Block
Copolymers

Block Copolymers

EDITED BY

Francisco J. Baltá Calleja
Instituto de Estructura de la Materia, CSIC
Madrid, Spain

Zbigniew Roslaniec
Institute of Materials Engineering
Technical University of Szczecin
Szczecin, Poland

MARCEL DEKKER, INC.   NEW YORK · BASEL

Chemistry Library

ISBN: 0-8247-0382-0

This book is printed on acid-free paper.

Headquarters
Marcel Dekker, Inc.
270 Madison Avenue, New York, NY 10016
tel. 212-696-9000; fax: 212-685-4540

Eastern Hemisphere Distribution
Marcel Dekker AG
Hutgasse 4, Postfach 812, CH-4001 Basel, Switzerland
tel: 41-61-261-8482; fax: 41-61-261-8896

World Wide Web
http://www.dekker.com

The publisher offers discounts on this book when ordered in bulk quantities. For more information, write to Special Sales/Professional Marketing at the headquarters address above.

Current printing (last digit):
10 9 8 7 6 5 4 3 2 1

PRINTED IN THE UNITED STATES OF AMERICA

Preface

In recent years, there has been increasing recognition that block copolymers are materials of outstanding scientific and technological importance. In the last two decades, two relevant monographs on block copolymers appeared: *Block Copolymers* by A. Noshay and J. E. McGrath, Academic Press, 1977, and *Processing, Structure and Properties of Block Copolymers* by M. J. Folkes, Elsevier, 1985. The former offers an important summary of the principles of chemical structure and methods of synthesis of block copolymers while the latter is a review volume focusing on properties and applications.

Rapid development of research activity on synthesis and physical properties of block copolymers has recently been observed. In addition, the present interest in this field is moving to more specific properties and applications where theoretical aspects and computer simulations are represented.

It has therefore been of considerable interest to select a few topics in which there has been particular progress and to present these together in one volume. We planned this volume to cover the field of block copolymers—chemistry, polymer structure, morphology, rheology, processing, physical properties, and typical applications—as of 1999. Among the authors are technical experts from leading research laboratories and universities that have developed new types of block copolymers and have used a variety of physical methods of characterization.

We offer a selection of scientific and technological aspects of block copolymers as multiphase polymer materials, and emphasize some of the characteristics that are of interest for topics in materials science and engineering. It is known that multiphase polymer systems include polymer blends, block copolymers, and IPNs. However, only block copolymers can be defined as multiphase systems having a stable microstructure, even after many processing cycles. Such multiphase systems are characterized by specific physical properties and can generate a variety of applications, such as thermoplastic elastomers, compatibilizers, high elastic fibers, and surfactant agents.

This book does not attempt to review all the scientific information in the field. Instead it focuses on selected results of the leading laboratories represented by the contributors.

The interest of the topic gave rise to the 1997 Europhysics Conference on Macromolecular Physics, "Structure-Physical Properties Relationship of Block Copolymers and Polymers Blends," which was held in Kolobrzeg, Poland.

Francisco J. Baltá Calleja
Zbigniew Roslaniec

Contents

Contents

Contributors

Volker Altstädt, Prof., Dr.-Ing. Polymer Engineering Department, Technical University Hamburg-Harburg, Hamburg, Germany

Francisco J. Baltá Calleja, Prof. Dr. Department of Macromolecular Physics, Instituto de Estructura de la Materia, CSIC, Madrid, Spain

Georg Broza, Ph.D.-Ing Polymer Composite Group, Technical University Hamburg-Harburg, Hamburg, Germany

Miroslawa El Fray, Ph.D., Ing. Department of Chemical Fibers and Physical Chemistry of Polymers, Technical University of Szczecin, Szczecin, Poland

T. A. Ezquerra, Ph.D. Instituto de Estructura de la Materia, CSIC, Madrid, Spain

Stoyko Fakirov, Prof. Dr. Laboratory on Polymers, University of Sofia, Sofia, Bulgaria

George Floudas, Ph.D. Institute of Electronic Structure and Laser, Foundation for Research and Technology-Hellas, Heraklion, Crete, Greece

Edward Grzywa, Prof., Dr.-Ing. Department of Polymers and Materials for Medicine, Industrial Chemistry Research Institute, Warsaw, Poland

G. P. Hellmann, Ph.D. Deutsches Kunststoff-Institut, Darmstadt, Germany

Helena Janik, Ph.D.   Polymer Technology Department, Technical University of Gdańsk, Gdańsk, Poland

Mariola Jastrzębska, Ph.D.   Department of Chemistry, Gdynia Maritime Academy, Gdynia, Poland

Robert J. E. Jérôme, Prof. Dr.   Center for Education and Research on Macromolecules, University of Liege, Liege, Belgium

Regina Jeziórska, Ph.D.   Department of Engineering Plastics and Specialty Polymers, Industrial Chemistry Research Institute, Warsaw, Poland

Nikos K. Kalfoglou, Prof., Ph.D.   Department of Chemistry, University of Patra, Patra, Greece

Ioannis K. Kallitsis, Prof., Ph.D.   Department of Chemistry, University of Patra, Patra, Greece

Thorsten Kirschnick   Polymer Engineering Department, Technical University Hamburg-Harburg, Hamburg, Germany

Efstathia G. Koulouri, Ph.D.   Department of Chemistry, University of Patra, Patra, Greece

Franciszek Lembicz, Ph.D.   Institute of Physics, Technical University of Szczecin, Szczecin, Poland

Jørgen Lyngaae-Jørgensen, Prof., Ph.D., H.D.   Department of Chemical Engineering, Technical University of Denmark, Lyngby, Denmark

Jerzy Majszczyk, Ph.D.   Institute of Physics, Technical University of Szczecin, Szczecin, Poland

Ernest Maréchal, Prof. Dr.   Laboratory of Macromolecular Synthesis, University P. M. Curie (Paris VI), Paris, France

Goerg H. Michler, Prof. Dr.   Institute of Materials Science, Martin Luther University Halle-Wittenberg, Halle, Germany

Harald Ott   Polymer Engineering Department, Technical University Hamburg-Harburg, Hamburg, Germany

Tadeusz Pakula, Prof. Dr.   Max-Planck-Institute for Polymer Research, Mainz, Germany

Jan Roda, Prof., Ph.D.   Department of Polymers, Institute of Chemical Technology, Prague, Czech Republic

Zbigniew Roslaniec, Ph.D., D.Sc.   Institute of Materials Engineering, Technical University of Szczecin, Szczecin, Poland

Maria Rutkowska, Ph.D., D.Sc.   Department of Chemistry, Gdynia Maritime Academy, Gdynia, Poland

Karl Schulte, Prof., Dr.-Ing.   Polymer Composite Group, Technical University Hamburg-Harburg, Hamburg, Germany

Jerzy Slonecki, Ph.D., D.Sc.   Department of Chemical Fibers and Physical Chemistry of Polymers, Technical University of Szczecin, Szczecin, Poland

Jiangdong Tong, Ph.D.   Chemical Department, University of Toronto, Toronto, Ontario, Canada

Ryszard Ukielski, Ph.D.   Department of Chemical Fibers and Physical Chemistry of Polymers, Technical University of Szczecin, Szczecin, Poland

Roland Weidisch, Dr.*   Institute of Materials Science, Martin Luther University Halle-Wittenberg, Halle, Germany

Witold Zielinski, Dr.   Industrial Chemistry Research Institute, Warsaw, Poland

---

* *Current affiliation*: Institute of Polymer Science and Engineering, University of Massachusetts at Amherst, Amherst, Massachusetts

1
Introduction and Plan

Francisco J. Baltá Calleja
Instituto de Estructura de la Materia, CSIC, Madrid, Spain

Zbigniew Roslaniec
Technical University of Szczecin, Szczecin, Poland

I. INTRODUCTORY REMARKS

The growth of the block copolymer industry has reached a high level of commercial importance, involving many new products and industry participants, as well as a large number of academic research groups. The increasing importance and interest in block copolymers arises mainly from their unique properties in solution and in the solid state, which are a consequence of their molecular structure. A block copolymer is a combination of two or more polymers joined end to end. Star-block or radial-block copolymers have branched structures and will not be considered here. In particular, sequences of different chemical composition are usually incompatible and therefore have a tendency to segregate. Amphiphilic properties in solution and microdomain formation in the solid state are directly related to this specific molecular architecture, which can be designed by using existing monomers and polymers.

As an indication of the interest and activity in block copolymers it is noteworthy that more than a thousand patents were filed between 1976 and 1982 and that the U.S. consumption of thermoplastic polymers, which represent only a part of the total block copolymer production, was approximately 800,000 metric tons per year by 1995, with a growth of 10%/year.

All block copolymers belong to the category of condensed matter often referred to as *soft materials,* which in contrast to crystalline solids are characterized by fluid disorder on the molecular scale and a high degree of order at longer-length scales. Their complex structure can give block copolymers mainly useful properties. The common poly(urethane) foams used in upholstery and bedding are com-

1

posed of multiblock copolymers known as *thermoplastic elastomers* that combine high-temperature resilience and low-temperature flexibility (1). The usual, inexpensive adhesive tape employs a different type of block copolymer, a linear triblock, to achieve pressure-sensitive adhesion. The addition of specific block copolymers to commodity plastics, such as polystyrene, can enhance their toughness or can modify surface properties for applications such as medical implantation and microelectronic fabrication.

II. APPLICATIONS

The unique properties and, thus, the application of block copolymer materials rely on their mesoscopic (10-nm scale) assembly in the molten and solid states. The patterns of assembly are referred to as *microphases, mesophases,* or *nanophases,* depending on the length scale. Phase separation is driven by chemical incompatibility between the different blocks that build up block copolymer molecules (2).

Several block copolymers are available commercially. Table 1 lists several examples of common trade names and typical applications, mainly of block copolymers produced in the United States, Europe, and Japan. The most common uses of block copolymers are as thermoplastic resins, membranes, polymer blends, and surfactants.

III. OBJECTIVES

The objectives of this book are twofold:

1. To record the recent research activity and development status of the block copolymer systems, up to the most recent research
2. To stimulate further research and development interest in these systems, since there are now many new ways of arriving at block copolymers, as we shall see later

There is now a great upsurge of interest, activity, and progress in this area, the consequences will be important, both technically and economically. The last two decades of theoretical development have culminated in predictive statistical theories that can account for the domain shapes, dimensions, connectivity, and ordered symmetry of many types of block copolymers (3–7). Recent advances in synthetic chemistry have exposed, in addition, new opportunities for using appropriate combinations of multiple blocks in novel molecular architectures to produce an unlimited number of structured materials endowed with tailored mechanical, optical, electrical, ionic, barrier, and other physical properties (7).

Table 1 Examples of Commercially Available Block Copolymers

Copolymer type	Trade names	Typical applications
Styrene-diene		
Diblock SB; SI	Solprene (Phillips)	Thermoplastic elastomers,
Triblock SBS; SIS	Kraton (Shell)	molding products,
Star block (SB)$_x$	Cariflex (Shell)	plastic modifiers,
	Tufprene (Asahi)	adhesives, films
	Europrene (Anic)	
	Buna BL (Bayer)	
	Eltar (Icechim–Romania)	
Hydrogenated styrene-diene		
Hydrogenated SBS	Kraton G (Shell)	Oxidation and weather-
Styrene-(ethylene-butylene)-styrene	Elaxar (Shell)	resistant thermoplastic elastomers, molding, cable insulation
Segmented polyester-polyurethane		
Thermoplastic polyurethanes (TPU)	Estane (Goodrich)	Thermoplastic elastomers, fibers
	Texin (Bayer, Mobay)	
	Roylar (Uniroyal)	
	Elastothane (Thiocol)	
	Pellethane (Upjohn)	
	Lycra (Du Pont)	
	Spandex (Du Pont)	
Segmented polyester-polyether		
	Hytrel (Du Pont)	Thermoplastic elastomers
	Pelprene (Toyobo)	
	Arnitel (Akzo-DSM)	
Segmented polyamide-polyether		
	Pebax (Atochem)	Molding extrusion,
	Vestamid E (C.W. Huls)	footwear, automotive parts, cable insulation
Polyolefinic block copolymers		
	TPR (Uniroyal)	Thermoplastic elastomers
	Somel (Du Pont)	
Ethylene oxide/propylene oxide block copolymers		
	Pluronics, Tetronics (BASF Wyandotte)	Surfactants

IV. PLAN

This book is devoted to the synthesis of block copolymers, to the properties that are relevant to their exploitation, and to the methodology of their preparation and measurement. The text is subtitled *Properties, Processing and Applications,* which is a description of its coverage; it is concerned with the physics and chemistry contributions to the science and technology of block copolymers and discusses the properties that determine the behavior processing and performance characteristics, together with some selected applications. This particular aspect, often underestimated by product designers, represents an important issue of this textbook.

Chapter 2 introduces new aspects of the synthesis of block copolymers and new catalysts and the combination of blocks with a widely varying chemical structure. These include: the synthesis, characterization, and properties of methacrylate butadiene–base triblock copolymers; the study of block copolymers with polyesters as flexible blocks; the description of new multiblock terpoly(ester-ether-amide)s with various amide blocks, and the applications of copolymers of poly(e-caprolactam)-block-(butadiene-containing elastomers) as engineering plastics.

Chapter 3 discusses the dynamic behavior of copolymers on the basis of computer simulations obtained for model block copolymer systems at various stages, from the disordered state to the fully segregated one. The simulation results are correlated with the viscoelastic behavior of real block copolymer melts of block copolymer blends and substituted block copolymers and with the flow behavior of diblock copolymers.

Chapter 4 discusses the use of physical methods, such as transmission electron microscopy, microhardness, dielectric spectroscopy, thermal mechanical relaxation, ultrasonic methods, and X-ray scattering, for the structural and chemical characterization of block copolymers. It offers new results on morphology and physical properties of segmented copoly(ether-ester)s and on morphology as applied to the study of micromechanical deformation mechanisms in weakly segregated diblock copolymers; it offers, for the first time, recent results on structure–microhardness correlations in condensation polymers and on the ultrasonic characterization of block copolymers. Finally, this chapter also brings up new dielectric studies on block copolymers relating to their relaxation behavior.

Chapter 5 offers new information on the microstructure–property correlation of block copolymers and their blends. In particular, the following three categories of blends are discussed: (a) polymer blends with block copolymers, (b) blends of two multiblock copolymers, and (c) blends of multiblock copolymers with homopolymers.

Chapter 6 discusses the rheological and processing parameters in multiphase polymer systems. In this chapter the structure/melt rheology relationship in block copolymers is highlighted and both the orientation behavior in axial tension and recent transmission electron microscopy studies of poly(ether-b-ester) copolymers are presented.

Chapter 7 offers new information on liquid crystalline and grafted copolymers. The specific examples of polyurethane grafted onto styrene-styrene sulfonic acid copolymers and of poly(ether-ester) block copolymers with liquid crystalline segments are discussed in detail.

Chapter 8 discusses recent new applications of block copolymers, in particular, the applications of triblock copolymers for toughness modification of immiscible engineering polymer blends and the application of blends and composites of poly(ether-*b*-ester copolymers as new materials of high mechanical, heat, and abrasion resistance and improved weatherability.

Chapter 9 surveys the latest developments and the expected future trends in synthesis, characterization, and applications.

In conclusion, the present volume gathers, for the first time, a variety of information on new block copolymer materials with specific properties, including high impact resistance, good processing, and high elasticity. The book presents, in addition, new views about the application of block copolymers as thermoplastic elastomers and modifying agents to polymer materials. Last but not least, it presents new ideas on molecular engineering aimed at developing block copolymers with enhanced properties.

ACKNOWLEDGMENTS

It is a pleasure for Z. Roslaniec to acknowledge the tenure of a sabbatical grant from DGICYT, Madrid, Spain.

REFERENCES

1. G Holden, NR Legge, R Quirk, HE Schröder, eds. Thermoplastic Elastomers—A Comprehensive Review. 2nd ed. Hanser, New York (1996).
2. FS Bates, GH Fredickson. Physics Today 52(2):32 (1999).
3. T Hashimoto, M Shibayama, H Kawai. Macromolecules 13:1237 (1980).
4. L Leibler. Macromolecules 13:1602 (1980).
5. E Helfand, ZR Wasserman. Macromolecules 13:994 (1980).
6. MD Whitmore, J Noolandi. J. Chem. Phys. 93:2946 (1990).
7. IW Hamley. Block Copolymers. Oxford University Press, Oxford (1999).

2
Methacrylate Butadiene-Based Triblock Copolymers: Synthesis, Characterization, and Properties

Robert J. E. Jérôme
University of Liege, Liege, Belgium

Jiangdong Tong
University of Toronto, Toronto, Ontario, Canada

I. INTRODUCTION

Since Szwarc's remarkable achievements in the 1950s (1), anionic polymerization has occupied a prominent position in polymer synthesis. Living anionic polyadditions allow synthetic polymers to be finely tailored (2). In addition to the precise control of molecular weight (MW), molecular architecture, and chain-end functionality, spectacular advances have been reported in the tailoring of multiphase block copolymers. As a representative achievement, Shell Chemical Company and Phillips Petroleum Company have successfully commercialized thermoplastic elastomers (TPEs), which are well-defined triblock and radial-block copolymers of styrene and butadiene (SBS) or isoprene (SIS) prepared via anionic process. As result of the thermodynamic immiscibility of their constitutive components, TPEs combine the intrinsic properties of the parent homopolymers with the additional benefit of spontaneous and thermoreversible crosslinking in strong relation to the phase morphology (3). The tensile strength of TPEs depends mainly on the ability of the hard blocks to resist plastic deformation under stress. Thus, the strength decreases sharply as the glass transition is approached, which explains why the upper service temperature (UST) of SBS is limited to ca. 70°C (4). It is thus worth substituting a hard block of a higher service temperature (or T_g) for polystyrene. Poly(α-methylstyrene) (5), poly(ethylene sulfide) (6), and poly(*tert*-butyl methacrylate) (7) have been proposed as possible substitutes for polystyrene.

Poly(methylmethacrylate) (PMMA) is another potential candidate, with T_g close to 130°C when the syndiotactic content is ca. 80% (8). PMMA is also known for stereocomplexation of syndiotactic and isotactic chains, the melting temperature of the stereocomplex being close to 190°C (9). However, the controlled synthesis of PMMA-b-polybutadiene (PBD)-b-PMMA triblock copolymers (MBM) is quite a problem. Indeed, living anionic polymerization of MMA is challenged by the difficult purification of MMA and the side reactions that occur during propagation of this highly reactive monomer above -65°C. Furthermore, the higher thermodynamic stability of the MMA enolate compared to the dienyl anion requires starting the synthesis with PBD. In order to meet this prerequisite, a difunctional initiator must be available, which additionally has to be soluble in hydrocarbons—otherwise the desired rubbery cis-1,4 microstructure is not formed. Finally, the polarity of the butadiene (BD) polymerization medium has to be increased before the addition of MMA, whose anionic polymerization is completely out of control in apolar solvents. This chapter is essentially a review of the research carried out on this type of block copolymers over the past 5 years.

II.   DIFUNCTIONAL INITIATORS FOR THE ANIONIC POLYMERIZATION OF BUTADIENE IN APOLAR SOLVENTS

So far, two general methods have been proposed to prepare such difunctional initiators, i.e., the coupling of radical anions (5), and the addition of 2 equivalents of monofunctional organolithium onto nonconjugated diene. Most problems met in the synthesis of difunctional initiators come from the solubility issue of dilithium compound in hydrocarbons, because the dipolar association of the organolithium pairs in solvent of low dielectric constant leads to insoluble aggregates (10,11).

A.   Overview of Recent Research on Difunctional Initiators

Recently, attention has been paid to the second strategy and to precursors of high ceiling temperature, so as to avoid competition between metalation and homopolymerization. (Scheme 1) Although bis-1,3-diphenylethylene (BDPE) compounds have been reported as successful precursors for efficient difunctional initiators (12–15), high cost and purification problems may limit industrial application. For these reasons, 1,3-diisopropenylbenzene (1,3-DIB) has received much attention. Rempp et al. were the first to consider the addition reaction of 2 mol of sec-butyllithium (s-BuLi) onto 1 mol of 1,3-DIB (16). Foss et al. reported on the favorable effect of triethylamine on this addition reaction (17). Cameron and Buchan, however found that oligomers were formed in addition to the expected diadduct (18). Ladd and Hogen-Esch concluded that the addition of s-BuLi to 1,3-DIB in a $^2/_1$ molar ratio allowed a nearly perfect difunctional initiator to be formed, when carried out in benzene at 50°C for 1h (19). However, our laboratory reported experimental observations indicating that this addition reaction was far

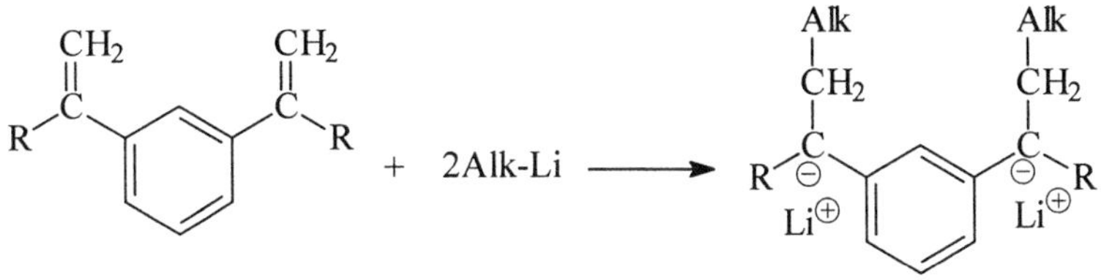

R = Phenyl; BDPE
R = Methyl; 1,3-DIB

Scheme 1

from being controlled. Clearly, the ceiling temperature of 1,3-DIB is higher than 50°C, since oligomers that bear approximately one double bond per monomeric unit are formed when 1,3-DIB is polymerized at 50°C in cyclohexane. Under the conditions used for the synthesis of the difunctional initiator, i.e., reaction of 1,3-DIB and s-BuLi in a ½ molar ratio in cyclohexane at 50°C, ^{1}H-NMR signals for unreacted vinylidene groups are detected at the early stage of the reaction; however, they disappear for longer reaction time. Nevertheless, a mixture of diadduct and oligomers is formed, as supported by size exclusion chromatography (SEC) (Fig. 1) and confirmed by ^{7}Li-NMR analysis, which detects a mixture of multi-lithiated species (Fig. 2). Furthermore, even when 1,3-DIB has been completely consumed, ca. 30% of s-BuLi remains unreacted.

1. Mechanism (20)

The addition reaction of s-BuLi onto 1,3-DIB is actually very complex, as a result of the difunctionality of 1,3-DIB and the occurrence of competing polymerization and depolymerization, which leads to a quasi-equilibrium state (Scheme 2). In the early stage of the reaction, propagation is faster than depropagation and the molecular weight (MW) of the oligomers increases. Then propagation is counterbalanced by depropagation that releases the monoadduct, and the average MW of the oligomers remains essentially constant. Finally, when 1,3-DIB is completely consumed, depropagation dominates and shorter oligomers are observed.

2. Initiator Efficiency (20,21)

When the s-BuLi/1,3-DIB diadduct is used as initiator in polar solvents, such as THF, initiation by at least monofunctional, difunctional, trifunctional, and tetrafunctional species is observed. Surprisingly enough, this mixture of mono- and multifunctional species typically behaves as a single anionic initiator in apolar solvents (e.g., cyclohexane, benzene, and toluene), yielding polymer of narrow MW distribution (MWD) and end-capped by an organolithium at both ends. A reasonable explanation is that the 1,3-DIB oligomers rapidly depolymerize upon monomer addition, in agreement with the formation of a unique low-MW com-

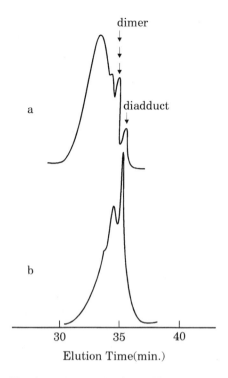

Fig. 1 SEC analysis of 1,3-DIB oligomers (s-BuLi/1,3-DIB molar ratio = 2; 50°C; in cyclohexane): (a) after 1 h; (b) after 20 h. (From Ref. 20.)

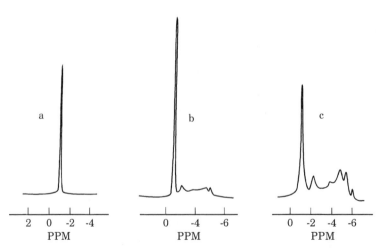

Fig. 2 ^{7}Li-NMR analysis of organolithium compounds in cyclohexane: (a) s-BuLi at 25°C; (b) s-BuLi/1,3-DIB in 2/1 molar ratio at 25°C for 8 days; (c) s-BuLi/1,3-DIB in 2/1 molar ratio at 50°C for 2 h. (From Ref. 20.)

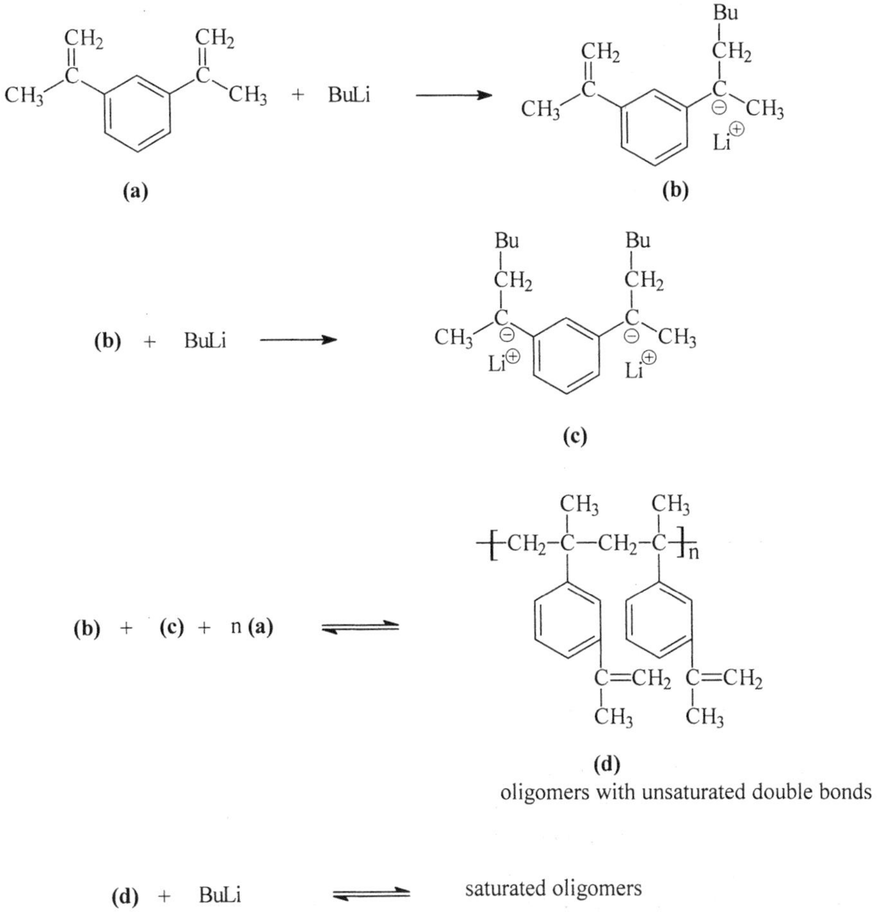

(a)

(b)

(b) + BuLi ⟶

(c)

(b) + (c) + n (a) ⇌

(d)
oligomers with unsaturated double bonds

(d) + BuLi ⇌ saturated oligomers

Scheme 2

pound when diphenylethylene (DPE), i.e., a nonpolymerizable monomer, is added to the "diadduct" solution (Fig. 3). Although this kind of initiator contains two organolithium pairs, it still behaves as a monofunctional initiator in apolar solvents, leading to the formation of diblock copolymers when butadiene is polymerized followed by styrene or MMA.

B. Synthesis of Pure Difunctional Initiator Based on *t*-BuLi and 1,3-DIB

In order to tackle this complex problem and to succeed in the synthesis of well-defined difunctional anionic initiator, the experimental conditions for *s*-BuLi/1,3-DIB diadduct formation have been changed. First, temperature was increased up

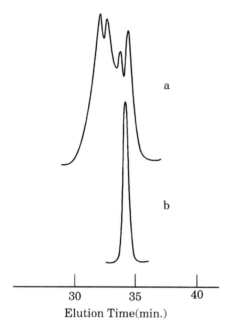

a

b

30   35   40

Elution Time(min.)

Fig. 3 SEC analysis of the addition product of *s*-BuLi onto 1,3-DIB in a 2/1 molar ratio at 50°C in toluene for 2 h and of the reaction product with DPE at 50°C for 20 h: (a) *s*-BuLi/1,3-DIB adduct; (b) reaction product with DPE. (From Ref. 20.)

to 80°C, closer to the ceiling temperature of 1,3-DIB. Figure 4 (a and b) shows that the diadduct is formed at the expense of high-MW oligomers. Unfortunately, the lithium diadduct is not thermally stable at 80°C, rapidly losing LiH (22). The activation energies of the addition reaction and the polymerization of the metalated 1,3-DIB are such that a decrease in temperature favors diadduct formation (Fig. 4c). A weakly polar additive, such as Et_3N, is then used in order to keep the reaction rate high enough. Finally, *tert*-butyllithium (*t*-BuLi), a more sterically hindered organolithium, was considered (22), making possible the selective formation of the dilithium diadduct (Fig. 4d). The reaction is conducted at −20°C, and 1,3-DIB in cyclohexane is added dropwise to the stirred cyclohexane solution of the 1:1 *t*-BuLi–Et_3N complex. The structure of the diadduct has been confirmed by SEC, ^{1}H-NMR, GC-MS, and ^{7}Li-NMR, with three stereoisomeric configurations being observed (22).

C. Initiation of Anionic Polymerization by the *t*-BuLi/1,3-DIB Diadduct

First of all, it is essential to prove that the *t*-BuLi/1,3-DIB diadduct is an actual difunctional anionic initiator suitable for the synthesis of living PBD chains capped

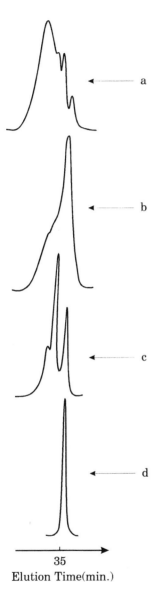

Elution Time(min.)

Fig. 4 SEC traces of BuLi/DIB (2/1) adducts prepared in cyclohexane. (a) *s*-BuLi/1,3-DIB at 50°C for 1 h; (b) *t*-BuLi/1,3-DIB at 50°C for 1 h; (c) *s*-BuLi/1,3-DIB at 80°C for 1 h; (d) *t*-BuLi/1,3-DIB in the presence of one equivalent of Et$_3$N at −20°C for 1 h. (From Ref. 22.)

by two isoreactive end groups. Indeed, in the case of uneven reactivity of the two organolithium end groups (particularly if one of them is "dormant"), the initiation of the second monomer (styrene, MMA, etc) may be incomplete, leading to diblock formation. That the expected SBS or MBM triblock copolymer (where S, B, and M stand for styrene, butadiene, and methylmethacrylate, respectively) is actually formed can be demonstrated via different methods, e.g., measurement of the stress–strain properties and selective degradation of the central polydiene block followed by determination of the MW of the second block. If the diadduct is an efficient difunctional initiator, the triblock copolymer (e.g., SBS) will show mechanical properties as high as the commercially available copolymer of the same MW and composition. In contrast, the parent diblock has very poor elastic properties. Moreover, the MW of the thermoplastic block (PS or PMMA) released upon the PBD degradation will fit the theoretical value calculated for a triblock. Otherwise, this MW will be two times higher than expected in the case of diblock formation. A mixture of diblock and triblock copolymers will provide the thermoplastic block with a bimodal MW distribution.

1. Apolar Solvent

The sequential polymerization of BD and styrene initiated by t-BuLi/1,3-DIB diadduct in cyclohexane yields a block copolymer of narrow monomodal MW distribution (1.1) and expected number average molecular weight (Mn) [100 K, 35% polystyrene (PS)]. However, the ultimate tensile strength is dramatically low (<2 MPa), and after selective degradation of the PBD block (21), Mn of the recovered PS block is about two times higher than expected, supporting qualitatively that a diblock copolymer has been formed. Therefore the pure diadduct is rather a monofunctional initiator in apolar solvent. Coordinative aggregation of the diadduct is at the origin of the problem, in agreement with previous suggestions by Quirk and Ma (15).

2. Polar Solvent

Although the MWD of polystyrene initiated by the s-BuLi/1,3-DIB "diadduct" in THF was multimodal, the SEC trace of polystyrene is monomodal when the t-BuLi/1,3-DIB "diadduct" is used as initiator in THF, which is additional evidence for the purity of this diadduct.

3. Apolar Solvent Modified by Polar Additives

Polar aprotic additives are known for their strong influence on the association of organolithium compounds in apolar solvent. Consistently, SBS triblock copolymers free from diblocks have been prepared by using the t-BuLi/1,3-DIB diadduct as initiator in cyclohexane added with THF or diethyl ether (Table 1). The PBD

Table 1 Characteristics of SBS Triblock Copolymers as Initiated by the t-BuLi/1,3-DIB Diadduct

Sample	Polar cosolvent	Content (vol %)	1,2-PBD[a] (%)	PS[a] (wt %)	Mn[b] ($\times 10^{-3}$)	Mw/Mn	Ultimate tensile strength (MPa)	Elongation at break (%)
1	THF	1.0	85	34	110	1.10	12.5	800
2	THF	3.0	85	34	105	1.10	12.0	800
3	Et$_2$O	10.0	46	33	105	1.10	31.5	1000
4	Et$_2$O	7.5	45	33	105	1.10	32.5	1000
5	Et$_2$O	5.0	45	35	105	1.10	30.0	1000
6	Et$_2$O	2.5	45	35	105	1.10	32.0	1000
7	Et$_2$O	1.0	33	35	110	1.35	19.0	1000

[a] ^{1}H-NMR analysis.
[b] From SEC of the first block and ^{1}H-NMR analysis of the final triblock.
Source: Ref. 22.

chains contain ca. 45% of 1,2-units when the amount of diethyl ether is in the range of 2.5–10%, compared to 85% when 1–3% THF is used as polar modifier. This modification of the PBD microstructure might, however, be deleterious to the elastomeric properties of the triblocks.

III. SYNTHESIS OF MBM TRIBLOCK COPOLYMERS INITIATED BY THE *t*-BuLi/1,3-DIB DIADDUCT

As mentioned in the introduction, the successful synthesis of MBM triblock copolymers requires that the diene polymerization be initiated in an apolar solvent by an effective difunctional initiator, although MMA has to be polymerized in a polar solvent at low temperature (−78°C). Early research in this laboratory (20–22) showed that when the BD polymerization is initiated by the pure *t*-BuLi/1,3 DIB diadduct in cyclohexane, followed by end-capping of the PBD chains by DPE in THF/cyclohexane (2/1, v/v) mixture and finally by the MMA polymerization at −78°C, no well-defined triblock copolymer is formed. As shown by Scheme 3, the living PBD chains are monofunctional rather than difunctional, being end-capped by organolithium species of completely different reactivity.

It appears that, on average, one of the two anionic sites of the diadduct does not contribute to BD initiation and forms a "dormant" organolithium species attached as an end group to the PBD. For steric reasons, this end group does not react with DPE, which leaves an anionic site that is overreactive to MMA, thus becoming prone to termination reactions (X end-group) (23). Furthermore, a small part of the original diadduct (ca. 10%) is completely inactive in the anionic poly-

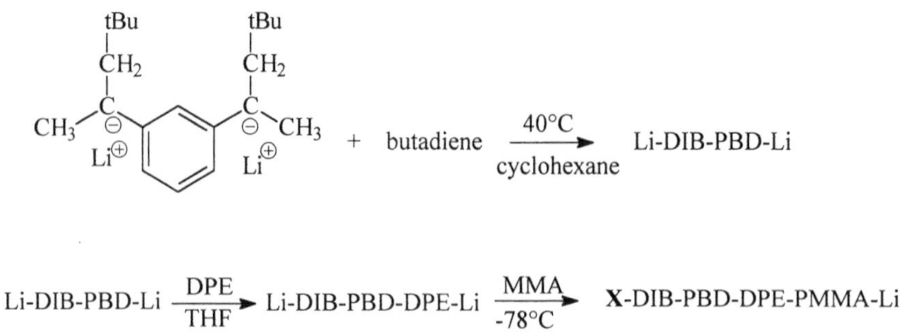
tBu
tBu
CH2
CH2
CH3
CH3
Li
Li
+ butadiene
40°C
cyclohexane
Li-DIB-PBD-Li
Li-DIB-PBD-Li
DPE
THF
Li-DIB-PBD-DPE-Li
MMA
-78°C
X-DIB-PBD-DPE-PMMA-Li

Scheme 3

merization of BD, which is, however, activated by the subsequent addition of THF and leads to homo-PMMA (24).

A. Synthesis of Poly(alkyl methacrylate)-*b*-PBD-*b*-Poly(alkyl methacrylate) Triblocks with PBD of Dominant 1,4-Microstructure (55%)

In order to dissociate the dormant anion of the diadduct in cyclohexane, some polar compounds have to be added so that the content of the 1,2-units of the PBD central block remains as low as possible (Scheme 4). Diethyl ether has proved to be a good candidate for the synthesis of MBM triblock (25–28) and MSBSM pentablock (29) copolymers of well-defined molecular structure and high tensile properties comparable to traditional SBS or SIS triblocks (σ_B: ~30MPa; ε_B: 1000%). The PBD central block then contains 55% 1,4-units. It must be mentioned that in Scheme 4, the *t*-BuLi/1,3-DIB diadduct may contain some multifunctional species, which will be very rapidly depolymerized into the diadduct as soon as butadiene is added, thus without detrimental effect on the structure and properties of the final MBM triblocks.

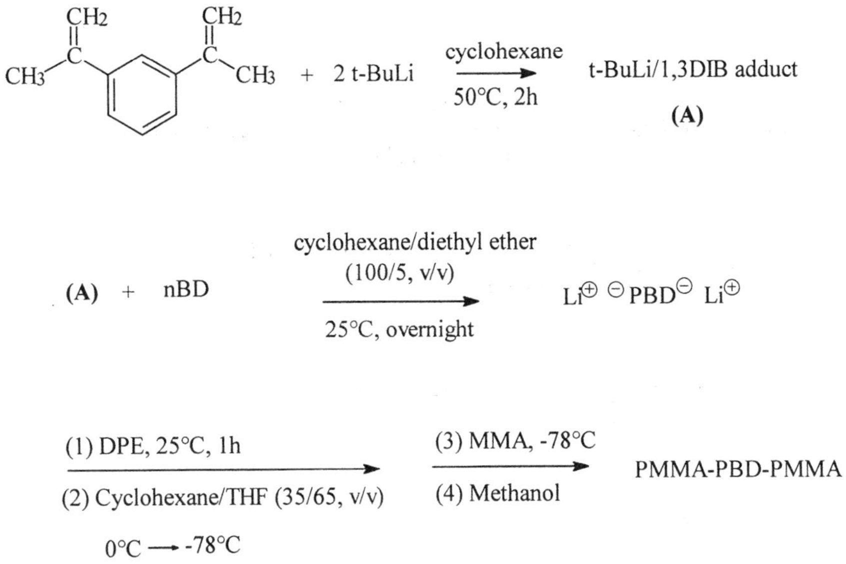

Scheme 4

B.   Synthesis of MBM Triblock Copolymers with PBD of High 1,4-Microstructure (>80%)

The synthesis of MBM triblock is quite a challenge when a high content of 1,4-units for the central PBD block (thus high elasticity and low T_g) is required. This complexity results from the severe requirements for the choice of the polymerization solvent. Indeed, a high 1,4-microstructure (>80%) for PBD is attainable only in hydrocarbon solvents, such as cyclohexane and toluene, in which organolithium species tend to aggregate. Triethylamine, anisole, and diphenyl ether are too weakly polar as additives to make the two organolithium sites of the diadduct available to initiation. In this respect, diethyl ether, *tert*-butyl methyl ether, tetramethylenediamine, and THF are efficient polar additives, which, however, result in too high a content of 1,2-units in the PBD central block. The seeding technique has been reported as an efficient method for the synthesis of polyisoprene with

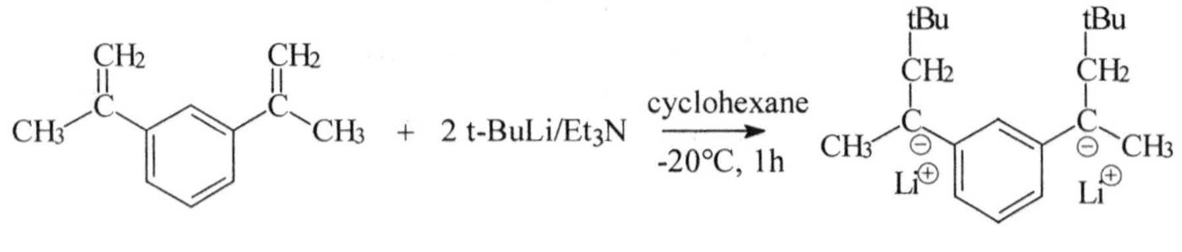

(A)

(A)   +   40BD $\xrightarrow[\text{25°C, overnight}]{\text{CHx, t-BuOLi, anisole}}$ seeded initiator

seeded initiator + nBD $\xrightarrow[\text{40°C, overnight}]{\text{CHx}}$ Li-PBD-DIB-Li(i),
Li-PBD-DIB-PBD-Li(ii)

$\xrightarrow[\text{(2) THF/CHx/DPE, -78°C, 0.5h}]{\text{(1) m St, 40°C, 4h}}$ Li-PS-PBD-PS-Li

$\xrightarrow[\text{(2) Methanol}]{\text{(1) MMA, -78°C}}$ PMMA-PS-PBD-PS-PMMA

Scheme 5

narrow MWD, although initiated by a difunctional initiator (13). In the case of the t-BuLi/1,3-DIB diadduct, although the seeded initiator prepared in the presence of anisole is efficient for the controlled synthesis of SBS triblock copolymers, it fails when the synthesis of MBM copolymers is envisioned, since the triblock is then contaminated by diblocks (30). These results suggest that two types of PBD chains are formed when the seeded initiator is used, i.e.,

Li-PBD-DIB-Li (i)   and   Li-PBD-DIB-PBD-Li (ii)

When styrene is added along with THF to these PBD species, the two types of organolithium species (i.e., PBD-Li and DIB-Li) initiate the styrene polymerization properly, and the expected SBS triblock copolymer is formed. The preliminary end-capping of the two active sites of PBD by DPE is prerequisite to the successful polymerization of MMA. However, the DIB-Li end group is unable to react with DPE, which explains the very early termination of the MMA polymerization and the formation of diblocks (or at least highly asymetric triblocks). Thus, the PBD chains have to be end-capped by a few styrene units followed by DPE, which provides them with two isoreactive styryl-Li end groups known for the efficient initiation of the MMA polymerization (Scheme 5).

IV. STRUCTURE–PROPERTY RELATIONSHIPS TYPICAL OF THE MBM TRIBLOCK COPOLYMERS

A. Service Temperature

Phase separation of the major central block and the minor outer blocks is the basic mechanism for SBS and MBM triblocks to build up a physical network of rubbery PBD chains. The measurement of glass transition temperatures (T_g) is a rapid, though not very sensitive, method to assess the two-phase structure of block copolymers. MBM triblock copolymers that contain the same PBD block (Mn:36,000) and two PMMA outer blocks of increasing Mn (7000–46,000), show two T_g's, which is a clear indication of phase separation. The T_g of PBD is essentially independent of the copolymer composition and, thus, of the PMMA molecular weight.

Figure 5 compares the MW dependence of the T_g for PMMA homopolymers and the PMMA outer blocks of the MBM triblocks. The unique relationship observed indicates a complete phase separation even for MBM copolymers, which contain low-MW PMMA blocks (<1000). In sharp contrast, the T_g of polystyrene (PS) blocks of 10,000 MW in SBS is smaller by ca. 40°C than the T_g of homo-PS of the same MW. This observation explains why the upper service temperature of commercially available SBS does not exceed ca. 70°C. Substitution of syndiotactic PMMA (sPMMA) for PS is thus an efficient way to increase further this service temperature.

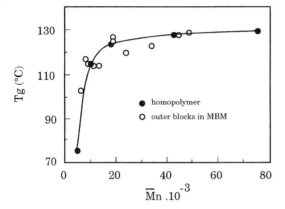

Fig. 5 Molecular weight dependence of the glass transition temperature for PMMA. (From Ref. 25.)

Another strategy has been proposed that consists of mixing the MBM triblocks with isotactic PMMA (iPMMA) (31). Solvent-cast films of these mixtures show a melting endotherm close to 180°C, which is a signature for stereocomplexation of the sPMMA blocks of the triblocks and iPMMA. Accordingly, the upper service temperature of neat MBM triblocks is increased by ca. 50°C as result of stereocomplexation of the outer blocks. As is the case for homopolymers, 2/1 is the best ratio between syndiotactic and isotactic partners for stereocomplexation to be the most extended. Decreasing tensile strength of MBM upon stereocomplexation with iPMMA (31) and very slow crystallization of the stereocomplex in the melt (2 days at 100°C) (32) are limitations to the stereocomplexation strategy. Still, with the purpose of increasing the upper service temperature of the traditional SBS thermoplastic elastomers, poly(isobornyl methacrylate) (PIBMA) has been considered as potential outer block (27). Compared to MMA that must be polymerized at −78°C, polymerization of isobornyl methacrylate can be conducted at room temperature, and the MWD remains very narrow (1.10). The additional advantage of PIBMA with respect to PMMA is much higher T_g, i.e., in the 170–206°C range, depending on the chain tacticity and MW. Figure 6 compares the dynamic mechanical properties of three different PBD-containing triblock copolymers. Clearly, the rubbery plateau extends until 90°C, 125°C, and 160°C when PS, sPMMA, and PIBMA, respectively, are the outer blocks.

B. Morphology

The effect of various parameters on the morphology of styrene-diene–based block copolymers has been extensively investigated (33). Highly periodic structures are

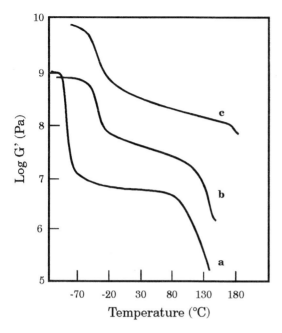

Fig. 6 Shear storage modulus (G') vs. temperature for (a) SBS, (b) MBM, (c) PIBMA-PBD-PIBMA.

known to be formed under equilibrium conditions, e.g., for films cast from suitable solvent and annealed properly. As a rule, spherical morphology is observed for polystyrene (PS) content in the range of 0–17 wt %. When this content is between 17 and 38 wt %, cylindrical morphology is reported, whereas lamellae are formed for 36–62 wt % PS. Phase morphology of poly(alkyl methacrylate)-polybutadiene–based triblock and pentablock copolymers has been analyzed by transmission electron microcopy (TEM) (25,27,29,34) and atomic force microscopy (AFM) (35,36). Figure 7 shows the typically observed phase morphologies, i.e., spheres (13% PMMA), cyclinders (36% PMMA), and lamellae (72% PMMA). It must be noted that the 6K-80K-6K MBM copolymer exhibits well-defined spherical morphology, although the PMMA block is very short, and no phase separation is observed for the SBS analogue (37), thus emphasizing a higher immiscibility for the PMMA/PBD pair compared to the PS/PBD one.

C. Mechanical Properties

The ultimate mechanical properties of a series of MBM triblocks are reported in Table 2 (25). The microstructure of the PBD central block consists of 45% of 1,4-

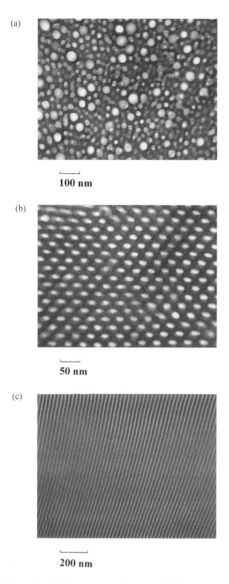

(a)

100 nm

(b)

50 nm

(c)

200 nm

Fig. 7 TEM micrographs for MBM samples: (a) 6K-80K-6K; (b) 20K-70K-20K; (c) 46K-36K-46K. Annealing at 140°C for 4 days. (From Ref. 25.)

Table 2 Mechanical Properties of MBM Triblock Copolymers

Sample	PMMA (wt %)	MW ($\times 10^{-3}$)	Mw/Mn	Ultimate tensile strength (MPa)	Elongation at break (%)
1	28	7-36-7	1.10	2.0	320
2	32	9-36-9	1.10	11.0	800
3	40	12-36-12	1.10	30.0	850
4	53	20-36-20	1.05	33.0	700
5	66	35-36-35	1.10	34.0	400
6	72	46-36-46	1.10	35.0	140
7	13	6-80-6	1.10	15.0	1300
8	20	10-80-10	1.10	22.0	1100
9	26	14-80-14	1.10	28.0	1000
10	33	20-80-20	1.10	33.0	850
11	38	25-80-25	1.10	34.0	800
12	56	50-80-50	1.10	34.0	600
13	39	13-40-13	1.10	34.0	850
14	35	16-60-16	1.10	34.0	900
15	36	20-70-20	1.10	34.0	890
16	38	40-130-40	1.10	34.0	850
17	32	14-60-14	1.10	26.0	700

units. As a rule, the tensile modulus is dependent mainly on the PMMA content and independent of the MW of the rubbery block. The ultimate tensile strength appears to be independent of either PMMA content or PMMA molecular weight, except for MBM containing very short PMMA blocks. Then the glassy domains of PMMA may be plasticized by PBD, which decreases their ability to withstand the applied stress. These observations are in good agreement with those for SBS and SIS triblock copolymers (38) and are explained by the filling effect of the hard microdomains in a highly entangled rubbery matrix.

1. Effect of PBD Microstructure

Modification of the butadiene polymerization by polar additives makes it possible to change the microstructure of the PBD central block. As the content of 1,2-units is increased, the glass transition temperature increases, and the elastic properties degrade (30). Figure 8 compares the stress–strain curves for MBM copolymers containing PBD with various 1,2-contents. Clearly the increase in 1,2-units decreases both the tensile strength and the elongation at break, particularly when the 1,2-content is predominant. Conversely, the initial modulus of these samples increases with the 1,2-content.

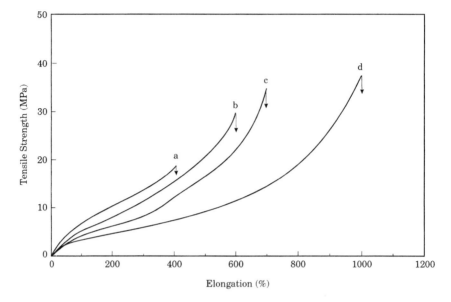

Fig. 8 Effect of the 1,2-content of the PBD block of MBM on the stress–strain curves: (a) 85% 1,2-units (MW 95000, PMMA content = 33 wt %; MWD = 1.15; prepared in cyclohexane with 1 vol % THF); (b) 66% 1,2-units (MW = 97,000, PMMA content = 33 wt %; MWD = 1.10; prepared in cyclohexane with TMEDA and TMEDA/Li of 0.45); (c) 45% 1,2-units (MW = 92000, PMMA content = 34 wt %; MWD = 1.10; prepared in cyclohexane with 7.5 vol % ether); (d) 17% 1,2-units (MW = 94000, PMMA content = 11 wt %; PS = 25 wt %; MWD = 1.10; prepared with seeded initiator in cyclohexane). (From Ref. 30.)

2. Tensile Properties at High Temperature

The physically crosslinked structure of the triblock copolymers results from the phase separation of the constitutive blocks on a microscopic scale. Although self-association of the hard blocks prevents viscous flow from occurring and provides these copolymers with rubberlike properties at room temperature (far below the T_g of the hard block), these characteristic features are lost upon heating in the vicinity of or above this T_g. Therefore, investigation of the mechanical properties at high temperature should provide useful information on the upper service temperature, or UST, of the triblock copolymers.

Table 3 lists the mechanical properties of poly(alkyl methacrylate)-based triblock copolymers at different temperatures. For the sake of comparison, the tensile properties of an SBS triblock are also reported. As a rule, the highest yield stress, ultimate tensile properties, and elongation at break are observed at room temperature. Mechanical properties for the SBS triblock decrease rapidly as the temperature is increased; i.e., more than 90% of the ultimate tensile strength is lost

Table 3 Mechanical Properties of Poly(alkyl methacrylate)-Based Triblock Copolymers

Sample	Tensile Properties[e]	T (°C)							
		25	50	70	90	110	125	140	150
SBS	σ_y (MPa)	—	—	—	—				
	σ_b (MPa)	33.0	13.0	2.4	0.8				
	ε_b (%)	1100	1200	600	200				
EBE[a]	σ_y (MPa)	4.0	3.2	2.3	—				
	σ_b (MPa)	18.7	14.0	4.0	1.4				
	ε_b (%)	1120	1280	850	530				
TBT[b]	σ_y (MPa)	—	—	—	—				
	σ_b (MPa)	24.0	16.1	4.8	1.8				
	ε_b (%)	1080	1270	910	570				
MBM	σ_y (MPa)	4.6	nd	3.8	3.2	2.4	1.7		
	σ_b (MPa)	32.0	nd	20.0	14.3	5.5	2.2		
	ε_b (%)	835	nd	1200	1240	970	1086		
MEBM[c]	σ_y (MPa)	6.3	nd	4.9	3.6	3.4	2.8		
	σ_b (MPa)	38.0	nd	24.0	18.2	8.4	3.8		
	ε_b (%)	700	nd	1150	1400	1100	970		
IBI[d]	σ_y (MPa)	9.6	nd	nd	3.5	2.9	2.6	2.5	1.9
	σ_b (MPa)	35.0	nd	nd	16.5	11.5	8.4	5.6	2.2
	ε_b (%)	650	nd	nd	1040	1050	1080	920	600

[a] PEMA-PBD-PEMA.
[b] PtBMA-PBD-PtBMA.
[c] Hydrogenated MBM.
[d] PIBMA-PBD-PIBMA.
[e] σ_y = yield stress, σ_b = ultimate tensile strength, ε_b = elongation at break.
Source: Ref. 34.

at 70°C, which corresponds to the lowest UST in the investigated series. PEMA (T_g = 90°C) and PtBMA (T_g = 110°C) containing triblocks show quite comparable behavior, at least for the composition and MW under investigation. The typical behavior of thermoplastic elastomer is maintained at 90°C for the MBM triblock that contains 38% of hard blocks, which is actually comparable to the SBS copolymer at 50°C. Thus an increase in T_g of the hard phase from 100°C (PS) to 130°C (PMMA) improves by ca. 40°C the upper service temperature of otherwise identical thermoplastic elastomers. The beneficial effect of hydrogenation of the PBD central block is worth being noted. Finally, the highest UST is observed for the PIBMA (T_g ~ 200°C) containing triblock copolymer, which shows an ultimate tensile strength higher than 2 MPa and elongation at break of 600% at 150°C.

V. CONCLUSION

Since MMA is much more reactive than butadiene, it is mandatory to polymerize butadiene first in the synthesis of MBM triblock copolymers, which requires the availability of a hydocarbon soluble α,ω-difunctional anionic initiator. Under optimized experimental conditions, pure t-BuLi/1,3-DIB diadduct can be prepared, which essentially acts as difunctional initiator in polar solvent (THF), in contrast to what happens in apolar solvents (toluene, cyclohexane), in which the initiator is apparently monofunctional, more likely as a result of μ-type aggregation. This conclusion has been drawn from the poor mechanical properties of the final copolymers and the length of the PS block released as a result of the selective degradation of the polybutadiene block. In order to activate the two anionic sites, polymerization of butadiene must be carried out in the presence of polar additives or be initiated by properly seeded difunctional butadiene oligomers. Depending on the strategy used, the content of 1,2-units is in the range of either 45% or 15%.

Alkyl methacrylate-butadiene–based triblock copolymers are clearly phase separated, with T_g of PBD ranging from $-90°C$ to $-60°C$ and T_g of the hard blocks ranging from $90°C$ to $190°C$. The tensile properties of MBM triblock copolymers are the same as those of SBS at room temperature ($\sigma_B = 30$ MPa, $\varepsilon_B = 800\%$), although they decrease much more rapidly upon increasing the temperature in the case of SBS. As result, the upper service temperature is higher by ca. $40°C$ for MBM compared to SBS. The largest service temperature range is, however, observed when PIBMA with T_g of $190°C$ is the hard block.

ACKNOWLEDGMENTS

R. Jérôme is deeply indebted to all his coworkers mentioned in the publications listed in the References. He is also grateful to the Services Fédéraux des Affaires Scientifiques, Techniques, et Culturelles for general support to CERM under the auspices of the PAI 4/11.

REFERENCES

1. M Szwarc. Living polymers: their discovery, characterization, and properties. J. Polym. Sci. Polym. Chem. Ed. 36:IX–XV, 1998.
2. R Jérôme, JD Tong. Recent developments in living anionic polymerization. Current Opinion. Solid State and Materials Science, 3:573–578, 1998.
3. G Holden, NR Legge. Styrenic thermoplastic elastomers. In: G Holden, NR Legge, R Quirk, HE Schroeder, eds. Thermoplastic Elastomers. 2nd ed. New York: Hanser, 1996, pp. 47–71.
4. G Kraus, W Rollman. J. Polym. Sci. Polym. Phys. Ed. 14:1133–1148, 1976.
5. LJ Fetters, M Morton. Macromolecules 2(5):453–458, 1969.
6. M Morton, SL Mikesell. J. Macromol. Sci. Chem. A7:1391–1397, 1973.
7. TE Long, AD Broske, DJ Bradly, JE McGrath. J. Polym. Sci. Polym. Chem. Ed. 27:4001–4012, 1989.

8. TE Long, RD Allen, JE McGrath. In: M Fontanille, A Guyot, eds. Recent Advances in Mechanistic and Synthetic Aspects of Polymerization. NATO ASI series. Dordrecht: Reidel, 1987, Vol. 215, pp. 79–100.
9. E Schomaker, G Challa. Macromolecules 21:3506–3510, 2195–2203, 1988.
10. RN Young, RP Quirk, LJ Fetters. Adv. Polym. Sci. 56:1–90, 1984.
11. F Bandermann, HD Speikamp, L Weigel. Makromol. Chem. 186:2017–2024, 1985.
12. LH Tung, GYS Lo, DE Beyer. Macromolecules 11:616–617, 1978.
13. LH Tung, GYS Lo. Studies on Dilithium initiators. 1. Hydrocarbon-soluble initiators, 1,3-phenylenebis(3-methyl-1-phenylpentylidene)dilithium and 1,3-phenylenebis[3-methyl-1-(methylphenyl)pentylidene]dilithium. Macromolecules 27:2219–2224, 1994.
14. CJ Bredeweg, AL Gatzke, GYS Lo, LH Tung. Studies on dilithium initiators. 2. The bimodal molecular weight distribution in polyisoprene. Macromolecules 27:2225–2232, 1994.
15. RP Quirk, JJ Ma. Dilithium initiator based on 1,3-bis(1-phenylethenyl)benzene. Tetrahydrofuran and lithium sec-butoxide effects. Polym. Int. 24:197–206, 1991.
16. G Beinert, P Lutz, E Franta, P Rempp. A bifunctional anionic initiator soluble in nonpolar solvents. Makromol. Chem. 179:551–555, 1978.
17. RP Foss, HW Jacobson, WH Sharkly. A new difunctional anionic initiator. Macromolecules 10:287–291, 1977.
18. GG Cameron, GM Buchan. Addition of sec-butyllithium to m-diisopropenylbenzene. Polymer 20:1129–1132, 1979.
19. BJ Ladd, TE Hogen-Esch. Synthesis of poly(butadiene) syndiotactic poly(methyl methacrylate) ABA block copolymers. Stereocomplex formation with isotactic PMMA. Polym. Prep. 30:261–262, 1989.
20. YS Yu, Ph Dubois, R Jérôme, Ph Teyssie. Difunctional initiator based on 1,3-diisopropenylbenzene. 2. Kinetics and mechanism of sec-butyllithium/1,3-diisopropenylbenzene reaction. Macromolecules 29:1753–1776, 1996.
21. YS Yu, R Jérôme, R Fayt, Ph Teyssie. Efficiency of the sec-butyllithium/m-diisopropenylbenzene diadduct as an anionic polymerization initiator in apolar solvents. Macromolecules 27:5957–5963, 1994.
22. YS Yu, Ph Dubois, R Jérôme, Ph Teyssie. Difunctional initiator based on 1,3-diisopropenylbenzene. 3. Synthesis of a pure dilithium adduct and its use as difunctional anionic polymerization initiator. Macromolecules 29:2738–2745, 1996.
23. TE Long, RD Allen, JE McGrath. Controlled synthesis of various poly(alkyl methacrylates) by anionic techniques. In: M Fontanille, A Guyot, eds. Recent Advances in Mechanistic and Synthetic Aspects of Polymerization. Dordrecht: Reidel, 1987, pp. 79–100.
24. YS Yu, Ph Dubois, R Jérôme, Ph Teyssie. Difunctional initiator based on 1,3-diisopropenylbenzene. 5. Effect of polar additives and initiator seeding on the synthesis of poly(styrene-b-butadiene-b-styrene) copolymers. Macromolecules 30:7356–7362, 1997.
25. JM Yu, Ph Dubois, R Jérôme, Ph Teyssie. Syndiotactic poly(methyl methacrylate) (sPMMA)-polybutadiene-sPMMA triblock copolymers: synthesis, morphology, and mechanical properties. Macromolecules 29:6090–6099, 1996.

26. JM Yu, Ph Dubois, R Jérôme. Synthesis of poly[isobornyl methacrylate (IBMA)-*b*-butadiene-*b*-IBMA] copolymers: new thermoplastic elastomers of a large service temperature range. Macromolecules 29:7316–7322, 1996.

27. JM Yu, Ph Dubois, R Jérôme. Poly[poly(isobornyl methacrylate-co-methyl methacrylate) (poly(IBMA-co-MMA))-b-polybutadiene-b-poly(IBMA-co-MMA)] copolymers: synthesis, morphology, and properties. Macromolecules 30:6536–6543, 1997.

28. JM Yu, Ph Dubois, R Jérôme. Poly[glycidyl methacrylate (GMA)/methyl methacrylate (MMA)-b-butadiene-b-GMA/MMA] reactive thermoplastic elastomers: synthesis and characterization. J. Polym. Sci. Polym. Chem. Ed. 35:3507–3515, 1997.

29. JM Yu, Ph Dubois, R Jérôme. Poly[methyl methacrylate (*M*)-*b*-styrene (*S*)-*b*-butadiene (*B*)-*b*-S-*b*-M] pentablock copolymers: synthesis, morphology, and mechanical properties. Macromolecules 30:4984–4994, 1997.

30. YS Yu, Ph Dubois, R Jérôme, Ph Teyssie. Difunctional initiator based on 1,3-diisopropenylbenzene. 6. Synthesis of methyl methacrylate-butadiene-methyl methacrylate triblock copolymers. Macromolecules 30:4254–4261, 1997.

31. JM Yu, YS Yu, Ph Dubois, R Jérôme. Sterocomplexation of sPMMA-PBD-sPMMA triblock copolymers with isotactic PMMA: 1. Thermal and mechanical properties of stereocomplexes. Polymer 38:2143–2154, 1997.

32. G Helary, G Belogy, TE Hogen-Esch. Stereocomplex formation in polybutadiene-syndiotactic poly(methyl methacrylate) block copolymers blended with isotactic poly(methyl methacrylate). Polymer 33:1953–1958, 1992.

33. BRM Gallot. Preparation and study of block copolymers with ordered structures. Adv. Polym. Sci. 29:85–161, 1985.

34. JM Yu, Ph Teyssie, R Jérôme. Poly[alkyl methacrylate (AMA)-*b*-butadiene (BD)-*b*-AMA] triblock copolymers: synthesis, morphology and mechanical properties at high temperatures. Macromolecules 29:8362–8370, 1996.

35. Ph Leclere, R Lazzaroni, JL Bredas, JM Yu, Ph Dubois, R Jérôme. Microdomain morphology analysis of block copolymers by atomic force microscopy with phase detection imaging. Langmuir 12:4317–4320, 1996.

36. Ph Leclere, JM Yu, R Lazzaroni, Ph Dubois, JL Bredas. Block copolymer microdomain morphologies by phase detection imaging. Mat. Res. Soc. Symp. Proc. 461:57–62, 1997.

37. H Gilman, FK Cartledge. The analysis of organolithium compounds. J. Organomet. Chem. 2:447–454, 1964.

38. RP Quirk, M Morton. Research on anionic triblock copolymers. In: G Holden, NR Legge, R Quirk, HE Schroeder, eds. Thermoplastic Elastomers. 2nd ed. New York: Hanser, 1996, pp. 71–100.

3
Block Copolymers with Polyethers as Flexible Blocks

Ernest Maréchal
University P. M. Curie (Paris VI), Paris, France

I. GENERAL CONSIDERATIONS

Many copolymers contain polyethers as flexible blocks, particularly poly(oxyethylene), poly(oxypropylene), and to a lesser extent poly(oxytetramethylene):

Polyamide-*block*-polyether (1–24)
Polyurethane-*block*-polyether (7,25)
Polyimide-*block*-polyether (26,27)
Polyester-*block*-polyether (28–37)
Polycarbonate-*block*-polyether (38)
Aromatic polyether-*block*-polyether (39)
Polystyrene-*block*-polyether (40–42)
Poly(methyl methacrylate)-*block*-polyether (43)
Polyacrylamide-*block*-polyether (43)
Polyacrylonitrile-*block*-polyether (44)
Polypeptide-*block*-polyether (45)

They can be prepared in very different ways, as described in the following sections.

A. Polycondensation

For instance, two α,ω-difunctionnal oligomers are polycondensed in bulk (Scheme 1).

n HOOC-POLYDECANAMIDE-COOH + n HO-POLY(OXYETHYLENE)-OH

HO $\left[\underset{O}{\overset{\parallel}{C}}\text{-POLYDECANAMIDE-COO-POLY(OXYETHYLENE)} \right]_{\tilde{n}}$ OH + 2n-1 H_2O

Scheme 1

B. Ring-Opening Polymerization

Several authors developed this method (46,47). Kobayashi et al. (48) prepared poly[polyimine-*block*-poly(oxytetramethylene)] with an amide as side group (Scheme 2).

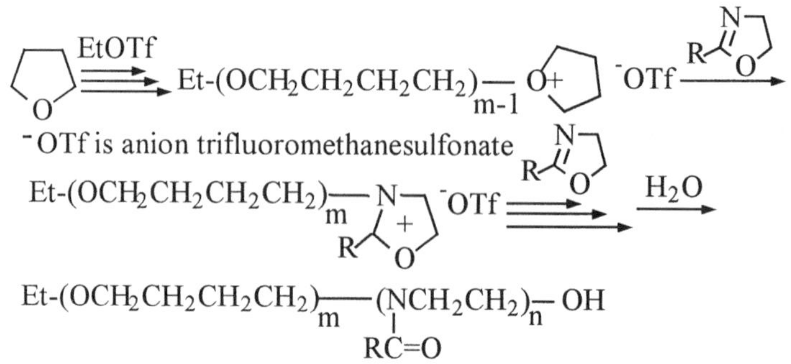

$-OTf$ is anion trifluoromethanesulfonate

Scheme 2

C. Radical Polymerization

Nagarajan et al. (49) prepared poly[poly(oxyethylene)-*block*-polyacrylonitrile] using ceric ions as catalyst (Scheme 3). The synthesis of some block copolymers associates several techniques, for instance, a cationic polymerization followed by an anionic polymerization or a cationic polymerization followed by a radical polymerization.

These techniques will be developed in the following sections.

POLY(OXYETHYLENE)-OCH$_2$CH$_2$OH $\xrightarrow{Ce^{4+}}$ POLY(OXYETHYLENE)-OCH$_2\overset{\cdot}{C}$H OH

$\xrightarrow{CH_2=CN}$ POLY(OXYETHYLENE)-OCH$_2$CH(OH)-POLYACRYLONITRILE

Scheme 3

II. BLOCK COPOLYMERS PREPARED BY POLYCONDENSATION

IUPAC recently published two separate definitions for polycondensation and polyaddition (50); however, in the following we will use the word *polycondensation* unless otherwise stated.

A. General Methods

1. Direct polycondensation of α,ω-difunctional oligomers:

X-[Oligomer 1]-X + Y-[Oligomer 2]-Y → ~X'-[Block 1]-X'Y'-[Block 2]-Y'~

2. Polycondensation of an α,ω-difunctional oligomer with the precursors of another block:

X-[Oligomer 1]-X + A-A + B-B → ~X'-[Block 1]-X'Y'-[Block 2]-Y"~

where Y' and Y" is either A'A' or B'B'

3. Oligomer coupling reactions

X-[Oligo 1]-X + Y-[Oligo 2]-Y + C-C → ~X'-[Block 1]-X'C'C'Y'-[Block 2]-Y'~

B. Direct Polycondensation of α,ω-Difunctional Oligomers

The direct polycondensation of α,ω-difunctional oligomers leads to well-defined block copolymers; however, this is less used than methods 2 and 3, due to two major difficulties: the oligomers must be soluble or molten in the reactional mixture, which is not the case for many aromatic polymers, and their functionality must be exactly 2.

The polycondensations of α,ω-diacid (or diester) polyamides with α,ω-dihydroxy polyethers (or copolyethers) were particularly studied. The following example (22,23) is relative to the preparation of poly[polydodecanamide-*block*-copolyether]s by polycondensation of an α,ω-dicarboxylic-poly(dodecamide) (PA) with different α,ω-dihydroxy-poly[poly(oxyethylene)-*block*-poly(oxy–propylene)] (PE) (Scheme 4).

n HOOC-PA-COOH + n HO-PE-OH

$\Big\downarrow$ Zr(OBu)$_4$; 240°C; vacuum

HO$\Big[$C-PA-C-O-PE-O$\Big]_n$H + 2n-1 H$_2$O
       $\overset{\|}{O}$ $\overset{\|}{O}$

Scheme 4

This is classical process, and Zr(OBu)$_4$ and Ti(OBu)$_4$ are efficient catalysts, even though they are very sensitive to hydrolysis and rapidly change into mixtures of various complexes, and aggregates are by far less active than the catalyst itself. The effect of water is increased when one of the blocks is hydrophilic, which is particularly the case with polyethers. Laporte et al. (20,21) studied the polycondensation of α,ω-dicarboxylic-polyundecanamide or of 11-dodecylamideundecanoic acid with α,ω-dihydroxy-poly(oxyethylene) or 1-dodecanol or 2-tridecanol. They showed that its kinetics drastically depends on the reaction medium; the reaction rate greatly decreases when the aliphatic alcohol is replaced by α,ω-dihydroxy-poly(oxyethylene). The water contained in the reactional medium reduces the efficiency of the catalyst, for only the surface of the aggregates is active; this effect is less important when the polyether block is poly(oxypropylene).

The elimination of water, whatever its origin (contained in the initial oligoether or resulting from the polycondensation), is a major problem, even when no catalyst is added. This was carefully analyzed by Michot et al. (27) when preparing poly(imide *seq*-polyether) by polycondensation of several dianhydrides with α,ω-diaminopropyl-oligo(oxyethylene) (Scheme 5):

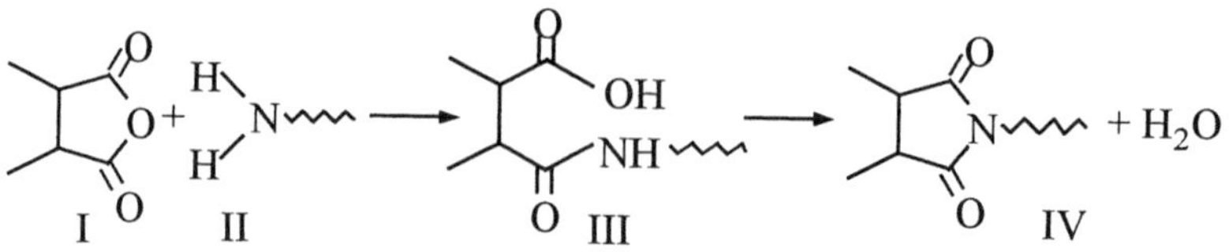

Scheme 5

Anhydride I derives from benzene, naphthalene, perylene, and benzophenone, and II is $H_2NCH(CH_3)CH_2[OCH_2CH_2]_nOCH_2CH(CH_3)NH_2$, dehydrated by anhydrous argon bubbling for 2 hours while kept at 80°C. The cyclization is carried out at 200°C in the presence of an acetic anhydride/pyridine mixture; the use of solvents such as dimethylacetamide at 120–140°C allows a one-pot synthesis from the starting ingredients. Molar mass (close to 20,000) was determined by SEC; however, standardization raises difficult problems and the values obtained are often only orders of magnitude. Determining an accurate value of cyclization extent is difficult, due to the insolubility of aromatic polyimides. The elimination of water is difficult, particularly when conversion is high, for the medium becomes very viscous and cyclization may be far from completion.

The molar mass and the structure of the polycondensate can be determined by different analytical processes, such as SEC, mass spectrometry, and concentration of end groups. Nuclear magnetic resonance (NMR) can sometimes provide a direct proof of the polycondensaction (see Sec. V.B, Refs. 22 and 23).

Eisenbach and Heinlein (51) prepared thermotrop block copolymers by polycondensing the oligomer in Scheme 6 with the oligomer in Scheme 7. The preparation of Scheme 6 requires several steps to obtain a rigorously bifunctional oligomer.

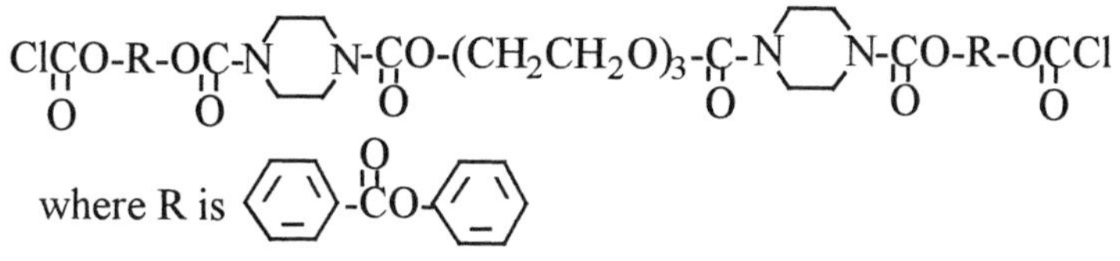

$$ClCO\text{-}R\text{-}OC\text{-}N \underset{\smile}{\frown} N\text{-}CO\text{-}(CH_2CH_2O)_3\text{-}C\text{-}N \underset{\smile}{\frown} N\text{-}CO\text{-}R\text{-}OCCl$$

where R is benzene ring–CO–benzene ring

Scheme 6

$$HN \underset{\smile}{\frown} N\text{-}C\text{-}O\text{—}(CH_2CH_2O)_n\text{—}C\text{-}N \underset{\smile}{\frown} NH$$

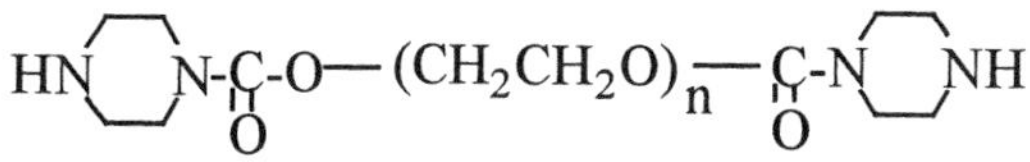

Scheme 7

This shows the difficulties linked to the direct polycondensation (Scheme 8):

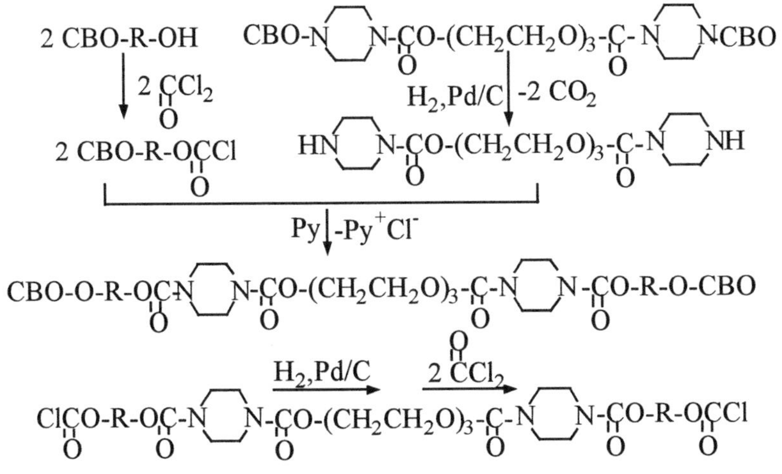

$$2\ CBO\text{-}R\text{-}OH \qquad CBO\text{-}N \underset{\smile}{\frown} N\text{-}CO\text{-}(CH_2CH_2O)_3\text{-}C\text{-}N \underset{\smile}{\frown} NCBO$$

$$\downarrow 2\ CCl_2 \qquad\qquad H_2,Pd/C \downarrow -2\ CO_2$$

$$2\ CBO\text{-}R\text{-}OCCl \qquad HN \underset{\smile}{\frown} N\text{-}CO\text{-}(CH_2CH_2O)_3\text{-}C\text{-}N \underset{\smile}{\frown} NH$$

$$Py\downarrow\text{-}Py^+Cl^-$$

$$CBO\text{-}O\text{-}R\text{-}OCN \underset{\smile}{\frown} N\text{-}CO\text{-}(CH_2CH_2O)_3\text{-}C\text{-}N \underset{\smile}{\frown} N\text{-}CO\text{-}R\text{-}O\text{-}CBO$$

$$\xrightarrow{\ H_2,Pd/C\quad 2\ CCl_2\ }$$

$$ClCO\text{-}R\text{-}OC\text{-}N \underset{\smile}{\frown} N\text{-}CO\text{-}(CH_2CH_2O)_3\text{-}C\text{-}N \underset{\smile}{\frown} N\text{-}CO\text{-}R\text{-}OCCl$$

CBO is carbonylbenzoloxy

Scheme 8

C. Polycondensation of an α,ω-Difunctional Oligomer with the Precursors of Another Block

This method is very useful when one of the initial oligomers is too fragile or too high melting to be directly polycondensed. Many block copolymers were prepared in that way; however, their structures are often ill defined because it is difficult to control the distribution of the units in the chains; this method requires a very careful identification of the product. In some cases it is possible to control the reaction or at least to obtain a good identification of the copolymer structure.

Gaymans et al. (3) prepared poly[polybutanamide-*block*-poly(oxytetramethylene)] by polycondensation of α,ω-diamino-poly(oxytetramethylene) with the salt prepared from adipic acid and 1,4-diamino-butane (Scheme 9):

$$H_2N-[(CH_2)_4O]_n-(CH_2)_4-NH_2 \,+_{\lfloor}H_2N-(CH_2)_4-NH_2,HOOC-(CH_2)_4-COOH_{\rfloor}$$

$$\text{Salt}$$

Scheme 9

The salt was independently prepared from the block polycondensation, and the diacid and the diamine remain linked in the reactional mixture, which leads to well-defined block copolymers, and the dispersity index of the rigid block is probably close to 2.

Polycondensations in the melt are often characterized by phase segregation between the precursors of the rigid blocks and the polyether; this may result in very heterogeneous copolymers. This difficulty can be overcome in different ways. Gaymans et al. (3) carried out a prepolycondensation in a solvent (pyrrolidone) leading to a homogeneous mixture of oligomers, and the prepolymer is polycondensed in the solid state (250–260°C). Deleens et al. (10–12) observed the same phenomenon when polycondensing various α,ω-diamino polyamides with different α,ω-dihydroxy-polyethers; the addition of diblock or triblock oligomers to the reactional mixture led to a homogeneous system. When no prepolymer is added the kinetics is erratic, whereas it obeys a well-defined second-order law when the medium is homogeneous.

Thuillier et al. (34,35) prepared thermotropic polyesters-*block*-polyethers by polycondensing an α,ω-dihydroxy-poly(oxytetramethylene) with an aromatic diacid and 1,5-pentanediol, which are the precursors of a mesogenic block

(Scheme 10):

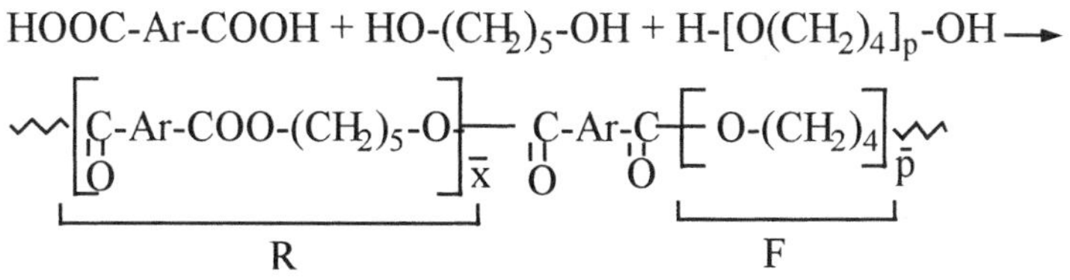

$$HOOC\text{-}Ar\text{-}COOH + HO\text{-}(CH_2)_5\text{-}OH + H\text{-}[O(CH_2)_4]_p\text{-}OH \longrightarrow$$

Scheme 10

Ar is either *trans*-stilbene or biphenyl; the following is relative to poly[poly(1,5-pentanediyl-4,4′-stilbenedicarboxylate)-*block*-poly(oxytetramethylene)]; the reaction in Scheme 10 was carried out in the bulk. R% is the rigid block content; $R_{th}\%$ is caculated from the composition of the feed; and $R_{exp}\%$ is determined directly from the copolymer. $R_{exp}\%$, $\overline{DP}_n$, and Mn of the rigid block were obtained from the integrations of methylene (F) and benzene (R) protons in [1]NMR; Table 1 shows that there is a good agreement between theoretical and experimental values.

The values given in Table 1 were obtained from the copolymer itself and not from the feed. The presence of double bonds in the chain permits a thorough analysis of the polymer structure using a 500-MHz ^{1}H-NMR spectrum; Fig. 1 shows the peaks relative to aromatic and ethylenic protons of copolymers with different R content. When R% decreases, the signals are progressively shifted toward low field and a significant split of the peaks, which is maximum when R% is 30. With the exception of the pure aliphatic polyester (R% = 0), all these spectra show a small peak at 0.05 ppm before the aromatic proton doublet and corresponding to small amounts of cyclic molecules. Any chain contains N_A units of A and N_B units

Table 1 Polycondensation of 4,4′-Dicarboxylic-stilbene with α,ω-Dihydroxy-poly (oxytetramethylene)

R_{th}	R_{exp}	$\overline{x}_{th}$	$\overline{x}_{NMR}$	$\overline{M}_{th}$	$\overline{M}_{NMR}$
76.3	73	18.8	16	6500	5540
50.0	52	5.3	6	2000	2210
30.0	28	1.8	1.5	850	770

R_{th}, R_{exp}, $DP_{n,th}$, $DP_{n,NMR}$, M_{th}, and M_{NMR} are, respectively, the theoretical (th) and experimental (NMR) values of the R content (molar %) in the copolymer and its degree of polymerization $\overline{DP}_n$ and its molar mass.

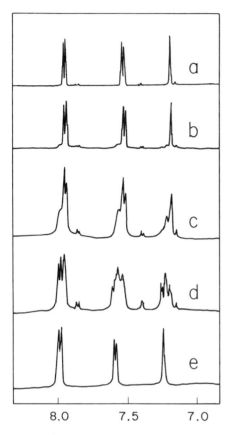

Fig. 1 ^{1}H-NMR spectra (500-MHz) of the block copolymer given in scheme 11 with different values of R%: (a) 100; (b) 66; (c) 50; (d) 30; (e) 0. CHCl$_3$, δ (ppm).

of B (Scheme 11) which are found in four arangements (Scheme 12) and N_A/N_B = $(I + III/2 + IV)/(II + III/2)$. Deconvolution of the double bond signal (Fig. 2) leads to the following assignments: peaks 5, 6, 7 to I; peak 2 to II; peaks 3 and 4 to III; and peak 8 to IV. Integration gives the following distribution (%): I 38.4; II 17.10; III 41.05; and IV 3.45. Assuming a statiscal distribution of polyether and

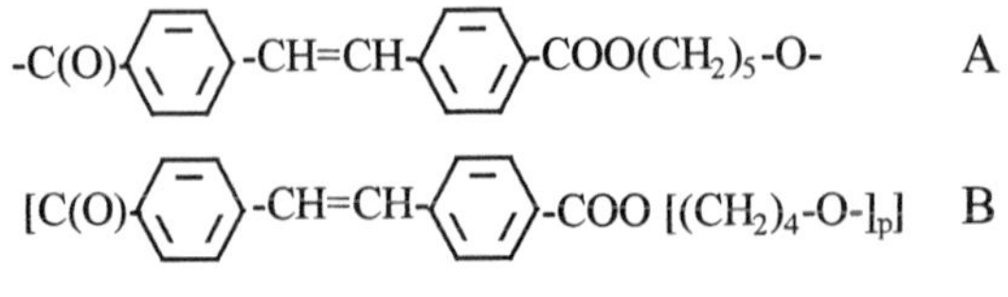

$$-C(O)\langle\overline{}\rangle-CH{=}CH{-}\langle\overline{}\rangle{-}COO(CH_2)_5{-}O{-} \qquad A$$

$$[C(O)\langle\overline{}\rangle-CH{=}CH{-}\langle\overline{}\rangle{-}COO\ [(CH_2)_4{-}O{-}]_p] \qquad B$$

Scheme 11

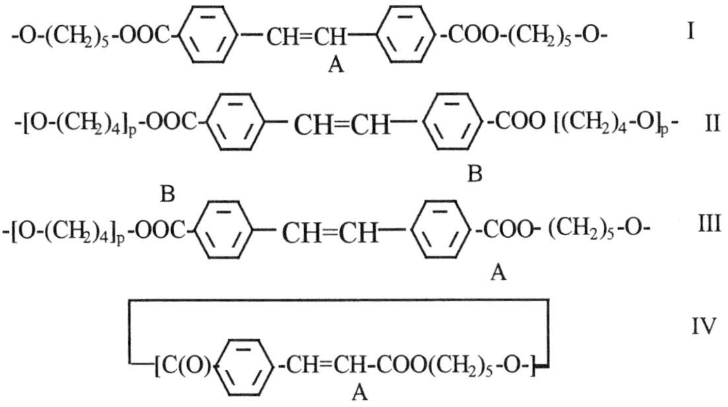

-O-(CH$_2$)$_5$-OOC-⟨‾⟩-CH=CH-⟨‾⟩-COO-(CH$_2$)$_5$-O-   I
                      A

-[O-(CH$_2$)$_4$]$_p$-OOC-⟨‾⟩-CH=CH-⟨‾⟩-COO [(CH$_2$)$_4$-O]$_p$-   II
                          B

    B
-[O-(CH$_2$)$_4$]$_p$-OOC-⟨‾⟩-CH=CH-⟨‾⟩-COO- (CH$_2$)$_5$-O-   III
                      A

                                IV
┌─────────────────────────────────────┐
└-[C(O)-⟨‾⟩-CH=CH-COO(CH$_2$)$_5$-O-]┘
                A

Scheme 12

pentanediyl units, the theoretical contents of I + IV, II, and III are 41.35, 12.75, and 45.9 respectively, and are close to experimental values: 42.85, 17.10, and 41.05.

This example shows that it is possible to obtain structural information on the chain itself even though rigid blocks were prepared during the block polycondensation.

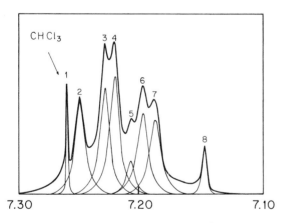

Fig. 2 Extending and deconvolution of ^{1}H-NMR spectrum (500-MHz) of the block copolymer given in scheme 11 with R% = 30. CHCl$_3$, δ (ppm).

D. Oligomer Coupling Reactions

Oligomer coupling reactions were reviewed by Fradet (52). This is the only way to prepare block copolycondensates from oligomers either with the same end groups or with different end groups unable to interreact. Contrary to the second method, coupling leads to well-defined blocks, since the functionnal oligomers are prepared and characterized before their polycondensation; on the other hand, their distribution is statistical, since one of the blocks can be repeated several times, one after another.

1. Oligomers with Hydroxy End Groups

a. Diisocyanates as Coupling Agents

In the same way, diacid chlorides can be used instead of diisocyanates (Scheme 13). Diisocyanates are efficient coupling agents as long as the temperature is not too high, when they would trimerize. When acid chlorides are used, hydrogen chloride is released, which often induces side reactions.

HO-(POLYETHER)-OH + HOCH$_2$-(POLYSILOXANE)- CH$_2$ OH + O=C=N-X-N=C=O

$\longrightarrow$ $\sim\!\sim\!\sim$(POLYETHER)-O$_7$C-NH -X-NH$_7$C-O-CH$_2$-(POLYSILOXANE)$\sim\!\sim\!\sim$
$\qquad\qquad\qquad$ O$\qquad\qquad\qquad$ O

Scheme 13

b. Bis(cyclic iminoesters) as Coupling Agents

Bis(cyclic iminoesters) are very efficient coupling agents and much less sensitive to temperature, which permits their polycondensation with molten poly(alkylene terepthalates) at 240°C, as shown in Scheme 14 (54). The following bis(cyclic iminoesters) are particularly efficient in coupling reactions (14):

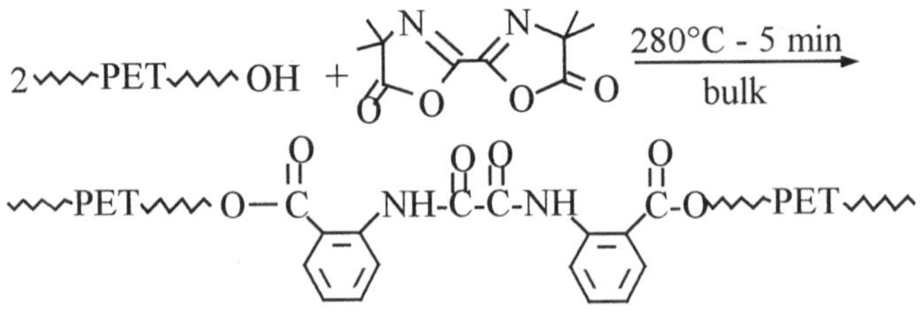

Scheme 14

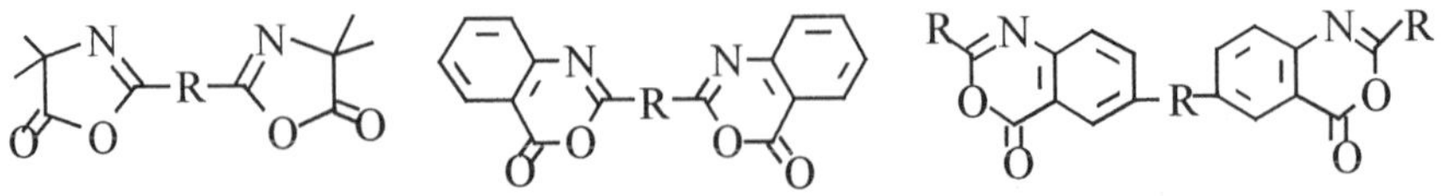

Scheme 15

Acevedo and Fradet (55) studied the polyaddition of bisoxazolinones with several α,ω-dihydroxy-polyethers, leading to diblock polyethers (Scheme 16):

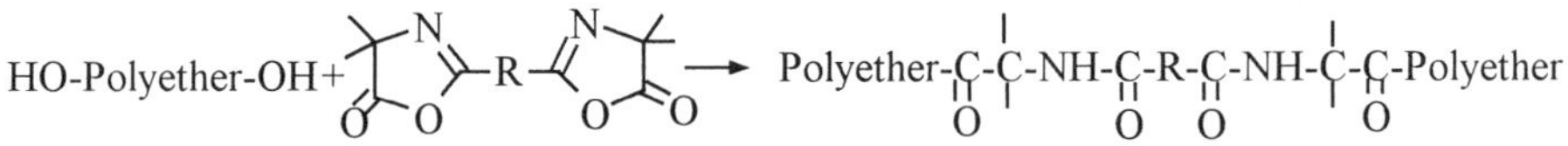

HO-Polyether-OH + [structure] ⟶ Polyether-$\overset{\text{O}}{\underset{}{\text{C}}}$-$\overset{}{\underset{}{\text{C}}}$-NH-$\overset{\text{O}}{\underset{}{\text{C}}}$-R-$\overset{\text{O}}{\underset{}{\text{C}}}$-NH-$\overset{}{\underset{}{\text{C}}}$-$\overset{\text{O}}{\underset{}{\text{C}}}$-Polyether

Scheme 16

The reaction is carried out in bulk and catalyzed by 4-dimethylaminopyridine. When the substrates are α,ω-dihydroxy-poly(oxytetramathylene) and 2,2'-bis[4,4'-dimethyl-5(4H)-oxazolinone], the conversion is 100% after 30 min at 184°C and M_W ranges from 7800 to 19,000. This reaction is particularly rapid when carried out in a double-screw extruder.

2. Oligomers with Amino End Groups

a. Diisocyanates as Coupling Agents

The remarks relative to Scheme 13 hold for Scheme 17.

H_2N-(POLYAMIDE)-NH_2 + $HOCH_2$-(POLYSILOXANE)-CH_2OH + O=C=N-X-N=C=O

⟶ ∿∿∿(POLYAMIDE)-NH-$\overset{\text{O}}{\underset{}{\text{C}}}$-NH -X-NH-$\overset{\text{O}}{\underset{}{\text{C}}}$-$OCH_2$-(POLYSILOXANE)∿∿∿

Scheme 17

b. Bis(cyclic iminoesters) as Coupling Agents

Reactions such as Scheme 16 can be applied to oligomers with amino end groups using bis(cyclic iminoethers), particularly bisoxazolones, which are stable and lead to linear condensation polymers by solution reaction with diamines (56). In both cases the nucleus contains several amide functions. Acevedo and Fradet (18) prepared polyether-*block*-polyamides by polyaddition of bisoxazolones with the mixture of an α,ω-diamino-polyether and an α,ω-diamino-polyamide (Scheme 18).

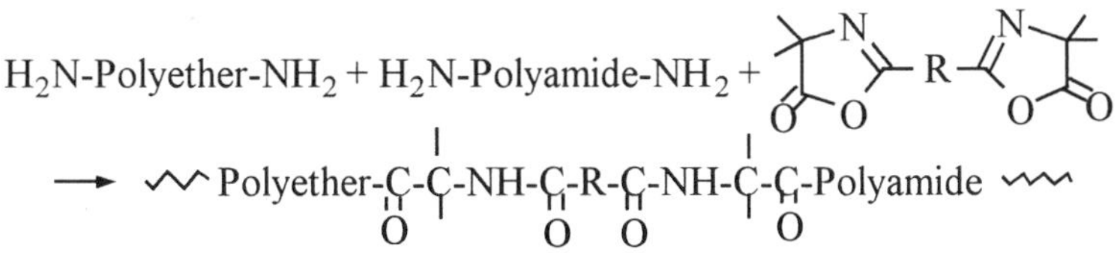

H_2N-Polyether-NH_2 + H_2N-Polyamide-NH_2 + [structure]

⟶ ∿∿Polyether-$\overset{\text{O}}{\underset{}{\text{C}}}$-$\overset{}{\underset{}{\text{C}}}$-NH-$\overset{\text{O}}{\underset{}{\text{C}}}$-R-$\overset{\text{O}}{\underset{}{\text{C}}}$-NH-$\overset{\text{O}}{\underset{}{\text{C}}}$-$\overset{}{\underset{}{\text{C}}}$-Polyamide ∿∿∿

Scheme 18

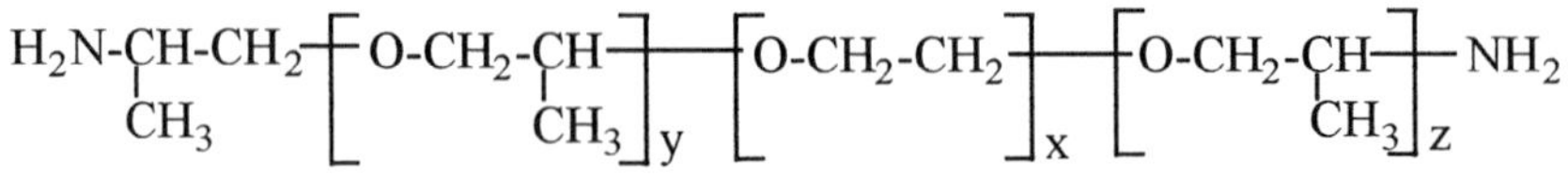

$$H_2N\text{-}\underset{CH_3}{CH}\text{-}CH_2\underset{}{\left[O\text{-}CH_2\text{-}\underset{CH_3}{CH}\right]_y}\left[O\text{-}CH_2\text{-}CH_2\right]_x\left[O\text{-}CH_2\text{-}\underset{CH_3}{CH}\right]_z NH_2$$

x+y = 3.0   x = 14.8

Scheme 19

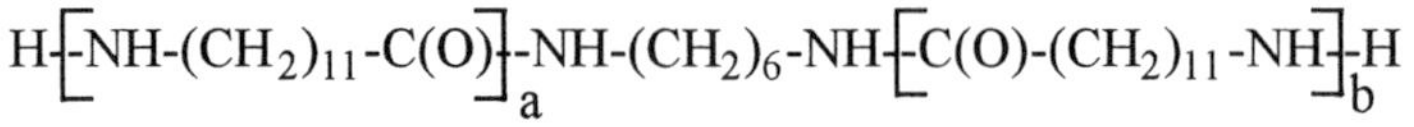

$$H\left[NH\text{-}(CH_2)_{11}\text{-}C(O)\right]_a NH\text{-}(CH_2)_6\text{-}NH\left[C(O)\text{-}(CH_2)_{11}\text{-}NH\right]_b H$$

a + b = 5.17

Scheme 20

When oligomers 19 and 20 are polycondensed with a bisonazolone in bulk at 200°C, conversion and M_{peak} are 100% and 13,900, respectively.

III. RING-OPENING POLYMERIZATION

The opening polymerizations used in the synthesis of block copolymers with polyethers as flexible blocks are ionic processes. However, they are often associated with other processes, such as polycondensation, chemical modification, and radical polymerization; for instance, the first step is a cationic polymerization and then the carbocation is changed into a free radical. In the following examples the first step is always ring-opening polymerization.

The same remark holds for the ring-opening polymerization of lactams, which are often accompanied by polycondensation. Hu et al. (2) prepared block copolyetheramides by melt polycondensation of a mixture of α,ω-diamino-oligo(oxypropylene) and piperazine with a mixture of diacid HOOC—R—COOH and ε-caprolactam, giving materials used as hot-melt adhesives. Their compositions and their characteristics are reproducible; however, their actual nature and the distribution of the rigid blocks are probably very different from those proposed in Scheme 21.

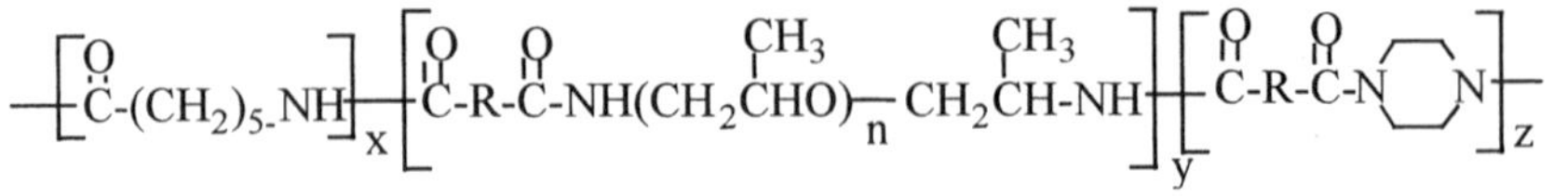

$$\left[\underset{}{\overset{O}{\underset{}{C}}}\text{-}(CH_2)_5\text{-}NH\right]_x\left[\overset{O}{\underset{}{C}}\text{-}R\text{-}\overset{O}{\underset{}{C}}\text{-}NH(CH_2\underset{CH_3}{CHO})_n CH_2\underset{CH_3}{CH}\text{-}NH\right]_y\left[\overset{O}{\underset{}{C}}\text{-}R\text{-}\overset{O}{\underset{}{C}}\text{-}N\diagup N\right]_z$$

Scheme 21

Biodegradable materials are the subject of many studies, particularly polyesters obtained by homo- or ring copolymerization of lactones, lactides, or cyclic carbonates. High-molar-mass poly(trimethylene carbonate) can be obtained by ring polymerization of 1,3-dioxan-2-one (trimethylene carbonate) initiated by

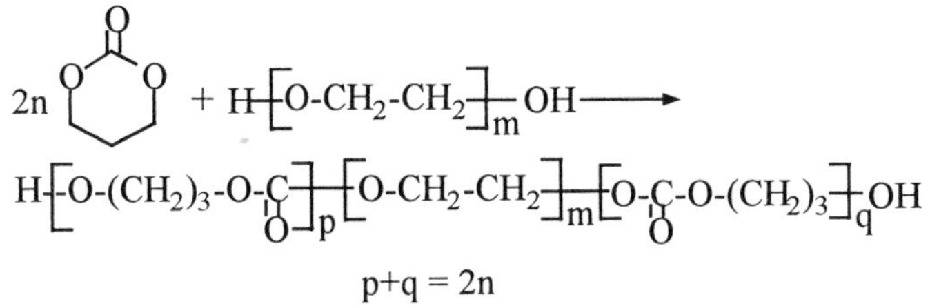

$$2n \quad \underset{O}{\overset{O}{\bigodot}}\overset{O}{} + H\text{-}[O\text{-}CH_2\text{-}CH_2]_m\text{-}OH \longrightarrow$$

$$H\text{-}[O\text{-}(CH_2)_3\text{-}O\text{-}\underset{O}{\overset{O}{C}}]_p\text{-}[O\text{-}CH_2\text{-}CH_2]_m\text{-}[O\text{-}\underset{O}{\overset{O}{C}}\text{-}O\text{-}(CH_2)_3]_q\text{-}OH$$

$$p+q = 2n$$

Scheme 22

$BF_3 \cdot O(C_2H_5)_2$ or $BuSnCl_3$, and its hydrolysis is slower than that of polylactones or polylactides; its copolymers are of great interest in medical applications (58,59). Wang et al. (38) prepared bioerodible ABA copolymers, where A is a poly(trimethylene carbonate) and B is polyoxyethylene; they were obtained by copolymerizing α,β-dihydroxy-polyoxyethylene and trimethylene carbonate in the presence of stannous octoate as catalyst and at 120°C (Scheme 22).

These copolymers (Scheme 21) were analyzed by infrared and NMR spectroscopies. Their 400-MHz ^{1}H-NMR spectra permitted determination of the poly(trimethylene carbonate) fraction in the copolymer calculated from the integrations of carbonate block methylene triplet (4.24 ppm) and poly(oxyethylene) methylene singlet (3.66 ppm). Their properties highly depend on their composition; they were purified via the addition of methanol to their solution in $CHCl_3$; however, when the polycarbonate content is below 20% (molar), no precipitation is observed.

Cho et al. (45) prepared and studied an ABA block copolymer, where A is poly(ε-benzyloxycarbonyl-L-lysine) and B is poly(oxyethylene). It was obtained by polymerization of the ε-benzyloxycarbonyl-L-lysine N-carboxyanhydride initiated by the amino groups located at both ends of the poly(oxyethylene) (Scheme 23).

Its $\overline{Mn}$ was determined from its ^{1}H-NMR by comparing the peak intensities

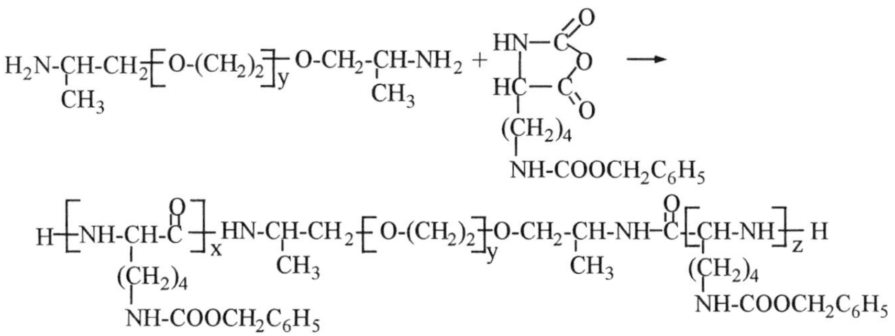

$$H_2N\text{-}\underset{CH_3}{\overset{|}{C}}H\text{-}CH_2\text{-}[O\text{-}(CH_2)_2]_y\text{-}O\text{-}CH_2\text{-}\underset{CH_3}{\overset{|}{C}}H\text{-}NH_2 + \underset{\underset{\underset{NH\text{-}COOCH_2C_6H_5}{|}}{(CH_2)_4}}{\overset{O}{\underset{HC\text{-}C}{HN\text{-}C}}}\overset{\nearrow O}{\underset{\diagdown O}{\diagup}} \longrightarrow$$

$$H\text{-}[NH\text{-}\underset{\underset{NH\text{-}COOCH_2C_6H_5}{|}}{\underset{(CH_2)_4}{\overset{|}{C}}}H\text{-}\overset{O}{\overset{\parallel}{C}}]_x\text{-}HN\text{-}\underset{CH_3}{\overset{|}{C}}H\text{-}CH_2\text{-}[O\text{-}(CH_2)_2]_y\text{-}O\text{-}CH_2\text{-}\underset{CH_3}{\overset{|}{C}}H\text{-}NH\text{-}\overset{O}{\overset{\parallel}{C}}\text{-}[\underset{\underset{NH\text{-}COOCH_2C_6H_5}{|}}{\underset{(CH_2)_4}{\overset{|}{C}}}H\text{-}NH]_z\text{-}H$$

Scheme 23

of benzyl protons (7.4 ppm) and polyoxyethylene block methylene protons (3.9 ppm). However, the diaminooligomer is an industrial sample whose functionality was "estimated" as 90%, which raises a doubt about the $\overline{Mn}$ value, because functionality is a determining parameter in the calculation of molar mass. The analysis of the functional oligomers raises problems, which are analyzed in Sec. V.A.

Hsiue et al. (60) prepared a multiblock copolymer of poly(oxyethylene) and poly(*N*-isovalerylethyleneimine) using a three-step process: (1) polymerization of 2-isobutyl-oxazoline, leading to a monodispersed oxazolonium end-capped poly(*N*-isovalerylethyleneimine); (2) its chemical modification to obtain lithium oxide end groups; (3) polycondensation of the dianion with polyoxyethylene ditosylate (Scheme 24).

Gallot collected interesting examples of block copolymers in a review devoted to comb and block liquid crystalline polymers for biological applications and containing polypeptide blocks (61); many of them contain polyethers and are prepared by ring-opening polymerization, often initiated by α,ω-diaminopolypeptides.

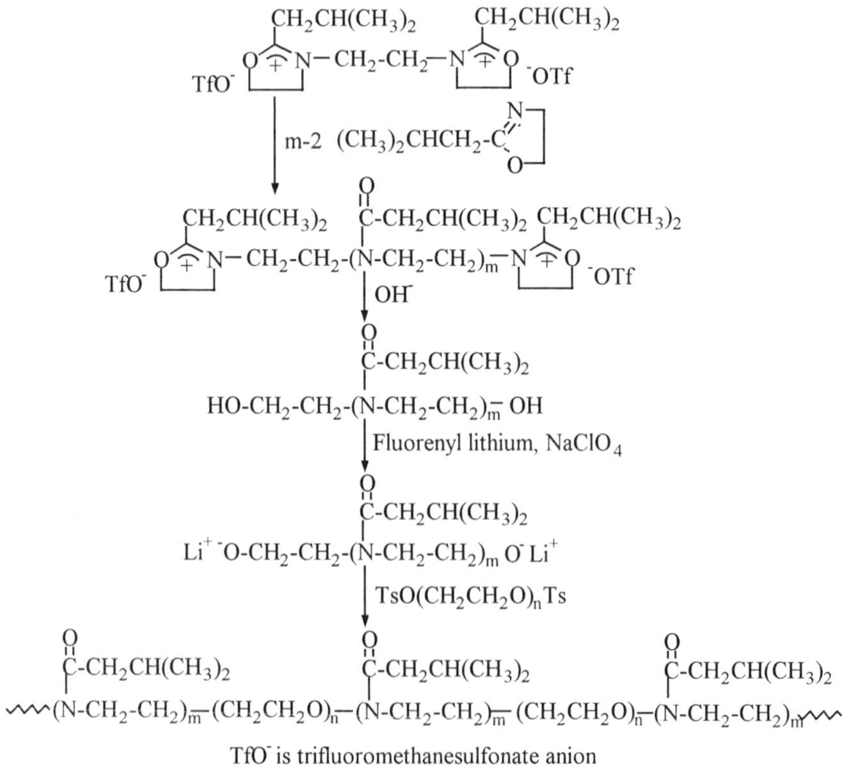
m-2 $(CH_3)_2CHCH_2$-C
OH⁻
Fluorenyl lithium, $NaClO_4$
$TsO(CH_2CH_2O)_nTs$
TfO⁻ is trifluoromethanesulfonate anion

Scheme 24

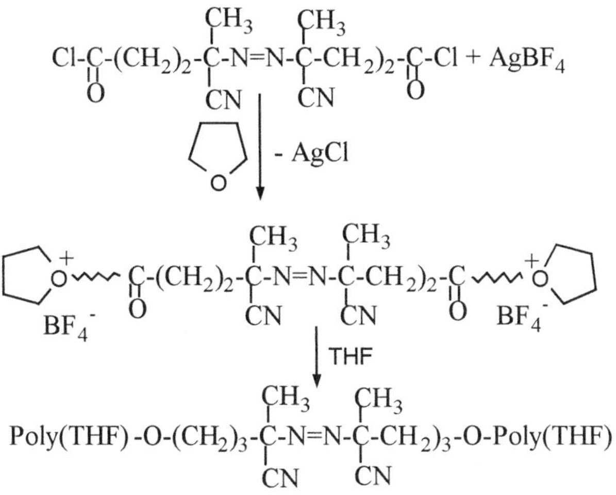

$$Cl-\underset{\underset{O}{\parallel}}{C}-(CH_2)_2-\underset{\underset{CN}{|}}{\overset{\overset{CH_3}{|}}{C}}-N=N-\underset{\underset{CN}{|}}{\overset{\overset{CH_3}{|}}{C}}-(CH_2)_2-\underset{\underset{O}{\parallel}}{C}-Cl + AgBF_4$$

$$\Big\downarrow - AgCl$$

$$\overset{+}{O}\!\sim\!\!\sim\!\!\sim \underset{\underset{O}{\parallel}}{C}-(CH_2)_2-\underset{\underset{CN}{|}}{\overset{\overset{CH_3}{|}}{C}}-N=N-\underset{\underset{CN}{|}}{\overset{\overset{CH_3}{|}}{C}}-(CH_2)_2-\underset{\underset{O}{\parallel}}{C}\!\sim\!\!\sim\!\!\sim\overset{+}{O}$$

$$BF_4^- \qquad\qquad\qquad\qquad\qquad\qquad\qquad BF_4^-$$

$$\Big\downarrow THF$$

$$Poly(THF)-O-(CH_2)_3-\underset{\underset{CN}{|}}{\overset{\overset{CH_3}{|}}{C}}-N=N-\underset{\underset{CN}{|}}{\overset{\overset{CH_3}{|}}{C}}-(CH_2)_3-O-Poly(THF)$$

Scheme 25

Yagci (40) prepared polymers containing thermolabile groups by associating cationic and radical polymerizations; initiating cations are formed at both ends of suitable azoinitiators, and the azo group contained in the chain initiates free-radical polymerization. This process was applied to tetrahydrofuran, cyclohexene oxide, and epichlorohydrin. The following example is relative to the preparation of poly(oxytetramethylene)-*block*-polystyrene-*block*-poly(oxytetramethylene).

Polytetrahydrofuran with a central azolinkage (62) was synthesized by cationic ring-opening polymerization of THF initiated by 4,4′-azobis(4-cyanopentanoyl chloride) in the presence of silver tetrafluoroborate with non-nucleophilic counteranion (Scheme 25). Kinetic results showed the absence of termination reactions (63); the molar mass of the macromolecular azo ranges from 3500 to 34,000, with a narrow distribution. It is used as a radical macroinitiator, with an azo group per molecule (Scheme 26).

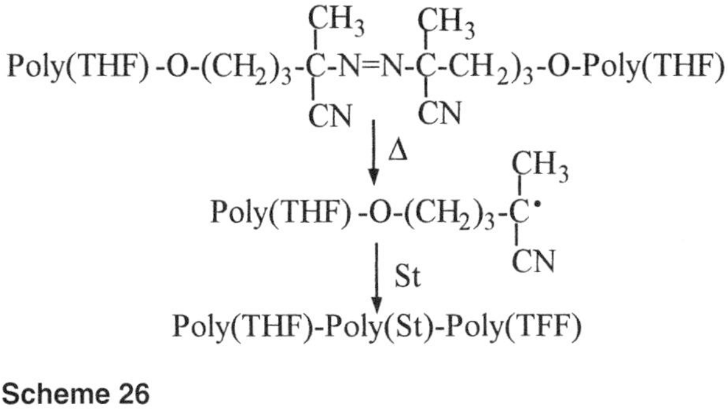

$$Poly(THF)-O-(CH_2)_3-\underset{\underset{CN}{|}}{\overset{\overset{CH_3}{|}}{C}}-N=N-\underset{\underset{CN}{|}}{\overset{\overset{CH_3}{|}}{C}}-(CH_2)_3-O-Poly(THF)$$

$$\Big\downarrow \Delta$$

$$Poly(THF)-O-(CH_2)_3-\underset{\underset{CN}{|}}{\overset{\overset{CH_3}{|}}{C}}{}^{\bullet}$$

$$\Big\downarrow St$$

$$Poly(THF)-Poly(St)-Poly(TFF)$$

Scheme 26

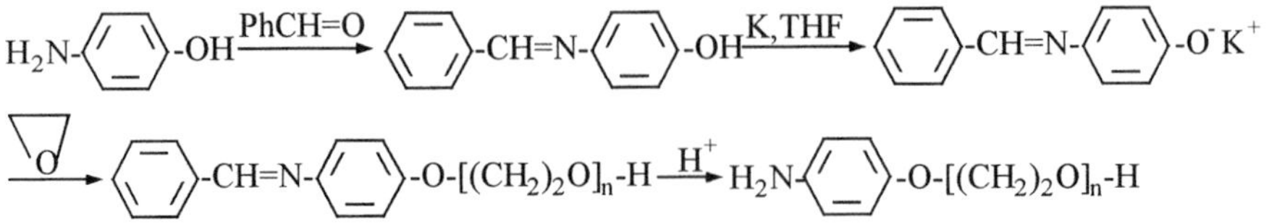

$$H_2N\text{-}\langle\bigcirc\rangle\text{-OH} \xrightarrow{\text{PhCH=O}} \langle\bigcirc\rangle\text{-CH=N-}\langle\bigcirc\rangle\text{-OH} \xrightarrow{\text{K,THF}} \langle\bigcirc\rangle\text{-CH=N-}\langle\bigcirc\rangle\text{-O}^-K^+$$

$$\xrightarrow{\overset{\text{O}}{\triangledown}} \langle\bigcirc\rangle\text{-CH=N-}\langle\bigcirc\rangle\text{-O-}[(CH_2)_2O]_n\text{-H} \xrightarrow{H^+} H_2N\text{-}\langle\bigcirc\rangle\text{-O-}[(CH_2)_2O]_n\text{-H}$$

Scheme 27

Living cationic polymerization could be extended to the preparation of azo polymers; Nguyen et al. synthesized azo polymers using azo-divinyl ethers (64). Huang and Huang (65) prepared α-(4-aminophenyl)-polyoxyethylene by anionic ring-opening polymerization and used it as initiator in methyl methacrylate charge-transfer polymerization (Scheme 27).

The charge-transfer polymerization of methyl methacrylate was carried out in the presence of benzophenone and under UV irradiation (Scheme 28). When the amine-benzophenone system is irradiated by 365-nm light, the benzophenone molecule absorbs a photon, giving an excited singlet, which then decays to an excited triplet by intersystem crossing, in this state it forms an exciplex with the aromatic amine. A proton transfer between the cation radical and the anion radical gives the corresponding free radicals. Diphenyl methanol radical is too stable to initiate methyl methacrylate polymerization, whereas the imino radical does. The polymerization and the polymer highly depend on the nature of the solvent, particularly the composition, the molar mass, and the dispersity index.

The block copolymer described in Scheme 28 was studied by ^{1}H-NMR and size exclusion chromatography. When the molar mass of the poly(oxyethylene) is 11,000 and a dispersity index is equal to 1.05, the corresponding block copolymer has a molar mass equal to 44,000 and its dispersity index is 1.40.

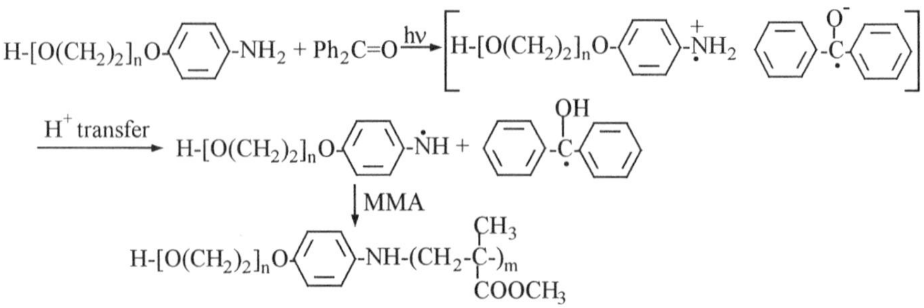

$$H\text{-}[O(CH_2)_2]_nO\text{-}\langle\bigcirc\rangle\text{-}NH_2 + Ph_2C=O \xrightarrow{h\nu} \left[H\text{-}[O(CH_2)_2]_nO\text{-}\langle\bigcirc\rangle\text{-}\overset{+}{N}H_2 \quad \langle\bigcirc\rangle\text{-}\overset{O^-}{\underset{\bullet}{C}}\text{-}\langle\bigcirc\rangle\right]$$

$$\xrightarrow{H^+ \text{ transfer}} H\text{-}[O(CH_2)_2]_nO\text{-}\langle\bigcirc\rangle\text{-}\dot{N}H + \langle\bigcirc\rangle\text{-}\overset{OH}{\underset{\bullet}{C}}\text{-}\langle\bigcirc\rangle$$

$$\downarrow MMA$$

$$H\text{-}[O(CH_2)_2]_nO\text{-}\langle\bigcirc\rangle\text{-}NH\text{-}(CH_2\text{-}\overset{CH_3}{\underset{COOCH_3}{C}}\text{-})_m$$

Scheme 28

IV. BLOCK COPOLYMERS PREPARED BY RADICAL POLYMERIZATION

Améduri et al. (66) reviewed the application of radical polymerization to the preparation of block copolymers. In the following sections, these three main methods are studied:

Polymerization of a monomer initiated by the end group of a polymer
The use of controlled radical polymerization
The use of macroinitiator

A. Polymerization of a Monomer Initiated by the End Group of a Polymer

Several authors (41,43) used manganese derivatives as initiators, such as $Mn_2(CO)_{10}$/α-methyl-ω-tribromoalkyl-poly(oxyethylene) (41), to prepare a multiblock of poly(oxyethylene) and polystyrene (Scheme 29). Oligomer I was obtained by reacting the sodium salt of ω-methoxy-polyoxyethylene with tribromoacetyl chloride.

Çakmak (43) prepared triblock copolymers by redox polymerization process (Scheme 30). Only triblock copolymers were formed, because α,ω-dihy-

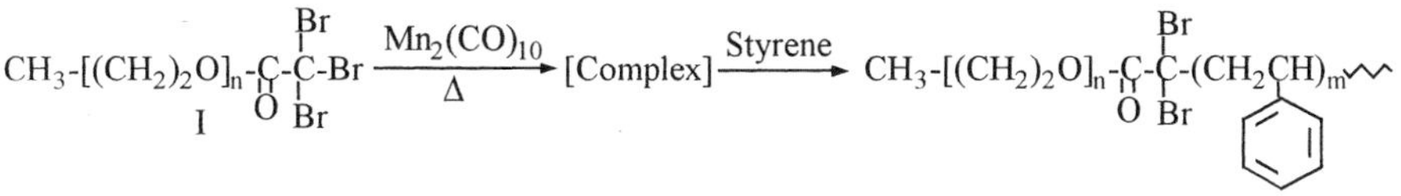

$$CH_3\text{-}[(CH_2)_2O]_n\text{-}\overset{Br}{\underset{\underset{Br}{O}}{C}}\text{-}\overset{}{C}\text{-}Br \xrightarrow[\Delta]{Mn_2(CO)_{10}} [Complex] \xrightarrow{Styrene} CH_3\text{-}[(CH_2)_2O]_n\text{-}\overset{Br}{\underset{\underset{Br}{O}}{C}}\text{-}\overset{}{C}\text{-}(CH_2CH)_m$$

I

Scheme 29

$$HOCH_2CH_2\text{-}(OCH_2CH_2)_{32}\text{-}OCH_2CH_2\,OH + 2Mn(III)$$

$$\updownarrow$$

Complex

$$HO\overset{\bullet}{C}HCH_2\text{-}(OCH_2CH_2)_{32}\text{-}OCH_2\overset{\bullet}{C}HOH + 2Mn(II) + 2H^+$$

$$\downarrow nM$$

Poly(M)-*block*-Polyoxyethylene-*block*-Poly(M)

Where M is methyl methacrylate or acrylamide

Scheme 30

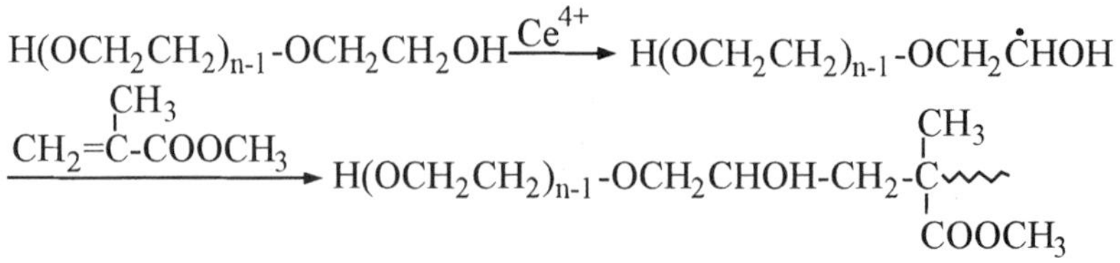

$$H(OCH_2CH_2)_{n-1}\text{-}OCH_2CH_2OH \xrightarrow{Ce^{4+}} H(OCH_2CH_2)_{n-1}\text{-}OCH_2\overset{\bullet}{C}HOH$$

$$\underset{CH_2=\overset{|}{C}\text{-}COOCH_3}{\xrightarrow{\hspace{2cm}}} H(OCH_2CH_2)_{n-1}\text{-}OCH_2CHOH\text{-}CH_2\text{-}\underset{\underset{COOCH_3}{|}}{\overset{\overset{CH_3}{|}}{C}}\text{\textasciitilde\textasciitilde\textasciitilde}$$

Scheme 31

droxy-polyoxyethylene gives a biradical that initiates the formation of two blocks, and their polymerization is stopped by reaction with Mn(III).

Nagarajan et al. (49,81) studied the copolymerization of methyl methacrylate initiated by the system Ce(IV)/α,ω-dihydroxy-polyoxyethylene (Scheme 31). When α,ω-dihydroxy-polyoxyethylene is maintained only in the presence of ceric ions, a long-chain aldehyde is formed. On the other hand, when ceric ions are added to methyl methacrylate, no polymerization takes place, showing that only the redox system is able to initiate the polymerization. The polymerization is carried out in water, and the medium is homogeneous, at least at the beginning of the reaction; however, when the length of the poly(methyl methacrylate) increases, the polymer separates from the water. The polymerization rate is proportional to [methyl methacrylate]^2 and [PEG] but does not depend on [Ce^{4+}] and [H$^+$]; this completely rules out the possibility of mutual combination, because that would require a 1.5 order in methyl methacrylate; moreover, this probably shows that the chains are terminated by reaction with Ce^{4+}.

B. Use of Controlled Radical Polymerization

Controlled radical polymerization has seen a rapid development in the preparation of well-defined homopolymers and block copolymers (82). Matyjaszewski (67,68) proposed several processes in controlled radical polymerization: unimolecular exchanges between growing radicals and dormant alkoxyamines or dormant organometallic species, and bimolecular exchanges between growing radicals and dormant species. Up to now this was very little used in the preparation of block copolymers with polyethers as flexible block; it is obvious that the "controlled radical polymerization" step must be prepared from a polyether whose end groups are modified into an α,ω-dinitroxyl oligomer. This was done by Yoshida and Sugita (69,70), who prepared polytetrahydrofuran with 2,2,6,6-tetramethyl-piperidinyl-1-oxyl (Tempo) as end group and used this polymer to initiate the controlled radical polymerization of styrene (Scheme 32).

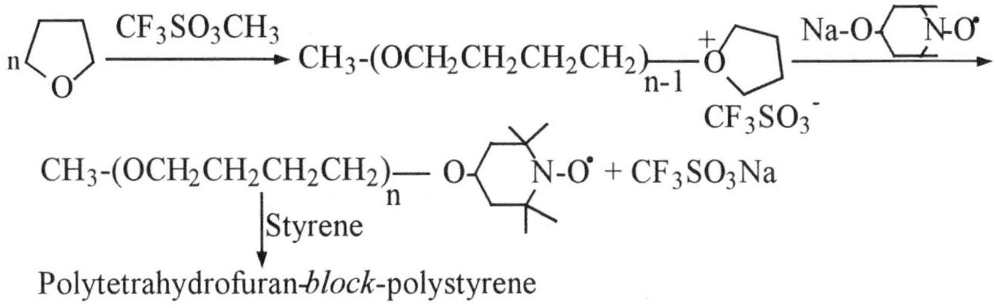

$$\underset{n}{\overset{}{\bigtriangleup}}_{O} \xrightarrow{CF_3SO_3CH_3} CH_3\text{-}(OCH_2CH_2CH_2CH_2)\overset{}{\underset{n-1}{}}\overset{+}{\bigtriangleup}_{O} \quad CF_3SO_3^- \qquad \xrightarrow{\text{Na-O}-\boxed{}\text{-N-O}^{\bullet}}$$

$$CH_3\text{-}(OCH_2CH_2CH_2CH_2)\underset{n}{}\!\!- O-\boxed{}\text{-N-O}^{\bullet} + CF_3SO_3Na$$

$$\Big\downarrow \text{Styrene}$$

Polytetrahydrofuran-*block*-polystyrene

Scheme 32

C. Use of Macroinitiator

This section is close to some of the examples given in Sec. III; however, the strategy is somewhat different.

Yagci et al. (71) prepared two identical samples of poly[polystyrene-*block*-poly(oxytetramethylene)] in two different ways. In the first method, a polystyrene radical is modified into a polystyrene cation, which initiates the polymerization of tetrahydrofuran (Scheme 33). In the second method, the same macroinitiator is irradiated by UV (350 nm), giving a macrobiradical, which is modified by EMP$^+$ (see Scheme 33), leading to a bication that initiates the cationic polymerization of tetrahydrofuran. The poly(oxytetramethylene) contains an azo group able to polymerize styrene (Scheme 34). The type of block copolymer depends greatly on the kinetics of these systems and on the nature of the monomers. In principle there is only one azo group per poly(oxytetramethylene) chain generating two growing

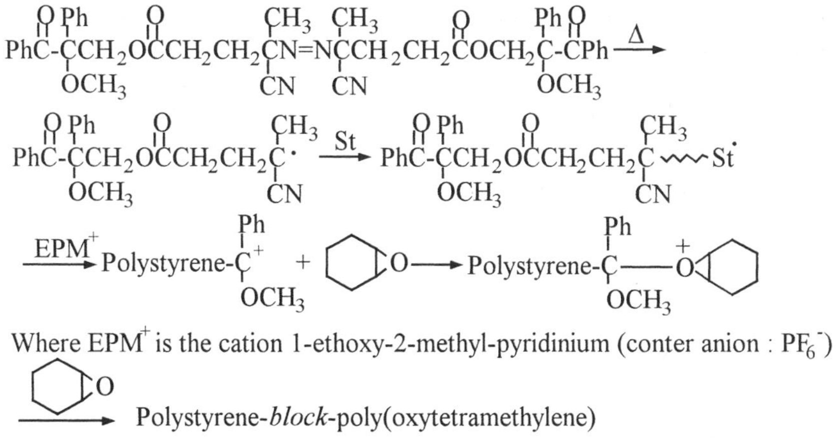

$$PhC\text{-}CCH_2OCCH_2CH_2CN=NCCH_2CH_2COCH_2C\text{-}CPh \xrightarrow{\Delta}$$

$$PhC\text{-}CCH_2OCCH_2CH_2C^{\bullet} \xrightarrow{St} PhC\text{-}CCH_2OCCH_2CH_2C\!\!\sim\!\!\sim\!St^{\bullet}$$

$$\xrightarrow{EPM^+} \text{Polystyrene-}C^+ + \boxed{}O \longrightarrow \text{Polystyrene-}C\text{---}O^+\boxed{}$$

Where EPM$^+$ is the cation 1-ethoxy-2-methyl-pyridinium (conter anion : PF$_6^-$)

$$\boxed{}O \xrightarrow{} \text{Polystyrene-}block\text{-poly(oxytetramethylene)}$$

Scheme 33

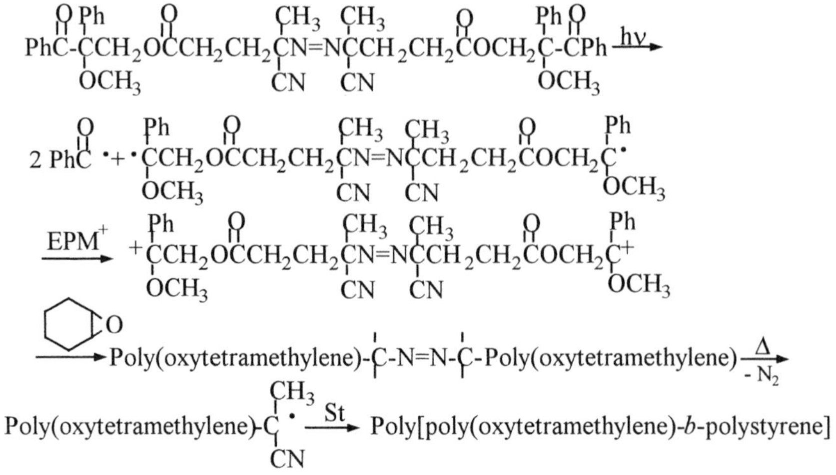

$$
\overset{O}{\underset{\underset{OCH_3}{|}}{PhC}}-\overset{Ph}{\underset{}{C}}CH_2O\overset{O}{\overset{\|}{C}}CH_2CH_2\overset{CH_3}{\underset{\underset{CN}{|}}{C}}N=N\overset{CH_3}{\underset{\underset{CN}{|}}{C}}CH_2CH_2\overset{O}{\overset{\|}{C}}OCH_2\overset{Ph}{\underset{\underset{OCH_3}{|}}{C}}-CPh \xrightarrow{h\nu}
$$

$$
2\ Ph\overset{O}{\overset{\|}{C}}\ \cdot+\cdot\overset{Ph}{\underset{\underset{OCH_3}{|}}{C}}CH_2O\overset{O}{\overset{\|}{C}}CH_2CH_2\overset{CH_3}{\underset{\underset{CN}{|}}{C}}N=N\overset{CH_3}{\underset{\underset{CN}{|}}{C}}CH_2CH_2\overset{O}{\overset{\|}{C}}OCH_2\overset{Ph}{\underset{\underset{OCH_3}{|}}{C}}\cdot
$$

$$
\xrightarrow{EPM^+}\ {}^+\overset{Ph}{\underset{\underset{OCH_3}{|}}{C}}CH_2O\overset{O}{\overset{\|}{C}}CH_2CH_2\overset{CH_3}{\underset{\underset{CN}{|}}{C}}N=N\overset{CH_3}{\underset{\underset{CN}{|}}{C}}CH_2CH_2\overset{O}{\overset{\|}{C}}OCH_2\overset{Ph}{\underset{\underset{OCH_3}{|}}{C}}{}^+
$$

$$
\xrightarrow{} Poly(oxytetramethylene)\text{-}\overset{|}{C}\text{-}N=N\text{-}\overset{|}{C}\text{-}Poly(oxytetramethylene)\xrightarrow[-N_2]{\Delta}
$$

$$
Poly(oxytetramethylene)\text{-}\overset{CH_3}{\underset{\underset{CN}{|}}{C}}\cdot \xrightarrow{\ St\ } Poly[poly(oxytetramethylene)\text{-}b\text{-}polystyrene]
$$

Scheme 34

polystyrene chains terminated by radical coupling. From the information given in Ref. 71 it is difficult to know wether disproportionation takes place or not; if it has a significant contribution to the process, this probably has a drastic influence on the length and distribution of the blocks.

Ropot et al. (72) published a large review on block and graft copolymers prepared by thermal and photochemical radical polymerizations in the presence of macromolecular initiators, including aliphatic polyethers. Galli et al. (73) prepared and studied liquid crystalline block copolymers using sequential cationic or promoted cationic and free-radical polymerizations; some of these copolymers contain aliphatic polyethers as flexible blocks. These authors elegantly illustrated their results with an interesting diagram, which is summarized in Scheme 35.

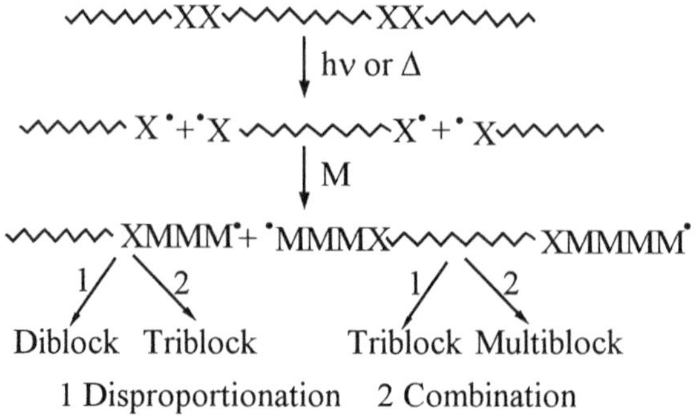

$$
\text{wwwXXwwwwwXXwww}
$$
$$
\downarrow h\nu\ or\ \Delta
$$
$$
\text{wwwX}^\bullet+{}^\bullet X\text{wwwwwwX}^\bullet+{}^\bullet X\text{www}
$$
$$
\downarrow M
$$
$$
\text{wwwXMMM}^\bullet+{}^\bullet MMMX\text{wwwww}XMMMM^\bullet
$$

Diblock Triblock   Triblock Multiblock

1 Disproportionation   2 Combination

Scheme 35

V. TECHNIQUES OF CHARACTERIZATION: SIDE REACTIONS

A. Functional Oligoethers

The characterization of oligomers is essential. It is necessary to control their exact composition (many of them are mixtures of homo- and copolymers), their functionality, which is particularly important in polycondensation, and their stability in the experimental conditions chosen for the blocking reaction.

Boularès et al. (22,23) prepared poly[polydodecanamide-*block*-copoly–ether] by polycondensation of an α,ω-dicarboxylic-polydodecanamide with different α,ω-dihydroxy-poly[poly(oxyethylene)-*block*-poly(oxypropylene)] (Scheme 4). In principle these copolymers contain two poly(oxyethylene) blocks (OE) and a central poly(oxypropylene) block (OP) (Scheme 36).

poly(oxyethylene)-*block*-poly(oxypropylene)-*block*-poly(oxyethylene)

Scheme 36

They are commercial products: Symperonics® (ICI) and Pluranics® (BASF). Their structures were carefully characterized, and to some extent it appeared that the experimental results does not fit Scheme 36. In the following, the characteristics of one of these oligomers (Symperonic L31®) and the analytical methods that were used are thoroughly studied. Mn^- was obtained by SEC, MS, and ^1H-NMR; Table 2 shows that the values are in good agreement.

Heatley et al. (74) provided very valuable NMR spectroscopic information on polyethers. The ^1H-NMR spectra of the copolyethers after trifluoration permitted determination of the ratio of oxyethylene and oxypropylene units in the chain and the concentration of primary (triplet at 4.42 ppm) and secondary (sextuplet at 5.16 ppm) hydroxy. Their ^{13}C-NMR spectra were compared to those of α,ω-dihydroxy-polyoxyethylene and α,ω-dihydroxy-poly(oxypropylene), which confirmed the results from the ^1H-NMR spectrum and permitted the identification of each configuration in the chain. The results given in Table 3 show that the actual values are somewhat some different from those proposed by the supplier, and knowledge of this information is essential to their polycondensation with α,ω-dicarboxylic-oligododecanamide. In principle the end groups are supposed to be primary alcohols; however, some of them are secondary alcohols, which changes completely the kinetics of the polycondensation. The ratio OE/OP is also very im-

Table 2 $\overline{Mn}$ of Copolyether L31

$\overline{Mn}_{th}$	$\overline{Mn}$ SEC	$\overline{Mn}$ MS	$\overline{Mn}$ ^1H-NMR
1100	1095	1000	1090

The values were obtained by size exclusion chromatography (SEC), mass spectrometry, and ^1H-NMR. $\overline{Mn}_{th}$ is the value given for the product.

Table 3 Characteristics of α,ω-Dihydroxy-copolyethers

Sample	OE/OP	%-pri-OH	%-sec-OH
S L31	0.25	50	50
S L35	1.26	90	10
S L42	0.26	75	25
S L43	0.77	80	20
S L44	1.09	85	15
S L61	0.23	60	40
P 4300	0.65	80	20
P 6100	0.14	45	55
P 3100	0.16	35	65

Symperonics® (S) and Pluronics® (P). % pri-OH, % sec-OH, and OE/OP are the molar contents of primary and secondary OH and the ratio of the numbers of oxyethylene and oxypropylene units, respectively.

portant, for it determines the hydrophily/hydrophoby balance and consequently the water content in the material, which can drastically modify the catalytic efficiency when it is an organometallic compound.

The correlation of NMR with mass spectroscopy (electrospray) showed that L31 is a mixture of polyoxypropylene (40%), diblock copolymer (20%), and triblock copolymer (40%), which was confirmed by its MALDI flight time spectrum.

Functionality is an essential parameter, but its value must be carefully analyzed. The global functionality of copolyethers (Scheme 36), is definitely 2 (22,23); however, it is necessary to state precisely the meaning of this value. In the case of L31, the functionality of secondary OH ranges from 0.2 to 1.3 and that of primary OH ranges from 0.7 to 1.8; the behavior of sec-OH and pri-OH is often similar, but it may be different, depending on the nature of the reaction.

It is amazing how little attention is paid to the composition, the impurity, and the structure of the initial oligomers. Thuillier et al. (34,35,75) studied the polycondensation of several aromatic diacids with α,ω-dihydroxy-polyoxytetramethylene and showed that the oligoethers contained a significant content of 1-butanol and butyrolactone, which changes the stoichiometric balance; moreover, they can be termination agents.

The presence of macrocycles concerns not only the polyether block but also the rigid one. Rozes (76) prepared liquid crystalline copolymers whose flexible blocks are aliphatic polyethers, and the rigid sequences were prepared by polytransesterification of aliphatic diols with aromatic diesters derived from biphenyl of stilbene. Oligomers (Scheme 37) always contain macrocycles where n ranges from 3 to 6.

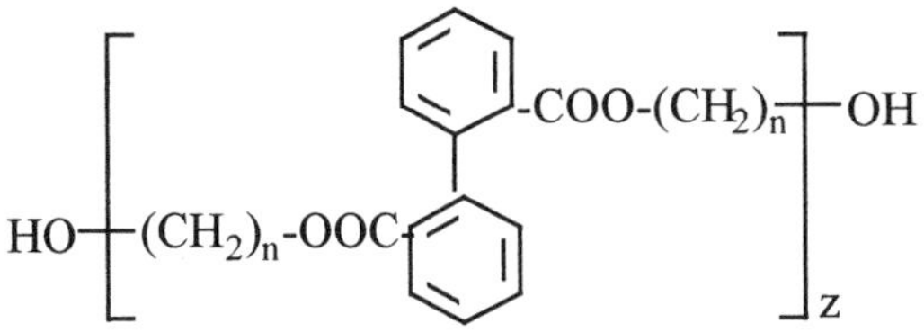

Scheme 37

(Scheme 38), whose formation is favored by the vicinity of esters in positions 2 and 2'.

The determination of the functionality of the oligomers is a difficult process. Many authors determine the molar mass of the oligomers from the end group concentration, implicitly assuming that the functionality is 2. It is absolutely necessary to compare the molar mass obtained by direct determination (chromatography, tonometry, mass spectrometry, osmometry) and from the concentration of the end groups. End group concentration can be determined by chemical titration, ^{1}H-NMR, or infrared spectroscopy, for instance; the use of several of these methods increases accuracy, but the actual functionality can be obtained only by comparing these values to those resulting from the direct determination of molar mass.

The use of ^{1}H- and ^{13}C-NMR gives deep insight into the structure of the end groups. Acevedo and Fradet (18) prepared poly[polyethers-*block*-polyamides] by polyaddition of bisoxazolones with mixtures of α,ω-di(2-aminopropyl)-poly(oxyethylene) (Jeffamine ED-900®) and α,ω-diamino-polydodecanamide (Scheme 18). The Jeffamine was extensively characterized; its functionality is very close to 2, determined by chemical titration, and ^{1}H- and ^{13}C-NMR spectra

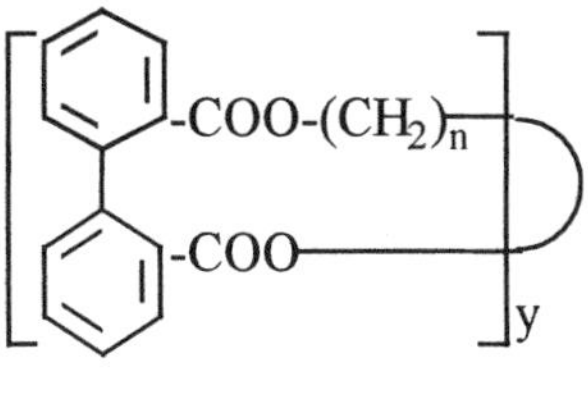

where n ranges from 4 to 6.

Scheme 38

show that chain ends are different (Scheme 39):

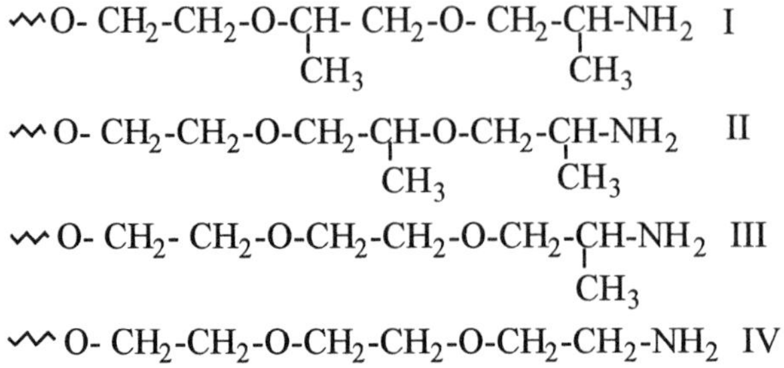

$\sim$O- CH$_2$-CH$_2$-O-CH- CH$_2$-O- CH$_2$-CH-NH$_2$ I
$\qquad\qquad\qquad$ |$\qquad\qquad\qquad\quad$|
$\qquad\qquad\quad$ CH$_3$$\qquad\qquad\qquad$ CH$_3$

$\sim$O- CH$_2$-CH$_2$-O-CH$_2$-CH-O-CH$_2$-CH-NH$_2$ II
$\qquad\qquad\qquad\qquad$ |$\qquad\qquad$|
$\qquad\qquad\qquad\quad$ CH$_3$$\qquad\quad$ CH$_3$

$\sim$O- CH$_2$- CH$_2$-O-CH$_2$-CH$_2$-O-CH$_2$-CH-NH$_2$ III
$\qquad\qquad\qquad\qquad\qquad\qquad\qquad$ |
$\qquad\qquad\qquad\qquad\qquad\qquad\quad$ CH$_3$

$\sim$O- CH$_2$-CH$_2$-O-CH$_2$-CH$_2$-O-CH$_2$-CH$_2$-NH$_2$ IV

Scheme 39

The content of the unexpected IV is 6%. The first three oligomers have probably very close reactivity; on the other hand, that of IV is somewhat different.

B. Block Copolymers

The preparation of block copolymers is often accompanied by side phenomena, such as:

The possible degradation of the oligomers during the blocking reaction, particularly the aliphatic oligoether.

The reactional medium can be heterogeneous, and the distribution of the reactants and of the catalyst between the phases is often unknown.

Contribution of secondary reactions such as cyclizations.

Formation of homopolymers and poor control of the distribution of the blocks in the chains and between the chains.

α,ω-Dihydroxy-oligoethers may contain numerous light components (see Sec. V.A), which may be eliminated during the polycondensation or may react with the chain end groups, which stops the propagation. On the other hand, the thermal stability of the oligomers may be very good in the reactional medium; Deleens (10–12) studied the polycondensation of α,ω-dihydroxy-oligo (oxyethylenes) with aliphatic α,ω-dicarboxylic polyamides at several temperatures and observed no significant degradation of the polyether, even though some of the reactions were carried out at 260°C. The formation of macrocycles was analyzed in Sec. II.C (34,35); some of them are already present in the precursors. Thuillier (75) showed that α,ω-dihydroxy-oligo(oxytetramethylenes) contain light fractions (see Sec. V.A), and often their contribution increases during the polycondensation. When α,ω-dihydroxy-oligo(oxytetramethylene) ($\overline{Mn}$ = 2030 g-mol^{-1}) is maintained at 250°C for 4 h, in the presence of tetrabutoxytitane (cat-

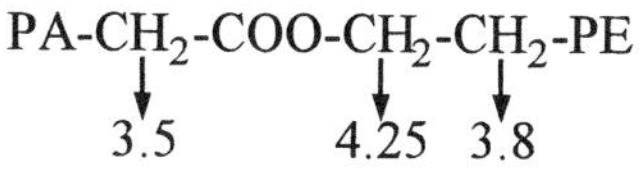

PA-CH$_2$-COO-CH$_2$-CH$_2$-PE

3.5   4.25   3.8

Scheme 40

alyst of the polycondensation) and under vacuum, *n*-butanol and butyrolactone are formed and eliminated under distillation; they probably result from cleavages and rearrangements of the chains.

There are very few absolute proofs of the formation of block copolymers free from homopolymers, and in some cases there is no copolymer at all. In most cases, elementary analysis and spectroscopy give characteristics that are common to the block copolymer and to the mixture of the corresponding homopolymers. In a very few cases it is possible to characterize the junction between the blocks. Boularès (22,23) prepared poly[polyamide-*block*-copolyether]s that were prepared by polycondensation of an α,ω-dicarboxylic-oligoamide (Mn = 1200) with different α,ω-dihydroxy-copolyethers (Mn around 1000); their ^{1}H-NMR spectra clearly show the signals (ppm) relative to the ester junction between the blocks (Scheme 40), even though the molar mass of the block copolymer is close to 20,000. Size exclusion chromatography often gives an unambigous proof of the formation of the copolymer. Otsuki et al. (1) prepared multiblock copolymers by reacting a mixture of α,ω-dicarboxylic-oligo(oxyethylene), isophthalic acid, and nonanedioic with di(4-isocyanatophenyl)-methane (Scheme 41). The synthesis can be a one-step process (B), in which the components are reacted all together, or a two-step process, in which the diacid mixture is reacted with the diisocyanate, leading to an α,ω-diisocyanate-polyamide, and then the latter is polycondensed with II (C).

Their elementary analyses give little information as to whether the compositions of the copolymer and the reactant mixture are identical, because Scheme 41 is a polyaddition. Their infrared spectra show that the reaction took place, since the absorptions of the amide group are observed at 3300 cm^{-1} (NH) and at 1655

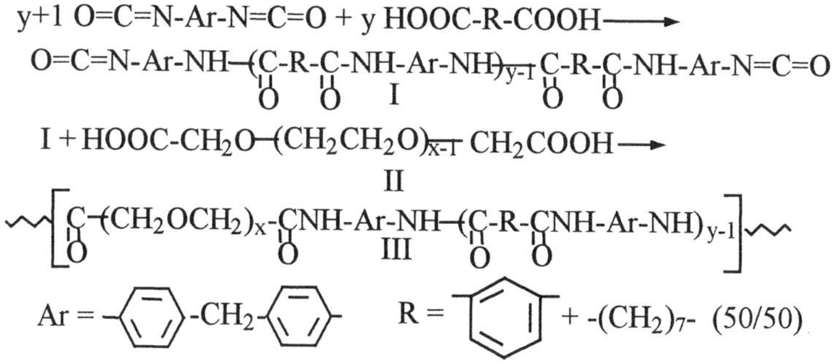

y+1 O=C=N-Ar-N=C=O + y HOOC-R-COOH ⟶

O=C=N-Ar-NH-(C-R-C-NH-Ar-NH)$_{y-1}$ C-R-C-NH-Ar-N=C=O
     O   O   I   O   O

I + HOOC-CH$_2$O-(CH$_2$CH$_2$O)$_{x-1}$ CH$_2$COOH ⟶
                          II

⟍⟍[C-(CH$_2$OCH$_2$)$_x$-CNH-Ar-NH-(C-R-CNH-Ar-NH)$_{y-1}$]⟋⟋
   O   O   III   O   O

Ar = -⟨◯⟩-CH$_2$-⟨◯⟩-   R = ◯ + -(CH$_2$)$_7$- (50/50)

Scheme 41

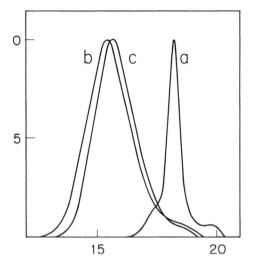

Fig. 3   Size exclusion chromatograms of the oligoether A(a) and of III (see text) obtained with the one-step process B(b) and with two-step process C(c).

cm^{-1} (CO); however, this does not prove that the copolymer does not contain homopolymers, since they have the same absorption bands. On the other hand, the size exclusion chromatograms of the products obtained via processes A and B (Fig. 3) are almost identical, which is very convincing.

The solubility of the block copolymers, of the initial oligomers, and of the corresponding homopolymers provides useful information, though it is mainly a qualitative property (Table 4). It shows that the copolymers are soluble in

Table 4   Solubility of the Poly[poly(oxyethylene) (POE)-*block*-polyamide (PA)]s

			Block copolymer			
Solvent	POE	PA	3c(I)	3c(II)	3g(I)	3g(II)
N-methyl-2-pyrrolidone	+ +	+ +	+ +	+ +	+ +	+ +
N,N'-dimethylacetamide	+ +	+ +	+ +	+ +	+ +	+ +
N,N'-dimethylformamide	+ +	+ +	+ +	+ +	+ +	+ +
Dimethylsulfoxide	+ +		+ +	+ +	+ +	+ +
m-cresol	+ +	+ +	+ +	+ +	+ +	+ +
Pyridine	+ +	–	+	–	±	±
Methanol	+ +	–	–	–	–	–

I and II are relative to one- and two-step processes, respectively (see text); compounds 3c are prepared from POE, whose molar mass is 3900, from PA, with molar mass of 3200; and polymers 3g are prepared from POE, whose molar mass is 7400, and from PA, whose molar mass is 8300. Solubility: + +, soluble at room temperature; +, soluble on heating; ± swelling; – insoluble.

dimethylsulfoxide, even at room temperature (+ +), whereas the polyamide is soluble only on heating (+); in the same way, copolymer **3c** (I) is soluble in pyridine on heating (+), whereas the polyamide is not (−). These observations clearly show that compounds are copolymers, if not, the polyamide would probably separate from the solution (1).

Table 4 gives additional information: **3c** is more soluble than the other copolymers; this could be expected, since the ratio [POE]/[PA] is 1.2 for **3c**(A) and **3c**(B) and only 0.9 for **3c**(A) and **3c**(B); the difference between **3c**(A) (soluble) and **3c**(B) probably results from different molar distributions: the block distribution in the copolymers prepared by process A is probably statistical, whereas the copolymers prepared by process B are probably more regular.

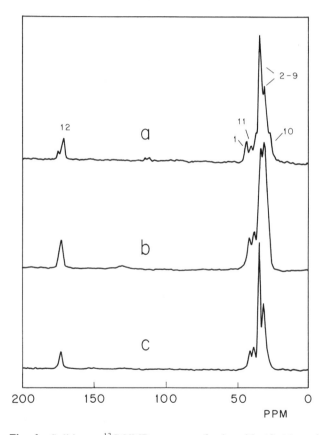

Fig. 4 Solid-state ^{13}C-NMR spectrum of polyamide-12. (a) α- plus γ-forms; (b) γ-form; and (c) γ′-form. Numbering of the carbons:

Solid-state ^{13}C-NMR provides information, particularly the comparison of the spectra of the block copolymers and those of the initial oligomers (77–79). Mathias and Johnson (80) studied the different crystalline forms observed in polydodecanamide: α-form (precipitated from phenol-ethanol solution); γ-form (by thermal annealing); γ′-form (quenched from the melt) using solid-state ^{13}C-NMR. The spectra of (α-plus γ-forms), γ-form, and γ′-form samples are shown in Fig. 4.

The spectrum (Fig. 5) of the polyamide sample used in the preparation of poly[polydodecanamide-*block*-copolyether] (22,23) shows that its crystalline parts are γ, which is confirmed by X-ray diffractometry, and this morphology is preserved in the block copolycondensates, even though they were prepared in the melt.

On the other hand, the literature provides much less information on polyether blocks. Different copolymers were prepared from α,ω-dihydroxy-polyoxytetramethylene and α,ω-dicarboxylic polydodecanamide; they are completely amorphous when the polyether molar mass is 1000 and partially crystallized when it is 2000 (Figs. 6 and 7).

Figure 8 is the spectrum of a multiblock copolymer prepared from a polyamide and a poly(oxyethylene); their molar masses are 2000 and 1000, re-

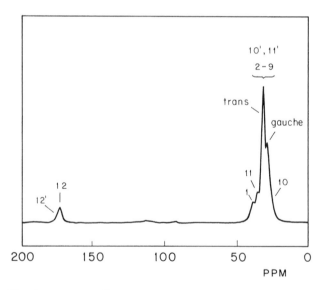

Fig. 5   Solid-state ^{13}C-NMR spectrum of α,ω-dicarboxylic-polyamide-12. Numbering of the carbons:

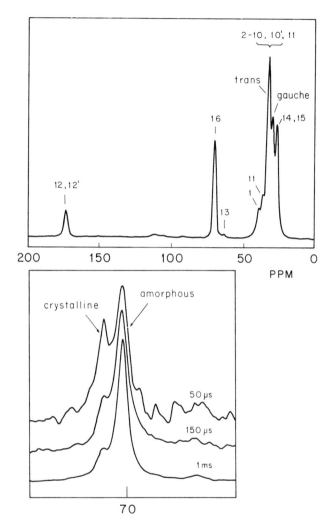

Fig. 6 Solid-state ^{13}C-NMR spectrum of poly[polyamide-12 (2100)-*block*-polyoxyte-tramethylene (2000)]. Numbering of the carbons:

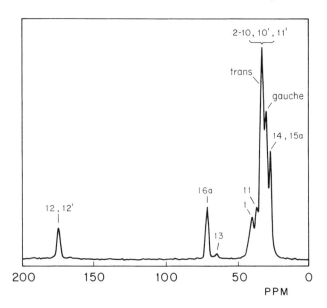

Fig. 7 Solid-state ^{13}C-NMR spectrum of poly[polyamide-12 (2100)-*block*-polyoxyte-tramethylene (1000)]. Same carbon numbering as in Fig. 6.

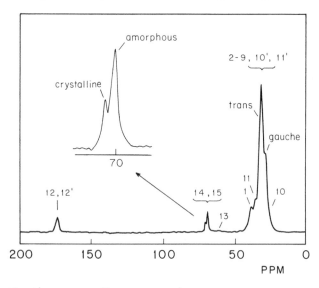

Fig. 8 Solid-state ^{13}C-NMR spectrum of poly[polyamide-12 (2100)-*block*-polyoxyethy-lene (1000)]. Numbering of the carbons:

spectively. The splitting of peaks 14 and 15 shows that polyether blocks contain crystalline and amorphous phases. On the other hand, this splitting is not observed when the flexible block is poly(oxypropylene), showing that the polyether phase is entirely amorphous. This holds for the copolycondensates prepared from copolyethers: the flexible blocks are amorphous even though some of the initial copolyethers are partially crystalline.

REFERENCES

1. T Otsuki, M-A Kakimoto, Y Imai. Synthesis and properties of multiblock copolymers based on polyoxyethlene and polyamides by diisocyanate method. J. Appl. Polym. Sci. 40:1433–1443, 1990.
2. AT Hu, R-S Tsai, Y-D Lee. Preparation of block copolyetheramides and their properties as hot melt adhesives. J. Appl. Polym. Sci. 37:1863–1876, 1989.
3. RJ Gaymans, P Schwering, JL de Haan. Nylon 46-polyteramethylene oxide segmented block copolymers. Polymer 30:974–977, 1989.
4. P Foy, C Jungblut, GE Deleens (ATO Chimie SA Fr.). U.S. Patent 4331786, 25 May 1982.
5. RJ Zdrahala, EM Firer, JF Fellers. Block copolymers of poly(*m*-phenylene isophthalamide) and poly(ethylene oxide) or polydimethylsiloxane: synthesis and characteristics. J. Pol. Sci. Polym. Chem. Ed. 15:689–705, 1977.
6. JD Garforth (ICI Ltd.). Br. Pat. 1 270 097, 1972.
7. LL Harrell. Segmented polyurethanes—properties as a function of segment size distribution. Macromolecules 2:607–612, 1969.
8. K Ateya. Fadenbildende polyesteramide aus oligomeren. Angew. Makromolecules Chem. 9:56–60, 1969.
9. L Castaldo, G Maglio, R Palumbo. Synthesis of polyamide-polyether block copolymers. J. Pol. Sci. Polym. Lett. Ed. 16:643–645 (1978).
10. G Deleens, P Foy, E Maréchal. Synthèse et caractérisation de copolycondensats séquencés poly(amide-seq-ether)—I. Synthèse et étude de divers oligomères $\omega\omega'$-difonctionnels du poly(amide-11). Europ. Polym. J. 13:337–342, 1977.
11. G Deleens, P Foy, E Maréchal. II. Polycondensation d'oligomères polyamides-11-$\omega\omega'$-diacide ou diester avec des oligomères polyethers $\omega\omega'$dihydroxy. Europ. Polym. J. 13:343–351, 1977.
12. G Deleens, P Foy, E Maréchal. III. Etude de la réaction de polycondensation du polyamide-11-ω,ω'-dicarboxylique et du polyoxyethylene-ω,ω'-dihydroxy. Détermination des constantes de vitesse et de l'énergie d'activation. Europ. Polym. J. 13:353–360, 1977.
13. B Masar, P Cefelin, Sebenda. Block copolymers of copolyamides with polyoxirane. J. Pol. Sci. Polym. Chem. Ed. 17:2317–2335, 1979.
14. JL Chen, DJ Shikkena, RWM van Berkel. Neth. Pat. 8:101–262, 1981.
15. M Xie, Y Camberlin. Etude morphologique de bloc-copoly(éther-amide)s. Makromol. Chem. 187:383–400, 1977.

16. SW Shalaby, EM Pearce, HK Reimschuessel. Nylon-12/poly(oxybutylene) block copolymers. Ind. Eng. Chem. Prod. Res. Develop. 12:128–32, 1973.

17. M Acevedo, A Fradet. Study of bulk chain coupling reactions. I. Reaction between bisoxazolones and amine-terminated polyamide 12. J. Polym. Sci. Polym. Chem. Ed. 31:817–830, 1993.

18. M Acevedo, A Fradet. Study of bulk chain coupling reactions. II. Reaction between bisoxazolones and amine-terminated polyether: synthesis of polyether-block-polyamides. J. Polym. Sci. Polym. Chem. Ed. 31:1579, 1993.

19. S Mumcu, K Burzin, R Feldmann, R Feinauer. Copolyetheramide aus laurinlactam, 1,10-decandicarbonsäre und α,ω-dihydroxy-(polytetrahydrofuran). Angew. Makromol. Chem. 74:49–60, 1978.

20. P Laporte, A Fradet, E Maréchal. Kinetics study on models of the noncatalyzed reaction between α,ω-dicarboxylpolyamide-11 and α,ω-dihydroxypolyoxyalkylenes. J. Macromol. Sci. Chem. A24:1269–1287, 1987.

21. P Laporte, A Fradet, E Maréchal. Kinetics study on models of the tetrabutoxy-zirconium-catalyzed reaction between α,ω-dicarboxylpoly-amide-11 and α,ω-dihydroxypolyoxyethylenes. J. Macromol. Sci. Chem. A24:1289–1302, 1987.

22. A Boularès, L Rozès, M Tessier, E Maréchal. Structure and properties of block-copolymers with polyethers as flexible blocks. Europhysics Conference Abstracts 21A:24, 1997.

23. A Boularès, L Rozès, M Tessier, E Maréchal. Synthesis and characterization of poly(copolyethers-block-polyamides). I. Structural study of polyether precursors. J. Macrom. Sci. Pure Appl. Chem. A35:933–953, 1998.

24. Ph Demont, D Chatain, C Lacabanne, M Glotin. Retardation and relaxation behavior of poly(ether-block-amide)s (PEBA) copolymers by thermally stimulated CREEP and current (TsCu). Makromol Chem. Macromol. Symp. 25:167–186, 1989.

25. WP Yang, CW Macosko. Macromol. Chem. Phase separation during fast (RIM) polyurethane polymerization. Macromol. Symp. 25:23–44, 1989.

26. P Honoré, G Deleens, E Maréchal. Synthesis and study of various reactive oligomers and of poly(ester-imide-ether)s. Europ. Polym. J. 16:909–916, 1980.

27. C Michot, D Baril, M Armand. Polyimide-polyether mixed conductors as switchable materials for electrochromic devices. Solar Energy Mater. Solar Cells 39:289–299, 1995.

28. D Pospiech, K Eckstein, H Komber, D Voigt, FR Böhme, HR Kricheldorf. The use of melt polycondensation for synthesis of LC multiblock copolymers. Polym. Prep. 38(2):398–399, 1997.

29. LT Hsiue, CCM Ma, HB Tsai. Hydrophilic thermotropic block copolyetheresters with poly(pentametylene p,p'-bibenzoate) segments. Macromol. Chem. Phys. 196: 3459–3468, 1995.

30. LT Hsiue, CCM Ma, HB Tsai. Synthesis and characterization of hydrophilic thermotropic block copolyetheresters. J. Pol. Sci. Polym. Chem. Ed. 33:1153–9, 1995.

31. HB Tsai, C Lee, NS Chang. Synthesis and thermal properties of thermotropic block copolyetheresters. Polymer J. 24:157–164, 1992.

32. HB Tsai, DL Lee, JL Liu, YS Tsao, RS Tsai, JW You. Block copolyetheresters—part 7. Polym. Bull. 35:743–749, 1995.

33. HB Tsai, DL Lee, JL Liu, YS Tsao, RS Tsai. Block copolyesters. V. Low-tempera-

ture properties of thermotropic block copolyetheresters. J. Appl. Polym. Sci. 59: 1027–1031, 1996.

34. P Thuillier, M Tessier, Maréchal. Synthesis and characterization of thermotropic polyester-block-polyethers. Makromol. Chem. Macromol. Symp. 70/71:37–45, 1993.

35. P Thuillier, M Tessier, E Maréchal. Synthesis and characterization of thermotropic polyester-block-polyesters. Molecular Crystals and Liquid Crystals Science and Technology Section A 254:1–16, 1994.

36. JP Leblanc, M Tessier, D Judas, C Friedrich, C Noël, E Maréchal. Aromatic copolyesters with stilbene mesogenic group. 1. Liquid crystalline properties of compounds containing a stilbene, terephthaloyl, or hydroquinone central group. Macromolecules 26:4391–4399, 1993.

37. JP Leblanc, M Tessier, D Judas, C Friedrich, C Noël, E Maréchal. Aromatic copolyesters with stilbene mesogenic groups. 2. Synthesis and thermal behavior. Macromolecules 28:4837–4850, 1995.

38. H Wang, JH Dong, KY Qiu. Synthesis and characterization of ABA-type block copolymer of poly(trimethylene carbonate) with poly(ethylene glycol): bioerodible copolymer. J. Pol. Sci. Polym. Chem. Ed. 36:695–702, 1998.

39. M Sano. Block copolymer epitaxy. Adv. Mater. 9:509–511, 1997.

40. Y Yagci, G Hizal, A Önen, E Serhalli. Synthetic routes to block copolymerization by changing mechanism from cationic polymerization to free radical polymerization. Macromol. Symp. 84:127–136, 1994.

41. JR Ascenso, AR Dias, PT Gomes, CC Romão, Q-T Pham, D Neibecker, I Tkatchenko. Reversible interpolymer complexation between poly(oxyethylene)-based amphiphilic block polymer and poly(acrylic acid) at the air–water interface. Macromolecules 22:1000–1002, 1989.

42. D Garg, S Höring, J Ulbricht. Initiation of anionic polymerization of methyl methacrylate by living poly(ethylene oxide) anions, a new way for the synthesis of poly(ethylene oxide)-*b*-poly(methyl methacrylate). Makromol. Chem. Rapid Commun. 5:615–618, 1984.

43. I Çakmak. Preparation of multiphase block copolymers by redox polymerization process. Angew. Makromol. Chem. 224:1–8, 1995.

44. I Çakmak. Synthesis of block copolymers by redox macro initiators. Macrom. Report A32:197–206, 1995.

45. CS Cho, BW Jo, JK Kwon, T Komoto. Synthesis and structural study of an ABA block copolymer consisting of poly(ε-benzyloxycarbonyl-L-lysine) as the A block and poly(ethylene oxide) as the B block. Macromol. Chem. Phys. 195:2195–2206, 1994.

46. EJ Goethals, PV Caeter, FE Du Prez, MF Dubreuil. Sophisticated macromolecular structures by cationic ring-opening polymerizations. Macrom. Symp. 98:185–192, 1995.

47. M Van de Velde, EJ Goethals. The synthesis of polyamine networks using cationic living poly(1-*tert*-butylaziridine). Makromol. Chem., Macromol. Symp. 6:271–276, 1986.

48. S Kobayashi, H Uyama, E Ihara, T Saegusa. Block copolymerization of tetrahydrofuran with cyclic imino ether: Synthesis of a new nonionic polymer surfactant. Macromolecules 23:1586–1589, 1990.

49. S Nagarajan, SS Sreeja, KSV Srinivasan. Kinetics and mechanistic studies on the block copolymerization of methyl methacryate initiated by Ce^{4+}-poly(ethylene glycol) redox system. J. Appl. Polym. Sci. 63:565–571, 1997.

50. AD Jenkins, P Kratchovíl, RFT Stepto, UW Suter. Glossary of basic terms in polymer science. Pure Appl. Sci. 68:2287–2311, 1996.

51. CD Eisenbach, J Heinlein. Synthesis and properties of block copolymers consisting of liquid crystalline/amorphous segments. Polym. Prepr. 37(1):75–76, 1996.

52. A Fradet. Coupling reactions in polymer synthesis. In: SL Aggarwal, S Russo, eds. Comprehensive Polymer Science. Oxford: Pergamon, 1996, pp 133–162.

53. Y Qing, H Jiasong. Thermotropic liquid crystalline block copolyesters with PET as flexible spacers. Polym. Bull. 29:633–638, 1992.

54. H Inata, S Matsumura. Chain extenders for polyesters. V. Reactivities of hydroxyl-addition-type chain extender: 2,2′-bis(4H-3,1-benzoxazin-4-one). J. Appl. Polym. Sci. 34:2609–2617, 1987.

55. M Acevedo, A Fradet. Chain coupling reaction of α,ω-dihydroxy-poly(oxytetramethylene) by bis oxazolones. Polym. Prepr. 34(1):457–458, 1993.

56. JK Ramunssen, LR Krepski, SM Heilmann, K Sakizadeh, HK Smith II, AR Katrizky. Polyimidazolinones via thermal cyclodehydration of polyamides containing α-aminoacids. Polym. Prep. 27(2):17–20, 1986.

57. Y Yagci, G Hizal, A Önen, E Serhati. Synthetic routes to block copolymerization by changing mechanism from cationic polymerization to free radical polymerization. Macromol. Symposia 84:127–136, 1994.

58. YX Li, XD Feng. Biodegradable polymeric matrix for long-acting and zero-order release drug delivery systems. Makrom. Chem. Macromol. Symp. 33:253–264, 1990.

59. AC Albertson, M Eklund. Synthesis of copolymers of 1,3-dioxan-2- and oxepan-2-one using coordination catalysts. J. Pol. Sci. Polym. Chem. Ed. 32:265–279, 1994.

60. GH Hsiue, AX Swamikannu, MH Litt. Synthesis and characterization of a multiblock copolymer of poly(N-Isovaleryl ethyleneimine) and poly(ethylene glycol). J. Pol. Sci. Polym. Chem. Ed. 26:3043–3069, 1988.

61. B Gallot. Comb and block liquid crystalline polymers for biological applications. Prog. Polym. Sci. 21:1035–1088, 1996.

62. Y Yagci. Block copolymers by combinations of cationic and radical routes—Part 1. Polym. Comm. 26:7–8, 1985.

63. G Hizal, Y Yagci. Block copolymers by cationic combination and radical routes: 4. Cationic polymerization of tetrahydrofuran initiated by difunctional oxacarbonium initiator. Polymer 30:722–731, 1989.

64. O Nguyen, H Kröner, S Aechner. Macro azo initiators via cationic polymerization. Synthesis and application. Makromol. Chem. Rapid Commun. 9:671–679, 1968.

65. J Huang, X Huang. Polyfunctional initiation system for preparation of block copolymer of ethylene oxide and methyl methacrylate by sequencial initiation of anion and charge transfer complex and the effect of polymerization conditions on the copolymerization. J. Macrom. Sci. Pure Appl. Chem. A34:685–694, 1997.

66. R Améduri, B Boutevin, Ph Gramain. Synthesis of block copolymers by radical polymerization an telomerization. Av. Polym. Sci. 127:88–142, 1997.

67. K Matyjaszewski, S Gaynor, JS Wang. Controlled radical polymerizations: the use of

alkyl iodides in degenerative transfer. Macromolecules, Communications 28:2093–2095, 1995.

68. K Matyjaszewski, S Gaynor, D Greszta, D Mardare, T Shigemoto. Synthesis of well-defined polymers by controlled radical polymerization. Macromol. Symp. 98, 73–89, 1995.

69. E Yoshida, A Sugita. Synthesis of poly(tetrahydrofuran) with a nitroxyl radical at the chain end and its application to living radical polymerization. Macromolecules 29:6422–6426, 1996.

70. E Yoshida, A Sugita. Synthesis of poly(styrene-*block*-tetrahydrofuran-*block*-styrene) triblock copolymers by transformation from living cationic into living radical polymerization using 4-hydroxy-2,2,6,6-tetramethylpiperidine-1-oxyl as a transforming agent. J. Pol. Sci. Part A, Polym. Chem. 36:2059–2068, 1998.

71. Y Yagci, A Önen, W Schnabel. Block copolymers by combination of radical and promoted cationic polymerization routes. Macromolecules 24:4620–4623, 1991.

72. M Ropot, V Harabagiu, BC Simonescu, C Simonescu. Macromolecular radical initiators in the synthesis of block and graft copolymers. Revue Roumaine de Chimie 40(9):937–956, 1995.

73. G Galli, E Chiellini, Y Yagci, EI Serhatli, M Laus, AS Angeloni. Liquid crystalline block copolymers by sequential cationic or promoted cationic and free-radical polymerizations. Macromol. Symp. 107:85–97, 1996.

74. F Heatley, YZ Luo, JF Ding, RH Mobbs, C Booth. A ^{13}C nuclear magnetic resonance study of the triad sequence structure of block and statistical copolymers of ethylene oxide and propylene oxide. Macromolecules 21:2713–2721, 1988.

75. P Thuillier. Synthèse et Caractérisation de polyesters-bloc-polyéthers contenant des motifs mésogènes. Thesis Paris, 1992.

76. L Rozes. Synthèse de poly(polyester-*block*-polyether)s thermotropes. Thesis Paris, 1996.

77. GH Hatfield, Y Guo, WE Killinger, RA Andrejack, PM Roubicek. Characterization of structure and morphology in two poly(ether-*block*-amide) copolymers. Macomolecules 26:6350–6353, 1993.

78. GH Hatfield, RW Bush, WE Killinger, RA Andrejack, PM Roubicek. Phase modification and polymorphism in two poly(ether-*block*-amide) copolymers. Polymer 35: 3943–3947, 1994.

79. A Boularès. Synthèse et caractérisation de poly(ethers-*block*-polyamides). Thesis Paris, 1995.

80. LJ Mathias, CG Johnson. Solid-state NMR investigation of Nylon 12. Macromolecules 24:6114–6118, 1991.

81. SS Nagarajan, KSV Srinivasan. Block copolymerization initiated by Ce(IV)-poly(ethylene glycol) redox system—kinetics and characterization. Europ. Polym. J. 30:113–119, 1994.

82. M Bouix, J Gouzi, B Charleux, JP Vairon, Ph Guinot. Synthesis of amphiphilic polyelectrolyte block copolymers using "living" polymerization. Application as stabilizers in emulsion polymerization. Macromol. Rapid Commun. 19:209–213, 1998.

4

New Multiblock Terpoly(ester-ether-amide) with Various Chemically Constitutive Amide Blocks

Ryszard Ukielski
Technical University of Szczecin, Szczecin, Poland

I. INTRODUCTION

The properties of thermoplastic elastomers (TPEs) are influenced by an appropriate phase structure and its thermal reproducibility in the heating–cooling cycle, processing properties (the possibility of multiple melting and solidification), and functional qualities (large, reversible deformations). From the point of view of physical constitution, thermoplastic elastomers are customarily considered polymeric materials, in which, as a result of the phase separation, at least two phases are distinguished soft (flexible) and hard. Hence, these are plastics possessing at least the two values of the physical transition temperatures, e.g., T_{g1} and T_{g2} or T_g and T_m. These temperatures determine a wide "plateau" of a small modulus of elasticity (likewise in rubber). Additionally, TPEs can be processed by methods analogous to those used for typical thermoplastics (1–3).

Some of the multiblock copolymers of the—$(A_x B_y)_{\overline{n}}$—type that have the heterophase structure are classified as thermoplastic elastomers. A macromolecule of block elastomers consists of flexible and hard blocks distributed alternately. These blocks differ considerably in their physical and chemical properties. The flexible blocks are capable of forming a soft-phase matrix. The hard blocks, as a result of aggregation, form the domains that constitute the hard phase. Such heterophase systems are unique in that the dispersed domain structures are thermodynamically stable in the dispersed state. The phase separation in block copolymers is restricted to molecular dimensions because the incompatible block components are joined together, thus preventing gross physical separation of the two components as would occur with their simple mixture (1).

In order to classify the block elastomers as TPEs, their internal structure must comply with the following conditions:

The soft phase (responsible for the elastic properties) must possess a relatively small modulus of elasticity, a relatively low glass transition temperature (T_g), and a low density value. Hence, the chemical constitution of blocks incorporated into the composition of this phase must ensure weak intermolecular interactions and a large ability for motion and rotation of the short sequences of chain (small cohesion energy).

The hard phase (responsible for the mechanical and processing properties) must possess a relatively large modulus of elasticity, a high glass transition temperature (T_g) and melting temperature (T_m), and a relatively high density. The blocks forming this phase must have a tendency to aggregate with the same kind of segments. Thus, strong intermolecular interactions occur, as does so-called "pseudo-crosslinking" reversible thermally (van der Waals bonds, hydrogen bonds, ionic bonds, the ability to crystallize, polar and dispersive interactions). The intermolecular interactions of hard blocks affect the stabilization of the phase structure of the entire polymeric system. The hard blocks must be characterized by a considerably larger cohesive energy density of matter than the flexible blocks, having thereby a higher thermodynamic potential. The potential difference comprises the driving force for the formation of a heterophase structure (1,3).

An important and a necessary feature of multiblock elastomers is a mutual thermodynamic incompatibility of the blocks, which depends on both the chemical constitution and the molecular weight. This feature determines the selection of blocks utilized for the manufacture of block TPEs. Summarizing: It can be concluded that the characteristic properties of block TPEs are a result of their specific molecular constitution and the phase structure (3–5).

The arduousness of manufacturing rubber goods (labor consumption, harmfulness, waste) is responsible for the development of such multiblock TPE copolymers as: ether-urethanes (PUE)—Estane Goodrich; ester-urethanes (PUA)—Desmopan Bayer; ether-esters (PEE)—Hytrel Du Pont; Arnitel AKZO and ether-amides (PAE)—Pebax Atochem, Vestamide-E Huls, Grilamid-ELY-Amer. Grilon (3,6).

The common objective of many worldwide investigations of TPE is the search for new materials with the specific functional properties of simplicity of processing and, contrary to rubber, the possibility of recycling. The author of the present chapter is also involved in this subject, and is still searching for readily available raw materials to be used to develop novel multiblock elastomers (7–17). As a result of the attainments achieved so far in the field of block copoly(ester-ether) syntheses and the research results obtained from the investigations of multiblock systems, a new idea has emerged for the preparation of poly(ester-ether-amide) block elastomer TPEEA, i.e., the terpolymer of the $—(A_xB_yC_z)_{\overline{n}}—$ type. Via prediction of intermolecular interactions in such a ternary system, the possibility of preparing a new type of polymers with the specific properties was assumed; this has been realized by combining polyamide with polyether and polyester using the method of block melt condensation polymerization. The three-blocks system creates greater control possibilities of the desired (expected) properties through a change in the quantitative composition or by preferential joining of segments. This also facilitates the synthesis process.

An initial point of TPEEA preparation was a typical synthesis of block copoly(ester-ether) obtained from dimethyl terephthalate, 1,4-butanodiol, and α,ω-oligo(oxytetramethylene)diol with the molecular weight of 1000 g/mol, where dicarboxylic oligoamid is additionally introduced. Therefore, it can be assumed that the synthesis of multiblock poly(ester-ether-amide) relies on a deep modification of copoly(ester-ether) by the dicarboxylic amide block (18,19).

In the present chapter the synthesis of TPEEA and the relationship between the chemical constitution of the amide block, the degree of polymerization (polycondensation) (DP) of the ester block and the properties in connection with phase structure of TPEEA block elastomers are described.

The following terpolymers were selected for this research study:

Poly[(tetramethylene terephthalate)-*block*-(oxytetramethylene)-*block*-(ε-caprolactam)] PBT-*b*-PTMO-*b*-PA6

Poly[(tetramethylene terephthalate)-*block*-(oxytetramethylene)-*block*-(laurolactam)] PBT-*b*-PTMO-*b*-PA12

II. EXPERIMENTAL PART

A. Materials

The synthesis of PBT-*b*-PTMO-*b*-PA multiblock elastomers was carried out using the commercially available substrates, namely: dimethyl terephtalate (DMT)—Chemical Plant "Elana"; 1,4-butanediol (BD)—BASF, α,ω-oligo (oxytetramethylene)diol (Terathane)—Du Pont with the number average molecular weight 1000 g/mol; ε-caprolactam—Chemical Plant Tarnów; dodecano-

12-lactam (laurolactam)—Aldrich Chemie; and sebacic acid—Aldrich Chemie. The two lactams and the dicarboxylic acid are the substrates prepared in our laboratory: α,ω-dicarboxylic oligo(laurolactam) (PA12) and α,ω-dicarboxylic oligo (ε-caprolactam) (PA6) with the number average molecular weight 2000 g/mol.

B. Synthesis of Oligoamide Blocks

The synthesis of oligoamides with the assumed molecular weight (ca. 2000 g/mol) double-side terminated with the carboxylic groups was carried out. These oligomers performed the role of hard segments intended for incorporation into the macromolecules of poly(ester-*b*-ether-*b*-amides) (20,21).

The methods for manufacture of different polyamides produced on a commercial scale as constructional and fiber-forming polymers are well known. However, in the available scientific literature, data has not been encountered concerning the synthesis of oligoamides terminated with the carboxylic groups, utilized as the blocks for incorporation into the macromolecules of other polymers (23–25). The stabilization of the molecular weight of polyamide-12 with sebacid acid in the process of its postpolycondensation was utilized (26). Reactions (1) and (2) illustrate the syntheses used for the oligoamides:

n HN(CH$_2$)$_5$CO + HOOC(CH$_2$)$_8$COOH $\xrightarrow{H_2O}$ HO[OC(CH$_2$)$_5$NH]$_n$O(CH$_2$)$_8$COOH

Oligoamide 6 (PA6)

Reaction 1

n HN(CH$_2$)$_{11}$CO + HOOC(CH$_2$)$_8$COOH $\xrightarrow{H_2O}$ HO[OC(CH$_2$)$_{11}$NH]$_n$O(CH$_2$)$_8$COOH

Oligoamide 12 (PA12)

Reaction 2

These reactions were carried out in a pressurized autoclave designed and constructed at the laboratory of Elastomers and Chemical Fibers Technology at the Technical University of Szczecin in Poland. The main unit of the equipment comprises a 6-dm^3 "pressure-vacuum" reactor made of stainless steel, having a cylindrical shape with a conical bottom. The ratio of height to reactor diameter is $h{:}d = 3.5$. The heating system comprises a set of three resistance heaters enabling control of temperature in the range of 20–400°C. The regulators are con-

Table 1 Composition of Raw Materials and Conditions of Synthesis of Dicarboxylic Oligoamides

Product symbol	Amide substrate Type	Amount, g	Auxiliary raw material Type	Amount, g	Pressure process Pressure, MPa	Temperature, °C	Time, h	Pressureless process Temperature, °C	Time, h
PA6	ε-Caprolactam	3000	Sebacic acid	337.5	0.8	260	2	260	2
PA12	Laurolactam	2600	Sebacic acid	303.0	1.6	320	5	300	5

trolled by Fe-constantan thermocouples and Pt-100 thermoresistors. The agitator drive permits a fluent variation of revolutions in the range of 100–800 rpm. Into the nitrogen purged-reactor appropriate lactam (ε-caprolactam or ω-laurolactam) and sebacic acid were introduced. A small amount of water was added to facilitate the initiation of the reaction, along with the same amount of sebacic acid, which allows one to obtain the assumed molecular weight of oligoamide, taking into account the double-side terminations of macromolecules with the carboxylic groups. The kinds and amounts of introduced initial compounds are compiled in Table 1.

The initial compounds were placed in the autoclave, previously purged three times with nitrogen at a pressure of ca. 0.5 MPa, at temperature of ca. 25°C directly before the synthesis. When an oxygen-free reaction environment was achieved, the reaction mixture was kept in the reactor under a pressure of ca. 0.1 MPa and then heated to a temperature of 270 or 300°C (depending on the kind of lactam), controlling the pressure in such a way that it would not exceed 0.8 or 1.6 MPa. The pressure was controlled by the removal of water vapor with a small amount of amide di- and trimers. The reaction was performed for 2 or 5 h (pressure stage) under the conditions of assumed temperature and pressure, and with a continuous stirring at the rotational speed of about 200 rpm. Subsequently, without decreasing the temperature, the pressure was reduced during 0.5 h to atmospheric pressure, and the polycondensation was carried out farther (pressureless stage) for 2 or 5 h under an atmosphere of anhydrous nitrogen. After the reaction was completed, the formed oligoamide was extruded by compressed nitrogen into the tub with water under vigorous stirring with bubbling air. The resulting product, after filtration, was rinsed three times with hot water and then with distilled water in order to extract the residual by-products or unreacted lactam. After drying successively in air and in a vacuum dryer at a temperature of 70°C, the oligoamide was ground and characterized.

C.   Synthesis of Multiblock PBT-*b-PTMO-b*-PA Polymers

The synthesis of multiblock poly(ester-ether-amides) (TPEEA) was performed in an apparatus designed and constructed in our laboratory: a set of two reactors composed of the following units and systems (Fig. 1):

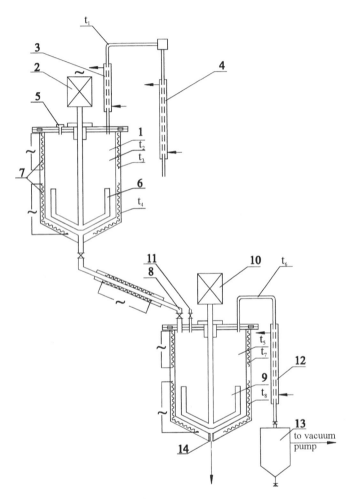

Fig. 1   Apparatus used in the synthesis of multiblock poly(ester-ether-amide)s: 1: reactor for transesterifacation; 2: variable-current motor; 3: dephlegmator; 4: condensor; 5: dosing pipe; 6: stirrer; 7: heating zones I and II; 8: overflow pipe; 9: reactor for polycondensation; 10: direct current motor; 11: nitrogen-introducing valve; 12: condensor; 13: pressure receiver; 14: bottom outlet; t_1, t_2, t_5, t_6: Fe-constantan thermocouples; t_3, t_4, t_7, t_8: Pt-100 thermoresistors.

Reactor I (for the ester exchange reaction): with a capacity of 3.6 dm³ (working capacity 1.2–1.6 dm³), made of stainless steel with a horseshoe agitator, equipped with a dephlegmator and a condensor, the system for drying and dosing of nitrogen, a thermostatically controlled heating system with the possibility of heating in the temperature range of 20–400°C, and a driving system composed of an ac motor of 1.1 kW and a gearbox (rotational speed of stirrer 120 rpm).

Reactor II (for the esterification and polycondensation) with a total capacity of 3.6 dm³ (working capacity 1.2–1.6 dm³), made of stainless steel with a helical stirrer sealed with a Teflon gland. The achievement of a high vacuum ensured the vacuum system, which was composed of a vacuum valve and a valve enabling a slow reduction of pressure and a vacuum pump. The reactor was equipped with a thermostated electrical heating system (20–400°C), a driving system, e.g., a dc motor of 0.3 kW, a gearbox, and a control system ensuring a fluent control of stirrer revolutions in the range of 0–100 rpm and a temperature with an accuracy to 2°C.

The preparation of the PTMO-*b*-PBT-*b*-PA12 and PTMO-*b*-PBT-*b*-PA6 block terpolymers proceeds as a three-step process (27–31). The initial stages comprise the transesterification (Reactor I) with a contribution of dimethyl terephthalate and 1,4-butanodiol (reaction 3), and the esterification reaction (Reactor II) of α,ω-dicarboxylic oligolaurolactam or α,ω-dicarboxylic oligo-ε-caprolactamwith α,ω-oligooxytetramethylenediol in the presence of tetrabutyl titanate as catalyst (reaction 4).

n H₃C-O-CO ⟨◯⟩-CO-O-CH₃ +2(n-1) HO(CH₂)₄OH $\overset{Ti(OR)_4}{\rightleftharpoons}$ (n -2) HO(CH₂)₄-OCO ⟨◯⟩- CO-O(CH₂)₄OH +

+   HO(CH₂)₄-OCO ⟨◯⟩-CO-O(CH₂)₄O-CO ⟨◯⟩-CO-O(CH₂)₄OH   + 2n CH₃OH↑

Reaction 3

Analyzing the quantity of methanol and water collected from reactions (3) and (4), respectively, it was concluded, that the conversion in the transesterification reaction (3) amounts to 98%, and the degree of esterification in reaction (4) to 88%.

n HO[OC(CH₂)₁₂NH]_N O(CH₂)₈COOH +2n HO[(CH₂)₄O]_n H) $\overset{Ti(OR)_4}{\rightleftharpoons}$

HO[(CH₂)₄O]_n[OC(CH₂)₁₁NH]_N O(CH₂)₈COO[(CH₂)₄O]_{n-1}(CH₂)₄OH + 2n H₂O↑

Reaction 4

The next stage of the synthesis comprises the proper condensation polymerization (Reactor II) of mixed intermediate products obtained in reactions (3) and (4). The course and the parameters of the entire synthesis are shown in the following block diagram:

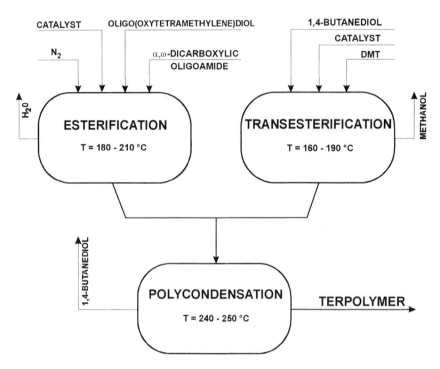

Structure 1   Structure flow chart.

The approximate chemical constitution of the obtained terpolymer is represented by the formula shown in Scheme 1. The kinds and molar ratios of the substrates used for the synthesis of terpolymers are given in upcoming Tables 3 and 4.

D. Techniques

The degree of conversion of esterification and transesterification reactions were expressed as a ratio of the mass of released water or methanol to the stoichiometric amounts of these products and reported as a percentage.

The relative viscosity η_{rel} of the α,ω-dicarboxylic oligoamides were determined via an Ubbelohde viscometer, according to PN 83/C-89039. The concen-

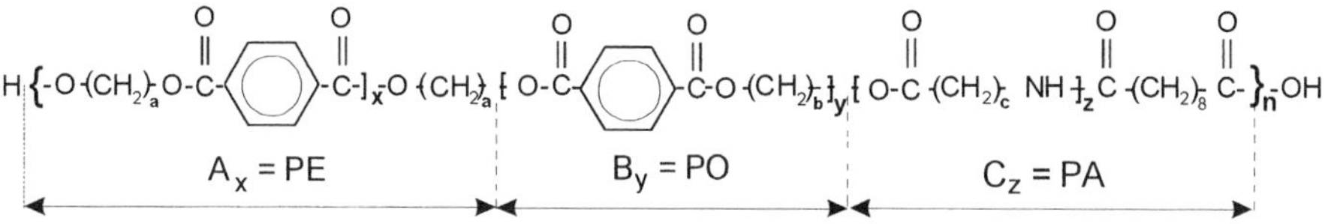

$$H\left\{-O\text{-}(CH_2)_a^{}\text{-}O\text{-}C\text{-}\underset{O}{\overset{O}{\|}}\text{-}C\text{-}\right]_x O\text{-}(CH_2)_a^{}\text{-}\left[O\text{-}C\text{-}\underset{O}{\overset{O}{\|}}\text{-}C\text{-}O\text{-}(CH_2)_b^{}\right]_y\left[O\text{-}C\text{-}(CH_2)_c^{}\text{-}NH\text{-}\underset{O}{\overset{O}{\|}}C\text{-}(CH_2)_8^{}\text{-}C\text{-}\right]_n\text{-}OH$$

$A_x = PE$ $\qquad$ $B_y = PO$ $\qquad$ $C_z = PA$

A_x = PE -polyester block, $\quad$ a = 2 to 6, $\quad$ x = 1 to 15 (m.w. ~200 to ~3000)

B_y = PO -polyether block, $\quad$ b = 2 or 4, $\quad$ y = 8 to 40 (m.w. ~400 to ~6000)

C_z = PA -polyamide block, $\quad$ c = 5 or 11 $\quad$ z = 6 to 30 (m.w. ~1000 to ~6000)

Scheme 1

tration of residual carboxylic end groups [—COOH] were determined by means of titration analysis. A solution of oligoamide in a phenol–ethanol mixture was titrated by 0.02 N HCl in the presence of bromophenol blue until the color changed to pink-violet.

The concentrations of the amine groups [—NH$_2$] were determined by titration of an oligoamide solution in benzyl alcohol with 0.02 N NaOH in the presence of phenolphthalein until the color changed to pink.

The limiting viscosity number [η] of the poly[ether-*block*-ester-*block*-amide] terpolymers in phenol–trichloroethylene (50:50 vol/vol) was determined by an Ubbelohde viscometer II at 30°C. The melt flow index (MFI) of terpolymers was examined by a plastometer (CEAST type 6841) at 180°C. Optical melting points T_m were determined using a Boetius microscope (HMK-type Franz Kustner Nacht KG) at a heating rate of 2°C/min.

Swelling in the benzol and water, p_b, p_{H2O}, were performed according to PN-66/C-08932. Hardness H measurements were performed on a Shore A and D apparatus (Zwick, type 3100) according to standard DIN 53505 (ISO 868, PN-80/C-04238). The tensile data were collected at room temperature with an Instron TM-M tensile tester at a crosshead speed of 20 cm·min^{-1}.

The microcalorimetric examinations were recorded on a DSC-2 (Perkin-Elmer) apparatus. The samples were examined in a triple cycle (heating, cooling, heating) in the temperature range −120 to +200°C. The rate of both heating and cooling was 10°C·min^{-1}. The glass transition temperature and melting temperature observed in the low-temperature region, T_{g1} and T_{m1}, respectively, refer to the soft phase. The endotherms observed in the high-temperature region, $T'_{m2} \approx T_{g2}$ and T_{m2}, refer to the hard phase.

The dynamic mechanical thermal analyses (DMTAs) were performed on a Rheovibron DDV-II viscoelastometer in the temperature range from −120°C to +200°C at 35 Hz. The storage modulus E', loss modulus E'', and loss tangent tan δ were also determined.

III. RESULTS AND DISCUSSION

A. Properties of α,ω-Dicarboxylic-oligoamides

Some properties of the obtained α,ω-dicarboxylic-oligoamides PA6 and PA12 are compiled in Table 2. A high content of carboxylic groups in these products attests to the reactivity of the sebacic acid utilized for the regulation of the molecular weight. A slightly higher amount of amine groups in the oligomers, for whose synthesis caprolactam was used, may be explained by a tendency toward the equilibrium reproduction of caprolactam rings.

The differences between the molecular weight values, enumerated from the amounts of the carboxylic groups and the viscometric testing, and the assumed theoretical values do not exceed 5% of the experimental error, which supports the correctness of the experimental assumptions.

The melting temperature values of the oligomers are lower than the respective melting temperatures of the polyamides. This is in agreement with the commonly known influence of the molecular weight of polymers on their melting temperature.

Multiple syntheses of the dicarboxylic oligoamides PA6 and PA12 were performed and a repeatability of results of 99% was achieved.

B. Properties of PTMO-*b*-PBT-*b*-PA6 and PTMO-*b*-PBT-*b*-PA12

1. Viscosity

An increase in the limiting viscosity number values [η] determined during the synthesis was used as a measure of the progress of the polycondensation reaction. In order to compare the progress of the polycondensation reaction of the PTMO-*b*-PBT-*b*-PA6 and PTMO-*b*-PBT-*b*-PA12 with typical condensation polymers,

Table 2   Properties of Oligoamides

Symbol	$\eta_{rel}{}^{a}$	[COOH], $\mu eq/g^{b}$	[NH$_2$], $\mu eq/g^{c}$	M g/mol^{d}	T_m, °C^{e}
PA6	1.178	969	21	2020	201–205
PA12	1.097	967	12	2040	176–180

^{a} According to PN 83/C-89039.
^{b} Solution of oligoamide in a phenol–ethanol mixture, titrated with 0.02 N HCl in the presence of bromophenol blue to change of color to pink-violet.
^{c} Solution of oligoamide in benzyl alcohol in the presence of phenolphthalein, titrated with 0.02 N NaOH to change of color to pink.
^{d} Calculated from the content of carboxylic groups.
^{e} Determined by means of the Boethius microscope, type HMK, heating rate 3°/min.

the syntheses of a poly(tetramethylene terephthalate) (PBT), a copoly(ether-*b*-ester) (PTMO-*b*-PBT), and a copoly(ether-*b*-amide) (PTMO-*b*-PA12) were also carried out. Figure 2 displays the progress of these reactions as viscosity varies. Figure 3 and 4 illustrate the changes of [η] values during the polycondensation reaction in the synthesis of poly[ether-*block*-ester-*block*-amide] terpolymers, differing in the chemical structure of the oligoamide block, in relation to the degree of polycondensation (DP) of the ester block. The most rapid, as evident from the increase in [η] values versus the molecular weight, is the polycondensation of PTMO-*b*-PBT-*b*-PA6 and PTMO-*b*-PBT-*b*-PA12 with a degree of polycondensation of PBT below 7 ($DP_{PBT} < 7$). Samples of elastomers exhibiting good mechanical properties with $DP_{PBT} = 14$ ([η] < 1.0) could not be prepared. This is probably due to the presence of a miscibility threshold of the reactants in the molten state.

Following the progress of the polycondensation reaction just described, two series of poly[ether-*block*-ester-*block*-amide] terpolymers (the first series: 1_{PA6}–5_{PA6}; the second series: 1_{PA12}–7_{PA12}) with the general formula shown in Scheme 1 were synthesized and examined. These series differ in the chemical structure of the oligoamide block. The obtained products have a light cream color; they resemble polyurethane elastomers (PUEs) in their appearance and texture.

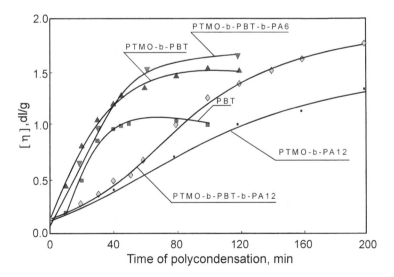

Fig. 2 Progress of the polycondensation reaction, expressed as an increase in [η] with time, for the following polymers: PBT; PTMO-*b*-PBT; PTMO-*b*-PA12; PTMO-*b*-PBT-*b*-PA6; PTMO-*b*-PBT-*b*-PA12 (MW block: PTMO = 1000; PBT = 880; PA12 = 2000; PA6 = 2000 g/mol).

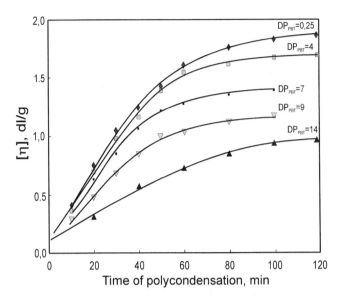

Fig. 3 Effect of PBT segment content on the progress of the polycondensation reaction of PTMO-*b*-PBT-*b*-PA6 terpolymers: $DP_{PBT} = 0.25$; $DP_{PBT} = 4$; $DP_{PBT} = 7$; $DP_{PBT} = 9$; $DP_{PBT} = 14$, molecular mass PTMO = 1000, PA6 = 2000 g/mol.

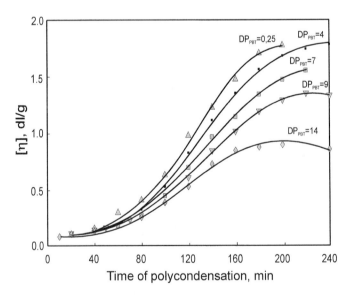

Fig. 4 Effect of PBT segment content on the progress of the polycondensation reaction of PTMO-*b*-PBT-*b*-PA12 terpolymers: $DP_{PBT} = 0.25$; $DP_{PBT} = 4$; $DP_{PBT} = 7$; $DP_{PBT} = 9$; $DP_{PBT} = 14$, molecular mass PTMO = 1000, PA12 = 2000 g/mol.

2. Mechanical Properties

This discussion is limited to the particular case where the PTMO-*b*-PBT-*b*-PA6 and PTMO-*b*-PBT-*b*-PA12 terpolymers have good mechanical properties. The "stress–strain" curves of these terpolymers are characteristic of thermoplastic elastomers. The relative molar composition and properties of the terpolymers are given in Tables 3 and 4. The tensile strength values σ_r of the samples of terpolymers from 1_{PA6} to 4_{PA6} are satisfactory and are similar to those of copoly(ester-ether) elastomers with the same hardness. Samples prepared from 2_{PA12} to 5_{PA12} exhibit a good elastic recovery, similar to urethane elastomers (PUEs) with the same hardness. The elongation ε is smaller than for PUEs.

The PTMO-*b*-PBT-*b*-PA6 terpolymers of the first series (Table 3) behave similar to the other elastomers. Therefore, in agreement with expectations, the melting point, hardness, and tensile data of these terpolymers rise, whereas the melt flow index and the elongation decrease with the DP of the PBT block, i.e., with an increase in the total fraction of PA6 and PBT hard blocks in the terpolymer. However, the PTMO-*b*-PBT-*b*-PA12 terpolymers of the second series behave in a different manner (Table 4): An increase in the degree of polycondensation of the ester block (PBT) causes a decrease in the melting point, hardness, and tensile data of these terpolymers. These unexpected results may be interpreted in the light of a change of morphology of the terpolymers as the DP of PBT increases. This is probably due to the intermolecular interactions between the three types of segments (PTMO, PBT, and PA12), with PBT causing a mutually disturbing effect on the phase-separated morphologies of the system.

The relaxation behavior of all samples was studied by dynamic mechanical thermal analysis (DMTA), measuring the storage modulus (E'), the loss modulus (E''), and the loss tangent (tan δ). The results are presented in Fig. 5 and 6. In both series of terpolymers the storage modulus E' shows rapidly decreasing values near T_g and a wide "plateau" of elastic state. After that, a quick decrease of E' values near the melting temperature is observed. The "plateau" of the elastic state decreases with increasing length of the ester block.

The loss modulus shows a wide maximum E'' for PTMO-*b*-PBT-*b*-PA6 and PTMO-*b*-PBT-*b*-PA12 terpolymers that corresponds to the glass transition of the amorphous phase. The position and the magnitude of these maxima for the PTMO-*b*-PBT-*b*-PA6 terpolymers vary slightly with an increase in DP of the PBT block. However, in PTMO-*b*-PBT-*b*-PA12 terpolymers, when the degree of polycondensation of the PBT block increases, the maxima flatten and shift toward higher temperatures, from -60 to $-40°C$. The tan δ curves show (Fig. 6) that the increasing degree of polycondensation of the ester block in PTMO-*b*-PBT-*b*-PA12 terpolymers causes an increase in tan δ in the range from -30 to $-10°C$ and a shift of the maximum temperature toward higher temperatures.

The tan δ curves of PTMO-*b*-PBT-*b*-PA6 terpolymers (Fig. 5) behave differently. Their maxima rise with the DP of the PET block, as with the PTMO-*b*-

Table 3 Relative Molar Composition and Properties of PTMO-b-PBT-b-PA6 Terpolymers

Polymer sample	PTMO/PA6/DMT/BD-1,4 molar ratio	DP	[η] (dL/g)	MFI (g/10 min)	H Shore D	σ_e (MPa)	σ_r (MPa)	ε (%)	Δl_{100} (%)	T_m (°C)
1$_{PA6}$	2/1/1.25/2.5	0.25	1.84	9.3	20	—	9.8	280	24	186,196
2$_{PA6}$	2/1/5/9	4	1.7	6.8	29	6.3	12.2	250	25	190,195
3$_{PA6}$	2/1/8/15	7	1.41	5.5	38	7.1	17.9	230	26	185,194
4$_{PA6}$	2/1/10/19	9	1.18	2.4	47	9.6	22.1	190	40	203,125
5$_{PA6}$	2/1/15/29	14	0.89	1.8	59	—	8.1	50	—	213,220

DP: Degree of polycondensation of ester block (PBT);
[η]: limiting viscosity number;
MFI: melt flow index;
H: hardness Shore D;
σ_e: yield stress;
σ_r: tensile strength;
ε: elongation;
T_m: melting point with Boëthius apparatus;
$\Delta l_{100} = [(l_k - l_p)/l_p]100\%$: residue after 100% elongation, where l_p is the initial length of the sample and l_k is the final length of the sample after the "elongation-recovery" cycle.

Table 4 Relative Molar Composition and Properties of PTMO-*b*-PBT-*b*-PA12 Terpolymers

Polymer sample	PTMO/PA12/DMT/BD-1,4 molar ratio	DP	$[\eta]$ (dL/g)	MFI (g/10 min)	H Sh. A	H Sh. D	σ_e (MPa)	σ_r (MPa)	ε (%)	Δl_{100} (%)	T_m (°C)
1_{PA12}	2/1/1.25/2.5	0.25	1.71	0.33	—	36	—	19.8	320	20	153,162
2_{PA12}	2/1/3/5	2	1.75	0.38	—	35	—	20.1	270	14	145,152
3_{PA12}	2/1/5/9	4	1.68	0.53	95	33	—	17.2	290	12	135,148
4_{PA12}	2/1/8/15	7	1.58	2.13	91	27	7.1	14.3	220	15	120,136
5_{PA12}	2/1/10/19	9	1.30	14.15	90	26	5.6	9.5	190	16	108,125
6_{PA12}	2/1/12/23	11	1.17	46.31	85	23	4.8	7.8	160	20	105,115
7_{PA12}	2/1/15/29	14	0.82	22.58	—	—	—	—	—	—	110,108

DP: degree of polycondensation of polyester block (PBT);
$[\eta]$: limiting viscosity number;
MFI: melt flow index;
H: hardness Shore A and D;
σ_e: yield stress;
σ_r: tensile strength;
ε: elongation
T_m: melting point with Boëthius apparatus;
$\Delta l_{100} = [(l_k - l_p)/l_p]100\%$: residue after 100% elongation, where l_p is the initial length of the sample and l_k is the final length of the sample after the "elongation-recovery" cycle.

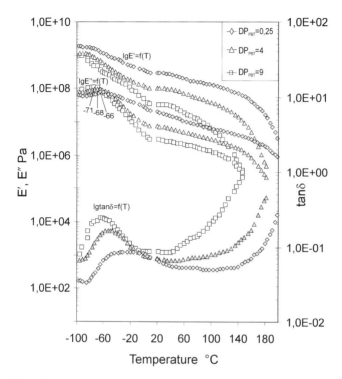

Fig. 5 Storage modulus (E'), loss modulus (E''), and loss tangent (tan δ) of PTMO-*b*-PBT-*b*-PA6 terpolymers vs. temperature.

PBT-b-PA terpolymers; however, they shift toward lower temperatures from −40 to −80°C. A shift of maxima E" and tan δ in PTMO-*b*-PBT-*b*-PA6 terpolymers along with an increase of DP of the PBT toward lower temperatures attests to a more complete isolation of the PBT and PA6 regions and their better separation from the PTMO regions. This means that in these terpolymers the multiphase structure is better formed than in PTMO-*b*-PBT-*b*-PA12. In the latter, both maxima E" and tan δ shift toward a higher temperature. This attests to a partial mixing of a respective region (phase) in PTMO-*b*-PBT-*b*-PA12 terpolymers. This is in compliance with an earlier conclusion resulting from the estimation of hardness and strength of these terpolymers.

3. Thermal Properties

The DSC (differential scanning calorimetry) results of PTMO-*b*-PBT-*b*-PA6 and PTMO-*b*-PBT-*b*-PA12 terpolymers are presented in Fig. 7–10 and Tables 5–6. The terpolymer samples were heated, cooled and reheated in the temperature range from $-100°C$ to $+250°C$. The glass transition temperature T_g, change of specific heat Δc_p., crystallization temperature T_c, enthalpy of crystallization ΔH_c, melting enthalpy ΔH_m, and melting temperature T_m for the soft phase as well as for the hard phase in both series of terpolymers were determined. The DSC data analysis for the respective blocks are presented in Table 7 for comparison.

The constant value of the glass transition temperature T_{g1} in the low-temperature region, which is independent of the content and the degree of polymerization (DP) of the PBT block in the range $0.25 \leq DP_{PBT} \leq 9$, is characteristic of PTMO-b-PBT-b-PA6 terpolymers. This proves the good separation of the soft phase from the hard phase in the samples of terpolymers from 1_{PA6} to 4_{PA6}. The enthalpy ΔH_{m1} of a low-temperature region for PTMO-b-PBT-b-PA6 decreases with the increase of DP_{PBT}. This means a reduction in the degree of crystallization in this region and, as a consequence, an increase in the amorphous soft phase. This is reasonable, since together with a decrease in DP of the PBT block, the terpolymer shows a higher concentration of the short sequences PBT capable of mixing with PTMO. The endothermic effect determining T'_{m2} (Fig. 5) is observed in all

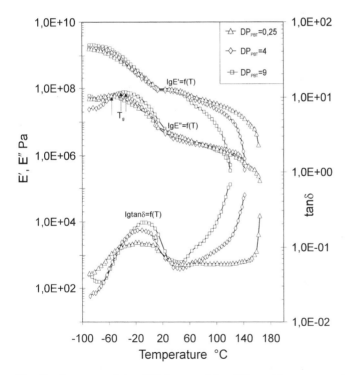

Fig. 6 Storage modulus (E'), loss modulus (E''), and loss tangent (tan δ) of PTMO-b-PBT-b-PA12 terpolymers vs. temperature.

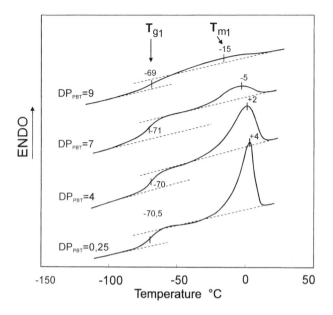

Fig. 7 DSC thermograms of PTMO-*b*-PBT-*b*-PA6 terpolymers in the low-temperature region.

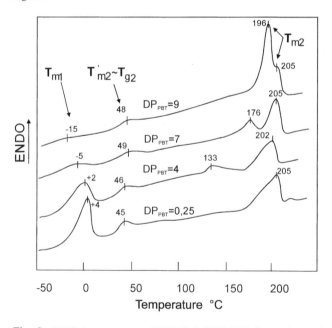

Fig. 8 DSC thermograms of PTMO-*b*-PBT-*b*-PA6 terpolymers in the high-temperature region.

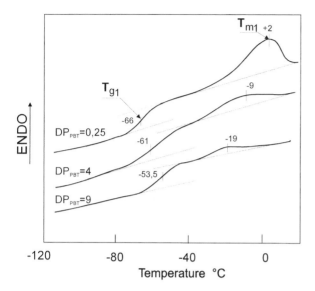

Fig. 9 DSC thermograms of PTMO-*b*-PBT-*b*-PA12 terpolymers in the low-temperature region.

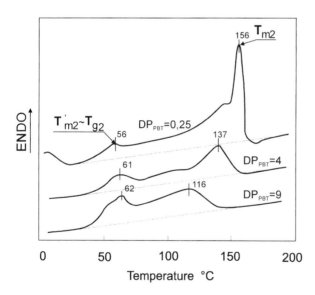

Fig. 10 DSC thermograms of PTMO-*b*-PBT-*b*-PA12 terpolymers in the high-temperature region.

Table 5 DSC Data for PTMO-b-PBT-b-PA6 Terpolymers

Polymer sample	T_{g1} (°C)	ΔC_{p1} J/gk	T_{c1} (°C)	ΔH_{c1} (J/g)	T_{m1} (°C)	ΔH_{m1} (J/g)	T_{g2}, T'_{m2} (°C)	ΔH_2 (J/g)	T_{c2} (°C)	ΔH_{c2} (J/g)	T_{m2}, (°C)	ΔH_{m2} (J/g)	$T_{m2} - T_{g1}$ (°C)
1$_{PA6}$	−70.5	0.259	−30	9.0	4	16.7	45	3.4	99	6.6	205	13.4	275.5
2$_{PA6}$	−70.0	0.228	−30	5.2	2	12.9	46	2.0	113	19.1	133,202	16.8	272
3$_{PA6}$	−71.0	0.205	−40	2.3	−5	4.7	49	1.9	136,145	33.4	176,205	33.9	276
4$_{PA6}$	−69.0	0.132	−36	−0.6	−15	2.2	48	2.0	151,169	42.4	196,205	47.0	274
5$_{PA6}$	−67.5	0.098	—	—	—	—	52	2.0	152,174	48.2	201,208	52.1	275.5

T_{g1}, T_{c1}, T_{m1} = glass transition, crystallization, and melting point temperatures, respectively, in low-temperature region;
ΔC_{p1} = heat capacity change in T_{g1};
ΔH_{m1} = heat of melting at T_{m1};
ΔH_{c1} = crystallization heat in T_{c1};
T_{g2}, T_{c2}, T_{m2} = glass transition, crystallization, and melting point temperatures, respectively, in high-temperature region;
ΔH_{c2} = crystallization heat in T_{c2};
ΔH_{m2} = heat of melting at T_{m2};
ΔH_2 = heat of melting at T'_{m2} (this endothermic effect was observed for many polymers crystallized from the melt).

Table 6 DSC Data for PTMO-b-PBT-b-PA12 Terpolymers

Polymer sample	T_{g1} (°C)	ΔC_{p1} J/gk	T_{c1} (°C)	ΔH_{c1} (J/g)	T_{m1} (°C)	ΔH_{m1} (J/g)	T_{g2}, T'_{m2} (°C)	ΔH_2 (J/g)	T_{c2} (°C)	ΔH_{c2} (J/g)	T_{m2} (°C)	ΔH_{m2} (J/g)	$T_{m2} - T_{g1}$ (°C)
1$_{PA12}$	−66.0	0.24	−19.5	11.7	2	11.7	56	2.7	125	42.4	156	43.3	222
2$_{PA12}$	−64.0	0.24	−29.0	5.8	−6	5.8	57	2.8	113	37.1	147	40.1	211
3$_{PA12}$	−61.0	0.28	−31.0	1.0	−9	1.0	61	4.0	92	28.4	137	31.9	198
4$_{PA12}$	−58.0	0.36	—	0.6	—	—	56	5.5	72	27.6	121	28.8	169
5$_{PA12}$	−53.5	0.27	—	—	—	—	62	5.2	59	20.0	116	25.0	169.5
6$_{PA12}$	−54.0	0.40	—	—	—	—	56	5.0	35	19.4	108	22.5	162
7$_{PA12}$	−40.5	0.31	—	—	—	—	59	5.1	80	25.1	104, 134	27.7	144.5; 174.5

T_{g1}, T_{c1}, T_{m1} = glass transition, crystallization, and melting point temperatures, respectively, in low-temperature region;
ΔC_{p1} = heat capacity change in T_{g1};
ΔH_{m1} = heat of melting at T_{m1};
ΔH_{c1} = crystallization heat in T_{c1};
T_{g2}, T_{c2}, T_{m2} = glass transition, crystallization, and melting point temperatures, respectively, in high-temperature region;
ΔH_{c2} = crystallization heat in T_{c2};
ΔH_{m2} = heat of melting at T_{m2};
ΔH_2 = heat of melting at T'_{m2}.

Table 7 DSC Study for Respective Blocks Incorporated into the Composition of Resulting Series of Terpolymers

Polymer sample	T_{g1} (°C)	ΔC_{p1} J/gk	T_{c1} (°C)	ΔH_{c1} (J/g)	T_{m1} (°C)	ΔH_{m1} (J/g)	T_{g2}, T_{m2} (°C)	ΔC_{p2} J/gk	T_{c2} (°C)	ΔH_{c2} (J/g)	T_{m2} (°C)	ΔH_{m2} (J/g)	$w_{c,H}$ (%)
PTMO$_{1000}$	−88	0.64	2	91.3	19	111							64.5
PBT$_{DP=4}$	—	—	—	—	—	—	60	0.18	127.17	83.9	149; 193	90.5	63.0
PA12$_{2000}$	—	—	—	—	—	—	10	0.17	134.00	75.6	134; 169	81.0	84.5

T_{g1}, T_{c1}, T_{m1} = glass transition, crystallization, and melting point temperatures, respectively, in low-temperature region;
ΔC_{p1} = heat capacity change in T_{g1};
ΔH_{m1} = heat of melting at T_{m1};
ΔH_{c1} = crystallization heat in T_{c1};
T_{g2}, T_{c2}, T_{m2} = glass transition, crystallization, and melting point temperatures, respectively, in high-temperature region;
ΔH_{c2} = crystallization heat in T_{c2};
ΔH_{m2} = heat of melting at T_{m2};
ΔH_2 = heat of melting at T_{m2};
$w_{c,H} = \dfrac{\Delta H_t}{\Delta H_{t,c}}$ = degree of crystallinity determined by calorimetric method;
ΔH_t = melt enthalpy of polymer sample = ΔH_{m2};
$\Delta H_{t,c}$ = melt enthalpy of completely crystalline polymer; $\Delta H_{t,c\text{PTMO}}$ = 172 J/g [2]; $\Delta H_{t,c\text{PBT}}$ = 144 J/g [32]; $\Delta H_{t,c\text{PA12}}$ = 96 J/g [2].

examined samples during the first heating. This effect overlaps with the effect of the specific heat variation on the amorphous part of the hard phase rich in PA12 and PBT blocks (T_{g2}). After cooling down it appears again after only 1 hour. This effect was observed for many polymers crystallized from the melt. It is admitted that this is responsible for the melting of microcrystalline species between the boundary layers of the large crystallites. The T'_{m2} and T_{g2} values were taken from the respective maxima observed, and they ranged from 45 to 49°C for the samples investigated (Table 5). The enthalpy of this effect achieves a constant value, independent of the chemical composition of the amide block, after 18 h. This may imply that the formation of the constant amount of fine crystallites is a part of the crystalline structure of aggregates. The changes in $T'_{m2} \approx T_{g2}$ are caused by the relaxation effects originating from the internal stress of this system or physical aging, before measurement, at room temperature.

The heat of melting and crystallization, ΔH_{m2} and ΔH_{c2}, respectively, and the melting temperature T_{m2} in a high-temperature region rises with increasing DP_{PBT} (an increase of PA6 + PBT fraction). An important feature of PTMO-b-PBT-b-PA6 is the occurrence of the two endothermic effects in the high-temperature region, associated with melting (two peaks T_{m2}, Fig. 5) and the two values of T_{c2} (Tab. 5) for the samples with $DP_{PBT} \geq 4$ (2_{PA6} to 5_{PA6}). Sample 1_{PA6}, with $DP_{PBT} = 0.25$, is characterized by a low PBT content and has one main endothermic effect connected with melting (single maximum T_{m2} occurs). The copoly(ester-ether)s of the PTMO-b-PBT type and copoly(ether-amide)s of the PTMO-b-PA6 type possess two maxima T_m2 (2,33), independent of the fraction and the molecular weight of the PBT block. Taking these facts into account and analyzing the remaining data in Table 5, and by comparison with the data in Table 7, it can be concluded that the soft phase in PTMO-b-PBT-b-PA6 terpolymers is composed mainly of PTMO, whereas the hard phase is probably composed from the two different regions, in which PA6 and/or PBT (multiphase structure) dominates. It turns out that $T_{m2} - T_{g1} = $ a constant (Table 5) for the investigated samples of PTMO-b-PBT-b-PA6. This is evidence of good separation of the hard phase from the soft phase in these terpolymers.

The DSC curves for the PTMO-b-PBT-b-PA12 terpolymers are characterized by completely distinctive runs from the previously described DSC curves for the PTMO-b-PBT-b-PA6 terpolymers. In PTMO-b-PBT-b-PA12 terpolymers, the glass transition temperature of a low-temperature region, T_{g1}, is a function of the degree of polymerization of the PBT block (Table 6). For sample 1_{PA12} with $DP_{PBT} = 0.25$, the glass transition temperature amounts to $T_{g1} = -66°C$; and for sample 7_{PA12} with $DP_{PBT} = 14$, it amounts to $T_{g1} = -40.5°C$. An increase in T_{g1} is a result of the reduction of the PTMO phase fraction in the terpolymer. However, a large difference in the glass transition temperatures between the extreme samples of this series of terpolymers, amounting to $\Delta T_{g1} = 25.5°C$, may reflect the increase in the fraction of PBT and PA12 blocks in the soft phase. A DSC curve (Fig. 9) showing an endothermic effect in the low-temperature region is seen only for a degree of polycondensation of PBT lower than 4 ($DP_{PBT} \leq 4$ in samples 1_{PA12}, 2_{PA12}, and 3_{PA12}). In samples 4_{PA12} and 5_{PA12}, with $DP_{PBT} > 4$, the endothermic effects are so small that it is difficult to estimated them. Therefore, it can be assumed that a partial crystallization in the low-temperature region occurs in the terpolymers containing short segments of PBT. In the high-temperature region, as in PTMO-*b*-PBT-*b*-PA6 terpolymers, an endothermic effect ($T'_{m2} \approx T_{g2}$) occurs in the range from 56 to 62°C, which is associated with the glass transition temperature T_{g2} of the hard phase, rich in PA and PBT. In this region, the decrease in both the heat of melting ΔH_{m2} and the melting temperature T_{m2}, together with an increase in DP_{PBT} and the sum of the fractions of PBT and PA12 in the terpolymers, is rather unexpected. This proves there is mixing of PA12 blocks with PBT and the formation of the high-temperature region (hard phase), which is responsible for T_{m2} from a mixture of these blocks. The two temperatures T_{m2} occur in sample 7_{PA12}. The following explanation can be put forward: The two temperatures T_{m2} are probably a result of the formation of the multiphase structure composed of an amorphous phase (constituted mostly of PTMO blocks) as well as a phase of PA12 and/or PBT blocks. Unfortunately, we could not obtain (in spite of many trials) PTMO-*b*-PBT-*b*-PA12 with a larger content of PBT, i.e., $DP_{PBT} > 14$. This would enable the verification of the foregoing hypothesis. An indirect evidence (by analogy) is the occurrence of the two temperatures T_{m2} in PTMO-*b*-PBT-*b*-PA6 terpolymers previously described. The approaching of maxima of endotherms determining T_{g1} and T_{m2} (Figs. 9 and 10), that is, the decrease of an interval between T_{g1} and T_{g2}, together with an increase in DP_{PBT} of the PTMO-*b*-PBT-*b*-PA12 terpolymers, supports the idea of partial mixing of the hard phase formed from PA12 and PBT blocks with the soft phase constituted from PTMO and, therefore, a poor phase separation of the entire system and the formation of a considerable interphase.

IV. CONCLUSIONS

The block terpolymers of the poly[oxytetramethylene-*block*-(tetramethylene terephthalate)-*block*-laurolactam] and poly[(oxytetramethylene)-*block*-(tetramethylene terephthalate)-*block*-(ε-caprolactam)] (PTMO-*b*-PBT-*b*-PA6 and PTMO-*b*-PBT-*b*-PA12) with different degrees of polycondensation of the ester block ($DP_{PBT} < 14$) were prepared by melt polycondensation. These terpolymers represent a novel class of polymeric material with potential application as elastomers. An analysis of the results indicate that both the course of the synthesis and the chemical composition of particular phases depend on the content of the segments and the length of the individual blocks. By changing the chemical structure and length of the blocks it is possible to prepare terpolymers with the desired properties.

The possibility of the synthesis of polymers comprising three immiscible blocks, namely, the PTMO, PBT, and PA6 or PA12 blocks, under appropriate conditions and with the use of a suitable catalyst, has been demonstrated. Two types of products were obtained: (1) PTMO-b-PBT-b-PA6 terpolymers with appearance and mechanical properties similar to those of copoly(ether-ester) elastomers of the HYTREL type, and (2) PTMO-b-PBT-b-PA12, transparent products of light cream color and with mechanical properties similar to those of PUEs.

Based on the results of differential scanning calorimetry and static and dynamic mechanical measurements, it is concluded that PTMO-b-PBT-b-PA terpolymers with $DP_{PBT} < 14$ are characterized by: physical changes T_{g1}, T_{m1}, T_{g2}, and T_{m2}, a wide "plateau" of a small modulus of elasticity, good reversible elasticity, and the ability for phase separation. Thus, the PTMO-b-PBT-b-PA terpolymers with the compositions given in Tables 3 and 4 can be classified as typical elastomers. In PTMO-b-PBT-b-PA6, the observed phase structure is better formed than in PTMO-b-PBT-b-PA12. However, the PTMO-b-PBT-b-PA12 terpolymers have a higher ability for elastic recovery (important property for use). The separation degree of PTMO-b-PBT-b-PA6 can be compared with the separation degree of copoly(ether-ester)s elastomers, whereas this degree in PTMO-b-PBT-b-PA12 is asimilar to the polyurethane elastomers with the same type of flexible blocks [33,34]. A small degree of separation in PTMO-b-PBT-b-PA12 allows us to assume that in these terpolymers, as in polyurethanes and copoly(ether-amide)s, a substantial interphase exists (6,34). This is confirmed by the low melting temperatures resulting from DMTA analysis. Poly[oxytetramethylene-$block$-(tetramethylene terephthalate)-$block$-laurolactam]s, blocks polyurethanes, and copoly(ether-amide)s with the high fraction of the interphase characterize a good elastic recovery, in contrast to the poly[(oxytetramethylene)-$block$-(tetramethylene terephthalate)-$block$-(ε-caprolactam)]s and blocks copoly(ether-ester)s, in which the residue interphase and large fractions of the crystalline and amorphous phases occur. Thus, it can be concluded that block elastomers that are characterized by an increased fraction of the interphase (i.e., with blocks partially mixing together) possess properties close to those of soft rubber. On the other hand, polymers with well-formed phases show properties closer to those of hard rubber. This may determine the direction of experiments for research searching for novel high-elastic plastics.

REFERENCES

1. DJ Meier. Block Copolymers: Morphological and Physical Properties. NATO ASI Ser., Ser E 89:173–194, 1985.
2. M Xie, Y Camberlin. Etude morphologique de $block$-copoly(ether-amide)s. Macromol. Chem. 187:383–400, 1986.

3. G Holden. Thermoplastic Elastomers. In: JC Salamone, ed. Polymeric Materials Encyclopedia. New York: CRC Press, 1996, vol. 11, pp 8343–8353.
4. Z Roslaniec. Polymeric systems with elastothermoplastic properties. Szczecin, Poland: Technical University of Szczecin Press, 1993. pp 15–50.
5. J Slonecki. Structure and some properties of copoly(ester-ether)s. Szczecin, Poland: Technical University of Szczecin Press, 1992. pp 11–35.
6. M Ash, J Ash. Handbook of Plastic Compounds, Elastomers, and Resins. New York: VCH, 1992, pp 623–640.
7. R Ukielski, H Wojcikiewicz. Synthesis and properties of copoly(ester-ether) elastomers. Effect of the chemical structure and length of the flexible segments. Int. Polym. Sci. Technol. 11:T/65–T/68, 1984.
8. R Ukielski, F Lembicz. Low-temperature relaxation processes and electrical properties of copoly(ester-ether) elastomer. Polimery (Polish ed.) 30:105–106, 1985.
9. R Ukielski, F Lembicz. Relaxation processes in a block copoly(ether-ester) elastomer investigated by the spin probe method. Makromol. Chem. 186:1679–1683, 1985.
10. R Ukielski. Block copoly(ester-ether)s modified by pentaerythritol. Polimery (Polish ed.) 33:9–12, 1988.
11. R Ukielski. Thermoplastic copolyesterether elastomers. Polish Technical Rev. Press Service IV, 1/1989.
12. R Ukielski, H Wojcikiewicz. Manufacturing method of thermoplastic copoly–esterether elastomers. Polish Pat. 108, 711 (1979).
13. R Ukielski, H Wojcikiewicz. Manufacturing method of thermoplastic copoly–esterether elastomers. Polish Pat. 115, 345 (1980).
14. R Ukielski. Manufacturing method of thermoplastic block copolyesterethers with elevated reversal deformability increased transversal viscosity in the molten state. Polish Pat. 136, 744 (1981).
15. R Ukielski, J Slonecki, H Zarzycka, Z Maćków. Manufacturing method of thermoplastic copolyesterether elastomers. Polish Pat. 137, 158 (1981).
16. R Ukielski, J Slonecki, H Wojcikiewicz, Z Roslaniec, P Kurek, A Kapelański, Z Maćków. Manufacturing method of thermoplastic block copolyesterethers. Polish Pat. 150, 278 (1987).
17. R Ukielski, K Pawlaczyk, J Slonecki, H Wojcikiewicz. Manufacturing method of thermoplastic block copolyesterethers with increased viscosity. Polish Pat. 151, 320 (1988).
18. R Ukielski. Synthesis of block polyesteretheramide terpolymers. Polimery 40:160–163, 1995.
19. R Ukielski. Synthesis of polyoxytetramethylene-*block*-tetramethylene terephthalate-*block*-laurolactam. Polimery (Polish ed.) 41:286–289, 1996.
20. K Pawlaczyk, R Ukielski. Synthesis of dikarboksyloligoamide blocks applied in production of terpolyesteretheramide elastomers. Euruphysics Conference on Macromolecular Physics, Kołobrzeg, Poland, 5–8 May, 1997, pp 171–175.
21. K Pawlaczyk, R Ukielski. Synthesis of α,ω-dicarboxyloamide oligomers. Polimery (Polish ed.) 42:680–681, 1997.
22. R Ukielski. Synthesis and properties of poly(ether-ester-amide) block terpolymers. International Polymer Seminar, Gliwice, Poland, 1997, pp 105–110.

23. J Zimmerman. Polyamides. In: Encyclopedia of Polymer Science and Engineering. 2nd ed. New York: Wiley, 1988, vol. 11, pp 315–381.
24. K Okajima, Ch Yamane, F Ise. Polyamide Fibers Aliphatic. In: JC Salamone, ed. Polymeric Materials Encyclopedia. New York: CRC Press, 1996, vol. 7, pp 5387–5397.
25. R Puffr, J Stehlíček. Polyamides Lactam-Based. In: JC Salamone, ed. Polymeric Materials Encyclopedia. New York: CRC Press, 1996, vol. 7, pp 5432–5440.
26. R Feldmann, R Feinauer. Nachkondensation von Polylaurinlactam in fester Phase. Angew. Makromol. Chem. 34:1–7, 1973.
27. R Ukielski, D Pietkiewicz. The effect of changes of the degree of condensation polymerization of oligo(tetramethylene terephthalane) block on properties of poly(ether-ester-amide)s. Polimery (Polish ed.) 43:39–43, 1998.
28. R Ukielski, D Pietkiewicz. Influence of changes in degree of polycondensation of polyester block on properties of poly[ether-*block*-ester-*block*-amide] terpolymers J. Macromol. Sci. Phys. B37(2):255–264, 1998.
29. R Ukielski, J Slonecki. Manufacturing method of thermoplastic block copoly(ester-ether-amide)s with branched constitution of macromolecular. Polish Pat. 161, 526 (1993).
30. R Ukielski. Manufacturing method of thermoplastic block copoly(ether-ester-amide)s. Polish Pat. 162, 304 (1993).
31. R Ukielski, Z Maćków. Manufacturing method of thermoplastic block terpoly(ester-ether-amide)s. Polish Pat. 165, 712 (1995).
32. M Gilbert, FJ Hybart. Effect of chemical structure on crystallization rates and melting of polymers. Part 1: aromatic polyesters. Polymer 13:327–332, 1972.
33. K Dijkstra, H Martens, M Soliman, RJM Borggreve. Morphology development and properties of segmented block co(poly-ether-esters). 4th International Symposium on Thermoplastic Elastomers, Europhysics Conference on Macromolecular Physics, Kołobrzeg, Poland, 1997, pp 43–46.
34. Y Camberlin, JP Pascault. Quantitative DSC evaluation of phase segregation rate in linear segmented polyurethanes and polyurethaneureas. J. Polym. Sci. 21:415–423, 1983.

5
Copolymers of Poly(ε-caprolactam)-Block-(Butadiene-containing elastomers)

Jan Roda
Institute of Chemical Technology, Prague, Czech Republic

I. INTRODUCTION

The application of polymers in fields dominated by traditional materials (silicates, metals) affects our everyday lives and offers a new field of research activity. The polymers substituting such classical materials are denoted as constructional-engineering plastics. They must possess sufficiently high moduli, strength, toughness, and a good resistance to heat. What is valued in these materials is their low density, resistance against wear, and, last but not least, relatively simple production, processing, and assembly.

Polyamides (1) derived from unsubstituted lactams (See Scheme 1), classified as aliphatic polyamides, are counted among the earliest engineering plastics. The aliphatic polyamides, above all the polymer derived from seven-membered ε-caprolactam, and in some cases also the polyamide derived from 13-membered ω-laurolactam, are—due to their mechanical properties—aimed at such applications (2). For special purposes, especially at temperatures below 0°C, it is necessary to increase the toughness, which is, in the case of certain heavy-duty applications, insufficient.

Generally, there are two ways to modify the physicochemical (and therefore also mechanical) properties of polyamides, especially poly(ε-caprolactam), or polyamide 6, which we shall focus on in the following sections.

1. Most frequently, by means of physical procedures (3), namely, by blending with other organic polymers, often using compatibilizing agents, or by mixing with proper inorganic fillers.

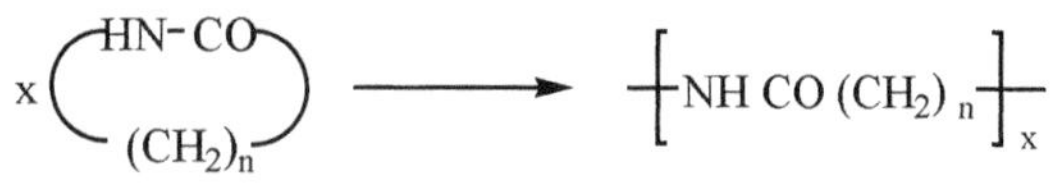

$$x \left(\begin{array}{c} \text{HN-CO} \\ (CH_2)_n \end{array} \right) \longrightarrow \left[NH\,CO\,(CH_2)_n \right]_x$$

n = 5   ε-caprolactam   polyamide 6, nylon 6, poly(ε-caprolactam)
                            poly[imino(1-oxohexamethylene)]

n = 11   ω-laurolactam   polyamide 12, nylon 12, poly(ω-laurolactam)
                            poly[imino(1-oxododecamethylene)]

Scheme 1   Commercially available homopolymers of lactams.

2. Alternatively, by means of chemical modification. One highly valued route is formation of block copolymers. When the synthesis is performed adequately, combining suitable polyamide and elastomeric blocks, elastic domains are formed in a continuous polyamide matrix that may increase substantially the toughness of the material.

The syntheses and properties of the block copolymers of aliphatic polyamides, particularly poly(ε-caprolactam), are discussed in detail elsewere (3). The synthetic routes were classified as follows:

1. Exchange reactions of homopolymers
2. Coupling of the prepolymer chains with reactive end (terminal) groups
3. Polymerization methods:

    a. Copolymerizations of lactams with nonlactam monomers
    b. Polymerizations of lactams in the ends of a prepolymer chain

The polymers having reactive functional groups in their main chains easily undergo exchange reactions at temperatures above their melting point. The exchange reactions are a random process, but the establishing of the equilibrium proceeds via the block copolymer stage.

The coupling of the prepolymer chains bearing reactive end groups is based on the reaction of a functionalized polyamide with another functionalized polymer. Obtaining a high-quality (i.e., sufficiently pure) block copolymer requires perfect functionalization of both polymers and a quantitative reaction of the end groups.

Theoretically, simultaneous copolymerization is possible only for the anionic copolymerization of lactams with lactones, due to the different mechanisms for opening the lactam and lactone rings (see Scheme 2). The alkoxide anion

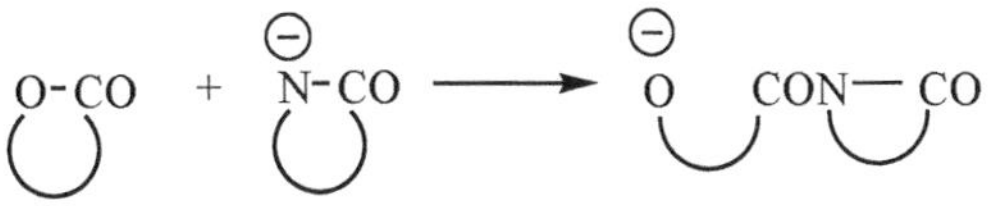

$$O\text{-}CO \;+\; \overset{\ominus}{N}\text{-}CO \;\longrightarrow\; \overset{\ominus}{O} \quad CON\text{---}CO$$

Scheme 2 Simultaneous initiation of lactam–lactone copolymerization.

forms a propagation center for a polyester chain, while the *N*-acyllactam group represents the center for incorporating the lactam. Due to exchange reactions (amide/ester), however, a copolymer having a rather random character is obtained, as was demonstrated for the copolymerizations of the pairs ε-caprolactam/ ε-caprolactone (4) and ω-laurolactam/ε-caprolactone (5).

For technological applications predominantly, anionic polymerization is used when lactams are to be polymerized on prepolymers, end-capped with various reactive functional groups. In this chapter, a detailed analysis of this technique is presented. A preparation of the block copolymers of lactams by anionic polymerization has been studied since the 1960s. Finally, an original technology (NYRIM) has been developed for the production of the polyamide 6/polyether block copolymer. To explain the chemistry of the process, it is necessary to mention briefly the main features of the anionic polymerization of lactams (6,7).

To initiate an anionic polymerization of lactams (ε-caprolactam being the most widely studied lactam), the presence of an initiator—a lactam anion—is necessary. Alkali or alkaline earth metal cations are used as counterions (see Scheme 3).

The salts of lactam can be prepared using strong nucleophilic agents (alkali metals, hydrides, alkoxides). The initiation system contains a further component, an activator, that is, a substance having the *N*-acyllactam (N-substituted carbamoyllactam) structure or its precursor (isocyanate) (see Scheme 4). By repeating the addition of the lactam anion on these *N*-acyllactam structures and cycleopening and neutralization reactions, accompanied by a regeneration of the lactam anion (ring-opening polymerization), the polyamide chains are formed.

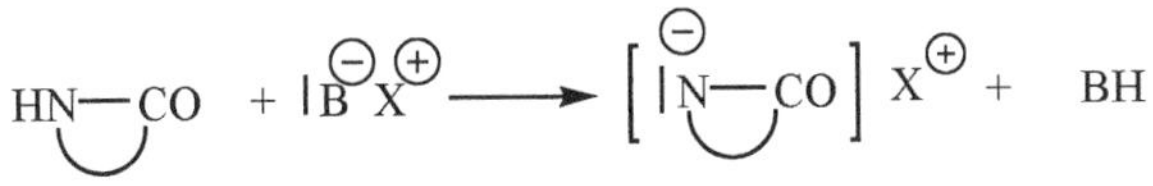

$$HN\text{---}CO \;+\; I\overset{\ominus}{B}\overset{\oplus}{X} \;\longrightarrow\; \left[\; I\overset{\ominus}{N}\text{---}CO \;\right] X^{\oplus} \;+\; BH$$

$$X = Na,\ K,\ Li,\ MgBr,\ Mg/2\ \text{etc.}$$

Scheme 3 Formation of initiator of anionic lactam polymerization.

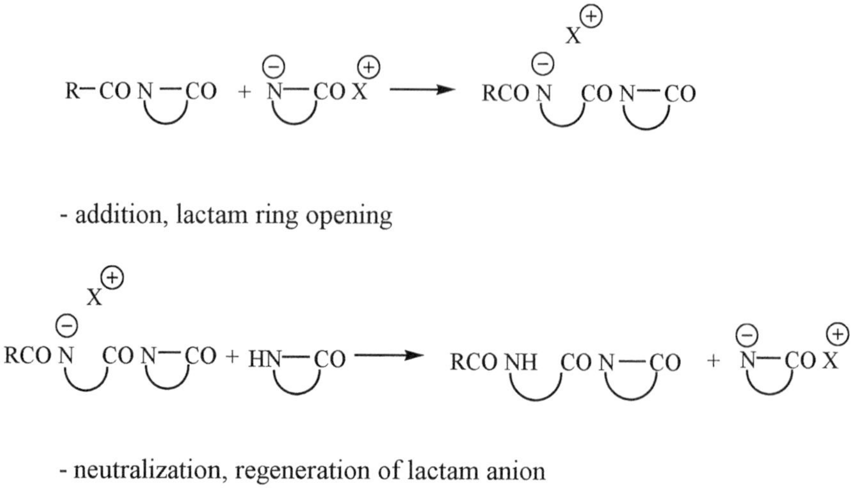

$$R\text{-}CO\,N\text{-}CO \;+\; \overset{\ominus}{N}\text{-}CO\,\overset{\oplus}{X} \;\longrightarrow\; RCO\,\overset{\ominus}{N}\quad CO\,N\text{-}CO \quad\overset{X^{\oplus}}{}$$

- addition, lactam ring opening

$$RCO\,\overset{\ominus}{N}\quad CO\,N\text{-}CO \;+\; HN\text{-}CO \;\longrightarrow\; RCO\,NH\quad CO\,N\text{-}CO \;+\; \overset{\ominus}{N}\text{-}CO\,\overset{\oplus}{X}$$

- neutralization, regeneration of lactam anion

Scheme 4   Basic steps of the anionic polymerization of lactams.

Polymerization of ε-caprolactam proceeds in a wide temperature range. Above the melting point (>220°C) and below it (140–200°C), where the polymerization is heterogeneous, polymerization is followed by crystallization. With respect to a low heat of polymerization (140 J/g), it is possible to carry out the process adiabatically below the melting point of polyamide 6.

Formation of the block copolymer poly(ε-caprolactam)/elastomer by an anionic polymerization of ε-caprolactam initiated on the ends of prepolymer chains requires that the prepolymer be soluble in the ε-caprolactam melt (mp 70°C) and bear suitable endgroups (hydroxy, amino, or carboxy). The first step of the synthesis is a transformation of the reactive end groups of the telechelic polymer into nonionic growing centers of the anionic polymerization. This functionalization yields the so-called macroactivator. The second step of the synthesis is the anionic polymerization of ε-caprolactam (the formation of the polyamide blocks) on the nonionic growing centers formed, in the presence of the initiator. The procedure can be illustrated by the NYRIM technology (8), the procedure giving the nylon 6 (polyamide 6) block copolymers by reaction injection molding (RIM). As the prepolymer, the hydroxy-terminated polyethers are used. The end groups are transformed into nonionic N-acyllactam growing centers by alcoholysis with bis-acyllactams (Scheme 5).

The success of the NYRIM technology is based primarily on the application of a special initiator, namely, ε-caprolactammagnesium bromide (CLMgBr). Owing to the high rate of the pseudoadiabatic polymerization, which proceeds below the melting point of the polymer and is followed by crystallization, molding of the

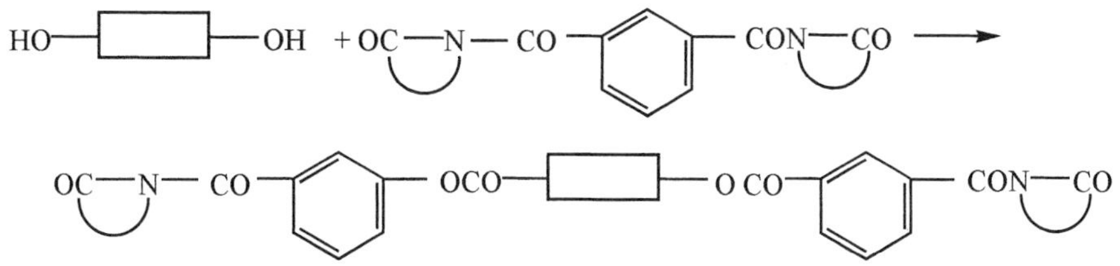

- functionalization, formation of macroactivator with N-acyllactam growth centres

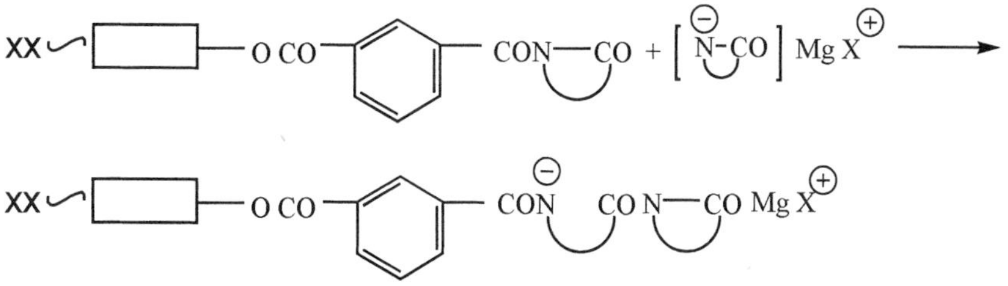

-propagation, repetition of addition and neutralization, formation of polyamide block

Scheme 5   Basic steps of NYRIM technology.

product can be completed within a few minutes, which is the main requirement of the RIM technology. Alternatively, the system can be used for the polymerization casting.

Most papers, and especially patents, focus on polyethers as precursors of the soft block copolymer segments. However, other types of the telechelic polymers (based on polybutadienes, silicons, polysulfides, etc.) were also used to modify the properties of the polyamides (3).

The aims of this chapter are to summarize the following:

The available literature concerning the preparation and properties of block copolymers poly(ε-caprolactam)-*block*-(butadiene-containing elastomers), especially poly(ε-caprolactam)-*block*-polybutadiene, aiming at the preparation of construction materials

The preparation procedures for less common block copolymers, where the poly(ε-caprolactam) blocks are formed by anionic polymerization of ε-caprolactam on prepolymers bearing reactive functional end groups (growth centers)

Unpublished results of the author's research group in optimizing the preparation of the mentioned copolymers where the improved toughness was reached

II. COPOLYMERS
POLY(ε-CAPROLACTAM)-*BLOCK*-POLYBUTADIENE

One of the early attempts at the preparation of copolymers with bonded poly(ε-caprolactam) and polybutadiene (PBD) blocks consisted in functionalizing living polybutadiene with 2,4(6)-toluylene diisocyanate (TDI) in a nonpolar solvent (hexane) (9). The fine suspension of the polymeric material produced in the same medium by an anionic polymerization could be processed at room temperature. The copolymers retained their rubberlike behavior, even at as a high content of incorporated ε-caprolactam as 70–80 wt%, and did not possess engineering plastic properties (Table 1). These properties are the consequence of the fact that polymerization proceeds in a solvent that is thermodynamically good for polybutadiene. According to transmission electron microscopy (TEM), the material consists of a continuous polybutadiene phase in which two types of poly(ε-caprolactam) particles are perfectly dispersed: the fine ones having dimensions of about 50–100 nm, and the coarse ones with a diameter of 0.3–3.0 μm (74% of ε-caprolactam incorporated in the material) (10). The extraction analysis as well as the fractionation of the material confirmed that it is composed of both homopolymers and a true block copolymer.

It was possible to make the material crosslinked by dicumylperoxide or by classical vulcanization sulphur system and thus to improve its mechanical properties (11). However, the parameters corresponding to a constructional plastic could not be achieved.

To prepare copolymers poly(ε-caprolactam)-polybutadiene, the commercially available α,ω-dihydroxypolybutadiene (Butarez) (Table 1) was functionalized by an extra situ reaction with TDI in boiling toluene (12) (see Scheme 6). The polyamide blocks joined to the ends of the functionalized prepolymer were formed by bulk anionic polymerization at 105°C. Alternatively, the synthesis was carried out in the presence of decahydronaphthalene (decaline). However, this solvent decreased not only the yield but also the degree of incorporation of the lactam into the material. The low temperature of the polymerization was derived from the kinetics of the model anionic polymerization of ε-caprolactam, which demonstrated that at 105°C the carbamate bonds joining the polyamide and polybutadiene blocks (Scheme 6) are not very sensitive to the nucleophilic attack in comparison with carbamoyl growth centers.

Table 1 Preparation of Copolymers Poly(ε-caprolactam) with Polybutadiene (PBD) Elastomers via Anionic Polymerization of ε-Caprolactam (CL) in the Presence of Functionalized Rubbers

Elastomers			Functionalization		Initiator		Polymerization			Copolymer Yield (%)	Mechanical properties[a]					Ref.
Type	$M_n \times 10^{-3}$	1,2-ad. (%)	Agent	In situ	Type	Mol (%)	Temp. (°C)	Time (h)	Solvent		E_t (GPa)	σ_z (MPa)	a_L (kJ·m^{-2})	ε_z (%)	PBD[b] (%)	
PBD[b]	~5		TDI	—	BuLi	~1.1	160	19	Hexane	98	—	Rubbery[d]		80–130	26	9
PBD(OH)₂		26	TDI	—	NaH	~1	105	16	—	87	—	22–39[f]	—	—	32	12
(Butarez) (Arco)	20.0[c]			—	NaH	~1	105	16	Decaline	26	—	—	—	—	59	12
PBD(OH)₂ (Arco)	6–24[c]		TDI	—	NaH	—	110	—	Decaline	87	2.1	56	15	57	10	13
PBD(OH)₂	6–24[c]	20	TDI	+	Na	~0.75	180	0.25	—	—	—	—	—	100	—	14
PBD(OH)₂	1		TDI	+	CLK[g]	0.5	150	0.5	—	97	1.5	53	24[h]	75	10	15
PBD(OH)₂	~5	65	MDI	+	CLK	0.5	150	0.5	—	97	1.2	39	22[h]	70	10	15
			MDI	+	CLNa[i]	0.75	160	0.5	—	98	0.9	36	39[h]	350	15	
SBR(OH)ₓ (Arco)	3.5		TDI	—	NaH	—	160	5	—	92	0.02	35–45[k]	—	—	20	20
NBR(NHR)ᵧ (Hycar)	3.7		TDI	+	CLNa	1.3	150	0.5	—	94	1.4[m]	70[m]	20	—	10	22

[a] E_t: tensile modulus;
σ_z: tensile strength;
ε_z: elongation at break;
a_L: Charpy notched impact resistance at 23°C.
[b] Content of elastic phase in material.
[c] Living polybutadiene.
[d] Poly(ε-caprolactam) as a discrete phase.
[e] After extension of the initial PBD chains ($M_n = 4.4 \times 10^3$, Ref. 12, or 3×10^3, Ref. 13) by reaction with TDI.
[f] Values determined by sample preparation (molding, solvent casting).
[g] CLK: potassium salt of ε-caprolactam.
[h] Determined at -20°C.
[i] CLNa: sodium salt of ε-caprolactam.
[j] Styrene-butadiene copolymer (SBR).
[k] Depending on concentration of initiator used.
[l] Butadiene-acrylonitrile copolymer with 17% of acrylonitrile (NBR).
[m] Flexural characteristic.
TDI: toluylene 2,4(6)-diisocyanate; MDI: 4,4'-diisocyanate diphenylmethane.

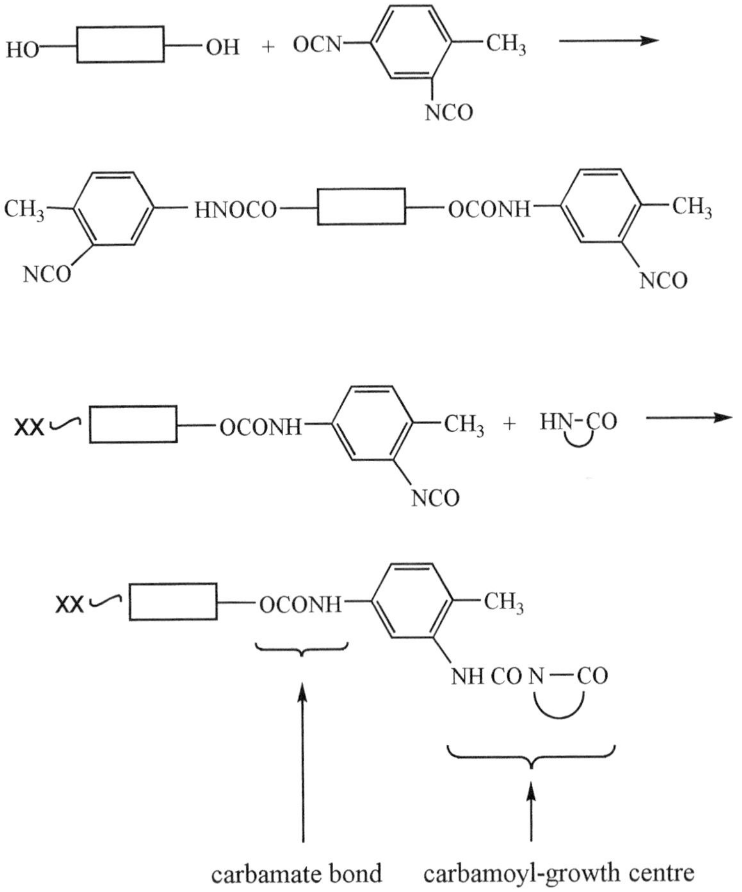

Scheme 6   Functionalization of α,ω-dihydroxypolybutadienes with diisocyanates.

In the product isolated after 16 h of polymerization, as low a degree of conversion of the lactam as 80% was reached. The copolymer contained 33% incorporated polybutadiene phase and 18% homopolymer. The mechanical properties were sensitive both to the thermal history of the sample and to the method of preparation (see Table 2). The foils prepared via casting revealed that spherical polybutadiene domains having the dimensions of about 10 nm were formed. The extraction analysis and morphology were used as a proof of the block character of the material.

Only a single study is available in the literature that also deals with the mechanical properties of the poly(ε-caprolactam)-*block*-polybutadiene (13) pre-

Table 2   Mechanical Properties of Poly(ε-caprolacatam)-*block*-polybutadiene Containing 33 wt% Incorporated Elastic Segments

Material	σ_z (MPa)	ε_r (%)
Isolated from melt	23	80
Isolated from solvent	39	130

Source: Ref. 12.

pared. In this study, α,ω-dihydroxypolybutadiene as a precursor of the elastomeric segment was employed, too. The functionalization was carried out in decalin at 95°C. The resulting molar mass of the elastomer was controlled by the value of the TDI/α,ω-dihydroxypolybutadiene ratio. Sodium hydride, present in an equimolar amount with respect to the unreacted NCO groups of the prepolymer, was used as an initiator of the dispersion polymerization in decalin at 110°C. The degree of conversion (max. 85%), as well as the mechanical properties measured with pressed test specimens, depends on the molar mass and the content of polybutadiene in the polymerization feed. Tensile and flexural moduli, as well as the strength and melting points, decreased with increasing concentration of polybutadiene in the copolymer; in contrast, the notch impact strength and the elongation at break increased with increasing polybutadiene content. All the measured quantities assumed their highest values for the lowest molar mass of polybutadiene used, i.e., $M_n = 6,000$ g/mol (soaking) (see Fig. 1). It was proved by TEM that the discrete particles of the elastic phase had spherical shape and that the samples prepared by pressing showed an orientation of the phases.

The finding that water sorption of the prepared materials decreased with increased content of the hydrophobic polybutadiene segments is rather controversial. When determining the sorption, the uncreated ε-caprolactam present in the starting material was probably not taken into account.

Through an anionic polymerization, initiated by a sodium salt of ε-caprolactam and performed in molds (15 min at 180°C), TDI-terminated polybutadiene segments having their molar mass of about 1,000 g/mol were incorporated into the poly(ε-caprolactam) structures (14). Via an extraction analysis, the content of the homopolymer of ε-caprolactam, the soluble block copolymer, and the insoluble fraction were determined (see Fig. 2). The presence of two structural modifications of the block copolymer is deduced from dynamic mechanical spectra in Ref. (14).

The copolymers poly(ε-caprolactam)-*block*-polybutadiene have long been the subject of research at the Department of Polymers (Institute of Chemical Tech-

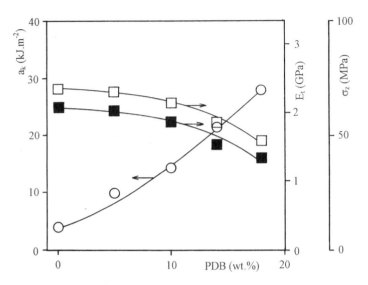

Fig. 1   Dependence of (○) notched impact strength a_k at 23°C, (□) tensile modulus E_t, and (■) tensile strength σ_z on the concentration of elastic polybutadiene segments (PBD) in the poly(ε-caprolactam)-*block*-polybutadiene copolymers prepared by the anionic polymerization of ε-caprolactam initiated with NaH in the presence of toluylene 2,4(6)-diisocyanate (TDI) and α,ω-dihydroxypolybutadiene ($\overline{M}_n \sim 6000$) as a precursor of elastic segments at 110°C in decaline. (Adapted from Ref. 13.)

nology, Prague) (15). The investigation was focused on a detailed analysis of parameters affecting the mechanical properties of the block copolymers, as well as on the optimization of the polymerization process. In this case, optimization meant finding a suitable compromise between the content of elastomer in the polymerization mixture composition and the chemical nature of the initiation system on the one hand, and the final mechanical properties, above all the toughness, on the other. A variety of α,ω-dihydroxypolybutadienes (PBD), prepared in the Research Institute for Synthetic Rubber (Kralupy, Czech Republic) by a solution polymerization with lithium initiators, was used. The molar-mass range was 2,000–5,500 g/mol and the average OH functionality was 1.7–1.9. The functionalization of these polybutadienes was carried out in situ in liquid ε-caprolactam at 110°C using isocyanates, mostly by the toluylene 2,4(6)-diisocyanate (TDI) mentioned earlier; an alkali salt of ε-caprolactam as initiator was introduced either as a solution or as a solid concentrate in ε-caprolactam. The polymerization mixture in a mold was immediately placed in an oil bath heated up to the polymerization temperature (150–170°C). This method of preparation simulates, to some extent, the conditions of industrial polymerization casting. The time period t_c, corresponding to a maximum on the temperature profile measured by a thermocouple directly inside the mold, was taken as a measure of the polymerization rate.

The TDI applied as a functionalization agent in all the available communications (9–13) (see Table 1) was chosen due to the fact that the reactivities of the two isocyanate groups in its molecule differ by an order of magnitude, which diminishes the probability of the coupling of the telechelic PBD molecules (Scheme 6).

A variety of other commercially available isocyanates were tested in Ref. 15. The criteria of their suitability as functionalization agents were, in the first place, the mechanical properties of the materials obtained. The characteristics of the materials prepared by an anionic polymerization in the presence of 10 wt% of polybutadiene having $M_n \approx 4,900$ g/mol are summarized in Table 3, where the data published in Ref. 15 are combined with supplemental data.

The increase by one order of magnitude of the toughness of the materials, which suggests the formation of block copolymers, was achieved with aromatic diisocyanates only. An important finding should be noted: The free diisocyanates can be substituted by the derivatives of isocyanates blocked by ε-caprolactam (carbamoyllactams).

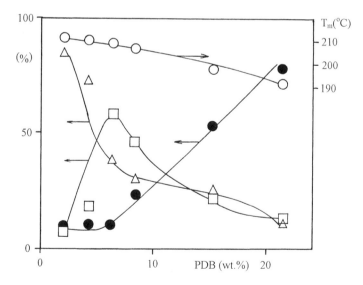

Fig. 2 Dependence of content of poly(ε-caprolactam) (△), linear copolymer poly(ε-caprolactam)-*block*-polybutadiene (□), and crosslinked copolymer poly(ε-caprolactam)-*block*-polybutadiene (●) and melting temperature of polyamide phase (○) on the initial concentration of polybutadiene (PBD) in materials prepared via the anionic polymerization of ε-caprolactam initiated with Na in the presence of toluylene 2,4(6)-diisocyanate (TDI), and α,ω-dihydroxypolybutadiene ($\overline{M}_n \sim 1\ 000$) at 180°C. (Adapted from Ref. 14.)

Table 3 Influence of Type of Isocyanate, Resp. Its Blocked Derivative as a Functionalization Agent (A) on the Properties of Copolymers Poly(ε-caprolactam)-*block*-polybutadiene Prepared via Anionic Polymerization of ε-Caprolactam Initiated with 0.5 mol% Potassium (CLK) or Sodium (CLNa) Salt of ε-Caprolactam in the Presence of 0.5 mol% (A) and 10 wt% α,ω-Dihydroxypolybutadiene (M_n = 4900 g/mol) at 150°C

| Type | | k_w | k_{wt} | E_t | σ_z | ε_r | a_k | x_c | t_c |
A	I	(%)	(%)	(Gpa)	(MPa)	(%)	(kJ-m^{-2})	(%)	(min)
TDI	CLK	96.9	95.9	1.47	53	101	24	49.1	12.7
TDICL	CLK	97.1	96.5	1.47	38	74	16	49.3	9.8
MDICL	CLK	97.4	—	1.53	51	76	24	43.0	9.7
MDI	CLK	96.9	97.1	1.19	39	75	22	40.0	9.3
HDI	CLK	—	—	—	—	—	6	—	3.3
HDICL	CLK	98.0	—	1.70	45	5	10	—	3.7
PIC	CLK	96.2	96.8	1.34	25	21	3	54.3	12.7
PIC3	CLK	96.8	95.8	1.27	24	17	4	52.2	10.3
TDI	CLNa	97.9	—	1.10	47	70	16	—	5.0
TDICL	CLNa	95.7	—	0.90	46	70	15	—	15.7
MDICL	CLNa	97.0	—	1.60	55	79	22	—	11.8
HDICL	CLNa	97.8	—	1.20	31	13	5	—	5.0
PIC3	CLNa	96.0	—	0.70	31	22	—	—	13.2

k_w: polymer yield determined by extraction with hot water; k_{wt}: polymer yield determined by extraction with hot toluene; E_t: tensile modulus; σ_z: tensile strength; ε_r: elongation at break; a_k: Charpy notched impact resistance at -20°C; x_c: crystallinity of polyamide phase; t_c: minimal polymerization time extrapolated as a maximum of time vs. temperature dependence measured in mold;
TDI: toluylene 2,4(6)-diisocyanate;
TDICL: toluylene 2,4(6)-diisocyanate blocked with CL;
MDI: 4,4'-diisocyanatodiphenylmethane;
MDICL: 4-4'-diisocyanatodiphenylmethane blocked with CL;
HDI: 1,6-diisocyanatohexane;
HDICL: 1,6-diisocyanatohexane blocked with CL;
PIC: phenylisocyanate;
PIC3: cyclic trimer of phenylisocyanate.

For all the materials, the content of the water-extractable, low-molar-mass fractions was convenient, i.e., less than 3%. The same is true for the extraction by boiling toluene, which means that PBD was incorporated into the polyamide matrix. In case of phenylisocyanate (PIC) and its trimer (PIC3), such an incorporation of the elastomer can be explained only by side reactions.

The results given in Table 3 indicate indirectly that the formation of the copolymer as well as its structure are controlled kinetically. These findings were also confirmed for PBD having lower molar mass ($\approx 3,100$ g/mol) (15).

The effect of molar mass of PBD on the basic parameters of the obtained block copolymers is illustrated in Fig. 3 for 10 wt% PBD in the polymerization feed. The use of PBDs having M_n between 4,000 and 5,000 g/mol guaranteed that the toughness of the modified polyamide increased by an order of magnitude, as compared with the homopolymer of ε-caprolactam. At the same time, as expected, the tensile strength decreased by some 40%.

A detailed analysis of the effect of the PBD concentration in the range of 0–20 wt% on the properties of the materials prepared by anionic polymerization initiated by potassium salt of ε-caprolactam (CLK) and by 2,4(6)-toluylenediiso-cyanate (TDI) as the functionalization agent was performed (15). For all PBDs having molar masses 2,500, 4,350, and 4,900 g/mol, the dependencies of mechanical properties showed similar trends, as illustrated in detail in Fig. 4 for PBD with molar mass of 4,350 g/mol. The toughness of the material increased with increasing PBD concentration: Even at as low a PBD concentration as 10 wt%, the toughness increased by an order of magnitude up to some 30 kJ/m^2 (determined at

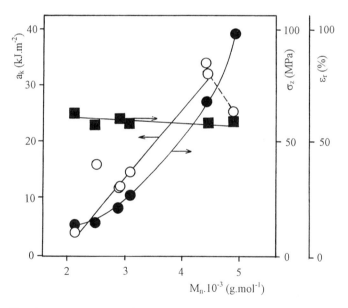

Fig. 3   Dependence of ($\bigcirc$) notched impact strength a_k at 20°C, ($\blacksquare$) tensile strength σ_z, and ($\bullet$) elongation at break ε_r on the molar mass (M_n) of α,ω-dihydroxypolybutadienes as precursors of the elastic polybutadiene segments (PBD) in the poly(ε-caprolactam)-*block*-polybutadiene copolymers prepared by the anionic polymerization of ε-caprolactam initiated with potassium salt of ε-caprolactam in the presence of toluylene 2,4(6)-diisocyanate (TDI) and 10 wt% precursor at 150°C. (From Ref. 15.)

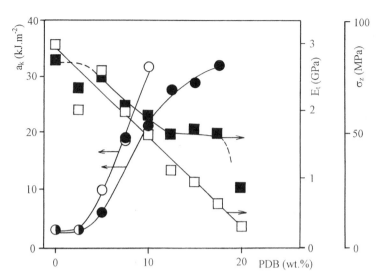

Fig. 4 Dependence of (○) notched impact strength a_k at 23°C, (●) resp. at −20°C, (□) tensile modulus E_t, and (■) tensile strength σ_z on the concentration of elastic polybutadiene segments (PBD) in the poly(ε-caprolactam)-*block*-polybutadiene copolymers prepared by the anionic polymerization of ε-caprolactam initiated with potassium salt of ε-caprolactam in the presence of toluylene 2,4(6)-diisocyanate (TDI) and α,ω-dihydroxypolybutadiene ($\overline{M}_n = 4350$) as a precursor of elastic segments at 150°C. (From Ref. 15.)

−20°C). The modulus of elasticity E_t decreased linearly with increasing PBD concentration. The tensile strength decreased as well, reaching a distinct plateau above 10 wt% PBD. At the same value of concentration, the elongation at break increased by an order of magnitude from the initial 5–10%.

The rate of polymerization, characterized by t_c, was a function of concentration and M_n of polybutadiene: with an increase in the latter quantities, the initial viscosity of the polymerization feed increases, thus retarding the polymerization process, as seen in Fig. 5. The degree of crystallinity of the polyamide phase in the block copolymer decreases distinctly only above 15 wt% PBD incorporated in the copolymer. The dependence of the melting point T_m of the polyamide phase on the PBD concentration dropped very slowly, splitting above 15 wt% PBD into two distinct endotherms of comparable intensities (Fig. 5).

An optimum PBD concentration in block copolymers was determined to be 10–15 wt%, depending on the molar mass of PBD and on the type of the functionalization agent. Such a content of the elastic phase causes a substantial increase in the impact strength of the polyamide material, as illustrated in the complexity of Fig. 6. For instance, substituting TDI for 4,4′-diisocyanatodiphenylmethane (MDI) in the polymerization feed leads to values of notched impact strength (a_k) exceeding those obtained with TDI. For a maximum increase of a_k, PBD with M_n = 4,900 g/mol proved to be the best (Fig. 6) (15). When compared with polyethers (PEO) (8), used commercially in the NYRIM technology, the optimum concentration of PBD constitutes one-half of that of PEO. In the given optimum range, the tensile strength of the material does not change substantially (40–50% of the value of the pure polyamide) (15) and the polymerization times are about 12 minutes. For all types of PBD, the E_t values decrease linearly with increasing PBD concentration, being located in a very narrow range (±0.2 GPa) and depending, therefore, on the polyamide phase content only, as seen in Fig. 7.

Another positive finding is that the content of the low-molar-mass fractions is constant and sufficiently low (2–3.5 wt%) in all materials prepared. Similar values obtained via toluene extraction confirmed that PBD was incorporated quantitatively into the polyamide matrix, which is consistent with published data (13).

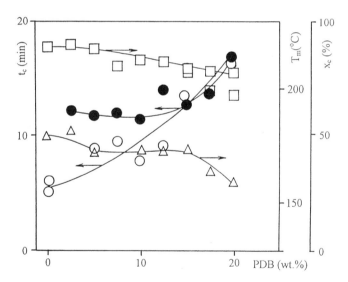

Fig. 5 Dependence of (○, ●) polymerization time t_c (corresponding to the maximum on the temperature profile), (△) degree of crystallinity x_c of polyamide phase, and (□) its melting temperature, determined by DSC, on the concentration of elastic polybutadiene segments (PBD) in the poly(ε-caprolactam)-*block*-polybutadiene copolymers prepared by the anionic polymerization of ε-caprolactam initiated with potassium salt of ε-caprolactam in the presence of toluylene 2,4(6)-diisocyanate (TDI) and α,ω-dihydroxypolybutadiene as a precursor of elastic segments at 150°C, $\overline{M}_n$ of polybutadiene: (○) 2500, (●, △, □) 4350. (Adapted from Ref. 15.)

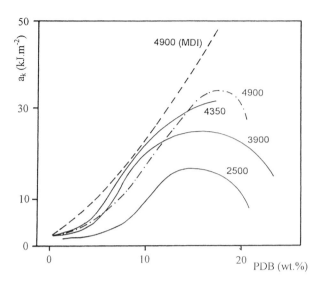

Fig. 6 Influence of molar mass (values displayed at curves) and concentration of elastic polybutadiene segments (PBD) on notched impact strength a_k of copolymers poly(ε-caprolactam)-*block*-polybutadiene prepared by the anionic polymerization of ε-caprolactam initiated with potassium salt of ε-caprolactam in the presence of α,ω-dihydroxypolybutadiene and toluylene 2,4(6)-diisocyanate (TDI) or 4,4'-diisocyanatodiphenylmethane (MDI) at 150°C.

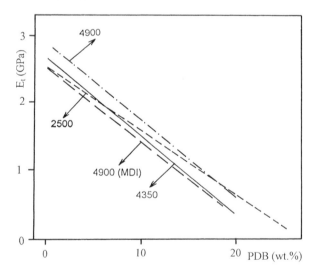

Fig. 7 Influence of molar mass (values displayed at curves) and concentration of elastic polybutadiene segments (PBD) on tensile modulus E_t of copolymers poly(ε-caprolactam)-*block*-polybutadiene prepared by the anionic polymerization of ε-caprolactam initiated with potassium salt of ε-caprolactam in the presence of α,ω-dihydroxypolybutadiene and toluylene 2,4(6)-diisocyanate (TDI) or 4,4'-diisocyanatodiphenylmethane (MDI) at 150°C.

108

As follows from Table 2, nontoxic diisocyanates blocked by ε-caprolactam can also easily be used for the functionalization of PBD. When polymerized at 150°C, at identical PBD concentrations and with MDI blocked by ε-caprolactam (MDICL) combined with potassium salt of ε-caprolactam, lower toughness is achieved, as compared with MDI alone. To achieve comparable toughness, it is sufficient to increase the polymerization temperature up to 160°C. Other mechanical properties, such as tensile modulus and strength, remain unchanged within the limits of experimental error (Fig. 8) (16).

Substituting the potassium salt of ε-caprolactam (CLK) for the sodium salt (CLNa) in combination with MDICL as the functionalization agent led to an improvement, as seen in Fig. 9. Surprisingly, with CLNa and at identical content of PBD, the toughness of the material is slightly but distinctly higher than with CLK. Other mechanical properties are comparable for both initiators and polymerization temperatures 150 and 160°C. For CLNa, which, according to theory (7), dissociates to a smaller extent than CLK, the copolymerizations proceed slowly (15,16). Thus, it can be recomended to use, at a previously determined optimum PBD concentration, the initiation system consisting of (cheaper) CLNa and nontoxic blocked MDI.

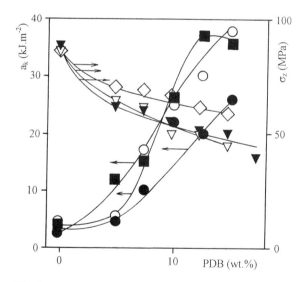

Fig. 8   Dependence of ($\bigcirc$, $\bullet$, $\square$) notched impact strength a_k at -23°C and ($\triangledown$, $\blacktriangledown$, $\lozenge$) tensile strength σ_z on the concentration of elastic polybutadiene segments (PBD) in copolymers poly(ε-caprolactam)-*block*-polybutadiene prepared by the anionic polymerization of ε-caprolactam (CL) initiated with potassium salt of ε-caprolactam in the presence of α,ω-hydroxypolybutadiene and 4,4'-diisocyanatodiphenylmethane (MDI) ($\blacksquare$, $\blacktriangledown$) or MDI blocked with ε-caprolactam (MDICL) ($\bullet$, $\triangledown$) at 150°C and MDICL at 160°C ($\bigcirc$, $\lozenge$).

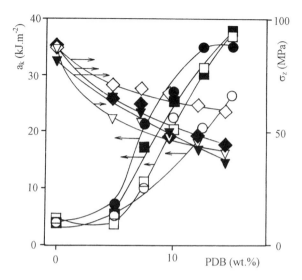

Fig. 9 Dependence of (○, □, ●, ■) notched impact strength a_k at 23°C and (▽, ◊, ▼, ◆) tensile strength σ_z on the concentration of elastic polybutadiene segments (PBD) in copolymers poly(ε-caprolactam)-*block*-polybutadiene prepared by the anionic polymerization ε-caprolactam initiated with potassium salt of ε-caprolactam at 150°C (○, ◆) resp. 160°C (○, ◊) or with sodium salt of ε-caprolactam at 150°C (□, ▽) resp. 160°C (■, ▼) in the presence of α,ω-hydroxypolybutadiene and 4,4′-diisocyanatodiphenylmethane blocked with ε-caprolactam (MDICL).

The copolymers poly(ε-caprolactam)-*block*-polybutadiene prepared using aromatic diisocyanates (TDI, MDI) as the functionalization agents are partially crosslinked (15), which was observed also in Ref. 14. The sol-gel analysis (17) using 2,2,2-trifluoroethanol reveals that the content of the gel increased (70–80%) with increasing concentration of the elastic segments (5–17.5%). A vast majority of the elastic segments is bound in the gel phase, because, according to the IR analysis (photoacoustic spectra), the soluble fraction contained 1 wt% PBD (18), at the most, and can therefore be taken as a homopolymer of ε-caprolactam.

The network density was determined on the basis of the compression modulus of equilibrious swollen gels. Network properties could formally be described using the model of the randomly crosslinked block copolymer having the most probable molar mass distribution. The number of crosslinks correlates with the initial PBD concentration and, therefore, also with the concentration of the functionalization agents (TDI, MDI), which are used for the in situ functionalization in an equimolar amount with respect to OH groups of PBDs (17).

Crosslinking is probably caused by well-known side reactions proceeding in the polyamide matrix (7), not in the elastomeric one, as is sometimes stated (14). By using a controlled heterogeneous hydrolysis of the block copolymers by methanolic KOH, which transforms the crosslinked product into a totally soluble one, the possibility of a coupling of the PBD blocks anionically through the C—C bonds was excluded. According to gel permeation chromatography (GPC), the isolated PBD had the same M_n as the pure starting PBD material (to be published). Details of the three-dimensional structure of the copolymer networks under study are not yet known.

A detailed morphological study on the tough copolymers poly(ε-caprolactam)-*block*-polybutadiene (prepared in Ref. 15) led to an unambiguous conclusion that the materials really have the block structure, as demonstrated by a typical phase separation.

When less than 10 wt% PBD is bound in the copolymer, the material shows the classical spherulitic structure with narrow border areas. Above 10 wt% PBD, the characteristic optical properties of the spherulites disappeared, and only the border areas remain visible (Fig. 10). Polybutadiene was located inside the spherulites, where it constituted a separated phase. According to TEM, with 5 wt% PBD content the elastomer forms isolated spherical particles, with some of them 20–40 nm in diameter. With PBD content of 10 wt%, a cellular structure appeared, the dimensions of the cells being about 100 nm and the thickness of PBD

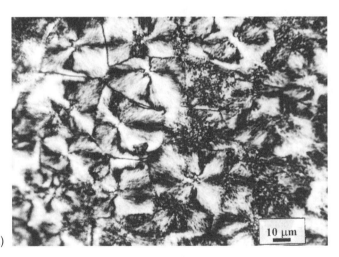

(a)

Fig. 10 Visible light microscopy of the copolymers poly(ε-caprolactam)-*block*-polybutadiene prepared by the anionic polymerization of ε-caprolactam initiated with potassium salt of ε-caprolactam in the presence of α,ω-dihydroxyterminated polybutadiene ($\overline{M}_n$ = 4900) at 150°C. Content of polybutadiene phase, in wt%: (a) 5, (b) 10, (c) 17.5.

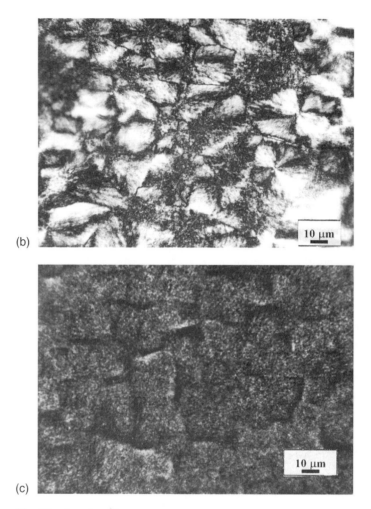

(b)

(c)

Fig. 10 *(continued)*

walls ranging from 7 to 20 nm (Fig. 11). When the content of PBD reached about 17 wt%, a distinct lamellar oriented structure is formed, having a lamellar thickness of about 20 nm. With increasing molar mass of the PBD segments, the structure of the PBD domains becomes finer, which resulted in a greater toughness of the materials. Conversely, a coarse morphology decreased the toughness (19).

Sorption isotherms of the copolymers poly(ε-caprolactam)-*block*-polybutadiene demonstrate that the maximum water content, as well as the rate of the sorption, increases with increasing content of the elastic (though hydrophobic) phase. This phenomenon is caused by the decrease of the crystalline phase in the polymer matrix with increasing content of the elastic component. Surprisingly enough, sorption isotherms show maxima that may indicate that the arrangement of the solid polyamide phase undergoes changes when water is present (20) (Fig. 12) (8).

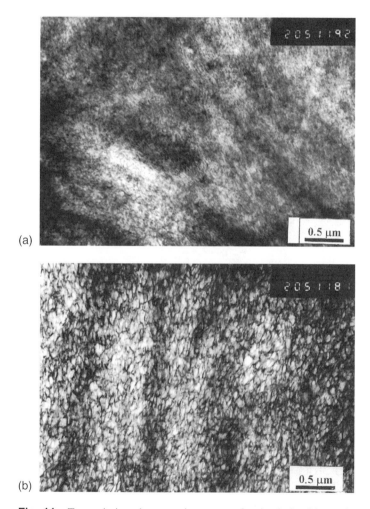

(a)

(b)

Fig. 11 Transmission electron microscopy of stained ultrathin sections of the copolymers poly(ε-caprolactam)-*block*-polybutadiene prepared by the anionic polymerization of ε-caprolactam initiated with potassium salt of ε-caprolactam in the presence of α,ω-dihydroxypolybutadiene ($\overline{M}_n$ = 4900) at 150°C. Content of polybutadiene phase, in wt%: (a) 5, (b) 10, (c) 17.5. *(figure continues)*

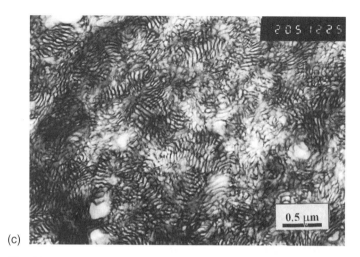

(c)

Fig. 11   *(continued)*

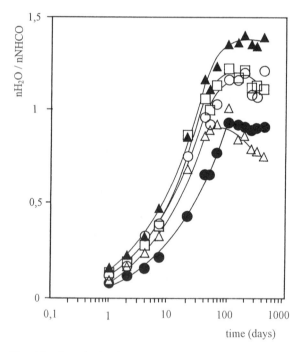

Fig. 12   Sorption isotherms of water in copolymers poly(ε-caprolactam)-*block*-polybutadiene with different content of elastic polybutadiene segments (PBD), in wt%: 0 (●), 5 (△), 10 (○), 15 (□), 20 (▲), at 95% relative humidity and at 25°C. Molar mass mg/mol of elastic PBD segments: 3000. Shape of samples: disc diameter = 25 mm, thickness = 2 mm.

Table 4 Mechanical Properties of Poly(ε-caprolactam)-*block*-polybutadiene/Wollastonite (Woll) Composites Prepared by Anionic Polymerization of ε-Caprolactam Initiated with Sodium Salt of ε-Caprolactam (CLNa) and Toluylene 2,4(6)-Diisocyanate (TDI) in the Presence of α,ω-Dihydroxy-Polybutadiene (PDB) (M_n = 4900 g/mol) and Wollastonite (Functionalized in situ with TDI) in Mold at 160°C, 30 min

Woll (%)	PBD (%)	$[CLNa]^a$ (mol %)	$[TDI]^a$ (mol %)	G^c (wt %-cm^{-1})	a_k^b (kJ-m^{-2})	E_t^b (GPa)	σ_z^b (MPa)	ε_r^b (%)
—	0	0.5	0.5	—	4	2.7	85	7
—	5	0.5	0.5	—	7	2.3	70	13
—	10	0.5	0.5	—	n.b.	1.4	48	52
—	15	0.7	0.7	—	15	1.3	28	127
10	5	1.0	0.5	0.34	5	2.5	—	5
10	10	1.0	0.5	0.14	9	2.3	62	5
10	15	1.4	0.7	0.01	7	2.1	44	5
15	10	1.0	0.5	0.03	8	2.5	62	5
15	15	1.4	0.7	0.12	7	2.1	44	5

^{a} Concentration relative to ε-caprolactam.
^{b} Dry specimens, 23°C
^{c} Sedimentation gradient of the filler.
n.b.: nonbreak;
a_k: Charpy notched impact resistance at 23°C;
E_t: tensile modulus;
σ_z: tensile strength;
ε_r: elongation at break.
Source: Adapted from Ref. 21.

Using the same procedure as described in Ref. 15, composites were prepared containing 5–15 wt% Wollastonite in the matrix constituted by the copolymers poly(ε-caprolactam)-*block*-polybutadiene (21). To attain a good dispersing of the filler and a good matrix/filler adhesion, it was necessary to modify Wollastonite by (3-amino-propyl)triethoxysilane. A functionalization in situ by diisocyanate (TDI) proves to be the most effective, preventing aggregation of the filler particles, which were the biggest obstacle during the preparation. The filler improves the tensile modulus as well as the strength and induces a decrease in toughness and a strong decrease in ductility, as can easily be observed in Table 4. Adhesion of the filler to the polyamide matrix was proved on the fractural surfaces by scanning electron microscopy (SEM).

III. COPOLYMERS POLY(ε-CAPROLACTAM)-*BLOCK*-POLY(BUTADIENE-CO-MONOMER)

Besides pure telechelic polybutadienes, copolymers containing butadiene were also utilized for a modification of poly(ε-caprolactam). Hydroxy-terminated poly(butadiene-*co*-styrene) (SBR) functionalized by an excess of TDI was incorporated into the polyamide matrix by an NaH-initiated anionic polymerization. The styrene content in the statistical copolymer, however, is not exactly known; the reported molar mass was about 3,500, functionality 2.2–2.4. Materials containing less than 70 wt% of ε-caprolactam are brittle and visually inhomogeneous. Therefore, it was even possible to measure the Shore D of both phases.

The polymeric materials prepared from a polymerization feed containing 75 and 80% of ε-caprolactam were tough and flexible. Their physical properties depend on the TDI/initiator mole ratio (Table 5). Low moduli indicate a rubberlike character of material and a lack of the properties of an engineering plastic. Unfortunately, experiments demonstrating unambiguously the block character of the material were not carried out.

Butadiene/acrylonitrile copolymers (Hycar) having molar masses 2,000–4,000 g/mol and average acrylonitrile content 16–17 wt% and bearing secondary amino end groups (ATBN) or hydroxy end groups (HTBN) were successfully applied to a preparation of a modified poly(ε-caprolactam). The precursors of the elastic blocks (Hycars) were functionalized in situ by N,N'-isophtaloyl-bis(ε-caprolactam) (IP), N,N'-hexamethylene-bis(carbamoyl-ε-caprolactam) (HC), or N,N'-2,4(6)-toluylene-bis(carbamoyl-ε-caprolactam) (TC). The functionalization was performed on the basis of model reactions: the rates of aminolysis and alcoholysis of N-acyllactams, and N-substituted carbamoyllactams in suitable solvents, as well as in the melt of ε-caprolactam (22,23) (Table 6). The aminolysis proceeds distinctly more rapidly and can thus be preferred for the functionalization (and thus for the preparation of block copolymers).

Table 5 Mechanical Properties of Modified Poly(ε-caprolactam) Prepared by Anionic Polymerization of ε-Caprolactam (CL) initiated with NaH in the Presence of Toluylene 2,4(6)-Diisocyanate (TDI) Functionalized Poly(butadiene-co-styrene) (SBR) (M_n = 3500 g/mol) at 160°C

CL (%)	NCO/NaH	y_L (%)	E_{100} (MPa)	σ_z (MPa)	ε_r (%)
75	2.7–4.3	90.3–92.3	12–16	34–43	340–440
80	2.2–3.7	92.3–92.8	17–23	35–45	290–370

CL: weight content of ε-caprolactam in initial polymerization mixture;
NCO/NaH: molar ration between isocyanate groups in SBR and initiator;
y_L: lactam conversion to polymer;
E_{100}: tensile modulus at 100% elongation;
σ_z: tensile strength;
ε_r: elongation at break.
Source: Adapted from Ref. 20.

Table 6   Second-Order Rate Constant k_2 of Lytic Reactions of N-acyllactams

Acyllactam	Reagent	Temp. (°C)	Medium	Catal. (1 mol %)	$k_2 \times 10^5$ (kg-mol^{-1}-s^{-1})
	Octylamine	50	THF	—	290
	Dibutylamine	80	Toluene	—	16
	Octanol	100	CL[a]	—	1.6
	Octanol	100	CL	CLNa[b]	1600
	Octanol	100	CL	CLMgBr[c]	22
	Octanol	130	CL	—	8.3
CH$_3$CH$_2$CON—CO	Octanol	100	CL	—	1.4
	Octanol	100	CL	CLNa	0.6
	Octanol	100	CL	CLMgBr	6.6
	Octanol	130	CL	—	4.4
	Octanol	100	CL	—	0.6
	Octanol	100	CL	CLNa	5800
	Octanol	100	CL	CLMgBr	18
	Octanol	130	CL	—	5.1

[a] ε-Caprolactam.
[b] 0.5 mol% sodium salt of ε-caprolactam.
[c] ε-Caprolactammagnesium bromide.
Source: Adapted from Refs. 22 and 23.

The effectiveness of a given functionalization of a prepolymer, as well as of a modification of poly(ε-caprolactam), was evaluated according to the mechanical properties of the product, mainly by measuring its toughness. In accordance with the model reactions, amino-terminated nitrile rubbers, functionalized in situ by IP and, apparently, also by TC, proved to be "successful." Conversely, the combination of hydroxy-terminated nitrile rubber and blocked aliphatic diisocyanate was less convenient. Nevertheless, in all cases, the extractions by water and toluene gave comparable results, which indicates that the rubber component is perfectly bound in the material. The bonding of the elastic phase may be of a different nature: true chemical bonds in a block copolymer, formation of crosslinked rubber particles, or interpenetrating networks (IPN). A certain role of the crosslinking was demonstrated by the fact that the materials were insoluble in cresol. It was confirmed that the molecular structure of the material—and thus also its mechanical properties—was controlled by a complex kinetic method. A participation of both main and side reactions is given by the polymerization conditions, such as the concentration of the elastomer, the type and concentration of the initiation system, and the polymerization temperature, as well as the changes of these quantities during all stages of the polymerization process. This affects the morphology of the material, the dimensions of the spherulites, and the character of the crystalline phase: α-form, γ-form, or a mixture of both these crystalline forms of poly(ε-caprolactam).

Reactive blending (24,25) represents an interesting and elegant procedure to simplify the preparation of modified poly(ε-caprolactam), especially to avoid the synthesis of a functionalized telechelic polymer (macroactivator). Again, an anionic polymerization of ε-caprolactam was utilized, this time in the presence of nitrile rubbers bearing no reactive end groups and having relatively high molar masses ($M_v \approx$ 160,000 g/mol), M_w/M_n of 3–5, and acrylonitrile content of 35–39 wt%. Due to a high acrylonitrile content, the elastomer dissolves easily in melted ε-caprolactam. The initiation was started by the sodium salt of ε-caprolactam combined with a relatively "slow" activator (N-acetyl-ε-caprolactam). The kinetics of the process were dramatically influenced by the preparation and the mode of stirring of the polymerization feed. The longer the time during which the feed was mixable (had sufficiently low viscosity), the shorter was the polymerization time and, surprisingly, the lower was the content of the unreacted lactam. Obviously, as with the poly(ε-caprolactam)-*block*-polybutadiene copolymer, the presence of the rubber (up to 12 wt%) increased the polymerization time and the content of unreacted monomer, as compared with the homopolymer.

The material consists of a polyamide matrix containing macrodomains (up to several tens of micrometers). Their size depends on the polymerization kinetics and on the rubber content. The macrodomains are covered by rubber shells. After the shells were removed by acetone, particles were distinctly visible having a relatively uniform diameter of about 4 μm and being bound together by an insoluble fraction of the nitrile rubber. This rubber constitutes a continuous crosslinked phase.

When a "rapid" activator (1,6-diisocyanatohexane blocked by ε-caprolactam) was applied, the polymerization times could be as low as 50 s (20% as compared with N-acetyl-ε-caprolactam), depending on the nitrile rubber concentration in the feed. Due to a high reaction rate, stirring was not a decisive parameter and high degrees of conversion were achieved. The arrangement of the phases was similar to that of the "slow" system, except for the fact that, in addition to spherical macrodomains, cylindrical macrodomains were present, again containing relatively uniform subdomains.

An attempt at a preparation of the tough polyamide 6 was made by an anionic polymerization of ε-caprolactam carried out in the presence of undissolved particles (26). These particles were prepared by an emulsion polymerization and had a core-shell structure. The inner part was constituted of a butadiene-acrylonitrile copolymer, whereas the shell consisted of either styrene-acrylonitrile copolymer or styrene-acrylonitrile-hydroxypropyl methacrylate terpolymer. The introduction of the OH group into the shell brought about the chemical bonding between the microparticles and the polyamide matrix, using the classical procedure with TDI as the functionalization agent. Mechanical properties of the product can be considered as comparable with those of the NYRIM materials.

IV. CONCLUSIONS

The anionic polymerization of ε-caprolactam on the functionalized chain ends of selected prepolymers might give the copolymers polyamide 6-*block*-elastomer, where elastomers are polybutadiene, poly(butadiene-*co*-acrylonitrile), and poly(butadiene-*co*-styrene).

Properties of copolymers depend on the experimental procedure of preparation, the type and composition of the initiation system, and the elastomer (polybutadiene prepolymer) concentration. The rubbery to hard materials can be synthesized.

A suitable composition of polymerization mixture also yields copolymers poly(ε-caprolactam)-*block*-polybutadiene with improved toughness at temperatures below zero. The most suitable prepolymers for block copolymer preparation are α,ω-dihydroxy-polybutadienes (PBD) with $M_n \sim 5.10$ (3), functionalized with aromatic diisocyanates in situ; the recommended polymerization temperature varies between 150 and 170°C. The optimal PBD concentration varies between 10 and 15 wt% for the improvement of toughness and the retaining of other mechanical properties in a reasonable range. Morphology study confirms the block character of the materials prepared.

Literature data confirm the possibility of preparing poly(ε-caprolactam)-*block*-poly(butadiene-*co*-acrylonitrile) by a polymerization method; "elegant" reactive blending can be used as well.

ACKNOWLEDGMENTS

The support of part this work by grant nos. 106/98/0699 and 106/94/1688 of the Grant Agency of the Czech Republic is highly appreciated.

REFERENCES

1. MI Kohan, ed. Nylon Plastics Handbook. Munich: Hanser Verlag, 1995, pp 2–11.
2. JCL Williams, SJ Watson, P Boydell. Properties. In: MI Kohan, ed. Nylon Plastics Handbook. Munich: Hanser Verlag, 1995, pp 291–360.
3. J Stehlíček. Modified polyamides. In: R Puffr, V Kubánek, eds. Lactam-Based Polyamides. Vol. 2. Boca Raton: CRC Press, 1991, pp 1–72.

4. I Goodmann, RN Vachon. Copolyesteramides—II. Anionic copolymers of ε-capro-
    lactam with ε-caprolactone. Preparation and general properties. Eur Polym J 20:529–
    537, 1984.
5. I Goodmann, A Valavanides. Copolyesteramides—I. Anionic copolymers of ω-lau-
    rolactam with ε-caprolactone. Eur Polym J 20:241–247, 1984.
6. J Šebenda. Lactams. In: HC Bamford, CFH Tipper, eds. Comprehensive Chemical
    Kinetics. Vol. 15. New York: Elsevier, 1976, pp 379–471.
7. J Šebenda. Lactam polymerization. J Macromol Sci Chem A6:1145–1199, 1972.
8. RM Hedrick, JH Gabbert, HM Wohl. Nylon 6 RIM. In: Reaction injection molding.
    ASC Symp Ser 270:135–162, 1985.
9. WL Hergenrother, RJ Ambrose. Block copolymers from isocyanate-terminated in-
    termediates. II. Preparation of butadiene-ε-caprolactam and styrene-ε-caprolactam
    block copolymers. J Polym Sci Polym Chem Ed 12:2613–2622, 1974.
10. RJ Ambrose, WL Hergenrother. Polar–nonpolar block polymers from isocyanate-ter-
    minated intermediates. J Polym Sci Polym Symp 60:15–27, 1977.
11. WL Hergenrother, RJ Ambrose. Block copolymers from isocyanate-terminated in-
    termediates. IV. Properties of cured butadiene-ε-caprolactam block polymers. J Appl
    Polym Sci 19:3225–3233, 1975.
12. D Petit, R Jerome, P Teyssie. Anionic block copolymerization of ε-caprolactam. J
    Polym Sci Polym Chem Ed. 17:2903–2916, 1979.
13. Yu-Der Lee, Ming-Jung Tsai, A Teh Hu. Preparation, characterization and physical
    properties of poly(ε-caprolactam)-b-butadiene tri-block copolymer. J Chin I Ch E 19:
    145–154, 1988.
14. NR Ashurov, ShG Sadykov, SV Voznecenskaja, LR Sajfutdinova, M Abdurazakov.
    Features of structure and properties of block copolymer polycaproamid-polybutadi-
    ene. Vysokomol Soedin 39B:2038–2041, 1977 (in Russian).
15. V Nováková, R Sobotík, J Matěnová, J Roda. Polymerization of lactams, 87. Block
    copolymers of poly(ε-caprolactam)-block-polybutadiene prepared by anionic poly-
    merization. Part 1. Preparation and properties. Angew Makromol Chem 237:123–
    141, 1996.
16. V Nováková, R Zýka, J Brožek, J Roda. Preparation of block copolymers poly(ω-
    caprolactam)-polybutadiene by the anionic polymerization, Proceedings of 6th EPF
    Symposium on Polymeric Materials, Aghia Pelaghia, Crete, 1996, PI 20.
17. B Meissner, V Nováková, R Pechová, J Roda. On the crosslinked structure of copoly-
    mers poly(ε-caprolactam)-polybutadiene. Rolduc Polymer Meeting 95, Kerkrade,
    Netherlands, 1995, P 15.
18. P Schmidt, J Roda, V Nováková, JM Pastor. Analysis of the chemical composition of
    poly(ε-caprolactam)-block-polybutadiene copolymers by photoacoustic FTIR spec-
    troscopy and by FT Raman spectroscopy. Angew Makromol Chem 245:113–123,
    1997.
19. J Roda, V Nováková, F Lednický. Morphology of block copolymers poly(ε-capro-
    lactam)-polybutadiene. EPS Conference on Macromolecular Physics, IMC, Prague,
    1995.
20. WT Allen, DE Eaves. Caprolactam based block copolymers using polymeric activa-
    tors. Angew Makromol Chem 58/59:321–343, 1977.

21. J Stehlíček, R Puffr, F Lednický, V Nováková, J Roda. Low-filled toughened composites poly(hexano-6-lactam)-*block*-polybutadiene-*block*-poly(hexano-6-lactam)/Wollastonite by polymerization casting. J Appl Polym Sci 69:2139–2146, 1998.

22. J Stehlíček, F Lednický, J Baldrian, J Šebenda. On the synthesis of poly(ε-caprolactam)-poly(butadiene-co-acrylonitrile) block copolymers for reaction injection molding process. Polym Eng Sci 31:422–431, 1991.

23. J Stehlíček, GS Chauhan, M Znášiková. Preparation of polymeric initiators of the anionic polymerization of lactams from polyetherdiols. J Appl Polym Sci 46:2146–2177, 1992.

24. GC Alfonso, G Dondero, S Russo, A Turturo. Nitrile rubber–modified poly(ε-caprolactam) by activated anionic polymerization: synthesis, molecular characterization and morphology. In: E Martuscelli, C Marchetta, eds. New polymeric materials. Reactive processing and physical properties. Utrecht: Elsevier, 1988, pp 101–128.

25. S Russo, G Bonta, A Imperato, F Parode. Chemical aspects in the synthesis of poly(caprolactam) for the RIM process. In: PJ Lemstra, LA Kleintjens, eds. Integration of Fundamental Polymer Science and Technology, 2. London: Elsevier, 1988, pp 17–29.

26. K Udipi. Particulate rubber modified Nylon 6 RIM. J Appl Polym Sci 36:117–127, 1988.

6
Dynamics and Viscoelastic Effects in Block Copolymers: Real and Simulated Systems

Tadeusz Pakula
Max-Planck-Institute for Polymer Research, Mainz, Germany

George Floudas
Foundation for Research and Technology-Hellas, Heraklion, Crete, Greece

I. INTRODUCTION

Polymer blends are the simplest polymer mixtures that can result in an enhancement of the desired properties that are exhibited by the individual components and thus to important applications. Block copolymers, on the other hand, are formed by the covalent bonding of two or more polymers. This combination is nontrivial and requires specialized chemistry, but it can be performed in different ways, resulting in linear and nonlinear copolymers (star, graft, miktoarm stars, etc.). In all cases, the most significant factor in determining the rich block copolymer phase behavior is the covalent bond that restricts the macrophase separation of chemically dissimilar blocks.

Upon microphase separation, a rich variety of periodic microstructures develops that strive to minimize the area of contacts between the unlike blocks. An interesting feature of these structures is that their length scale is comparable to the size of the block copolymer molecules—typically 5–50 nm. Therefore, the microstructure is highly coupled to the physical and chemical characteristics of the molecule. This observation has fueled continued research—both theoretical (1–6) and experimental (7,8)—attempting to understand the physical processes governing the order-to-disorder transition. Contrast this behavior with that observed in phase-separated polymer blends, in which the domains are typically several microns in size. In these systems, the size of the phase-separated structures is nearly independent of the detailed features of the molecules.

For linear block copolymers, two extreme cases have been considered the most: (1) block copolymers composed of two incompatible polymer chains, and (2) random copolymers with monomers of different types more or less randomly distributed along the chain. More sophisticated distributions of monomers have been studied in the so-called "tapered-block copolymers" (9–11), multiblock copolymers (12,13), and, more recently, the gradient copolymers (14,15). All these kinds of comonomer distributions can essentially be considered as building blocks in more complex macromolecules, illustrated in Fig. 1. However, most of synthetic efforts have up to now been made to synthesize various branched structures consisting of macromolecular blocks (16,17).

Although copolymers have been a subject of intensive research for many years, there are many aspects of their behavior, concerning both structure and dynamics and their relation to the thermodynamic state, that are still not sufficiently understood. A rich variety of phase behavior makes these systems an attractive subject of studies but also a difficult one (e.g., Ref. 1). Many parameters, such as the overall degree of polymerization (N), the composition (f), and the already-mentioned type of composition distribution, the architecture of macromolecules, and the segment–segment interaction parameter (χ_{AB}, ε_{AB}), control the thermodynamics and the structure of copolymers. Whereas mainly the equilibrium-phase behavior and related morphologies for linear-block copolymers have been studied in the past (1–7), more effort is now devoted to studies of complex polymer architectures and more sophisticated comonomer arrangements (16,17).

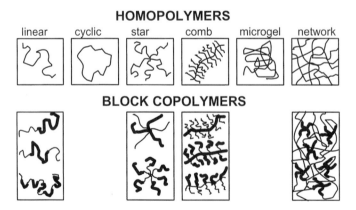

Fig. 1   Schematic illustration of various polymer topologies and some corresponding architectures of block copolymer macromolecules.

Furthermore, the dynamics of block copolymer melts and the effects of the order–disorder transition on the dynamics are being studied with ever-increasing interest (18–21). Experimentally, a variety of techniques, including rheology, dielectric relaxation, dynamic light scattering, NMR, and forced Rayleigh scattering, have recently been applied in investigations of this problem. Some of these methods have appeared to be a very useful tool for studying the phase transition temperatures between microphases and the phase transition kinetics in various block copolymers. The knowledge about the dynamics of block copolymers became very important in relation also to morphological changes accompanying large-amplitude oscillatory shear flow involving various orientations of morphological elements with respect to the shear direction, a problem investigated recently by several research groups that, however, lacks a satisfactory explanation (22–26). Only recently, the dynamics of diblock copolymers, and the influence of the order-to-disorder transition on the dynamics have attracted the increased interest of theoreticians (27–35). Computer simulations have provided results concerning mainly the static properties of block copolymers (36–39), but dynamics has also been considered recently (40–42).

In this chapter, the dynamic behavior of copolymers is discussed on the bases of computer simulation results and real experiments obtained for model block copolymer systems at various states ranging from the disordered state to the strongly segregated one. The simulation results are correlated with the viscoelastic behavior observed for real block copolymer systems. The simulation method used is the *cooperative motion algorithm,* which has appeared very effective in application to various complex dense-polymer systems. The chapter is organized in the following way: In Sec. II we provide a brief presentation of the simulation methods, and in Sec. III we discuss the correlation between simulated dynamics and the viscoelastic behavior of homopolymer melts. In Sec. IV, the thermodynamics and structure of simulated block copolymer systems are presented. Section V, deals with the experimental investigation of the viscoelastic properties of linear and nonlinear block copolymer melts, of block copolymer/homopolymer blends, and of substituted block copolymers with strongly interacting groups. The equilibrium order-to-disorder transition and the phase transformation kinetics are discussed in detail. Finally, in Sec. VI, simulations of the flow of symmetric diblock copolymers are presented, and the detected morphological changes are correlated to those observed in real systems.

II. METHODS

A. Cooperative Motion Algorithm

Among various simulation methods used to represent the behavior of polymers on the coarse-grained scale, the cooperative-motion algorithm (CMA) (38,40,43) is a unique one, because it allows one to simulate really dense systems corresponding to polymer melts. In this type of simulation, ensembles of beads located at lattice sites are connected by nonbreakable bonds to form structures representing macromolecules of various topologies. An fcc lattice is used, with all lattice sites occupied in order to mimic dense systems such as polymer melts. The systems are considered under the excluded-volume condition, which means that each lattice site can be occupied by only a single molecular element (bead). In such systems, strictly cooperative dynamics are used consisting of rearrangements satisfying local continuity of the simulated system (no empty lattice sites are generated). This is realized by local motions consisting of displacements of a certain number of molecular elements along closed loops so that each element replaces one of its neighbors, with the sum of displacements of elements taking part in the rearrangement being zero (continuity condition). During such rearrangements the models of the macromolecules undergo conformational transformations preserving, however, their identities given by the number and sequences of elements in the polymer.

Figure 2 shows an example of the evolution in time of a dense system of linear chains. The two-dimensional case is taken here only for illustration. This demonstrates that the local rearrangements (rearranged chain elements are shown by thicker lines) appear randomly in space and time and can bring the whole dense system to motion (in the last state shown, each chain element has been moved at least once). Monitoring positions and/or orientations of system elements in time allows one to detect the dynamics in such a system. The simulation of this type is very efficient and allows, for example, the observation of a system consisting of 32,000 beads (320 linear chains of length $N = 100$) within the time period by factor 10^7 longer than the relaxation time of a single bead, using only a personal computer.

In the case of copolymers, two types of monomers, A and B, are considered. The two types of monomers are partially compatible, which is characterized by direct interaction parameters ε_{ij}. The energy of mixing is given only by the interactions of monomers of different types; therefore, we assume here that $\varepsilon_{AA} = \varepsilon_{BB} = 0$ and $\varepsilon_{AB} = 1$. The effective energy of one monomer E_m, given by the sum of ε_{AB} over z nearest neighbors, will depend on the local structure.

In order to generate equilibrium states, a dense system of chains is subjected to motion at a given temperature. Moving a chain element alters the local energy, because the monomers contact new neighbors. An attempt to move a single monomer is assumed as one Monte Carlo step, and the probability of motion is related to the interaction energy of the monomer in the attempted position. At a given temperature, T, the Boltzmann factor $p = \exp(-E_{m,\text{final}}/k_B T)$ is compared with a random number r, $0 < r < 1$. If $p > r$, the move is performed and a motion of another monomer is attempted. Under such conditions, at low temperatures, the different types of monomeric units tend to separate from each other in order to reduce the number of A-B contacts, and, consequently, to reduce the energy. Simu-

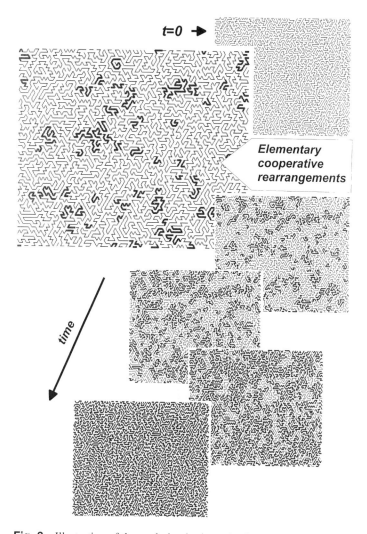

t=0 →

Elementary
cooperative
rearrangements

time

Fig. 2 Illustration of the evolution in time of a dense system of linear chains simulated by the CMA, starting from an arbitrary configuration (at $t = 0$). A sequence of states with increasing rearrangement number is shown within the time interval necessary to move every element at least once. The rearranged fraction of the system is plotted by thicker lines. The magnified snapshot illustrates the state at which individual rearrangements can be distinguished.

lations are usually performed in a broad temperature range including both the homogeneous and the microphase-separated states. The simulation allows one to get information about the structure, dynamics, and thermodynamic properties of the system.

B. Measuring Viscoelastic Properties

Mechanical spectroscopy can be used to investigate the dynamic properties of the polymer chains in the melt. Dynamic mechanical measurements are usually performed using a mechanical spectrometer, with the sample between two parallel plates or with a cone-and-plate geometry. Various diameters of plates, ranging between 6 and 25 mm, have to be used in order to adjust the sensitivity of the instrument to variable properties of samples that are changing over a broad range with temperature. Measurements should usually be performed under nitrogen atmosphere. Master curves, characterizing frequency dependencies of the storage (G') and loss (G'') moduli at a reference temperature, can be constructed merely by a horizontal shift of frequency sweeps measured at various temperatures.

An example of the master curve for an polyisoprene melt with molecular weight $M_W = 130,000$ is shown in Fig. 3. Two characteristic relaxation processes

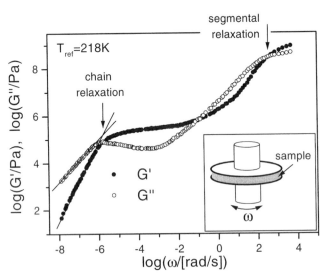

Fig. 3 Example of master frequency dependencies of the storage (G') and loss (G'') moduli determined for a polyisoprene melt of linear chains ($M_w = 130,000$), at the reference temperature $T = 218$ K. Dependencies of $G' \sim \omega^2$ and $G'' \sim \omega$, characteristic of the Newtonian flow range at low frequencies, are represented by lines with slopes 2 and 1, respectively. The arrows indicate frequencies corresponding to the two distinguishable relaxation processes.

can be detected in such dependencies. The segmental relaxation at high frequency and the terminal-chain relaxation at low frequency. The relaxation times of the two processes can be determined by fitting of Fourier transformed stretched exponential functions to the frequency dependencies of G' and G'', in corresponding frequency ranges. Mean relaxation times are used

$$<\tau> = \left(\frac{\tau_0}{\beta}\right) \cdot \Gamma\left(\frac{1}{\beta}\right) \qquad (1)$$

where τ_0 and β are parameters of the stretched exponential function

$$G(t) = \exp\left[-\left(\frac{t}{\tau_0}\right)^{\beta}\right] \qquad (2)$$

and Γ denotes the gamma function. It has been observed that the mean relaxation time determined in this way coincides well with the reciprocal value of the frequency at which G' and G'' dependencies cross each other (see arrows in Fig. 3).

III. SIMULATED AND MEASURED RELAXATIONS IN HOMOPOLYMER MELTS

Before discussing results concerning more complex polymer systems such as copolymers, we present here a summary of the main results obtained by means of the discussed methods for homopolymer melts of linear chains.

The dynamic properties of the simulated models can, for example, be characterized by the following quantities: (a) the autocorrelation function of a bond

$$\rho_b(t) = \frac{1}{Nn} \sum_{n}^{N} \sum_{i} b_i(t) b_i(0) \qquad (3)$$

where b_i are unit vectors representing bond orientation, and (b) the autocorrelation function of the end-to-end vector of chains

$$\rho_R(t) = \frac{1}{n} \sum_{n} R(0) \cdot R(t) \qquad (4)$$

with end-to-end vectors $R(0)$ and $R(t)$ at time $t = 0$ and t, respectively.

The bond and the end-to-end autocorrelation functions allow one to determine relaxation times of corresponding objects (bonds and chains) in a polymer model system. Examples of time dependencies of the bond and chain end-to-end vector correlation functions are shown in Fig. 4a. Various chain lengths, ranging from $N = 5$ to $N = 800$, are considered.

Typical experimental results concerning the dynamics of linear chains obtained for polyisoprene melts with chains of various length (M_W) are shown in Fig. 4b (44). The two characteristic regimes of segmental and terminal relaxation

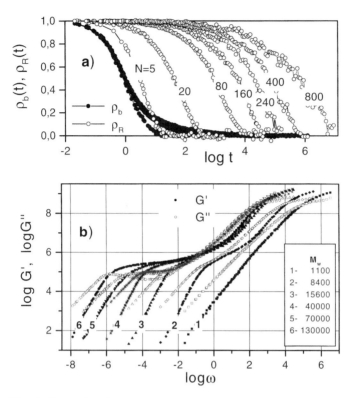

Fig. 4 Examples of dependencies characterizing relaxations in (a) simulated and (b) real polymer melts with linear chains of various lengths. (a) Time dependencies of the bond and the end-to-end autocorrelation functions. (b) Frequency dependencies of the real (G′) and imaginary (G″) components of the complex shear modulus for polyisoprene melts at the reference temperature $T = 218$ K.

of chains are well distinguishable, especially for longer chains, and allow a determination of characteristic relaxation times τ_s and τ_t.

Figure 5 shows chain-length dependencies of the terminal relaxation times for polyisoprene melts together with the relaxation times of bonds and chains determined for simulated systems. These results show that the dynamic behavior of simulated systems corresponds well to the behavior of real polymer melts, as detected experimentally. The relaxation times of chains in simulated systems reach the scaling law $\tau \sim N^{\omega}$, with $\omega > 3$, as observed in chain-length dependencies of melt viscosities for long chains. The ratio of the terminal and segmental relaxation times (τ_t/τ_s) is considered a reduced relaxation time, reflecting chain dynamics in polyisoprene independent of other effects, such as temperature. This reduced relaxation time should be used for comparison with theoretical models in which some chain-length-dependent effects are included in a friction coefficient (more

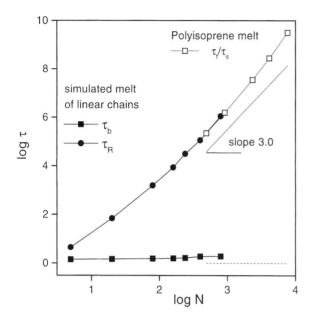

Fig. 5 Comparison of the bond and end-to-end vector relaxation times determined for the simulated systems with the reduced relaxation times determined from viscoelastic spectra of polyisoprene melts. The lengths of chains in simulated systems are given by the number of beads (N), whereas the chain length of polyisoprene samples is expressed by the number of bonds in the backbone (4 per monomer unit).

important for short chains). The results for polyisoprene reflect the known scaling low for long linear chains $\tau \sim N^{3.7}$ (determined here for systems with chains longer than $M_w = 15,000$).

Both simulation and experimental results form a continuous dependence of terminal relaxation times of chains when the lengths of polyisoprene chains are expressed by the number of bonds in the backbone. The observed coincidence between experimental and simulation results can be considered a consistency proof between the two methods. More detailed analysis of the static and dynamic behavior of polymer melts simulated by means of the CMA has been presented in other publications (43,44).

IV. SIMULATED BLOCK COPOLYMER SYSTEMS

A. Thermodynamics and Structure

Various types of copolymers differing in the distribution of comonomers and by chain architectures have been simulated using the cooperative-motion algorithm (15,42,44). Dense systems consisting of model macromolecules of a given structure (see Fig. 1) are usually equilibrated at various temperatures corresponding both to the disordered and to the microphase-separated states. In this section, the thermodynamic and structural characterization of some simulated block copolymer systems, taken as examples, is presented.

Examples of temperature dependencies of the thermodynamic quantities recorded during heating of the initially microphase-separated systems are shown in Fig. 6. The following quantities have been determined: (a) the energy of interaction of a monomer E_m determined as the average of interactions of all monomer

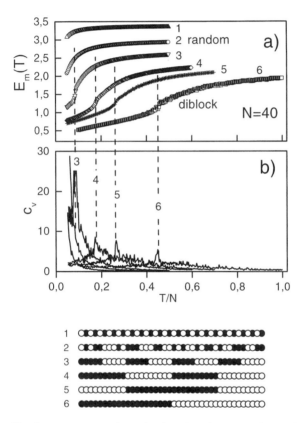

Fig. 6   Temperature dependencies of (a) the average interaction energy per monomer and (b) the specific heat, determined using Eq. (6), for copolymer systems with various distributions of monomers along the chain (illustrated in the bottom of the figure): (1) alternating copolymer, (2) random copolymer, (3) and (4) multiblock copolymers, (5) triblock and (6) diblock copolymer. Vertical dashed lines indicate temperatures of the microphase separation in systems (3) to (6).

pairs at a given temperature:

$$E_m = \sum_{i=1}^{z} \varepsilon_{kl}(i) \tag{5}$$

and (b) the specific heat calculated via fluctuations in the total energy:

$$c_v = \frac{\langle E^2 \rangle - \langle E \rangle^2}{k_B T^2} \tag{6}$$

where the brackets denote averages over the energy of subsequent states sampled during simulation of the system at constant temperature. The temperature at which a step-wise change in the energy and the corresponding peak in the specific heat is observed is regarded as the temperature of the order-to-disorder transition, T_{ODT}. In Fig. 6, the behavior of linear copolymers with various distributions of comonomers along chains are compared. The results indicate a high sensitivity of the system to changes of the comonomer distribution when other parameters (chain length and composition) are kept constant. The observed range of variation of the temperature of the ODT is very broad. In accord with the theoretical predictions (45), the order-to-disorder transition in the triblock copolymer (#5 in Fig. 6) is lower as compared to the diblock copolymer (from a critical value of the product χN of 10.5 in the diblock to 18 in the triblock copolymer). Moreover, the more dispersed the comonomers, the lower is the T_{ODT} due to the increasing number of contacts between blocks, which result in a broader interface. Similar results have also been obtained for other types of comonomer distributions (15).

The nature of the transitions corresponding to structural changes in copolymers can be well established from an analysis of distributions of local concentrations, which are directly related to the free energy. An example of such distributions for a symmetric diblock copolymer, in a broad temperature range, is shown in Fig. 7, by means of contour lines of equal composition probability projected on the composition–temperature plane. Such contour plots reflect many details of the thermodynamics and structure of the system. It is easily seen that, at high temperatures, the system can be considered as homogeneous, because locally the most probable concentration corresponds to the nominal composition in the diblock. This changes at temperatures close to T_{ODT}, where at first a plateau and later two maxima corresponding to two coexisting phases are detected. At T_{ODT}, a sudden change transforms the system to a state with well-defined microphases, indicated by the most probable local concentrations corresponding to pure components.

These results indicate three characteristic ranges of thermodynamic behavior of the system assigned as (1) disordered, (2) weakly segregated, and (3) strongly segregated regimes appearing with decreasing temperature. Structures of simulated systems corresponding to these regimes are illustrated in Fig. 7 by assuming different colors for copolymer constituents (black and gray for A and B, respectively).

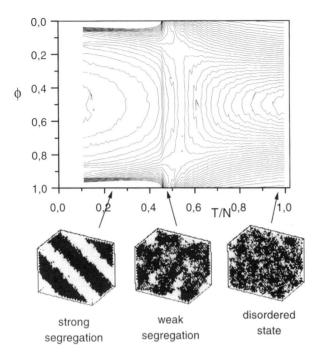

Fig. 7 Temperature dependence of concentration distributions in small-volume elements consisting of the nearest neighbors of each chain segment. Some representative structures corresponding to various temperature ranges are illustrated.

The structure of the simulated block copolymer systems is usually characterized by the following quantities: (a) mean squared end-to-end distance of chains

$$\langle R^2 \rangle = \langle (r_N - r_1)^2 \rangle \tag{7}$$

where r_1 and r_N are space coordinates of chain ends, and (b) two orientation correlation functions of the end-to-end vectors: (i) the local chain orientation correlation factor

$$f_1 = \langle \cos[\theta_{RR}(r_{ij})] \rangle \tag{8}$$

where $\theta_{RR}(r_{ij})$ is the angle between end-to-end vectors of two chains whose centers of mass are separated by r_{ij}, smaller than the root mean squared radius of gyration of corresponding unperturbed linear chains, and (ii) the global orientation correlation parameter

$$f_{RR} = \tfrac{1}{2}(3\langle \cos^2\theta_{ij} \rangle - 1) \tag{9}$$

where the averaging is performed over end-to-end vector pairs belonging to many states of the systems in equilibrium. The temperature dependence of these quantities for the symmetric diblock copolymer system is shown in Fig. 8. It is evident that all of foregoing quantities change in a characteristic way at T_{ODT}.

The microphase separation in the diblock copolymer system is accompanied by chain extension. This is easily seen in Fig. 8a. The chains of the diblock copolymer start to extend well above the transition to the strongly segregated regime. As

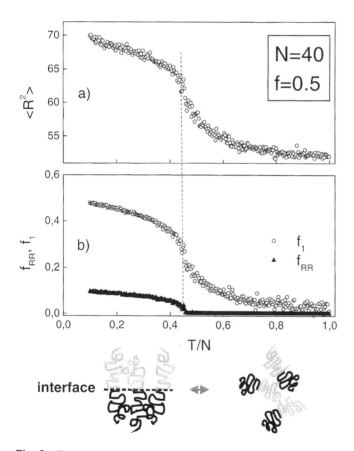

Fig. 8 Temperature dependencies of (a) mean squared end-to-end distance of a symmetric diblock copolymer, and (b) two orientation factors: local (f_1) and global (f_{RR}), determined according to Eqs. (8) and (9), respectively. The vertical dashed line corresponds to the order-to-disorder transition temperature. At the bottom, the noncorrelated orientations of chains in the disordered state (right-hand side) and the correlated orientation of chains in the ordered state (left-hand side) are illustrated.

seen in Fig. 8b, this extension of chains is also related to an increase in local orientation correlations, which appear well above the transition. On the other hand, the global orientation correlation factor f_{RR} remains zero above the microphase separation transition and jumps to a finite value at the transition. Structures shown schematically in the bottom of Fig. 8 illustrate the most important changes taking place in a diblock copolymer on the molecular level in relation to the microphase separation, i.e., the extension of chains and the orientation correlation of chain axes with respect to the interface.

The microstructure in copolymers is often studied experimentally by scattering methods. In order to get corresponding results for simulated copolymers, static collective structure factors can be determined:

$$S(q) = \sum_{ij} e^{iqr_{ij}} f_c(r_{ij}) \tag{10}$$

where q is the scattering vector and $f_c(r_{ij}) = (1/n_p N)\langle c_k(r_i) \cdot c_l(r_j)\rangle$ is the monomer–monomer correlation function of monomers separated by $r_{ij} = r_i - r_j$ and c_k and c_l are contrast operators, assuming the value 1 or -1 when the monomer at a given position is of type A or B, respectively. The results obtained for the symmetric diblock copolymers are shown in Fig. 9a and are compared with experimental results obtained for poly(styrene-*b*-butadiene) (P(S-*b*-B)) block

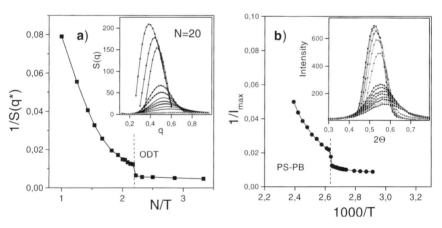

Fig. 9 (a) Reciprocal maximum value of the structure factor for the simulated diblock copolymer system, and (b) the reciprocal maximum scattered intensity for the poly(styrene-*b*-butadiene) copolymer plotted vs. reciprocal temperature. Corresponding static collective structure factors vs. the wave vector q and the angular distributions of scattered intensities at various temperatures are shown in the inserts. The jumps in their temperature dependence correspond to the order-to-disorder transition.

copolymers (46) shown in Fig. 9b. A very good qualitative agreement between the two kinds of results can be noticed.

B. Dynamic Properties

In order to get information about the dynamic properties of simulated systems, the relaxations of various structural elements (as was described earlier for homopolymer melts) are monitored with time at equilibrium states corresponding to various temperatures: (1) the mean squared displacement of monomers, (2) the mean squared displacement of the center of mass of chains, (3) the single bond autocorrelation function (Eq. 3), and (4) the end-to-end vector autocorrelation function (Eq. 4) Additionally, for block copolymers we can record (5) the autocorrelation of the end-to-end vector of the block

$$\rho_{bl}(t) = \frac{1}{n} \sum_n R_{bl}(0)R_{bl}(t) \tag{11}$$

and (6) the single point concentration correlation

$$\rho_c(t) = \frac{1}{n} \sum_n C(0) \cdot C(t) \tag{12}$$

with C assuming the value 1 or -1 when the lattice site is occupied by monomer A or B, respectively.

Based on these correlation functions, various quantities characterizing the dynamic properties of the systems can be determined, i.e., the diffusion constant of chains and various relaxation times corresponding to considered correlations. The relaxation times are determined by fitting a sum of stretched exponential functions to the simulated correlation functions:

$$\rho(t) = \sum_i A_i \exp\left[-\left(\frac{t}{\tau_i}\right)^{\beta_i}\right] \tag{13}$$

where i is the number of components.

In Fig. 10, various correlation functions, as specified earlier, are shown for the diblock copolymer system at high and low temperatures. At high temperature, the system behaves like a homogeneous melt. All correlation functions show a single-step relaxation. The fastest is the bond relaxation; the slowest is the chain relaxation described by the end-to-end vector autocorrelation function. The relaxation of the block is faster than the whole chain relaxation by approximately a factor of 2. Such relations between various relaxation times in the disordered state of the copolymer can be regarded as confirmed experimentally for some real systems, in which the dielectric spectroscopy allows distinction of the different relaxation modes (44). At low temperatures, drastic changes can be noticed for the

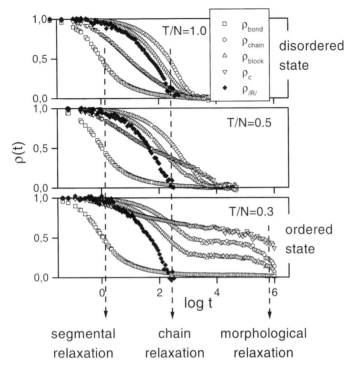

Fig. 10 Various correlation functions determined at various temperatures for the symmetric diblock copolymer case. The two extreme temperatures correspond to edges of the temperature range within which the system has been simulated. The high temperature ($T/N = 1.0$) corresponds to the homogeneous regime and the low temperature ($T/N = 0.3$) to the strongly segregated limit. The intermediate temperature ($T/N = 0.5$) is only slightly higher than the temperature of the order-to-disorder transition for this system.

dynamics of the block copolymer. At temperatures $T/N < 0.45$ (Fig. 10c), the diblock system is in the microphase-separated regime and most of the correlation functions determined show a bifurcation of the relaxation processes into fast and slow components. The fast components of chain, block, and concentration relaxations are attributed to the almost-unchanged-in-rate, but limited, relaxation of chains when fixed at the A-B interface, and the slow component indicates the part of the relaxation coupled to the relaxation of the interface within uniformly ordered grains with the lamellar morphology. The concentration relaxation becomes the slowest one in such a state of the system. The dynamic behavior of diblock copolymers, as simulated by the CMA, is presented in detail and discussed in Ref. 40, where the spectra of various relaxation modes have been determined in order to compare simulation results with dielectric spectra determined for real copolymer systems in the vicinity of the order-to-disorder transition (18).

A more detailed comparison of the dynamic behavior of various simulated systems is presented in Fig. 11, where changes of the different correlation functions with temperature are shown. The correlation functions shown by solid lines correspond to a temperature only slightly above the order-to-disorder transition of the system. Actually, they separate the faster-relaxing correlation functions corre-

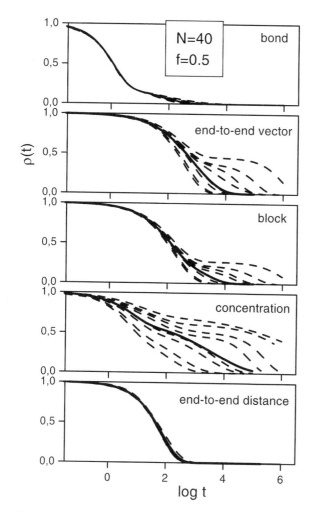

Fig. 11 Different correlation functions plotted at various temperatures as obtained by fitting the simulated data via the function given by Eq. (13). The correlation functions shown by solid lines correspond to the temperature just above the order-to-disorder transition and separate the faster relaxations at higher temperatures from the slower relaxations at temperatures below the transition. Correlation functions recorded at temperatures $T/N = 0.3$, 0.35, 0.4, 0.43, 0.45, 0.47, 0.5, 0.7, and 1.0 are shown.

sponding to high temperatures, and consequently to the homogeneous regime, from the correlation functions, in which clearly a new slow component is observed, related to the relaxation of the morphology existing at low temperatures (strong segregation regime). As will be shown in Sec. V, this slow relaxation can be detected as an extra viscoelastic relaxation mode and can be used as an indication of the ordered microstructure.

The preceding effects are better discussed when the temperature dependence of the relaxation times of various structural elements are compared, as presented in Fig. 12. The characteristic bifurcation of the chain relaxation in the phase-separated regime results in two relaxation times, which can be detected below the microphase separation temperature. Similar effects as these for the chain relaxation can be observed in temperature dependencies of other relaxations considered here, i.e., the block or concentration relaxation.

The diffusion in the systems studied has been detected by monitoring the mean squared displacements of monomers and centers of mass of chains as a function of time. Typical results for a diblock copolymer system are shown in Fig. 13. They indicate that the short time-displacement rates are not sensitive to temperature but the long time displacements are influenced slightly by the microphase

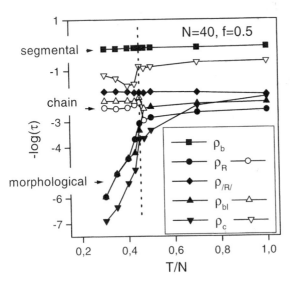

Fig. 12   Temperature dependencies of relaxation times determined from various correlation functions for the diblock copolymer system. The vertical dotted line corresponds to the temperature of the order-to-disorder transition for this system. Filled and open symbols of the same shape correspond to two components of the double-step correlation functions describing slow and fast relaxation modes, respectively.

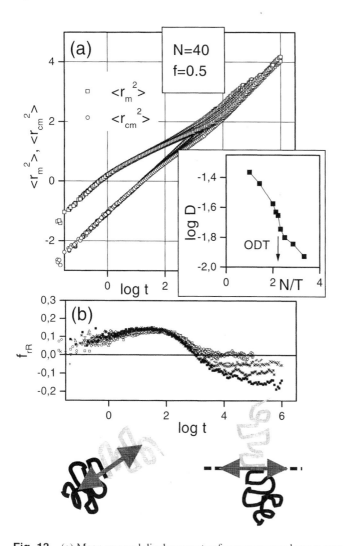

Fig. 13 (a) Mean squared displacements of monomers and mean squared displacements of chain centers of mass vs. time for the diblock copolymer system at various temperatures (as specified in Fig. 11). The temperature dependence of the self-diffusion constant of chains in the simulated melt is shown in the inset. (b) Orientation correlation factor between the end-to-end vector and the center-of-mass displacement vector of copolymer chains at various temperatures above and below the ODT. At the bottom, preferential directions of chain motions in the disordered and ordered states are illustrated.

separation. The self-diffusion constants of chains determined at the long time limit are shown in the inset of Fig. 13a, where the effects of the microphase separation in the diblock can be clearly noticed. The slowing down observed at the microphase separation of the system is, however, rather small, in agreement with experimental observations (47,48) and indicates a considerable mobility of chains left even when the chains are confined at interfaces. The nature of this mobility has been analyzed by monitoring the correlation between the orientation of chain axes and directions of chain displacements (Fig. 13b). It is established that in the phase-separated regime the chains diffuse easily in directions parallel to the interface. This is indicated by negative values of the orientation correlation factor and means that the limitations in the mobility of diblock chains imposed by the morphology concern only the diffusion through the interface. Preferential directions of motion of chains in disordered and ordered states of the diblock copolymer are illustrated at the bottom of Fig. 13.

V.   VISCOELASTIC EFFECTS IN BLOCK COPOLYMERS

A.   Viscoelastic Properties of the Ordered Phases

For disordered states of block copolymers, the viscoelastic behavior is usually similar to that observed for homopolymer melts. The two relaxation processes, as described in Sec. II.B, i.e., the chain relaxation at low frequencies and the segmental relaxation at high frequencies, can be observed, however, in some cases with a broadening or bifurcation in the range of segmental relaxation, depending on the nature of the comonomers. At $T > T_{ODT}$, the time–temperature superposition (TTS) also works well and the moduli exhibit typical terminal behavior ($G' \approx \omega^2$ and $G'' \approx \omega$). When examined over a broad temperature range, however, the TTS is violated (49–52) due to the order-to-disorder transition, which drives the system from the disordered state to a microphase-separated state. At $T < T_{ODT}$ and at low frequencies, the moduli exhibit weak frequency dependencies of the order of $\omega^{1/2}$ (for symmetric block copolymers) to $\omega^{1/4}$. This results, in our opinion, from the appearance of the new ultraslow relaxation process related to morphological rearrangements, as discussed in Sec. IV.B for the simulated systems. Unfortunately, quite often a complete relaxation of the microstructure copolymers cannot be observed within the accessible frequency and temperature window.

An example of the influence of the order-to-disorder transition on the frequency dependencies of G' and G'' is shown in Fig. 14 (53). It demonstrates the breakdown of the TTS; the main effect is observed at low frequencies, where the Newtonian behavior of the disordered state is replaced by a rubbery state related to the unrelaxed morphology.

The linear viscoelastic properties of microphase-separated block copolymers have been the subject of recent theoretical studies. In the study by Rubinstein and Obukhov (54), both microscopic and mesoscopic mechanisms have been in-

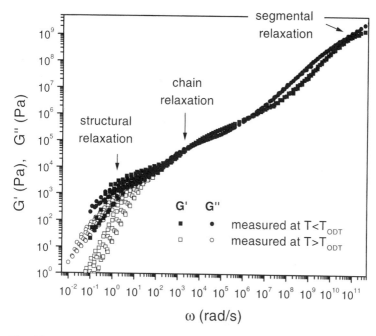

Fig. 14 Frequency dependencies of G' and G'' for poly(methylphenylsiloxane-b-styrene) [P(MPS-b-S)] copolymer measured at various temperatures and shifted horizontally to form master curves for the reference temperature $T_{ref} = 383$ K ($T_{ODT} = 403$ K, $f_{PS} = 0.36$). Below ODT, the hexagonal cylindrical morphology has been detected by small-angle x-ray scattering (SAXS).

voked, which were attributed, respectively, to the dispersion in the number of entanglements of a chain with the opposite brush (high-frequency response) and to the collective diffusion of copolymer chains along the interface. The latter mechanism is controlled by defects in lamellar orientation and contributes to the low-frequency side. For the disordered lamellar mesophase they predict: $G'(\omega) \approx G''(\omega) \approx \omega^{\frac{1}{2}}$, whereas for the cylindrical mesophase: $G'(\omega) \approx G''(\omega) \approx \omega^{\frac{1}{2}}$. On the other hand, Kawasaki and Onuki (55) reached the same conclusion through a completely different approach: They proposed that overdamped second-sound modes in an orientationally disordered lamellar phase could result in a complex shear modulus proportional to $(i\omega)^{\frac{1}{2}}$.

Recently, the moduli in bicontinuous ($I a\bar{3}d$) and spherical (Im3m) morphologies in linear (56) and nonlinear (57,58) block copolymers were found to exhibit a maximum, which signifies a new characteristic frequency for the relaxation of the ordered phase. We attribute the maximum in $G''(\omega)$ to the grain relaxation. Such dependence of G'' on frequency has been reported in asymmetric super-H-shaped block copolymer colloids (57). Similarly, in the asymmetric miktoarm stars $(PS)_2(PI)_2$, there was an upturn of G'' at low frequencies, which was found to depend on the annealing history and therefore on the coherence of grains (58).

An advantage of rheology in the study of block copolymer microstructures is that each phase has a different viscoelastic contrast that can be used as a dynamic signature of the phase (59). To illustrate the distinct frequency dependence of the storage and loss moduli for the different phases, we show in Fig. 15 the moduli of a polyisoprene-poly(ethylene oxide) (PI-PEO) diblock copolymer (M_n = 9800, f_{PI} = 0.61) at three temperatures (59). This diblock copolymer un-

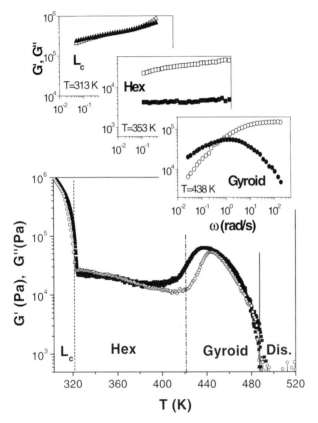

Fig. 15 Temperature dependence of the storage modulus for PI-PEO 5-4 (M_n = 9800, f_{PI} = 0.61) obtained at 1 rad/s with a low strain amplitude of 0.5% shown for two heating rates; (□) 1K/min, (○) 4 K/min. The vertical lines separate the high-temperature disordered phase (Dis.), the bicontinuous cubic phase (Gyroid), the hexagonally packed cylinders (Hex), and crystalline lamellar phases (L_c). Notice the difference in the $G'(T)$ values in the vicinity of the Hex-to-Gyroid transition. The insets show frequency dependencies of the storage (open symbols) and loss (filled symbols) moduli at three temperatures: T = 313 K, T = 353 K, and T = 438 K, corresponding to L_c, Hex, and the Gyroid phases, respectively [59].

dergoes several order-to-order transitions (see also later): from a low-temperature crystalline lamellar (L_c) to a hexagonally packed cylinders (Hex) to the Gyroid phase and finally to the disordered phase. The L_c phase has the highest moduli, followed by the Gyroid and finally the hexagonal phase. The L_c phase is formed by a semicrystalline block (PEO) and an elastomeric block (PI), both giving rise to a spherulitic superstructure. Therefore the storage and loss moduli below the equilibrium melting temperature T_m° correspond to the spherulitic superstructure and the relatively low values of G' and G'', as compared to bulk semicrystalline PEO, are caused by the unentangled PI block. The hexagonal phase at 353 K is characterized by an elastic response with weak frequency dependence. In contrast, in the Gyroid phase the moduli show a strong frequency dependence with a crossover in $G'(\omega)$ and $G''(\omega)$ to a nonterminal regime. Near the crossover, G'' develops a maximum characteristic of all cubic structures.

B. Localization of the (Apparent) Order-to-Disorder Transition

Because of the distinctly different frequency dependencies of the ordered and disordered states, low-frequency rheology is very sensitive for detecting the dissolution of ordered microstructures (49). Isochronal measurements of the storage modulus performed at low frequencies by heating the specimen provide a good way of locating the T_{ODT}. As we will see shortly with respect to the kinetic studies, the temperature extracted in this way reflects some apparent temperature (T'_{ODT}) that is lower than the true equilibrium ODT (T_{ODT}°) (60).

Figure 16a shows the result of the isochronal measurements of the storage (G') and loss (G'') moduli at $\omega = 1$ rad/s obtained on heating (symbols) and subsequent cooling (lines) of a symmetric SI diblock copolymer ($M_n = 12200$ and $f_{PS} = 0.51$). The changes of the moduli within the indicated range signify an apparent ODT. There is a pronounced hysteresis on cooling that extends to some 10°C below the transition and that is the signature of metastability.

Figure 16b shows the result of a similar experiment for the asymmetric and nonlinear miktoarm star–block copolymer SI$_3$ ($M_n = 24000$, $f_{PS} = 0.3$). The isochronal ($\omega = 1$ rad/s) temperature run reveals an apparent ODT and a pronounced hysteresis for about 15°C below the (apparent) transition. The well-defined loop calls for a pronounced supercooling effect. The hysteresis observed in all block copolymers around the ODT indicate a first-order transition at the ODT and will be used later as a tool to investigate the ordering kinetics and to localize the equilibrium ODT.

C. Phase Transformation Kinetics

The ordering process in block copolymers bears some similarities to the crystallization process in semicrystalline materials. A variety of different experiments

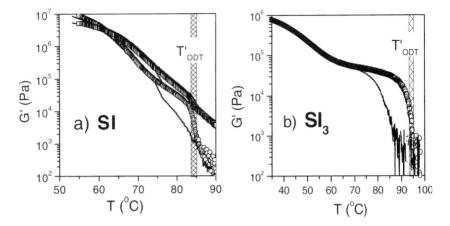

Fig. 16 (a) Isochronal measurement of the storage ($\bigcirc$) and loss ($\square$) moduli for the symmetric diblock copolymer SI ($M_n = 12200$, $f_{PS} = 0.51$). (b) The isochronal temperature scan of the storage modulus for the miktoarm star SI_3, ($M_n = 24,000$, $f_{PS} = 0.3$). The dependencies are taken with $\omega = 1$ rad/s and a strain amplitude of 2% on heating (symbols) and subsequent cooling (line). The heating rate and the cooling rate are 0.2 K/min in all cases.

can be used to monitor the crystallization process, e.g., dilatometry (61), x-ray diffraction (62), rheology (63,64), and dielectric spectroscopy (65), but undoubtedly the most direct evidence is provided by dilatometry via recording the discontinuous change of the specific volume and of the isothermal compressibility as the material passes through the melting point. Our earlier work on this subject has shown, however, that this method is not a sensitive probe of the ordering process in diblock copolymers, since both the density and the compressibility are continuous functions of T as a result of the weak first-order transition at the T_{ODT}. Herein we review the results from rheology on the ordering kinetics of block copolymers. Small-angle x-ray scattering (SAXS) has also been used to monitor the ordering process (51,67–70), and recently transmission electron microscopy (TEM) has been employed for the same purpose (71,72). The main results from these experiments are in agreement.

1. Detection and Analysis of the Ordering Kinetics

The time evolution of G' and G'', which is monitored after the temperature jumps, is used to analyze the ordering kinetics (51,57,60,73). Some typical data are shown in Fig. 17 for a symmetric SI diblock copolymer ($M_n = 12,200$, $f_{PS} = 0.51$). Both moduli show a sigmoidal shape with distinct plateaus at short and long times. The plateaus of G' and G'' at short times are assumed to describe the mechanical properties of the disordered phase of the system at the quenched temperature. On the other hand, the plateaus at longer times are regarded as describing the properties of the final microphase-separated state. In the intermediate time range, the system is regarded as a composite material made of two phases (ordered and disordered), the proportion of which changes with time, causing the observed changes in the rheological response. The long "incubation" time observed for the shallow quenches as well as the overall shape of the curves shown in Fig. 17 point toward a *nucleation-and-growth* ordering mechanism (51,57,60,73).

In order to analyze the ordering kinetics in terms of a nucleation-and-growth mechanism, the time dependence of the volume fractions of the constituent phases is needed. Several models have been developed to describe the properties of composite materials, but they do not provide any precise and unique solution to this problem. Therefore, we have used here the simplest "series" and "parallel" models, which provide the limits for the mechanical response of a two-phase system as a function of the properties of the constituent components and the composition (74). In the "series" model, the same stress is applied to both phases (ordered and disordered); this results in different displacements. The modulus of the mixed phase is expressed as a linear combination of the compliances of the constituent

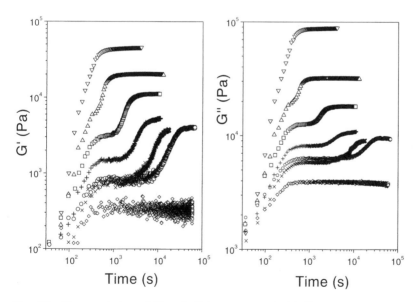

Fig. 17 Time evolution of G' and G'' for a symmetric SI diblock copolymer (M_n = 12,200, f_{PS} = 0.51) following quenches from the disordered state (T_i = 363 K) to different final temperatures: ($\Diamond$) 359, ($\circ$) 358, ($\times$) 356.5, ($+$) 356, ($\bullet$) 355, ($\triangle$) 353, and ($\triangledown$) 347.6 K [51].

phases:

$$\frac{1}{G(t)} = \frac{1 - \varphi(t)}{G_0} + \frac{\varphi(t)}{G_\infty} \tag{14}$$

where $G(t) \{= [G'^2(t) + G''^2(t)]^{\frac{1}{2}}\}$ is the absolute value of the complex modulus and G_0, G_∞ are the moduli of the initial ($t = 0$) disordered and final ($t = \infty$) ordered phases, respectively. In the "parallel" model, the two phases have the same displacement but different stresses and the modulus of the mixed phase can be expressed as a linear combination of the moduli of the constituent phases:

$$G(t) = [1 - \varphi(t)] G_0 + \varphi(t)G_\infty \tag{15}$$

The time dependence of φ obtained from both models at every temperature is analyzed by fitting the Avrami equation (75):

$$\varphi(t) = 1 - \exp(-\alpha t^n) \tag{16}$$

where α is the rate constant and n is the Avrami exponent. The former gives a quantitative information on the course of "crystallization," and it is usually expressed in terms of the half-time (or completion time):

$$t_{1/2} = \left(\ln \frac{2}{\alpha}\right)^{1/n} \tag{17}$$

The Avrami exponent n is a combined function of the growth dimensionality and the time dependence of the nucleation process and provides qualitative information on the nature of the nucleation-and-growth process. The Avrami parameters are usually extracted from a plot of $\log[-\log(1 - \varphi)]$ vs. $\log t$, from which it is possible to derive n and α from the slope and intercept, respectively.

2. Equilibrium Order-to-Disorder Transition Temperature

After the long-time plateaus have been reached in the previously described experiments, we performed isochronal temperature scans (at $\omega = 1$ rad/s), aiming to disorder the system. Should there be a single T_{ODT}, all curves would overlap. The results of the isochronal measurements is shown in Fig. 18 for a miktoarm-star-block copolymer of the SI_3 type (the corresponding isochronal temperature scan for the same material is shown in Fig. 16b). In Fig. 16b the isochronal ($\omega = 1$ rad/s) temperature run revealed a pronounced hysteresis for about 15°C below the (apparent) transition. The well-defined loop calls for a pronounced supercooling effect. As shown in Fig. 18, for every final ordered state there is a different T_{ODT}; the lower the ordering temperature, the lower the apparent ODT. The pronounced dependence of the ODT on the ordering temperature is shown in the inset to Fig. 18, where we plot the apparent transition temperature (T'_{ODT}) as a function of the corresponding ordering temperature (T_{ord}). The corresponding plot for semicrystalline polymers is known as the Hoffman–Weeks plot, where the apparent melting temperature (T'_m) is plotted as a function of the crystallization temperature

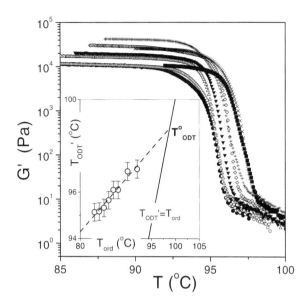

Fig. 18   Isochronal (ω = 1 rad/s) temperature scans (heating rate = 0.2 K/min) for the miktoarm-star-block copolymer SI_3 (M_n = 24,000, f_{PS} = 0.61) obtained by heating the ordered phase formed at the kinetic experiments: (●) 83, (○) 84, (△) 85, (▽) 86, (◊) 87, (+) 88, (X) 90, (■) 92°C. In the inset, the corresponding Hoffman–Weeks plot is shown and the $T°_{ODT}$ is obtained by extrapolation [60].

(T_c). The solid line in the figure signifies $T_{ord} = T'_{ODT}$, and the equilibrium ODT can be obtained by extrapolation (dashed line) (60). The extrapolation used in Fig. 18 is based on the underlying Gibbs–Thompson equation, which, for

$$T'_m = T°_m \left(1 - \frac{2\sigma_e}{\Delta H_f} \frac{1}{d} \right) \tag{18}$$

semicrystalline polymers, gives the melting point depression due to the finite thickness (d) of the crystal (60). In Eq. (18), σ_e is the surface free energy and ΔH_f is the heat of fusion. Some typical values of these parameters for a semicrystalline polymer (e.g., PEO) are 20 erg/cm^2 and 200J/g, respectively. It is worth noticing that the equilibrium ODT is at about 99°C, that is, some 4°C higher than the apparent transition one would obtain by ordering the system at 85°C ($\Delta T = T°_{ODT}$ − T'_{ODT} = 14°C).

Motivated by the similarities between the order/disorder process and the crystallization/melting process, we first simply rewrite Eq. (18) for the former process as

$$L = \left(\frac{2\sigma}{\Delta H} \right) \left(\frac{T°_{ODT}}{T°_{ODT} - T'_{ODT}} \right) \tag{19}$$

where σ is now the interfacial tension and ΔH is the change of enthalpy upon mixing. Equation (19) implies a strong effect of supercooling on sizes (L) of ordered regions.

In applying Eq. (19) we should keep in mind that it has been derived for lamellar crystallites, where only the interactions at the lamellar surfaces play a role. In the case of grains, however, the whole intergrain surface should be considered. In the following we generalize Eq. (19) by reformulating the problem for the case of grains in ordered copolymers. The transition temperature is defined as $T_{ODT} = \Delta H/\Delta S$, where ΔH (ΔS) is the enthalpy (entropy) difference between the ordered and disordered states. Under the assumption that the finite grain size causes the enthalpy reduction related to the disorder at the intergrain surfaces (constant entropy approximation), $\Delta H_0/T^{\circ}_{ODT} = \Delta H/T'_{ODT}$, where ΔH_0 refers to the enthalpy of an infinitely large grain. The enthalpy reduction is related to a kind of disorder (defects) at the grain boundaries. This effect should be proportional to the surface-to-volume ratio of individual grains, and the remaining enthalpy limiting the stability of grains can be expressed as $\Delta H_0 [1 - (s/v^*)l]$, where s and v^* are, respectively, the grain surface and volume and l is the intergrain layer thickness. Based on these considerations the grain dimensions, L, are:

$$L = \beta \, \frac{T^{\circ}_{ODT}}{T^{\circ}_{ODT} - T'_{ODT}} \tag{20}$$

where β is a function of the geometrical characteristics of the grain and of the intergrain layer thickness. Equation (20) can be considered as a generalized Gibbs–Thomson equation (60,63) for grain sizes, and knowledge of β would allow a determination of the grain size. A pronounced grain size dependence on supercooling has been observed in experiments that probe the development of birefringence (76) and optical anisotropy (77) following quenches from the disordered phase. From the former study and the dependence shown in the inset to Fig. 18 we deduce $\beta \sim 1.3$ µm, which gives reasonable grain sizes on supercooling.

Finally, we comment on the consequences of these findings for block copolymers. For experiments made far away from the T°_{ODT} there should be no considerable effects; however, for experiments made in the vicinity of the transition (weak segregation), strong coupling is expected. The equilibrium transition temperature, for example, is needed if kinetic experiments are performed, since knowledge of the quench depth, ΔT, is essential. Furthermore, shear- and pressure-induced transitions might also be affected in a narrow temperature interval below T°_{ODT}.

3. Effect of Fluctuations

The mean-field theory (MFT) (1) predicts for a symmetric diblock copolymer a continuous transition (second order) from the disordered to a lamellar phase at a critical value

$$(\chi N)_c = 10.5 \tag{21}$$

where χ is the Flory interaction parameter and N is the degree of polymerization. The MFT, however, cannot account for the metastability observed in Fig. 17. The reason is that, within the MFT approach, the free energy density below the spinodal develops two symmetric minima that describe the stable lamellar phase, and the disordered phase becomes completely unstable. Therefore, within the MFT it is not possible to account for the existence of metastable states below the ODT. The problem arises from the fact that the MFT neglects fluctuations of the order parameter. Fredrickson and Helfand (FH) (2) have demonstrated that with the introduction of fluctuation corrections, the critical point is suppressed and is replaced by a weakly first-order transition at

$$(\chi N)_{ODT} = 10.5 + 41\bar{N}^{-\frac{1}{3}} \tag{22}$$

where $\bar{N}$ is the Ginzburg parameter ($= N\alpha^6/u^2$, where α and u are the statistical segment length and volume, respectively). Furthermore, a new ordering mechanism is now expected near the ODT, since the FH theory allows for the existence of such metastable states as observed in the experiments.

The theoretical work of Fredrickson and Binder (78) predicted the existence of such metastable states and described the nucleation and growth of a lamellar phase from an undercooled disordered phase. The nucleation barrier for $f = 0.5$ was found to be unusually small when compared to polymer blends and equal to:

$$\frac{\Delta F^*}{k_B T}\Big|_s \approx \bar{N}^{-\frac{1}{3}} \delta^{-2} \tag{23}$$

and the characteristic ordering time is given by:

$$t_{\frac{1}{2}} \approx \exp\left(\frac{\Delta F^*}{k_B T}\right) \tag{24}$$

A pertinent feature of the ordering kinetics as described by Eq. (23) is that the characteristic ordering time-scales with $\bar{N}^{-\frac{1}{3}}$.

Testing this theoretical prediction requires a large number of exactly symmetric diblock copolymers. Instead, here we have tested the theoretical prediction (79) by employing just two symmetric diblock copolymers of polystyrene-b-polyisoprene (SI-85/65: $M_n = 12,200$ and SI-115/85: $M_n = 16,400$) and by preparing mixtures of the two at different compositions. Because of the small difference in molecular weights of the parent copolymers, all mixtures formed a single lamellar phase (as revealed by SAXS). The corresponding ODTs were extracted from rheology, as shown in Fig. 19. The (apparent) ODT is plotted in the inset to Fig. 19 as a function of the number-averaged degree of polymerization. The linear dependence can be parameterized as: T_{ODT}^{SI} (K) $= 213 + 1.0074N_n$. By forcing the

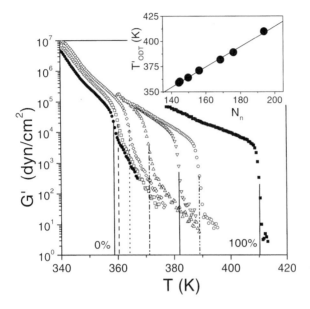

Fig. 19 Temperature dependence of the storage modulus G' obtained at 1 rad/s while heating the pure diblock copolymers and their blends. Vertical lines indicate the different T_{ODT}'s. The inset gives the dependence of the order-to-disorder transition temperatures on the number average degree of polymerization N_n for all copolymers (79).

MFT and FH approach (Eqs. 21 and 22) to the T_{ODT} we obtain $\chi_{MFT} = 54/T - 0.079$ and $\chi_{FH} = 89/T - 0.14$, respectively, for the interaction parameter in the mean-field and fluctuation approaches.

Next, we performed kinetic studies for each one of the new copolymers, and the kinetics have been analyzed as described in Sec. V.C.1. The characteristic ordering times were then plotted as a function of $N^{-1/3}$ for two supercooling temperatures in Fig. 20. The linearity shown constitutes a proof of the theoretical predictions and shows that the ordering kinetics in symmetric diblocks is controlled by the composition fluctuations (79). As already discussed, the kinetic studies provide a direct proof for the inadequacy of the MFT near the transition, and their existence and molecular weight dependence is in agreement with the fluctuation approach.

4. Effect of Block Copolymer Architecture

The synthesis of model nonlinear block copolymers (16,17) allows the investigation of the effect of copolymer architecture on the phase state and on the ordering kinetics. We have studied the ordering kinetics in: linear symmetric SI diblock copolymers (s-SI) (51,60), linear asymmetric (a-SI) (73), miktoarm star-block copolymers (SI_2, SIB: B: polybutadiene (67) and SI_3 (80)), and four-arm-star-block copolymers (($SI)_4$ (73)). We found that all block copolymers order with the same mechanism (nucleation and growth). Figure 21 shows Avrami plots for the volume fraction of the ordered phase for some of the foregoing cases. However, in comparing the ordering kinetics as a function of supercooling for the different architectures (Fig. 22) we find a strong influence of the architecture on the ordering times. The most intriguing feature of Fig. 22 is the freezing of the kinetics in the star diblock. Under the same undercooling, the star diblock needs about three additional decades of time—compared to the equivalent linear diblock—to develop the ordered structure. The ordering kinetics in block copolymers are related to the mobility of polymer chains and therefore to the presence of entanglements, which can result in the quenching of the star dynamics through the prefactor in Eq. (24).

5. Transitions Between Different Ordered States

Although the ordering kinetics have been well investigated, the transformation dynamics between ordered phases are much less explored. On the theoretical side, a time-dependent Landau–Ginzburg approach (81) has been employed and was successful in identifying the kinetic pathways of some order-to-order transitions and order-to-disorder transitions. Recently, the a self-consistent field theory (SCFT) (82) has also been applied to study the cylinder-to-Gyroid epitaxial transition.

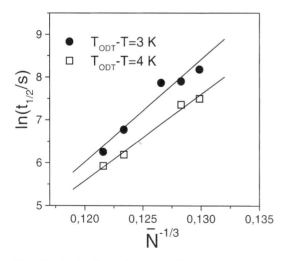

Fig. 20 Molecular weight dependence of the ordering times in the symmetric diblock copolymers. The characteristic time $t_{1/2}$ is plotted vs. $\overline{N}^{-1/3}$ at two temperatures, as indicated (79).

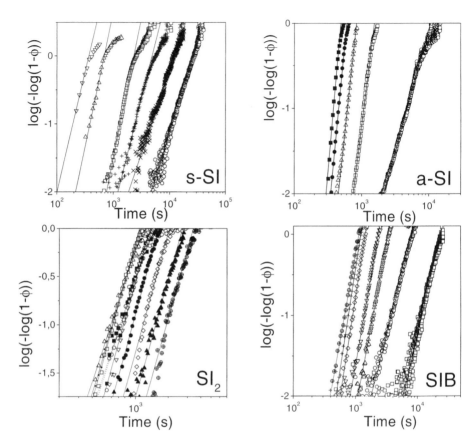

Fig. 21 Time evolution of the volume fraction of the ordered phase calculated using the "series" and "parallel" models (see text). The solid lines are fits to the Avrami equation. The different samples are: s-SI: symmetric SI diblock copolymer; a-SI: asymmetric SI diblock copolymer ($M_n = 23,800$, $f_{PS} = 0.22$); SI_2: miktoarm star-block copolymer ($M_n^{star} = 23,800$, $w_{PS} = 32\%$); SIB: miktoarm star-block copolymer of PS, PI, and PB ($M_n^{star} = 23,800$, $w_{PS} = 34\%$) (67,73).

However, to our knowledge there are very few (59,83) direct experiments that probe the kinetic pathways between amorphous ordered phases after sudden temperature changes.

    Ordered microdomains in block copolymers show a marked variation in elasticity, and this property has been used to identify order-to-order transitions via rheology. On the other hand, the structure factor for each phase is distinctly different, therefore SAXS and SANS can also be used for the structure investigation. Of key importance in the processing of block copolymers is the time period one can undercool or overheat ordered structures ("incubation time"). For example, as we will show later, we can process a block copolymer in the overheated "soft" hexagonal phase for long times before it is transformed to the "hard" Gyroid phase. Similarly, we can process a "soft" undercooled semicrystalline block below the equilibrium melting temperature (T_m°) for long times before it starts to crystallize. The present investigation (59) focuses on a series of recently synthesized poly(isoprene-b-ethylene oxide) diblock copolymers with a strong $\chi(T)$, which makes possible the observation of a variety of order-to-order transitions within an accessible temperature range.

In Fig. 15, the results of two heating runs with different heating rates (1 and 4 K/min) for a PI-PEO diblock copolymer with $M_n = 9800$ and $f_{PI} = 0.61$ have been shown. This copolymer shows the following succession of phases (verified by independent SAXS measurements): L_c (crystalline lamellar phase) $\rightarrow$ Hex (hexagonally packed cylinders) $\rightarrow$ Gyroid (bicontinuous structure with the $I a\bar{3}d$ group symmetry) $\rightarrow$ Dis. The first drop of the storage modulus in Fig. 15 signifies the (apparent) melting point of PEO crystals. In the hexagonal phase there is only a weak $G'(T)$ dependence. At higher temperatures (in the range 380–420 K depending on the heating rate), the modulus starts to increase and develops a maximum corresponding to the Gyroid phase. At even higher temperatures, the storage

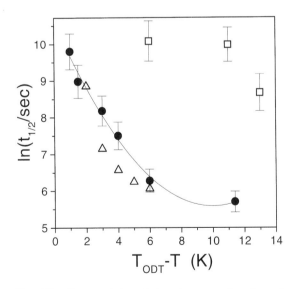

Fig. 22 Characteristic ordering times $t_{1/2}$ plotted vs. the temperature difference from T_{ODT}, for three block copolymers: ($\bullet$) SI-85/65 symmetric diblock ($M_n = 12,200$, $f_{PS} = 0.51$), ($\triangle$) a-SI asymmetric diblock ($M_n = 23,800$, $f_{PS} = 0.22$), and ($\square$) (SI)$_4$ star diblock copolymer ($M_n^{star} = 98,000$, $f_{PS} = 0.25$) (73).

modulus drops at the Gyroid-to-disordered-phase transition at about 486 K, in good agreement with the SAXS results. As discussed earlier, the drop of $G'(T)$ at the $L_c \rightarrow$ Hex and Gyroid $\rightarrow$ Dis. transitions gives only an apparent melting (T'_m) and order-to-disorder (T'_{ODT}) temperatures, respectively.

First we concentrate on the kinetics of the hexagonal-to-L_c transition, aiming to identify the equilibrium melting point (T°_m) and structure of PEO crystals. The experiments involved measuring the storage and loss moduli following T-jumps from an initial temperature of 328 K to four temperatures below T°_m. The evolution of G' and G'' following quenches to 316 and 318 K are shown in Fig. 23 (top). The data show an initial plateau with moduli corresponding to the hexagonal phase. This demonstrates that it is possible to undercool the hexagonal phase from seconds to some hours, in just the same way that we can undercool the disordered phase. On the other hand, the moduli at long times reflect the properties of the final spherulitic material. At intermediate times, the system is regarded as a

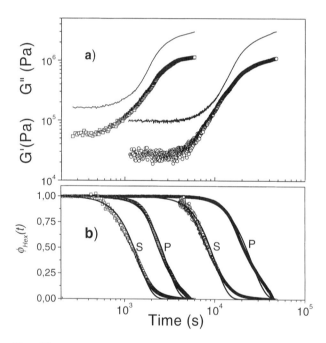

Fig. 23 (a) Time evolution of the storage (symbols) and loss (lines) moduli during isothermal crystallization of PEO in the diblock PI-PEO 5-4. The sample was initially heated to 328 K and quenched to different final temperatures: (□) 316; (○) 318 K, where it was set to crystallize. The frequency was 1 rad/s, and the strain amplitude was 1%. (b) Volume fraction of the hexagonal phase obtained assuming a two-phase system using the series (S) and parallel (P) models. The lines are fits to the Avrami equation (see text) (59).

composite made of two *ordered* phases (amorphous hexagonal microstructure and spherulitic superstructure), the proportions of which change with time, causing the observed "S" shape. As with the ordering kinetics, we need to obtain the evolution of the volume fraction ϕ of the final spherulitic structure (or of the Hex microstructure). We again use the two limiting cases, but this time for the different microstructures. According to the parallel model:

$$G^* = \varphi_{Sph}G^*_{Sph} + \varphi_{Hex}G^*_{Hex} \tag{25}$$

where "Sph" and "Hex" stand for the spherulitic and Hexagonal microstructures, respectively. In the case of the series model:

$$J^* = \varphi_{Sph}J^*_{Sph} + \varphi_{Hex}J^*_{Hex} \tag{26}$$

The $\phi_{Hex}(t)$ obtained from the two models are analyzed independently by fitting them to the Avrami equation $\varphi_{Hex}(t) = e^{-\alpha t''}$. The result from the fits to the series and parallel models are shown in Fig. 23 (bottom). Irrespective of the model, the value of the Avrami exponent is 3.0 ± 0.2, which in principle implies either a spherulitic growth from heterogeneous nuclei or a disclike growth from homogeneous nuclei. In the present case, where crystallization starts from the microphase-separated melt, the process is heterogeneous and the Avrami exponent reflects solely the formation of a three-dimensional spherulitic structure composed from stacks of lamellar crystallites. Based on the information from the kinetics, the equilibrium melting temperature and crystal structure were investigated (59).

Second, we have studied the kinetics of the hexagonal-to-Gyroid transition, by both SAXS and rheology, by imposing sudden temperature changes from an initial temperature corresponding to the hexagonal structure (353 K) to different temperatures corresponding to the Gyroid phase. Here we review the results of the rheological investigation. In this study we take advantage of the large viscoelastic contrast between the hexagonal and Gyroid phases (Fig. 15). The results for $G'(t)$ from two isothermal/isochronal kinetic experiments are shown in Fig. 24. At each temperature, a two-step structure is evident in the $G'(t)$ data. The initial plateau value around 10^4 Pa corresponds to the properties of the initial hexagonal phase (Fig. 15). The long-time plateau is attributed to the equilibrium final (Gyroid) structure, and the increase of the final plateau values with increasing temperature is consistent with the isochronal heating runs of Fig. 15. Temperature stabilization from the initial temperature (353 K) to the different final temperatures takes some 200 s. At about 200 s, the $G'(t)$ data develop an intermediate plateau that is not directly associated with a quenched hexagonal phase, since the $G'(t)$ values at this plateau are higher than 10^4 Pa and display a strong T dependence that contrasts with the weak $G'(t)$ dependence in the hexagonal phase. The formation of this structure is very fast, and part of the kinetics is coupled with the T stabilization and will not be discussed further. The kinetics of the second process, however, are much slower and can be investigated well with rheology.

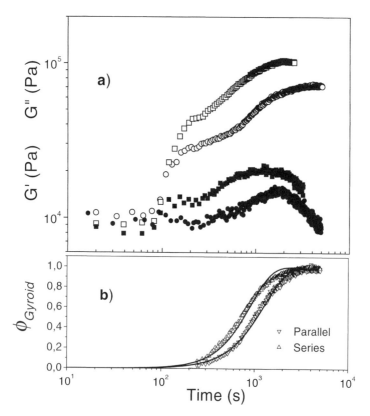

Fig. 24 (a) Isothermal (ω = 1 rad/s, strain amplitude = 0.5%) kinetic experiments of the hexagonal-to-Gyroid transition in the PI-PEO 5-4: (open symbols) G', (filled symbols) G''. The initial temperature was 353 K and the storage modulus was monitored for the different jumps to final temperatures: ($\bigcirc$) 418; ($\square$) 428 K. (b) Evolution of the volume fraction of the Gyroid phase calculated at 418 K using the parallel ($\triangledown$) and series ($\triangle$) models. The lines are fits to the Avrami equation (59).

In order to extract the characteristic times from this process, we have used the series and parallel models (Eqs. 25 and 26) for the Hex and Gyroid phases, and the evolution of the resulting volume fraction of the Gyroid phase is plotted in Fig. 24 (bottom). The $\phi_{Gyroid}(t)$ is then fitted to the Avrami equation for the "slow" process, and the results of the fit are shown in Fig. 24 with the lines. The Avrami exponent corresponding to the "slow" process is 2 $\pm$ 0.1, which implies anisotropic growth of either a disclike object from heterogeneous nuclei or a rod-like object from homogeneous nuclei. If the Gyroid phase was grown from the disordered phase, then the expected exponent would be 3 or 4, in view of the 3-dimensional structure. The lower exponent can be qualitatively understood if we consider that the Gyroid phase is formed epitaxially on an existing ordered structure (heterogeneous nucleation). The characteristic half-times of the order-to-order transition, obtained as $t_{l_i} = (\ln 2/\alpha)^{1/n}$, were very similar to the SAXS times with an activation energy of 47 kcal/mol characteristic of a collective process. The similarity of the activation energies obtained in the two kinetic experiments (~47 kcal/mol) with the energy involved in the viscoelastic properties of the Gyroid phase (~60 kcal/mol) may suggest that the dynamics of the phase transformation (Hex → Gyroid) are controlled by the same collective motions associated with the bicontinuous cubic phase.

Synchrotron x-ray experiments revealed that the dynamics of the "fast" process are likely to originate from fluctuations and/or modulations of the hexagonal phase (84). Furthermore, the nucleation sites of the "slow" nucleation-and-growth process were identified with the help of the recent SCFT (82) treating the Hex-to-Gyroid transition. The theory identifies as nucleation sites the formation of a single fivefold junction. The formation of such junction points involves a very cooperative motion of three neighboring cylinders, which may account for the high activation energy found in the present study. Furthermore, the growth mechanism proposed by the theory is highly anisotropic, since the Gyroid phase grows predominantly along the cylinder axis, which is also in agreement with our experimental findings. In the present study, beyond the description of the type of process, the associated time scale and activation energy of the phase transformation are also determined.

In conclusion, the investigation of the dynamics of the order-to-order transitions was facilitated by the marked variation in elasticity (rheology) and structure (SAXS) of the three phases involved (L_c, Hex, Gyroid). We have studied the dynamics of the Hex-to-L_c and Hex-to-Gyroid transitions. We find that it is possible to undercool and overheat the Hex phase. More specifically:

1. The Hex-to-L_c transition proceeds via a heterogeneous nucleation-and-growth process and results in the formation of a spherulitic superstructure composed from stacks of lamellar crystals.
2. The Hex-to-Gyroid transition occurs in two steps. The "fast" process involves fluctuations/modulations of the Hex phase, and the time scale is too fast to be studied by rheology. The "slow" process can be investigated well with SAXS and rheology and is described as a nearly epitaxial process involving the nucleation and growth of elongated (anisotropic) objects. The time scale and activation energy have been identified.

D. Block Copolymer/Homopolymer Blends

Recent experiments (85) in diblock copolymer/homopolymer blends have shown that the order-to-disorder transition temperature of a diblock copolymer can be shifted by adding a small amount of homopolymer A or B. Herein, we review the process of microphase separation in diblock copolymer/homopolymer blends that have been investigated both theoretically and experimentally (85). The system is an asymmetric diblock copolymer (SI, $M_n = 23,800$, $f_{PS} = 022$), and the homopolymer concentration was kept to below 25%. The degree of polymerization of the added homopolymer (N_h) covered all possible cases: $N \approx N_h$, $N > N_h$, and $N < N_h$ (N is the degree of polymerization for the diblock).

First, the transition point (temperature) was obtained theoretically (in collaboration with A. E. Likhtman and A. N. Semenov (85)) by calculating the free energies of the system in the disordered (F_{dis}) and microphase-separated states (F_{ord}). Neglecting macrophase-separation effects, the transition temperature is defined by:

$$F_{dis} = F_{ord} \tag{27}$$

These energies for a block copolymer/homopolymer mixture were calculated considering an *ideal* system (with no excluded volume and other interactions between monomers) as a reference state. Using the standard mean field Flory–Huggins model for the monomer–monomer interactions (thus neglecting fluctuation effects), the excess free energy (per monomer) in the disordered state is:

$$F_{dis} = f(1 - \varphi)(1 - f(1 - \varphi))(\chi N) \tag{28}$$

where $f(1 - \varphi)$ is the total volume fraction of A monomers, f is the volume fraction of A blocks in the diblock, and φ is the volume fraction of the added homopolymer B. The free energy of the microphase-separated state cannot be calculated exactly analytically, even within the Flory–Huggins model. The free energy in the strong segregation limit (SSL) (when the thickness of interfacial layers is much smaller than the period of the structure) can be represented as the sum of an interfacial (or surface) term, an elastic (or conformation) term, and an ideal-gas term:

$$F_{ord} = F_{surface} + F_{conformation} + F_h \tag{29}$$

Different cases have been considered (addition of the minority or of the majority phase), and the result for the critical values of χN have been calculated for the specific cases studied in the experiment (see later) and are depicted in Fig. 25. These theoretical predictions for $(\chi N)_c(\varphi)$ are then compared with the following experimental results.

On the experimental side, we performed isochronal measurement of the storage modulus G' with small-strain amplitude in order to identify the T_{ODT}. The result of such measurements, at $\omega = 1$ rad/s, for the SI diblock and the four blends with homopolymers PS ($M_n = 3300$) and PI ($M_n = 4700$) are shown in Fig. 26a. There is a discontinuous drop of G' at the order–disorder transition for all systems except for the blend SI/IB (blend composition 75/25 in SI/PI), where $G'(T)$ change continuously with T. A rough estimate of the T_{ODT} for this system is about

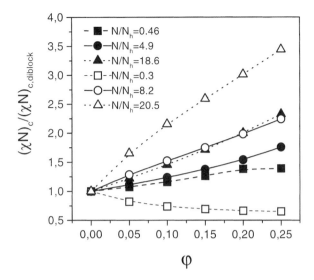

Fig. 25 Critical value of $_{\chi}N$ in diblock copolymer/homopolymer blends relative to the same product in the neat diblock copolymer shown for different ratios N/N_h, which correspond to the experiment: open symbols correspond to $f = 0.75$, solid symbols to $f = 0.25$ (85).

345 K, and the absence of a steplike decrease in $G'(T)$ is explained by the closeness to the polystyrene T_g. The isochronal measurements are in agreement with the SAXS results (not shown here) and show that addition of the minority (majority) phase increases (decreases) the T_{ODT} for the case $N \geq N_h$. Moreover, there is a quantitative agreement in the transition temperatures from the two experiments (SAXS and rheology), with only the exception of the blend SI/IA (blend composition 87/13 in SI/PI), where a difference of about 8 K is found.

Blends of the diblock SI with homopolymers of different molecular weights and for a homopolymer volume fraction up to 0.25 have been investigated via rheology. The order–disorder transition temperatures are summarized in Fig. 26b as a function of the homopolymer volume fraction φ. Independent of the homopolymer length, addition of the majority component brings disorder (lowers T_{ODT}) into the system, and this is in good agreement with the theoretical predictions (Fig. 25). Therefore, the addition of small amounts of the majority component to a diblock copolymer is an efficient way to increase the block compatibility. The same trend is observed, in both experiment and theory, when the minority component is added with $N \gg N_h$. On the other hand, the addition of the minority component with $N < N_h$ brings order into the system. There is again agreement between theory (Fig. 25) and experiment. The only case where a disagreement is found is

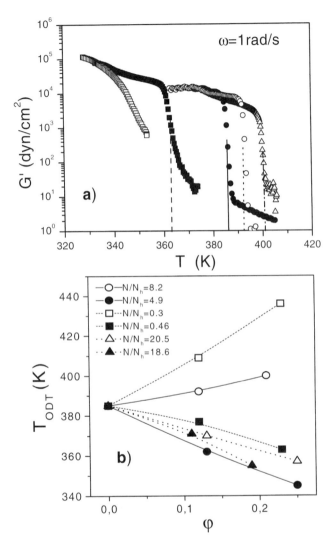

Fig. 26 (a) Temperature dependence of the storage modulus G' obtained at $\omega = 1$ rad/s while heating the diblock copolymer SI (●) and the mixtures with PI: SI/IA (■) (blend composition 88/12 in SI/PI), SI/IB(□) (blend composition 75/25 in SI/PI), and PS: SI/SA(○) (blend composition 88/12 in SI/PS), SI/SB(△) (blend composition 79/21 in SI/PS). The ODTs are indicated by the vertical lines. (b) Order–disorder transition temperatures for diblock copolymer/homopolymer blends with $\varphi \leq 0.25$ and for different ratios of N/N_h as indicated. The points are obtained from rheology (see text). Open symbols: addition of PS; solid symbols: addition of PI (85).

when $N \sim 8N_h$. As expected, the theoretical predictions are in good agreement with the experiment in the limiting cases: $N >> N_h$ and $N << N_h$.

In conclusion, the order-to-disorder transition for an asymmetric diblock copolymer mixed with homopolymer was found to shift systematically and the *sign* and *amount* of shift were found to depend on the *type* and *length* of added homopolymer. In general, addition of the *majority* component *decreases* the T_{ODT}. This is a composite effect owing to an increase in the free energy of the ordered diblock and to a decrease in the disordered free energy. Addition of the minority component can decrease ($N > N_h$) or increase ($N < N_h$) the T_{ODT}. The latter results from the sharp decrease of the ordered diblock free energy with the addition of homopolymer. These experimental results are in good agreement with theoretical calculations and demonstrate that the degree of compatibility between the two blocks can be effectively controlled by adding small amounts of homopolymers.

E. Block Copolymers Substituted with Strongly Interacting Groups

End-functionalized diblock copolymers with short but strongly interacting groups (which can be considered as model ABC block copolymers) provide a new way of manipulating the phase state by triggering the interaction parameter and thus the intrinsic compatibility between the different blocks. The first experiments (86,87) and theoretical considerations (88,89) on the phase behavior have already appeared. The model system that we have considered (86) is an ω-functionalized asymmetric SI diblock copolymer with a zwitterion group at either end of the chain. The molecular weights were in the range 0.62–2.44×10^4 g/mol. Depending on the temperature, the SAXS results revealed two separate levels of microphase separation, one between the polystyrene (PS) and polyisoprene (PI) blocks forming the microdomain structure and another one between ionic and nonionic material. The latter process creates sufficient contrast, notwithstanding the small fraction of the zwitterionic groups. When the zwitterion is linked to the PI chain end, aggregates are formed at low temperatures within the PI phase. When the zwitterion is located on the PS chain end, association takes place, at high temperatures, within the PS phase and acts as to stabilize the new microdomain up to very high temperatures. Figure 27 shows a schematic of the microstructures obtained with the ω-functionalized SI block copolymers, which is based on the SAXS study, for the cases where the zwitterion (Zw) is linked to the I block or to the S block.

The SAXS picture of aggregates formed within the "soft" phase, when the functional group is attached to the I-block (Zw-IS), or within the "hard" phase, when the functional group is attached to the S-block (Zw-SI), is supported by dy-

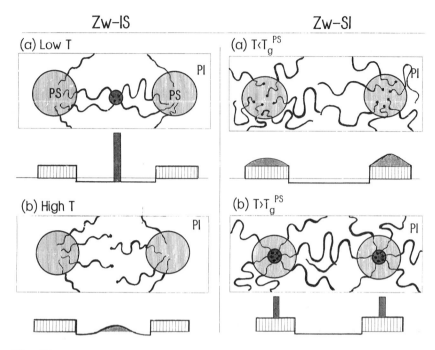

Fig. 27 Schematic illustration of the microstructures in ω-functionalized SI block copolymers, showing the Zw-IS (left) and Zw-SI (right) cases at low (upper) and high (lower) temperatures. The corresponding electron density distributions are also shown (86).

namic studies (86). Figure 28 gives the result of the isochronal measurements for the dimethylamino- (N-IS#5) and zwitterionic- (Zw-IS#5) substituted copolymer IS#5 (M_n = 12,200, f_{PS} = 0.25). Notice that the two samples have the same composition and molecular weight. As expected, the isoprene segmental relaxation at low T is identical in the two copolymers. However, the plateau zone in Zw-IS is extended due to ionic aggregates, which act as physical crosslinks, and the flow zone is shifted by about 50 K. Furthermore, the T dependence of the storage and loss moduli in the terminal zone is distinctly different, with a weak T dependence in the case of Zw-IS and with no sign for an order-to-disorder transition within the investigated temperature range.

When compared with neutral diblocks, the ω-functionalized diblock copolymers constitute a new class of materials that provide the possibility of altering the phase behavior by introducing a small amount of a polar group at one chain end.

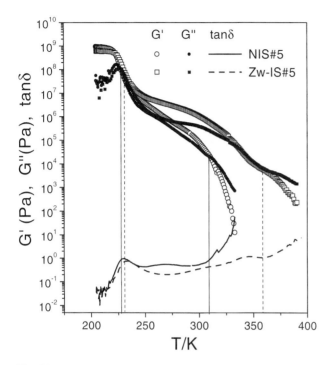

Fig. 28 Temperature dependence of the storage (open symbols) and loss (filled symbols) moduli at 1 rad/s as well as of the loss tangent for the end-functionalized block copolymer ω-IS#5 (M_n = 12,200, f_{PS} = 0.25) with the dimethylamino and zwitterionic groups.

VI. SHEAR-INDUCED ORIENTATIONAL ORDER

A. Experimental Observations

It is well known that shear (22–26,90–93) is very effective in inducing macroscopic orientation in block copolymers. When a symmetric diblock copolymer, microphase-separated to lamellar microdomains, is subjected to a high-amplitude oscillatory shear between parallel plates, a macroscopic alignment of lamellae can be observed (see Fig. 29). The orientation of the structure is usually detected by means of small-angle x-ray scattering performed with the primary beam incident along various directions with respect to the characteristic directions of the deformation geometry. Figure 29 shows schematically characteristic scattering patterns that can be observed for a sample element (deformed between parallel plates) when analyzed with the x-ray incident beam directed along the radial (X), tangential (Y), and normal (Z) directions, as indicated. In the case of lamellae

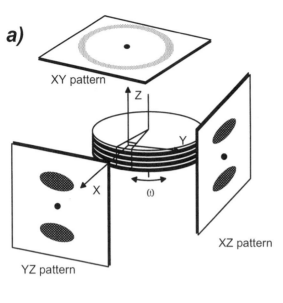

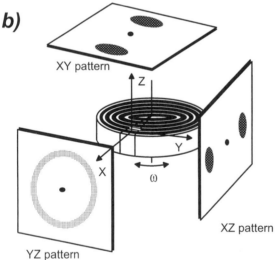

Fig. 29 Schematic illustration of structures detected in a microphase-separated copolymer melt subjected to large-amplitude oscillatory shear flow between parallel plates in a rotation rheometer. SAXS patterns recorded with various directions of the incident beam are shown schematically for: (a) "parallel" alignment observed at low- and high-deformation frequencies; (b) "perpendicular" alignment observed at intermediate deformation frequencies.

aligned parallel to the rheometer plates ("parallel" alignment, Fig. 29a), the scattering patterns YZ and XZ show two point diagrams with intensive maxima, indicating periodicity of oriented lamellae, whereas the XY scattering pattern shows only an isotropic, low-intensity halo. The structure with lamellae perpendicular to the rheometer plates ("perpendicular alignment," Fig. 29b) is indicated by two point-scattering patterns recorded with the incident beam along the Z and Y directions.

The two kinds of lamellar orientations have been observed in diblock copolymer melts as appearing in different ranges of deformation frequencies (22–26). In SI diblock copolymers, for example, at low and high deformation frequencies the "parallel" alignment of lamellae was detected; in an intermediate-frequency range, extending over almost two decades of frequency, the "perpendicular" alignment was observed (23,94) (Fig. 30). Although the dependence of lamellar alignment on frequency was correlated with the dynamic behavior of the copolymers, studied by means of mechanical spectroscopy methods, the explanation of the observed behavior was an open problem, and therefore simulations of the block copolymer flow have been performed (41).

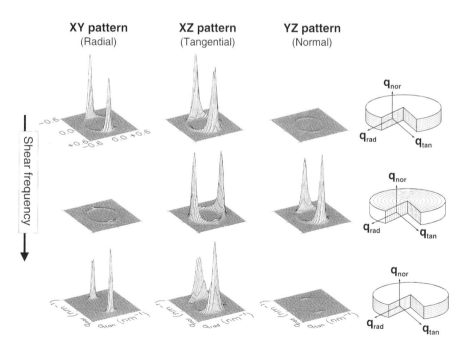

Fig. 30 SAXS patterns recorded at various directions showing the sequence of parallel-perpendicular-parallel orientations of lamellae with increasing deformation frequency. (From Ref. 23.)

B. Simulated Systems

In order to simulate the shear flow, the systems were placed between two parallel, impenetrable walls normal to the Z direction (Fig. 31). In the two other directions (X and Y), the systems remained artificially unlimited due to periodic boundary conditions. Shear stress is applied by assuming an asymmetry of probabilities of segmental motions along the Y direction with a linear gradient along the Z direction, as illustrated in Fig. 31a. Probabilities of segmental displacements in the positive and negative directions of the Y axis are assumed to be

$$p_{\pm} = p_0 \left[1 \pm \sigma\left(\frac{2z - d}{d}\right) \right] \tag{30}$$

where p_0 is a probability of a displacement in a given direction in a nonstressed state, d is the layer thickness, and σ is considered an external stress. The probabilities of displacements in other directions are left symmetric, as in the nonstressed state. It has been proven that under such conditions in a system of nonbounded beads (simple liquid), a Newtonian simple shear flow is achieved. Structures of systems under flow are illustrated by means of projections of chains on three planes perpendicular to the X, Y, and Z coordinates, as illustrated in Fig. 31b. A color contrast between elements constituting different blocks is used.

Two cases of different wall properties have been considered (41): (a) neutral walls and (b) walls interacting selectively with the two types of beads, A and B, constituting the copolymer. In the second case, walls have been assumed as having properties of component A, which means that they interact repulsively with component B.

The systems simulated under shear flow have been characterized by the orientation factors of end-to-end vectors of diblock chains with respect to various coordinates of the system (i: x,y,z):

$$f_i = \tfrac{1}{2}(3 \langle \cos^2 \theta_{Rri} \rangle - 1) \tag{31}$$

Figure 31c and d shows dependencies of these quantities on the shear stress for a system subjected to shear flow between neutral walls, at temperature $T/N = 1$, i.e., far above the order-to-disorder transition temperature. Results in Fig. 31c show that the initially nearly isotropic system becomes oriented under shear flow with chain axes aligning toward the shear direction (the small difference between f_z and f_x for other directions is caused by a slight orientation of chains at surfaces). The system remains in a disordered state, as indicated by E_m values (Fig. 31d) considerably higher then the E_m corresponding to the temperature of the order-to-disorder transition.

Figure 32 shows the same quantities for the diblock copolymer system subjected to shear flow between neutral walls but at the temperature $T/N = 0.3$, below the order-to-disorder transition, i.e., in a state with well-developed lamellar

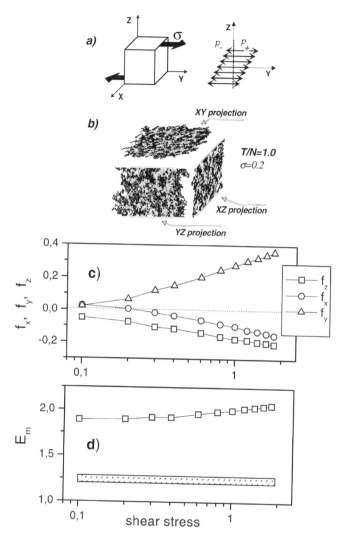

Fig. 31 (a) Illustration of the shear flow geometry and the distribution of motion probabilities along the Y direction. (b) Projections of polymer chains on three planes XY, XZ, and YZ, illustrating the structure in a model system at a stationary state under flow at $T/N = 1.0$. The copolymer constituents A and B are plotted in different colors, black and gray, respectively. (c) and (d) represent shear stress dependencies of (c) orientation factors of end-to-end vectors of diblock copolymer chains with respect to axes of the coordinate system and (d) mean interaction energy per monomer. The horizontal bar in (d) corresponds to the interaction energy level at which the order–disorder transition takes place. Results are obtained for stationary states of the model system subjected to flow between neutral walls at $T/N = 1.0$, i.e., far above the order-to-disorder transition temperature (41).

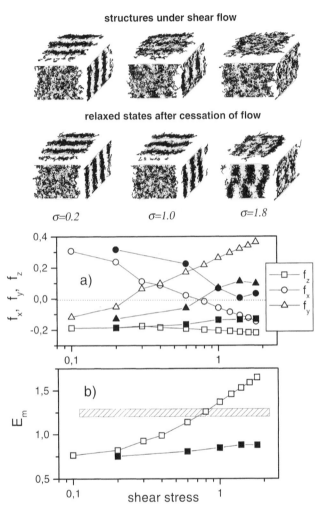

Fig. 32 Shear stress dependencies of (a) orientation factors of end-to-end vectors of diblock copolymer chains with respect to axes of the coordinate system, and (b) mean interaction energy per monomer. The horizontal bar in (b) corresponds to the interaction energy level at which the order-to-disorder transition takes place. The results obtained for stationary states of the model system subjected to flow are represented by open symbols; the results describing the states of the system obtained after cessation of flow are represented by filled symbols. The system subjected to flow between neutral walls, at $T/N = 0.3$, i.e., within the microphase-separated regime, is presented. Morphologies in the block copolymer systems under flow between neutral walls and after cessation of flow are shown. For the orientations of models with respect to flow geometry, compare with Fig. 31. The copolymer constituents A and B are plotted in different colors, black and gray, respectively. Results for $T/N = 0.3$, and various shear stresses are shown.

morphology. The observed behavior of the system under such conditions is considerably different than in the former case. At small shear stresses, independent of the orientation of lamellae before deformation, the system assumes a unique orientation, with chain axes aligned preferentially along the X direction. It indicates that the lamellar structure becomes oriented with interfaces perpendicular to the walls and aligned with the shear direction (the "perpendicular" alignment). In this low-stress range, the interaction energy remains at the level characteristic for a nondeformed system, indicating that the perfection of the morphology is not changed by the low shear flow rate.

At higher stress, the situation changes considerably. The orientation of chains toward the shear direction starts to increase and becomes dominating. The interaction energy increases with increasing stress, which indicates that the system becomes less ordered. At the stress values at which the interaction energy exceeds the level characteristic for the order-to-disorder transition temperature, the system can be considered as disordered. In this range the orientation correlation of chain axes with the X direction disappears. These results show that, when the system is subjected to shear flow between neutral walls, there is only one characteristic orientation at low shear stress, which is spontaneously assumed. This orientation and the microstructure can be distorted by high shear stresses, when the alignment of chains along the flow direction becomes dominating. A necessary translation of oriented flowing chains prohibits the formation of a lamellar morphology with another orientation. Structures of systems at various shear stresses are illustrated in the upper part of Fig. 32.

A considerable change of behavior of diblock copolymer systems under flow is observed when selective interactions of components with walls are taken into account (Fig. 33). It has been observed qualitatively that even a nondeformed diblock copolymer melt, when annealed between selectively interacting walls, relaxes to a state with lamellae parallel to the walls. Under shear flow with small stresses (low shear rates), such a system preserves the parallel orientation of the lamellae with respect to the walls. This is clearly indicated by the positive values of f_z and the negative values of both f_x and f_y. At a certain shear stress, however, a spontaneous reorientation in the system takes place to a state with the lamellae perpendicular to the walls and aligned with the shear direction. This is observed at relatively low shear stresses, which still do not involve any remarkable orientation of chains along the direction of shear. Further increase of the shear stress destroys the morphology by increasing the alignment of chains in the shear direction, as indicated both by the increase of E_m above the level characteristic for the order-to-disorder transition and by the considerable increase in f_y. Structures corresponding to stationary states of systems flowing under various shear stresses are illustrated in the upper part of Fig. 33.

All states just described are stationary, i.e., stable under applied stress. If

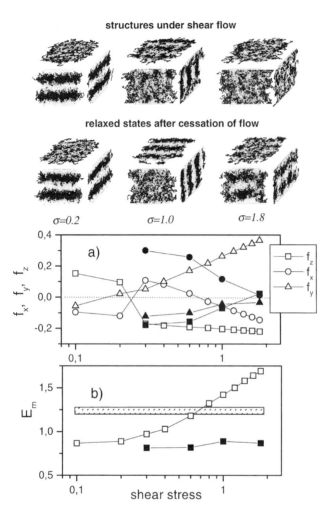

Fig. 33 Shear stress dependencies of (a) orientation factors of end-to-end vectors of diblock copolymer chains with respect to axes of the coordinate system, and (b) mean interaction energy per monomer. The horizontal bar in (b) corresponds to the interaction energy
level at which the order-to-disorder transition takes place. Results obtained for stationary
states of the model system subjected to flow are represented by open symbols; results describing states of the system obtained after cessation of flow are represented by filled symbols. The system subjected to flow between walls having properties of component A, i.e.,
interacting repulsively with component B, at $T/N = 0.3$, is presented. Morphologies in the
simulated block copolymer system under flow and after cessation of flow are shown by
means of projections of polymer chains on three planes XY, XZ, and YZ. For the orientations of models with respect to flow geometry, compare with Fig. 31. The copolymer constituents A and B are plotted in different colors, black and gray, respectively. Results for
$T/N = 0.3$, and various shear stresses are considered.

stresses are removed, the systems relax to other states, which, in the range of high stresses, differ considerably from these under flow. The relaxed states after cessation of flow with various stresses are characterized in Figs. 32 and 33, by solid symbols, with the shapes corresponding to open symbols characterizing the systems before the cessation of flow. It is interesting to notice that different orientational situations are obtained, depending on both the shear stress under which the system was previously deformed and the type of interactions with the walls. In the case of neutral walls, a new type of orientation appears with lamellae nearly perpendicular to the flow direction (Fig. 32). In the case of interacting walls, two different ranges of the behavior can be distinguished: the intermediate shear stress range, with the lamellae perpendicular to the walls but aligned with the shear direction ("perpendicular" alignment), and the high stress range, with the lamellae parallel to the walls (Fig. 33). In all cases just considered, a constant period of time, comparable to the relaxation time of the copolymer chain in a homogeneous melt, was left to the systems to relax after cessation of flow. The structures observed might, therefore, still be unstable to some extent and can further relax to more perfect morphologies when annealed over longer time periods.

The results presented here have shown that, in simulated systems, morphological rearrangements similar to those observed experimentally can be obtained when systems are subjected to shear flow with various shear rates. It is demonstrated that the interaction of copolymer constituents with the walls can considerably influence the observed structure orientations. This interaction can control the structure at low shear rates as well as the structure to which a system, disordered under high shear rate, relaxes after cessation of flow. Only between neutral walls (a case that is not realistic for experiments) or at intermediate deformation rates is the perpendicular orientation observed, which can be considered as controlled by the anisotropic dynamics of chains within lamellar microdomains. In this case, the higher mobility of chains in the direction parallel to the interfaces enables the microphase-separated system to flow easily only when lamellae become aligned with the flow direction and are perpendicular to the walls. Under high shear stresses leading to high flow rates, the microphase-separated system can be disordered. After cessation of flow, the fast relaxation of orientation leads to microstructures with orientations dependent on interactions of the copolymers with the walls. The foregoing observations can be summarized by means of the schematic diagram shown in Fig. 34, where the appearance of lamellar morphologies with various orientations with respect to flow direction is illustrated within the field of two parameters: the shear stress, and the strength of the selective interaction of one block with the layer surface. Positions of transitions between different structures represented by dashed lines can additionally be influenced by the thickness of the layer and by the chain length, as indicated. In this scheme, structures after cessation of flow and relaxation are considered. This scheme indicates that the possibility of observing various morphological orienta-

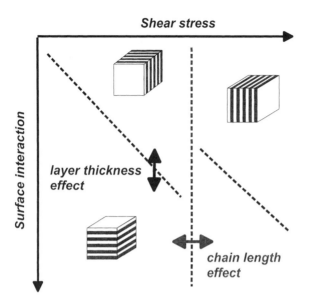

Fig. 34 Schematic representation of areas of appearance of different lamellar orientations induced by shear within a parameter field of the shear stress, and the strength of selective interaction of one component with the surface. Transitions between various orientations are shown by the dashed lines.

tions depends on many parameters, therefore, various sequences of structural rearrangements should be possible, depending on the experimental conditions, on the type of copolymer studied, as well as on the relaxation conditions after cessation of flow.

VII. CONCLUSIONS

We believe we have demonstrated in this chapter that the dynamics and related viscoelastic properties of block copolymers are very sensitive to properties of the extremely rich variety of structural states. The latter are controlled by a large number of parameters, both intrinsic, such as block dissimilarity, composition, and architecture, and extrinsic, such as temperature and flow fields. Due to this sensitivity, the analysis of the viscoelastic properties of copolymers provides details about the thermodynamic state, the location of the phase transitions, and the phase transformation dynamics. When these results are combined with structural investigations, a complete picture of the phase state and the associated dynamics emerges.

It has also been shown that the simulation of such complex systems, when performed using an appropriate simulation method, not only can mimic the structure and dynamic behavior of real systems but in addition can bring insight on a molecular level. This is possible through the selection of parameters and the monitoring of quantities that are not always accessible in the experiments.

ACKNOWLEDGMENTS

We would like to thank: N. Hadjichristidis, S. Pispas, U. Wiesner, R. Ulrich, A. E. Likhtman, A. N. Semenov, I. Erukhimovich, and A. Khokhlov for participation in parts of this work.

REFERENCES

1. L Leibler. Macromolecules 13, 1602, 1980.
2. GH Fredrickson, E. Helfand. J. Chem. Phys. 87, 697, 1987.
3. M Olvera de la Cruz, IC Sanchez. Macromolecules 19, 2501, 1986.
4. AN Semenov. Sov. Phys. JETP 61, 733, 1985.
5. AV Dobrynin, IYa. Erukhimovich. Macromolecules 26, 276, 1993.
6. MW Matsen, FS Bates. Macromolecules 29, 1091, 1996.
7. FS Bates, GH Fredrickson. Annu. Rev. Chem. 41, 525, 1990.
8. RH Colby. Curr. Opin. Coll. Int. Sci. 1, 454, 1996.
9. RP Zielinski, CW Childers. Rubber Chem. Technol. 161, 41, 1968.
10. G Kraus, KW Rollmann. Angew. Macromol. Chem. 271, 16/17, 1971.
11. T Hashimoto, Y Tsukahara, Y Tachi, H Kawai. Macromolecules 16, 648, 1983.
12. NR Legge, G Holden, HH Schroeder, eds. Thermoplastic Elastomers. Hanser Verlag, Munich, 1987.
13. PE Gibson, MA Vallance, SL Cooper. In: J Goodman, ed., Developments in Block Copolymers—I. Applied Science, London, 1982.
14. D Greszta, K Matyjaszewski, T Pakula. Polym. Prepr. (Amer. Chem. Soc.) 37(1), 569, 1996.
15. T Pakula, K. Matyjaszewski. Macromol. Theory Simul. 5, 987, 1996.
16. H Iatrou, N. Hadjichristidis. Macromolecules 25, 4649, 1992; 26, 2479, 1993.
17. A Avgeropoulos, BJ Dair, N Hadjichristidis, EL Thomas. Macromolecules 30, 5634, 1997.
18. K Karatasos, S Anastasiadis, G Floudas, G Fytas, S Pispas, N Hadjichristidis, T Pakula. Macromolecules 29, 1326, 1996.
19. G Fleischer, F Fujara, B Stühn. Macromolecules 26, 2340, 1993.
20. TP Lodge, MC Dalvi. Phys. Rev. Lett. 75, 657, 1995.
21. MW Hamersky, MA Hillmyer, M Tirrell, FS Bates, TP Lodge, ED von Meerwall. Macromolecules 31, 5363, 1998.
22. KA Koppi, M Tirrell, FS Bates, K Almdal, RH Colby. J Phys. II 2, 1941, 1992.

23. Y Zhang, U Wiesner, HW Spiess. Macromolecules 28, 778, 1995.
24. VK Gupta, R Krishnamoorti, JA Kornfield, SD Smith. Macromolecules 28, 4464, 1995.
25. KI Winey, SS Patel, RG Larson, H Watanabe. Macromolecules 26, 4373, 1993.
26. D Maring, U Wiesner. Macromolecules 30, 660, 1997.
27. T Jian, AN Semenov, S Anastasiadis, G Fytas, F-J Yeh, B Chu, S Vogt, F Wang, JEL Roovers. J. Chem. Phys. 100, 3286, 1994.
28. AZ Akcasu, M Benmouna, H Benoit. Polymer 27, 1935, 1986. AZ Akcasu, M Tombakoglu. Macromolecules 23, 607, 1990. AZ Akcasu. Macromolecules 24, 2109, 1991.
29. R Borsali, TA Vilgis. J. Chem. Phys. 93, 3610, 1990.
30. GH Fredrickson, E Helfand. J. Chem. Phys. 89, 5890, 1988.
31. RG Larson, GH Fredrickson. J. Chem. Phys. 86, 1553, 1987.
32. M Benmouna, TA Vilgis. Makromol. Chem. Theory Simul. 1, 333, 1992. U. Genz, TA Vilgis. J. Chem. Phys. 101, 7111, 1994.
33. TA Vilgis, A Weyersberg, MG Brereton. Phys. Rev. E 49, 3031, 1994.
34. MG Brereton, TA Vilgis. Macromolecules 29, 7588, 1996.
35. H Tang, K Schweizer. J. Chem. Phys. 103, 6296, 1995.
36. H Fried, K Binder. Europhys. Lett. 1991, 16, 237; J. Chem. Phys. 1991, 94, 8349.
37. K Binder, H Fried. Macromolecules 1993, 26, 6878.
38. A Gauger, A Weyersberg, T Pakula. Makromol. Chem. Theory Simul. 1987, 2, 531. A Weyersberg, TA Vilgis. Phys. Rev. E 1993, 48, 377.
39. RG Larson. Macromolecules 27, 4198, 1994.
40. T Pakula, K Karatasos, SH Anastasiadis, G Fytas. Macromolecules 30, 1326, 1996.
41. T Pakula. J. Macromol. Sci. Phys. B37(2), 181, 1998.
42. T Pakula. J. Comp. Aided Mat. Des. 3, 329, 1996.
43. T Pakula. Macromolecules 20, 679, 1987. T Pakula, S Geyler. Macromolecules 20, 2909, 1987.
44. T Pakula, S Geyler, T Edling, D Boese. Rheol. Acta 35, 631, 1996.
45. AM Mayes, MO de la Cruz. J. Chem. Phys. 95, 4670, 1991.
46. H Ladynski. W De Odorico, M Stamm, J Non-Cryst. Solids 491, 235–237, 1998.
47. K Shull, E Kramer, FS Bates, JH Rosedale. Macromolecules 24, 1383, 1991.
48. D Ehlich, M Takenaka, S Okamoto, T Hashimoto. Macromolecules 26, 189, 1993.
49. JH Rosedale, FS Bates. Macromolecules 23, 2329, 1990.
50. CD Han, DM Baek, JK Kim, T Ogawa, N Sakamoto, T Hashimoto. Macromolecules 28, 5043, 1995.
51. G Floudas, T Pakula, EW Fischer, N Hadjichristidis, S Pispas. Acta Polymer. 45, 176, 1994.
52. NP Balsara, HJ Dai, H Watanabe, T Sato, K Osaki. Macromolecules 29, 3507, 1996.
53. B Gerharz. Static and dynamic small angle x-ray scattering studies of the phase behavior of semicompatible, newly synthesized poly(styrene-6-methylphenylsiloxane)copolymers. PhD dissertation, University of Mainz, Mainz, 1991.
54. M Rubinstein, SP Obukhov. Macromolecules 26, 1740, 1993.
55. K Kawasaki, A Onuki. Phys. Rev. A 42, 3664, 1990.
56. GH Fredrickson, FS Bates. Annu. Rev. Mater. Sci. 26, 501, 1996.
57. G Floudas, N Hadjichristidis, H Iatrou, A Avgeropoulos, T Pakula. Macromolecules 31, 6943, 1998.
58. JM Johnson, JB Allgaier, SJ Wright, RN Young, M Buzza, TCB McLeish. J. Chem. Soc. Faraday Trans. 91, 2403, 1995.

59. G Floudas, R Ulrich, U Wiesner. J. Chem. Phys. 110, 652, 1999.
60. G Floudas, T Pakula, G Velis, S Sioula, N Hadjichristidis. J. Chem. Phys. 108, 6498, 1998.
61. B Lotz, AJ Kovacs. Polym. Prepr. (Am. Chem. Soc., Polym. Chem. Div) 10, 820, 1969.
62. DJ Quiram, RA Register, GR Marchand, AJ Ryan. Macromolecules 30, 8338, 1997.
63. G Floudas, C Tsitsilianis. Macromolecules 30, 4381, 1997.
64. I Alig, S Tadjbakhsch, G Floudas, C Tsitsilianis. Macromolecules 31, 6917, 1998.
65. TA Ezquerra, J Majszczyk, FJ Balta Calleja, E Lopez-Cabarcos, KH Gardner, B Hsiao. Phys. Rev. B 50, 6023, 1994.
66. G Floudas, S Vogt, T Pakula, EW Fischer. Macromolecules 26, 7210, 1993.
67. G Floudas, N Hadjichristidis, H Iatrou, T Pakula, EW Fischer. Macromolecules 27, 7735, 1994.
68. B Stühn, A Vilesov, HG Zachmann. Macromolecules 27, 3560, 1994.
69. T Hashimoto, N Sakamoto. Macromolecules 28, 4779, 1995.
70. JL Adams, DJ Quiram, WW Graessley, RA Register, GR Marchand. Macromolecules 29, 2929, 1996.
71. N Sakamoto, T Hashimoto. Macromolecules 31, 3815, 1998.
72. N Sakamoto, T Hashimoto. Macromolecules 31, 8493, 1998.
73. G Floudas, S Pispas, N Hadjichristidis, T Pakula, I Erukhimovich. Macromolecules 29, 4142, 1996.
74. JD Ferry. Viscoelastic Properties of Polymers. 3rd ed. Wiley, New York, 1980.
75. MJ Avrami. J. Chem. Phys. 7, 1103, 1939; 8, 212, 1940; 9, 177, 1941.
76. NP Balsara, BA Garetz, HJ Dai. Macromolecules 25, 6072, 1992.
77. G Floudas, G Fytas, N Hadjichristidis, M Pitsikalis. Macromolecules 28, 2359, 1995.
78. GH Fredrickson, K Binder. J. Chem. Phys. 91, 7265, 1989.
79. G Floudas, D Vlassopoulos, M Pitsikalis, N Hadjichristidis, M Stamm. J. Chem. Phys. 104, 2083, 1996.
80. G Floudas, N Hadjichristidis, Y Tselikas, I Erukhimovich. Macromolecules 30, 3090, 1997.
81. S Qi, Z-G Wang. Phys. Rev. Lett. 76, 1679, 1996.
82. MW Matsen. Phys. Rev. Lett. 80, 4470, 1998.
83. S Sakurai, H Umeda, C Furukawa, H Irie, S Nomura, HH Lee, JK Kim. J. Chem. Phys. 108, 4333, 1998.
84. G Floudas, R Ulrich, U Wiesner, B Chu. In preparation.
85. G Floudas, N Hadjichristidis, M Stamm, AE Likhtman, AN Semenov. J. Chem. Phys. 106, 3318, 1997.
86. G Floudas, G Fytas, S Pispas, N Hadjichristidis, T Pakula, AR Khokhlov. Macromolecules 28, 5109, 1995.
87. V Schädler, U Wiesner. Macromolecules 21, 6698, 1997.
88. EE Dormidontova, AR Khokhlov. Macromolecules 30, 1980, 1997.
89. I Erukhimovich, V Abetz, R Stadler. Macromolecules 30, 7435, 1997.
90. A Keller, E Pedemonte, FM Willmouth. Kolloid Z Z Polym. 238, 385, 1970.
91. G Hadziioannou, A Mathis, A Skoulios. Colloid Polym. Sci. 257, 136, 1979.
92. T Pakula, K Saijo, H Kawai, T Hashimoto. Macromolecules 18, 1294, 1985.
93. ZR Chen, JA Kornfield. Polymer 19, 4679, 1998.
94. U Wiesner. Macromol. Phys. 198, 3319, 1997.

7

Structure: Microhardness Correlations in Condensation Copolymers

Francisco J. Baltá Calleja
Instituto de Estructura de la Materia, CSIC, Madrid, Spain

Stoyko Fakirov
University of Sofia, Sofia, Bulgaria

I. INTRODUCTION

The microhardness technique was used for many years for the characterization of such "classical" materials as metals, alloys, and inorganic glasses. Its application to polymeric materials developed during the last several decades. Nowadays, the microhardness technique, being a nondestructive, sensitive, and relatively simple method, enjoys wide application, as can be concluded from the publications on the topic that have appeared during only the last five years—their number is more than 100, as shown by a routine computer-aided literature search. In addition to some methodological contributions to the technique, the microhardness method has been further successfully used to gain a deeper understanding of the microhardness–structure correlation of polymers, copolymers, polymer blends, and composites. A very attractive feature of this technique is its ability for micromechanical characterization of some components, phases, or morphological entities that are otherwise inaccessible to direct determination of their microhardness.

In the last few years the value of microhardness as a technique capable of detecting a variety of morphological and textural changes in crystalline polymers has been amply emphasized, leading to an extensive research. This is so because microindentation hardness is based on plastic straining and consequently is correlated directly with molecular and supermolecular deformation mechanisms occurring locally at the polymer surface. These mechanisms critically depend on the specific morphology of the material. The fact that crystalline polymers are multiphase materials has prompted a new route to identifying their internal structure and relating it to the resistance against local deformation (microhardness).

It seems important to note in these introductory remarks that, like many mechanical properties of solids, microhardness obeys the additivity law (1):

$$H = \sum_i H_i \varphi_i \tag{1}$$

where H_i and φ_i are the macrohardness and mass fraction, respectively, of each component and/or phase. This law can be applied to multicomponent and/or multiphase systems, provided each component and/or phase is characterized by its own H. Equation (1) is frequently used for semicrystalline polymers for one or another purpose, with the H-value of the crystalline H_c or amorphous H_a phase.

The application of the additivity law (Eq. 1) presumes a very important requirement: each component and/or phase of the complex system should have the T_g value above room temperature; i.e., it should be a solid at room temperature and thus capable of producing an indentation after removing the indenter. If this is not the case, it is frequently assumed that $H = 0$, although this may not be the best solution.

The presence in the complex system of a very soft, liquidlike component and/or phase (not having measurable H at room temperature) can affect the deformation mechanism of the entire system in such a way that it deviates from the additivity law. The aim of this chapter is to present some recent applications of the microhardness technique to study the influence of temperature and composition of condensation copolymers on mechanical properties and also to highlight the value of this technique to the study of polymorphic transitions of block copolymers.

II. MEASUREMENT OF MICROHARDNESS: BASIC CONCEPTS

Microindentation with a point indenter involving a deformation on a very small scale is one of the simplest methods of determining the mechanical properties of a material. The method uses a diamond pyramid indenter that penetrates the surface of a specimen upon application of a given load at a constant rate. A convenient measure of the microhardness (H) may be obtained by dividing the peak contact load, P, by the projected area of impression, A, $H = P/A$. The contact pressure when using a pyramid indenter is independent of indent size, and thus affords a convenient measure of the hardness (1). The strain boundaries for plastic deformation below the indenter have been shown to depend on the morphology of the polymer material (crystal size and perfection, degree of crystallinity, etc.). Typical loads of 10–50 mN, using a square-based diamond (Vickers) when applied to the surface of a polymer like poly(ethylene terephthalate) (PET) produce penetration depths of about 1–2 μm. From a macroscopic point of view, H is correlated directly to the yield stress, σ_y, following Tabor's relation $H \sim 3\sigma_y$. This approximation is well substantiated for amorphous and crystallized unoriented polymers when creep is minimized. To minimize the creep of the material surface under the indenter, loading cycles of 0.1–0.2 min are commonly used.

III. MICROHARDNESS OF CONDENSATION COPOLYMERS PREPARED IN VARIOUS WAYS

A. Microhardness of Copolymers Prepared from Monomers: Effect of Composition and Aging

In order to study the effect of composition on microhardness H, a series of random copolymers of poly(ethylene terephthalate) (PET) and poly(ethylene naphthalene-2,6-dicarboxilate) (PEN) over a wide range of compositions having 10%, 20%, 30%, 50%, 80%, and 100% PEN mole composition have been synthesized, starting with the monomers (2).

It should be noted that the whole-series of PET/PEN copolyesters is completely amorphous when quenched from the melt, as revealed by the wide-angle x-ray scattering (WAXS) patterns (3). However, when annealed at high temperature, some of the samples are capable of crystallizing. Figure 1 illustrates the WAXS patterns of the annealed copolyester series as a function of composition. For the samples containing from 0–30 mol % PEN, the PET sequences crystallize while the PEN segments remain in the amorphous regions. In contrast, in the samples containing 80% and 100% PEN, the PEN sequences crystallize in the α-polymorphic form (4). The PET segments are excluded here in the noncrystalline regions. Only the samples containing 50% and 60% PEN (the latter is not shown in Fig. 1), annealed at high temperature, cannot crystallize.

Figure 2 shows the variation of the microhardness of the quenched amorphous samples, H_a, and that of the crystallized samples, H, as a function of PEN content. It is seen that H_a increases linearly with the increasing concentration of PET units, w_{PET}, according to the predictions of the additivity law (parallel model) (Eq. 1):

$$H_a = H_a^{PET}\, w_{PET} + H_a^{PEN}\, w_{PEN} \tag{2}$$

where w_{PEN} is the total concentration of PEN units within the copolymer. The larger H values shown in Fig. 2 for the crystallized samples are related either to the presence of PET (left) or PEN (right) crystals. When the concentration of PET units prevails, the probability for bundles of PET sequences to agglomerate, forming crystallites, also increases. On the other hand, if the concentration of PEN segments is larger, then the probability that these sequences will pack in the form of crystalline aggregates is also larger. Larger crystallinity and crystal thickness values consequently give rise to larger H values (1).

In addition, the hardness of the crystals, H_c, was calculated using the additivity relation of crystalline and amorphous hardness values for the PET/PEN compositions 100/0, 90/10, 20/80, and 0/100 (5,6):

$$H = H_c w_c + H_a(1 - w_c) \tag{3}$$

where w_c represents the degree of crystallinity. Equation (3) has been shown to apply for PET samples crystallized at different temperatures and times of crystallization (7).

It is noteworthy that the materials with the 80/20 and 70/30 PET/PEN compositions do not show any small-angle x-ray scattering (SAXS) maxima (3). However, these copolyesters present a well-defined semicrystalline WAXS pattern (see Fig. 1). These results suggest the presence of a very small proportion (0.17 and 0.05 volume fraction) of crystallites that do not form diffracting units and that are embedded in a predominantly amorphous matrix.

In order to calculate H_c values for the materials with the 80/20 and 70/30 PET/PEN compositions, it is necessary to use the dependence of H_c on l_c, which

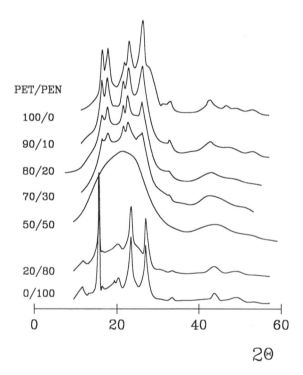

Fig. 1 Wide-angle x-ray diffraction patterns as a function of scattering angle for some of the investigated samples. (From Ref. 3.)

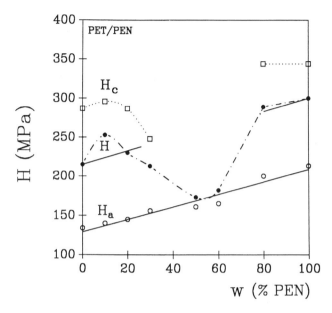

Fig. 2 Experimental microhardness values as a function of PEN mole content for the quenched materials, H_a, annealed samples, H. The crystal hardness value, H_c, was derived using Eq. (4). (From Ref. 3.)

has been derived in previous studies (5):

$$H_c = \frac{H_c^\infty}{1 + b/l_c} \qquad (4)$$

where $H_c^\infty \sim 400$ MPa is the hardness value for the infinitely large PET crystals and $b = 3$ nm is a parameter that characterizes the hardness depression from H_c^∞ due to the finite thickness of the crystals (8). The b-parameter is shown to be related to the ratio of the surface free energy to the enthalpy for plastic deformation of the crystals (9). It is worth noting that all the H_c and l_c values obtained for PET and PEN crystals contained within these copolyesters fit the predictions of Eq. (4) found for PET and PEN crystallites, and for the PET/p-hydroxybenzoic acid (PHB) and PEN/PHB copolyesters (8), using the above mentioned H_c^∞ and b values for PET, and $H_c^\infty = 820$ MPa and $b = 6.0$ nm values for PEN.

While in the case of PHB copolymers, samples with a high concentration of PHB units could not be quenched into the amorphous state, in the present flexible copolyesters, glassy amorphous materials can be produced over the whole range of compositions. In the case of the amorphous PET/PEN copolymers, the microhardness for the whole range of compositions follows simple additive behavior. Similar to the results obtained in PET/PHB and PEN/PHB systems, the annealed samples show the lowest microhardness values for concentrations near 50%. This finding is consistent with the fact that, even after annealing, the samples are always amorphous at these compositions. If, however, one increases either the PEN or the PET content, microhardness will always increase due to the hardening contribution of the developing crystalline regions. The influence of crystal thickness, crystallinity, and the hardness of the crystals can be quantitatively accounted for.

The microhardness of films of poly(butylene terephthalate)-cyclo-aliphatic carbonate (PBT-PC_c) block polymers prepared from the monomers has also been subject to study. The PBT-PC_c compositions investigated were 100/0, 95/5, 80/20, 70/30, 60/40, 50/50, and 40/60. For this purpose, dimethyl terephthalate, 1,4-butanediol, and dihydroxy-polycarbonate [cyclo-aliphatic carbonate (PC_c), Duracarb 140] were polymerized by a three-stage polycondensation method, as described in Refs. 10 and 11. Copolymers with a varying content of PC_c segments (between 40% and 100% in weight were prepared. The obtained polymers were extruded from the reactor and granulated. The granulated polymers were dried for 8 h at 70°C under reduced pressure and injection molded at $\approx$50 MPa. For x-ray differential scanning calorimetry (DSC) and microhardness characterization, 300-μm-thick films were prepared by compression molding between thin Teflon plates at a temperature 20° higher than the corresponding melting point of each polymer. Then the films were quenched at room temperature in the form of glassy films. The films were finally stored, at room temperature, for a period of 6 months.

When quenched from the melt, the PBT-PC_c copolymers containing more than 20 wt % PC_c are completely amorphous. After annealing at appropriate temperature, the PBT segments crystallize (12). On the other hand, pure PBT and the PBT-PC_c sample containing 5 wt % PC_c are not amorphous even after quenching. From the WAXS patterns it turns out that mainly the PBT structure retains its form in all the copolymers (10). The DSC scans of the copolymer samples immediately after quenching show the presence of a typical step, in the range 40–50°C, depending on the composition corresponding to the glass transition temperature (T_g) (12).

Figure 3 illustrates the variation of microhardness (H) as a function of PC_c content for the quenched amorphous samples measured immediately after solidification (free samples, H_a^{fresh}) and after storage for 6 months at room temperature (aged samples, H_a^{aged}). The aged samples clearly show larger microhardness values than the freshly prepared samples. For all amorphous blends, the microhardness increases linearly with the concentration of PC_c units according to:

$$H_a = H_0 + kw_{PCc} \tag{5}$$

where the intercept H_0 gives the H value of purely amorphous PBT and w_{PCc} is the content (wt %) of the PC_c component. For the fresh and the aged samples, two different values of H_0 are obtained, 54 MPa and 94 MPa, as the H_a values of the fully amorphous fresh PBT and aged PBT samples, respectively.

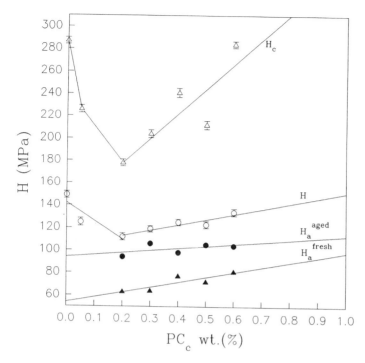

Fig. 3 Plot of microhardness H vs. composition PC_c for freshly prepared amorphous (H_a^{fresh}), physically aged (H_a^{aged}), semicrystalline (H), and fully crystalline (H_c) PBT-PC_c copolymers. (From Ref. 10.)

From the thermograms (12) it has been found that for the amorphous samples (PC_c content greater than 20 wt %), T_g increases roughly with increasing PC_c content. The parallel variation of microhardness and T_g with PC_c content immediately suggests that the microhardness increase is caused by the shift of T_g toward higher temperatures. The shift of T_g increases the $\Delta T = T_g - T$ values, ΔT being the difference between the temperature of measurements T and the T_g of the copolymer. Consequently, microhardness increases due to the fact that hardness increases with increasing ΔT.

The microhardness of the crystallized samples (H) shows first a decrease with PC_c content up to 20 wt % and then a linear increase with $PC_c \geq 20$ wt % (Fig. 3). This decrease in microhardness value can be attributed partly to the corresponding decrease in crystallinity from the PBT homopolymer ($\alpha = 0.28_5$) to the PBT_{80}/PC_{c20} copolymer ($\alpha = 0.17_6$). The hardness of a semicrystalline polymer additionally depends on the hardness of the crystals (H_c) themselves. Assuming the crystals are randomly distributed and oriented, the value of H_c can be calculated from the additivity relation (Eq. 3).

From Fig. 3 it is also observed that H_c for the PBT crystals decreases with increasing PC_c concentration until $w_{PC_c} \simeq 20$ wt %. It is clearly seen from Fig. 4 (plot of $1/H_c$ vs. $1/l_c$) that the decrease in H_c can be correlated with the decrease in crystal thickness of the PBT domains. The dependence of H_c on l_c is given by Eq. (4) (13,14), where H_c^{∞} is in this case the hardness value for infinitely large PBT crystals and $b = 2\sigma_e/\delta h$ is a parameter, as we have already seen, that is related to the surface free energy σ_e of the crystalline lamellae and to the energy δh required for plastic destruction of the crystals.

For the intercept of the straight line of Fig. 4 one can obtain a value for the microhardness of infinitely thick PBT crystals as $H_c^{\infty} = 369$ MPa, and from the slope, the value of $b = 15$ Å. At this point we must mention that for infinitely thick PET crystals, the reported microhardness value is ~400 MPa (14), which is not far from the obtained value earlier for PBT. In addition, we have to consider that the values of l_c used in this evaluation represent only a lower bound. The upper bound is obtained if one lets $l_c = L$, assuming that all amorphous material lies outside of the crystal lamellar stacks. With this assumption, and performing the same extrapolation, we obtain a microhardness value of infinitely large crystals as $H_c^{\infty} \simeq 511$ MPa and $b = 15.3$ nm.

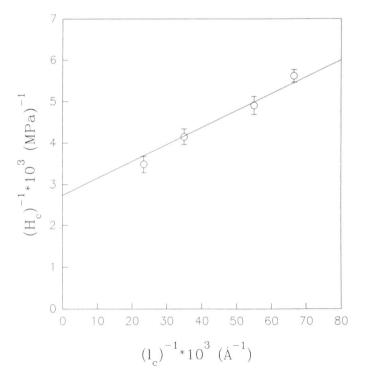

Fig. 4 Plot of $1/H_c$ vs. $1/l_c$. (From Ref. 10.)

In Fig. 4 we used only the l_c value for the copolymers with a PC_c content less than 50%. The samples with higher PC_c concentration do not show any discrete long spacing. Hence, the occurrence in these samples of wide-angle x-ray diffraction maxima could be explained by the absence of crystal stacks of minimum size to give coherently diffracting domains.

In conclusion, the microhardness of amorphous PBT/PC_c block copolymers can be explained in terms of a simple additive parallel model of two independent components. Physical aging, after several months, induces a significant hardening of the samples. Upon annealing, the hardness of the random copolymers increases owing to the reinforcing contribution of the developing crystalline regions. Composition, crystal thickness, and surface free energy play an important role in determining the final hardness value and can be quantitatively accounted for.

B. Microhardness of Condensation Copolymers from Reactive Blending: Relationship Between Amorphous Blends and Random Copolymers

The microhardness of corrective blends of PET and bisphenol A polycarbonate (PC) was also investigated over the whole range of compositions (15). Due to its significant commercial interest, the PET/PC blends have been investigated rather intensively, particularly with respect to their miscibility. In preceding studies, complete miscibility for all compositions to only partial miscibility have been reported (16). For example, blends prepared with an extruder at a die temperature of 290°C led to two amorphous phases over the whole composition range, suggesting that neither miscibility nor transreactions take place (17). On the other hand, the exchange reaction in molten PET/PC blends (in the range 240–274°C and in the presence of a transesterification catalyst) gave rise to four-component copolycondensates (18).

More recently, it was reported (19,20) that the same system is entirely miscible only after transesterification, as per melt mixing at high temperature (300°C), or under long processing time. It is therefore not surprising that immiscible blends are obtained on blending after a single extrusion due to insufficient time for reaction (21). Without reaction, this pair is immiscible.

Very recently, the sequential reordering in copolymers from a PET/PC blend was also investigated in detail (22). In agreement with previous reports (16,23), it was found that annealing at 280°C for hours (without an extra added catalyst) results in a complete disappearance of crystallization or melting of the blends. Such an amorphization is attributed to the formation of random copolymers. This statement has been confirmed by nuclear magnetic resonance (NMR) measurements, by the observation of one glass transition temperature in the range of the initial two T_{gs}', and by solubility tests. Because of the outlined reasons, it was of interest to supplement our earlier microhardness studies of blends of condensation polymers (12,24), in which evidence for chemical interactions was observed related to the microhardness additivity of the two components. In the present case, the same chemical reactions are in a more advanced stage and result in a profound change of the chemical composition of the starting blend and eventually its phase state.

The blends used were prepared from commercial grades of PET and PC with varying contents of PET, between 0 and 100 wt %, by means of coreactive mixing. For this purpose, PET and PC pellets were mixed in the reactor in the temperature range of 275–290°C and a vacuum of 0.1–0.2 mm Hg pressure in the presence of magnesium hydrohexabutoxy-*o*-titanate as a catalyst (1.5 cm^3 on 400 g of the mixture). After 45 min, the blend was extruded and granulated.

For further thermal and mechanical characterization of the blends, films were prepared by compression molding. The films so prepared, differing in their starting PET/PC ratio, were studied by WAXS and found to be completely amorphous. Such a result has been reported earlier (25) for the same blends even without performing chemical reactions between the partners.

Figure 5 shows the DSC curves for the homopolymers PET and PC and their blends after coreactive blending and film preparation. The homo-PET indicates

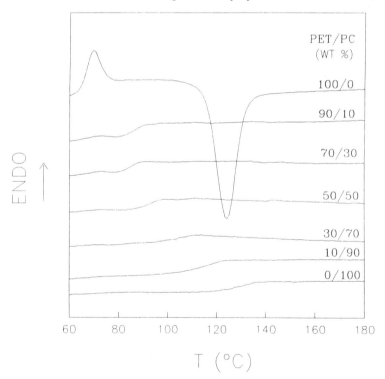

Fig. 5 DSC curves for homo-PET and homo-PC and their coreactive blends of various compositions as indicated on the respective curves. (From Ref. 15.)

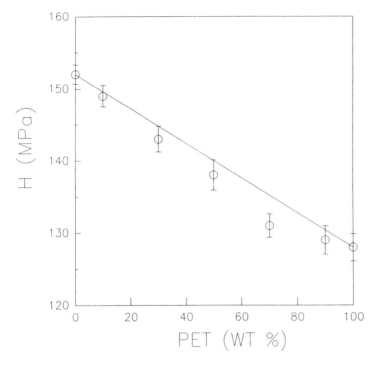

Fig. 6 Dependence of microhardness H on the composition of coreactive PET/PC blends. The values of the homo-PET and homo-PC are also given. The solid line reflects the H values calculated according to Eq. (9). (From Ref. 15.)

both a well-defined glass transition and a crystallization peak at around 65°C and 120°C, respectively. The small endothermic peak at 65°C is typical of an aged PET material and is due to an enthalpy relaxation of the glass toward equilibrium, as shown previously (26). The homo-PC shows only a glass transition around 130°C. More striking is the observation that all the blends show only one glass transition temperature, which increases with increasing PC content in the starting blends. There is no sign for crystallization in the interval studied (60–180°C), and this observation supports the WAXS results and findings reported for the same blends (25) that the samples are in the glassy state. This peculiarity of the blends is very important for the subsequent microhardness measurements performed at room temperature.

Figure 6 shows the dependence of microhardness on the composition of the coreactive blends. The H values for the two homopolymers are also given. One observes an almost linear decrease of microhardness with increasing content of the PET in the blends, i.e., with increasing concentration of the component with the lower H value.

The solid straight line in Fig. 6 reflects the H values calculated according to the mechanical parallel model (Eq. 1):

$$H = wH_a^{PET} + (1 - w)H^{PC} \tag{6}$$

where w and $1 - w$ are, respectively, the weight fractions of PET and PC, and H_a^{PET} and H^{PC} are the microhardness values for amorphous PET and for PC, respectively. Rather good agreement between the experimentally measured and calculated values is observed for the coreactive blends with different compositions (Fig. 6).

For crystalline polymers and copolymers, it is known that the microhardness depends primarily on crystal characteristics such as size, perfection, and chain conformation within the crystals (24). When dealing with completely amorphous samples, these parameters obviously cannot be used. The only quantity left is the glass transition temperature. For this reason, an attempt was made to look for a relationship between H and T_g, bearing in mind that they should be sensitive in a similar way to the blend composition and the occurrence of chemical reactions leading to the formation of copolymers.

From the plot of the relationship between the microhardness depression $\Delta H = H^{PC} - H^{BL}$ and the glass transition depression $\Delta T_g = T_g^{PC} - T_g^{BL}$ for coreactive blends with various compositions for which H^{PC} and H^{BL} are the microhardness of the PC homopolymer and of the coreactive blend, respectively, and T_g^{PC} and T_g^{BL} are the glass transition temperatures of the PC homopolymer and the coreactive blend, respectively (18–23), a linear correlation between ΔH and ΔT_g has been obtained (15).

Considering the reports (18–23) and taking into account the experimental conditions used for the coreactive blending in the present case (275–290°C, treatment time of 45 min, presence of a transesterification catalyst), one has to assume the occurrence of intensive chemical interactions. It is also to be expected that these reactions not only will lead to the formation of copolymers but will result in a more or less complete randomization of the sequential order of the repeating units. If this is the case, the initially two-component blends should be converted into one-component ones. In addition, the one-component ones will also represent one-phase systems so far as randomization (i.e., amorphization) is taking place. This is really the case, as can be concluded from the DSC curves shown in Fig. 5. The blends do not crystallize, in contrast to the homopolymer-PET, and they exhibit only one glass transition temperature.

The conclusion that the starting PET/PC blends are converted via coreactive blending into one-component material with only one phase (amorphous) has important consequences when the "blend" is characterized with respect to its mechanical properties. Unfortunately, this peculiarity of such blends is frequently disregarded or underestimated, which results in erroneous conclusions, for example, concerning the miscibility of the starting partners (16).

A second detail concerns how to express the blend composition. Dealing with blends of condensation polymers that have been reacted and converted into copolymers, the copolymers being uniform with respect to the number of components (as well as with respect to the number of phases provided no phase separation via crystallization or defacing has occurred), it seems more reasonable to express the ratio of the components in mol % rather than in wt %. This reflects more realistically the composition of the system and, at the same time, particularly, the expression as a mole ratio offers an idea about the character of the sequential order in the chains, assuming complete randomization has taken place. The fact that the mole ratio reflects the block length when complete randomization is achieved allows one to make direct conclusions about the crystallization ability of the copolymers obtained. For example, in the present case only the richest PET blend (90/10) is potentially crystallizable. For the rest of the blend, the PET "blocks" are too short to form lamellae with a thickness of 50–60 Å, which represents the lowest limit (22). It should be emphasized that, in the case studied, the differences between wt % and mol % are not that large, because the molecular weights of the two repeating units are rather close to each other, but there are cases when the difference is significant.

Bearing in mind the outlined peculiarities of blends of condensation polymers and particularly when they consist of one component and one phase (this case is more the exception rather than a general rule, since block copolymers usually consist of two, three, or more phases), the application of the additive law for evaluation of their characteristics seems not to be completely justified. The observed good agreement between the measured microhardness values and the calculated ones according to Eq. (6) (Fig. 6) allows one to make an important conclusion in this respect.

Basically, the application of the additive law for blends assumes the presence of spatially well-defined regions of chemically and structurally uniform moieties, which is not the case for the present amorphous PET/PC copolymers, since only a single T_g is observed. Consequently, it follows that the contribution of the two species to the microhardness of the copolymers is transferred via the repeating units constituting the copolymer molecules. Since in the present system there are no crystalline phases that can independently contribute to the microhardness of the copolymer, the observed microhardness can be regarded as arising from and depending only on the chemical composition of the copolymer itself. Possibly, one can speak about the intrinsic contribution of the two repeating units to the microhardness of the copolymeric solid. One can conclude that the additivity law can be applied at the molecular level; i.e., the microhardness of amorphous copolymers differing in their composition obeys the additivity law based on the H values of the respective amorphous homo-polymers, provided no other factors affect the microhardness of the copolymers.

It is important to note here that the additivity law is applicable to blends of miscible pairs of polymers. This was demonstrated for the blends of polymethyl methacrylate and polyvinylidene fluoride (27).

In conclusion, when working with blends of condensation polymers, one always has to consider the possibility of chemical interactions and formation of copolymers. The extent of these reactions is also important, because it is possible to obtain a one-component, as well as a one-phase, system when the blocky sequential order is converted to a random one. Such systems are very appropriate for the verification of relationships reflecting the effect of composition on various properties, since they are independent from other factors. Finally, in such cases one deals with copolymers distinguished by the creation of new chemical bonds, but not anymore with blends, although one mixes two or more homopolycondensates at the very beginning. Nevertheless such noncrystalliz-able copolymers obey the additivity law (Eq. 1) in the same way as the mechanical mixture of the respective homopolymers, provided they are also in a completely amorphous state. In other words, the microhardness of the mixture of two or more amorphous homopolymers is the same as their random (non-crystallizable) copolymer, with the same composition. In both cases of amorphous systems, the mixture and the copolymer, the microhardness can be calculated by means of the additivity law (Eq. 1) using the independent H values of the amorphous homopolymers involved.

C. Microhardness of Condensation Copolymers Prepared by Blend Compression: Effect of Composition and Processing Conditions

Previous studies on blends of PET and PEN indicate that the thermal behavior of the blends strongly depends on the time during compression molding (t_m) before quenching the films in ice water (28). For t_m values between 0.2 and 0.5 min, two glass transition (T_g) values were observed by means of dynamic mechanical analysis, indicating the presence of two phases, an amorphous PEN-rich phase and an amorphous PET-rich phase. For t_m in the range of 2–45 min, only a single value of T_g and, consequently, a single phase is found to exist. Within this time range, transesterification of the two compounds also takes place, resulting in a copolyester of PET and PEN. It looked interesting to examine the variation of the mechanical properties (microhardness) of these PET/PEN blends as a function of both composition and melt pressing time t_m. In particular, it is important to compare these results with those obtained in the case of random copolyesters of PET and PEN (3).

PET and PEN were obtained by coprecipitation from the solution in hexafluoroisopropanol. Amorphous films were, then, prepared from the precipitated powder, by melt pressing in vacuo for different times t_m, varying from 0.2 min to 45 min followed by quenching in ice water. PET/PEN blends with weight compositions 90/10, 70/30, 60/40, 44/56, 30/70, and 10/90 were prepared (24).

The PET and PEN blends are completely amorphous after quenching from the melt, as revealed by the DSC experiments (28). In all cases, H shows first a rapid initial increase with t_m, exhibiting a maximum just before $t_m = 10$ min, and then, for longer times, a gradual decrease down to values that can be even lower than the starting ones.

In order to explain the variation of H vs. t_m, it is convenient to analyze the results of hardness as a function of composition. Figure 7 shows the experimental values of H as a function of composition, taking t_m as a parameter. It is seen that for $t_m \sim 0.2$ min, H increases linearly with increasing concentration of PEN content according to the prediction of the mechanical parallel model (Eq. 1) given by

$$H = H_a^{PET} w_{PET} + H_a^{PEN} w_{PEN} \tag{7}$$

where $HPET\backslash a$ and $HPEN\backslash a$ are the hardness of amorphous PET and PEN and w_{PET} and w_{PEN} are the weight fractions of PET and PEN, respectively. These results indicate that the microhardness of the PET/PEN blends shows similar values as those obtained for PET/PEN amorphous random copolyesters (3) in accordance

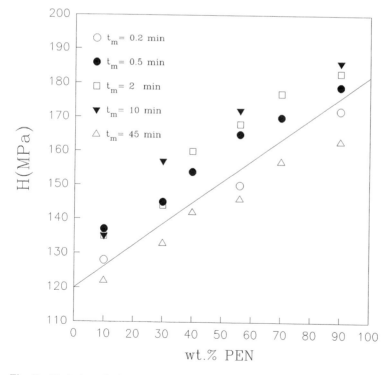

Fig. 7 Variation of microhardness as a function of wt % PEN for different melt-pressing times (t_m). The dotted straight line follows the additivity predictions of the single components (Eq. 1). (From Ref. 24.)

with the conclusion drawn in the previous section for the system PET/PC. With increasing t_m up to 10 min, one observes a shift of the straight line toward higher values. Finally, for t_m = 45 min, one observes the lowest values of H.

Let us recall that when the melt-pressing time is about 0.2–05 min, two T_g values are observed, indicating that there are two phases present (28). In the case of t_m = 2 min, a single T_g value and, thus, a single phase is found. For $t_m \sim$ 10–45 min, no crystallization and melting during heating in the differential scanning calorimeter (DSC) at 10°C min^{-1} is observed, indicating that an amorphous copolyester has been obtained by transesterification of PET and PEN during melt pressing. The first initial increase in hardness up to t_m = 0.5 min could be attributed to the corresponding shift of T_g toward higher temperatures. It is known that in the case of amorphous blends, temperature is the dominant parameter in determining the yield behavior of the glassy material (29). The further increase in H up to t_m = 2 min could be associated to the change from a two-phase system into one single amorphous phase composed of interpenetrating molecules of both polymers. Such a homogeneous system should offer a higher mechanical resistance to yield and to plastic deformation. Finally, H increases further up to $t_m \sim$ 10 min, where the copolyesters of PET and PEN have been formed by transesterification. The variation of the viscosity as a function of t_m for the 44/56 blend showed that no detectable changes in this property are observed (24).

One may ask at this stage: Why does H gradually decrease with increasing t_m if the molecular weight and the viscosity remain practically constant? One possible explanation could be that at the beginning of the transesterification process the copolyester has a rather blocklike character. Only after longer times does it become a statistical copolymer. Our results, therefore, indicate that the microhardness of the block copolyester is larger than that of the statistical copolymer. The existence of blocks may lead to a microphase separation between PEN and PET blocks. It seems, then, reasonable to assume that parallely packed sequences of blocks with the same chemical compositions would yield less easily than parallel copolymer sequences of statistical composition.

IV. MICROHARDNESS OF LIQUID CRYSTALLINE CONDENSATION COPOLYMERS: EFFECT OF COMPOSITION AND TEMPERATURE DEPENDENCE

Copolyester of 4-hydroxybenzoic (HBA) and 2-hydroxy-6-naphthoic acid (HNA) have lately been receiving particular attention, for they exhibit a thermotropic liquid crystalline phase from which highly oriented fibers can be produced.

Recent studies highlight the dependence of microhardness upon crystallinity, crystal thickness, and polymorphic crystal forms in random liquid crystalline copolyesters containing flexible spacers (3,8). In the present section we wish to re-examine these concepts in the case of rigid copolyesters that show substantial differences from flexible polymers. It is known that SAXS maxima have been detected only in P(HBA/HNA) copolyesters of lower molecular weight (degree of polymerization DP ≥ 125) (6). However, for high molecular weights (DP ~ 250), the packing of rigid molecular rods in the material does not give rise to any SAXS maxima. In addition, the measured crystallinity of the copolyesters shows very little dependence upon thermal treatments. A value of crystallinity close to 0.2 has been obtained in all cases. P(HBA/HNA) samples of high molecular weight (DP ~ 250), with molar ratios 30/70, 58/42, and 75/25, were compression molded and quenched from the melt, as described elsewhere (30). Films 0.25 mm thick were obtained. The microcrystals are disoriented in the plane of the film with the molecular axis parallel to the film surface. Samples were annealed for 24 h at a temperature of 240°C for P(HBA/HNA) 30/70 and 75/25 and 220°C for the molar ratio 58/42. The microhardness of the homopolymers PHB and PHN was also measured at room temperature to examine the influence of the comonomer ratio.

Figure 8 represents the plot of microhardness as a function of comonomer composition for a given loading time of 0.1 min. Hardness and degree of crys-

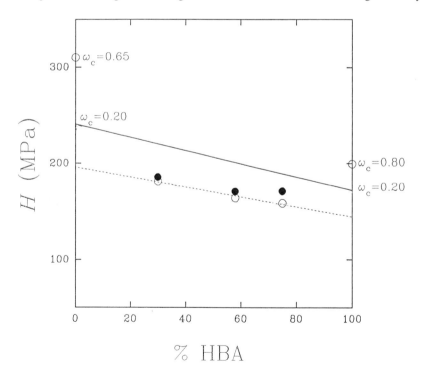

Fig. 8 Experimental values of microhardness as a function of comonomer composition for the P(HB/HN) copolyesters. Open symbols, quenched samples; solid symbols, annealed samples. The solid straight line corresponds to the additivity law of single homopolymers normalized to a crystallinity of $w_c = 0.2$ (From Ref. 30.)

tallinity, ω_c, are usually related according to the following expression (1):

$$H = \omega_c H_c + (1 - \omega_c)H_{nc} \tag{8}$$

where H_c and H_{nc} are the intrinsic hardness values of the crystalline and noncrystalline phases, respectively. In most cases, a controlled variation of the crystallinity permits the calculation of H_c and H_{nc} through Eq. (8). However, as mentioned before, the crystallinity of our samples cannot be substantially changed even if we use a widely different preparation method, such as quenching from the melt or long annealing time. Irrespective of the treatment used, ω_c values near 0.2 are always obtained. In order to plot comparable results, we have normalized the values of microhardness of the homopolymers to a crystallinity value of 0.2. To do this, one of the two missing values, H_c or H_{nc}, should be known in Eq. (8). Let us consider the H_{nc} values of PET and PEN [homopolymers of quite similar monomer units but with a higher degree of molecular flexibility (31)] as a lower limit for our PHB and PHN homopolymers. Using these H_{nc} values together with Eq. (1), the corresponding H values for $\omega_c = 0.2$ were computed for both homopolymers. These H values are represented in Fig. 8 as the intersection of the solid straight line with the vertical axes. This solid line corresponds to the linear additivity law of microhardness values given by the two homopolymers, with an assumed crystallinity of 0.2. It is clear that the H values for the copolyesters are lower than those calculated from the additivity law (solid line). After annealing at high temperature, a slightly increased microhardness value is observed for all samples. This is more evident in the 75/25 copolyester (see Fig. 8).

Figure 9 shows the dependence of microhardness with temperature, T, for the quenched copolyesters P(HB/HN) 30/70, 58/42, and 75/25 and for the thermally annealed 75/25. At temperatures close to the glass transition temperature (T_g) (above 100°C), there is a large viscoelastic recovery of the material and no indentations are observed after load removal. Above T_g, the microcrystals are probably mechanically stable at the strains involved in microindentation, and the noncrystalline regions have enough molecular mobility, in the nematic-like state, to contribute to a complete recovery of deformation. In the temperature range below 125°C, H follows an exponential decrease as a function of T given by

$$H = H_0 e^{-\beta T} \tag{9}$$

where H_0 is the hardness of the material at a temperature of 0 K and β is the so-called coefficient of thermal softening. It is to be noted that these values (ranging from 9.9 to 6.1 $\times$ 10^{-3} K^{-1} for 58/42 and 75/25, respectively) are of the same order of magnitude as that of copper measured at high temperature (>400°C) (32).

When discussing the microhardness of P(HB/HN) copolyesters, it is convenient to emphasize the fact that 80% of the material is located in the noncrystalline regions. However, by taking into account that many molecules have to be attached to one or more crystalline segments, it seems reasonable, owing to chain rigidity,

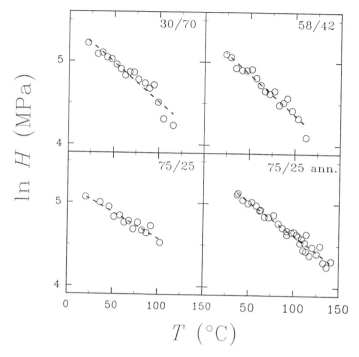

Fig. 9 Temperature dependence of microhardness for the quenched samples, 30/70, 58/42, 75/25, and the annealed one, 75/25. (From Ref. 30.)

that the overall packing of the molecular chains across the material is influenced by the three-dimensional distribution of crystallites.

Once the influence of crystallinity and crystal thickness is discarded, one is bound to consider that changes in microhardness are due mainly to variations of the average molecular packing within the crystalline and the noncrystalline phases. The packing of the chains in polymers and, therefore, the changes in cohesive energy of the crystals, have been shown to play a role in determining the hardness value (33). Thus the closer molecular packing of crystalline polymers (PE, PEEK) with increasing crystallization temperature leads to an increase in H (33,34).

The dashed curve in Fig. 8 represents the experimental values of microhardness as a function of comonomer composition of constant crystallinity and crystal thickness. Our results show that the H values for the copolyesters are lower than those calculated from the additivity law (solid line), while those of the annealed samples show a slight tendency to smaller deviations.

The molecular packing in the crystal phase can be described in terms of the molecular cross-sectional area ($a \times b/2$), a and b being the intermolecular parameters of the orthorhombic unit cell (30).

The deviation of microhardness values from the additivity law can now be related directly to the corresponding deviation of the values of the cross-sectional area of the crystalline phase from a linear plot. The deviation from the hexagonal symmetry (decrease in γ) when annealing the samples is reflected in a small increase in the microhardness values (see Fig. 8). The largest H increase, corresponding to the HBA-rich copolymer, corresponds to a clear decrease in the cross-sectional area.

In order to account for the double contribution of crystalline and noncrystalline phases, let us next introduce an average cross-sectional area of the molecules, A:

$$A = w_c \frac{a \times b}{2} + (1 - w_c)\pi \left(\frac{d}{2}\right)^2 \tag{10}$$

In this way, the dependence of microhardness on temperature can, as a first approximation, be related directly to the variation of the defined average cross-sectional area. However, a unique curve has not been found, probably due to the simplicity of the model and the errors in the calculation of the intermolecular noncrystalline distance (30).

In conclusion, microhardness values, at room temperature, of nonoriented P(HB/HN) copolyesters as a function of composition show a clear deviation from the linear additivity law of single components. Such a deviation is shown to be related mainly to changes in the packing of the rigid molecules. Degree of crystallinity and crystal size, on the contrary, remain constant with composition and do not add any significant contribution to the microhardness value. The packing of the molecules, and therefore H, can be characterized by an average cross-sectional area, which includes crystalline and noncrystalline regions. The slightly higher microhardness values found in annealed materials as compared with the quenched ones is proposed to be due to an increase of H_c together with a concurrent decrease of H_{nc}. Finally, the parallel temperature dependence of both microhardness and average cross-sectional area leads to an analytical expression that relates the coefficient of thermal softening to the thermal expansion coefficient of the cross-sectional area of the materials (30).

V. POLYMORPHIC TRANSITIONS AS REVEALED BY MICROHARDNESS IN PBT POLYETHERESTER BLOCK COPOLYMERS

A. Preliminary Aspects

When polymers are crystallized from melt or solution, their crystalline region may exhibit various types of polymorphic modifications, depending on the cooling rate, evaporation rate of solvent, temperature, and other conditions. These crystal modifications are different in their molecular and crystal structures as well as in their physical properties. Many types of crystalline modifications have been reported (35).

Some polymorphic modifications can be converted from one to another by a change in temperature. Phase transitions can be also induced by an external stress field. Phase transitions under tensile stress can be observed in natural rubber when it orients and crystallizes under tension and reverts to its original amorphous state by relaxation (36). Stress-induced transitions ar also observed in some crystalline polymers, e.g., poly(butylene terephthalate) (PBT) (37,38) and its block copolymers with poly(tetramethylene oxide) (PTMO) (39), poly(ethylene oxide) (PEO) (40,41), nylon-6 (42), keratin (43), and others. These stress-induced phase transitions are either reversible, where the crystal structure reverts to the original structure on relaxation, or irreversible, where the newly formed structure does not revert after relaxation. Examples of the former include PBT, PEO, and keratin.

Two crystalline phases of PBT can appear under tension and relaxation: the so-called β-form and the α-form, respectively (43,44). The α-form was found in a relaxed sample, whereas the β-form could be observed only when the sample was held under strain (Scheme 1).

There have been a number of attempts to determine the unit cell parameters for the two crystalline forms (38,45,46), and there is still some degree of controversy. However, the general consensus is that in the α-form the molecular chain is not fully extended, probably with the glycol residue in a gauche-trans-gauche conformation, whereas in the β-form the chain is fully extended with the glycol residue in the all trans conformation. There has been considerable interest in the mechanisms of the α–β transition, and this has been modeled for static and dynamic measurements. X-ray diffraction patterns and infrared–Raman spectra show specific changes through this $\alpha \angle \beta$ transition (37–39,47–49). The stress and strain dependence of the molar fraction of the β-form, X_β, can be evaluated by quantitative analysis of the infrared spectra (47). The characteristic behavior of this phase transition observed in a uniaxially oriented sample is as follows: the X_β increases drastically above the critical stress f^*; the fraction X_β is almost linearly proportional to the strain; the transition is reversible; and the stress–strain curve

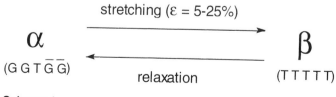

Scheme 1

has a plateau at the critical stress f^*; that is, the curve is divided into three regions: the elastic deformation of the α-phase (0–4% strain), the $\alpha \angle \beta$ transition (plateau region, 4–12% strain), and the elastic deformation of the β-phase (>12% strain). These experimental results indicate that this stress-induced phase transition is a thermodynamic first-order transition.

It has been also pointed out (39) that in block copolymers of PBT and PTMO, the transition is smeared over a wide range of stresses due to the effect of the PTMO soft segments (39).

As we have seen, crystal hardness H_c has been shown to depend principally on the average thickness ℓ of the crystalline lamellae (13). In systems, however, where the degree of crystallinity w_c and ℓ remain constant, H_c is a function of chain packing within the crystalline phase (51). Crystal hardness—the critical stress required to plastically deform the crystal—hence, reflects its response to the intermolecular forces holding the chains within the lattice and is a function of co-hesion energy (51).

This peculiarity of this technique makes possible to distinguish between polymorphic modifications of the same polymer, as convincingly demonstrated for polypropylene (52). In a very recent study (53) on PBT, known as a polymer capable of undergoing a reversible stress-induced polymorphic transition, it was demonstrated that the microhardness technique can also be applied successfully for detecting and following this transition. The observed very sharp change in H values (within 2–4% of external overall deformation) makes the method compet-itive with respect to the sensitivity of such other commonly used techniques as wide-angle x-ray scattering (WAXS), IR- and Raman spectroscopy, among others (37–39,44,47–49). Furthermore, by means of the additive law (Eq. 1) it was pos-sible to evaluate the H value of completely crystalline PBT comprising crystallites of the β-modification. It looked attractive to examine the microhardness behavior during the stress-induced polymorphic transition of PBT in its block copolymers, for some of which the transition is very smeared (39). An additional reason for performing such a study is the fact that the copolymers of PBT with poly(ethylene oxide) (PEO) and/or polycarbonate (PC) were not so far studied with reference to polymorphic transitions.

B. Experimental Details

The starting material represents a polyblock poly(ether ester) PEE comprising PBT as hard segments and PEO with a molecular weight of 1000 and polydis-persity of 1.3, according to gel permeation chromatography analysis, as soft segments in a ratio of 57/43 wt %. The synthesis was carried out on a semi-commercial scale, as described elsewhere (54). The sample was shaped as a bris-tle, drawn at room temperature to five times its initial length, and then annealed with fixed ends for 6 h in vacuum at a temperature of 170°C. The initial material used for measurements was highly oriented with a dominating α-modification of PBT (55).

Microhardness was measured at room temperature. Deformation measurements up to 80% overall relative deformation ε, after which the sample breaks, were performed by means of a stretching device. The elongation ε is defined as ε $= (\ell - \ell_0)/\ell_0$, where ℓ_0 and ℓ are the starting length and the length of the sample after a given deformation, respectively.

Although the stress-induced α → β polymorphic transition in PBT is well documented, comparative WAXS measurements were carried out on the same samples in the same deformation range.

C. Stress-Induced Polymorphic Transition

The dependence of the microhardness H on the deformation ε for drawn and annealed bristles of PEE (PBT/PEO = 57/43) is plotted in Fig. 10. One can see that the H variation presents several regimes, depending on the stress applied. At the lowest deformations (ε up to 25%), H is nearly constant (~33 MPa), thereafter, in a very narrow deformation interval (ε = 2–3%) the H value suddenly drops by 30%, reaching the value of 24 MPa, which is maintained in the ε-range between

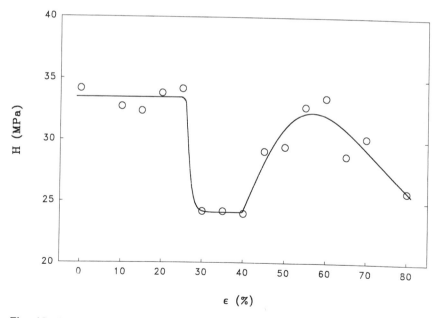

Fig. 10 Microhardness H vs. overall relative tensile deformation ε of drawn and annealed PEE (PBT/PEO = 57/43) bristles. (From Ref. 55.)

30 and 40%. With further ε increase (between 40 and 60%), H then increases up to nearly its starting value, followed by decrease again to 25 MPa for ε ~ 60–80%.

The most striking change in the H behavior is the sharp decrease at ε = 25–27%. Taking into account the recent similar study on homo-PBT (53), where a similar change has been observed (a drop in H by 20%) due to the well-documented stress-induced polymorphic transition in PBT, one can assume that in the present case also the same transition takes place.

The fact that the α–β polymorphic transition in the PEE under investigation occurs at a much higher deformation (around ε = 27%) in comparison to the homo-PBT, where it was observed at ε between 5 and 10% (53), can be explained by the peculiar behavior of the thermoplastic elastomers.

In a preceding study (56–62) it was shown that the deformation of PEE with varying PBT/PEO ratio deformed in the 25–50% range is related to the conformational changes in the amorphous regions, the crystallites remaining untouched. The stretching and relaxation of these amorphous segments causes an affine and reversible change of the measured x-ray long spacing (56–62). With further increase of deformation (for PEE with compositions as in the present case—starting at ε = 25–30%), the external load is transferred to the crystallites contributing to the observed polymorphic transition [Fig. 10, (53)].

The drop in the microhardness is due to the fact that the β-phase has a lower density in comparison to the α-phase (45), as concluded in our previous study (53).

Additional support for the assumption concerning the stress-induced polymorphic transition around ε = 27% can be found in our WAXS study on the same samples in the same deformation interval.

Figure 11 illustrates the radial (near meridional) scans of the ($\bar{1}$04) reflection of PBT taken from drawn and annealed PEE(PBT/PEO = 57/43) at various tensile deformations. One sees that up to ε = 12%, the angular position of the ($\bar{1}$04) peak remains unchanged. The same is also true for the diffractogram taken at ε = 28.8%, though, in contrast to the previous ones, here one sees indication for another peak at a lower angle (~27°). This tendency is very well defined on the diffractogram taken at ε = 58.8%, where the intensity of the new peak is higher than that of the starting one. This observation on PEE is in accordance with previous (48,63,64) WAXS studies of homo-PBT during the stress-induced polymorphic transition.

Similar changes can be found on equatorial scans of drawn and annealed PEE (PBT/PEO = 57/43) taken at various tensile deformations (55), where drastic changes, both with respect to the peak shape and peak angular position, can be detected at ε = 28.8% and 58.8%. These diffractograms (55) were used for evaluation of the "relative crystallinity" w_c (WAXS) at various stages of sample deformation, since calorimetry or density measurements cannot be used for strained samples.

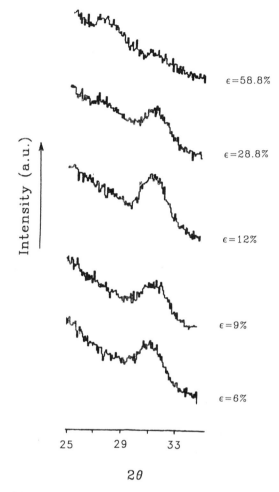

Fig. 11 Radial scans of the ($\overline{1}$04) reflection of drawn and annealed PEE (PBT/PEO = 57/43) bristles taken at various elongations ε (the ε values are given in the respective diffractograms). (From Ref. 55.)

Let us come back to Fig. 10 and follow the H variation as a function of ε. The observed H increase resulting in restoration of the initial microhardness cannot be related to a regeneration of the α-phase, distinguished by a higher microhardness H^{α}, because the sample is under stress. In order to explain the observed H increase, one has to remember a peculiarity of the system: PEE is a physical network where the crystallites are the crosslinks. In addition, the crystallites are embedded in a soft amorphous matrix; i.e., they are "floating" in a relatively low-vis-

Table 1 Structural Data and Mechanical Properties of Poly(ester-ether carbonate) Block Terpolymers

Composition (wt %) (PBT/PO4/PC)	WAXS			SAXS			Microhardness (MPa)		
	w_c (%)	D_{100} (Å)	D_{010} (Å)	L (Å)	l_c (Å)	l_φ (Å)	H_{exp}	H_{cal}	ΔH
100/0/0	59.0	—	—	—	—	—	—	—	—
60/40/0	37.7	84.7	116.6	146.0	55.0	83.7	29.6	85.1	55.5
60/32/8	25.8	87.8	93.8	130.8	33.8	77.2	22.8	68.5	45.7
60/20/20	24.3	105.0	113.9	118.5	28.8	89.6	18.6	66.4	47.8
60/12/28	21.2	86.2	119.6	118.5	25.1	85.3	17.5	62.0	44.5
60/0/40	13.0	77.9	104.0	116.3	15.1	77.3	15.5	50.6	35.1

Source: Ref. 66.

cosity matrix. During the deformation, the network will stretch further, depending on the relative "softness" of the various components of the system. The stretching of the amorphous segments restricts the mobility of the crystallites and makes the system harder. In other words, the decrease of H caused by replacement of crystallites of a higher H by others of lower H is compensated by this hardening effect caused by stretching of the network.

In addition to the assumed hardening of the noncrystalline phases, in the same direction simultaneously the crystallization of the PEO soft segments in acts PEE in the same deformation range.

The subsequent decrease of H in Fig. 10 down again to 25 MPa at the highest deformation ($\varepsilon = 60$–80%) can be explained by a partial destruction of crystallites due to the pulling out of hard segments. This mechanism has been demonstrated by SAXS (56–61) and WAXS (62) measurements. The introduction in this way of defects in the crystallites may induce a decrease in their H value.

A characteristic feature of the material under investigation is its very low microhardness—between 25 and 35 MPa, depending on the crystalline modification present. These values are up to 5–6 times lower than those for semicrystalline homo-PBT regardless of the crystalline modification. Even more, the obtained values for H of PEE (Table 1, Fig. 1) are about two times lower than the amorphous hardness, H_a, of PBT, being 54 MPa (38). This means that there should be other factors responsible for the very low H values of the copolymer, as will be demonstrated in the following section.

VI. MICROHARDNESS OF CONDENSATION COPOLYMERS CONTAINING A SOFT (LIQUID-LIKE) PHASE

As we have seen, microhardness is determined primarily by the crystalline phase, which in the case of polymers is always dispersed in an amorphous matrix (1,50).

In the present case the PBT crystallites are embedded in a two-phase amorphous matrix, the amorphous PBT segments and the soft PEO segments. The latter are distinguished by very low viscosity at room temperature—PEO of mol. wt 1000 melts around 30°C and being incorporated in a polymer chain its T_m is even lower—around 0°C (65).

This means that during the first H decrease (see Fig. 10), PEO is in a molten state. It is worth mentioning that for the same reason the observed stress-induced polymorphic transition cannot arise from PEO, even though basically it is known that PEO is capable to undergo such a transition (40,41).

By means of the additivity hardness law (Eq. 1) one can try to evaluate the contribution of the soft segments' amorphous phase to the overall microhardness. For a four-phase system, as in the present case (two crystalline modifications and two amorphous phases), Eq. (1) can be presented in the following way:

$$H = \varphi\, [H_c^\alpha w_c^\alpha + H_c^\beta w_c^\beta + (1 - w_c^\alpha - w_c^\beta H_a] + (1 - \varphi)H_{\text{soft}} \qquad (11)$$

where φ is the mass fraction of the hard PBT segments, w_c^α and w_c^β are the degree of crystallinity values of both α and β modifications of PBT, H_c^α and H_c^β are their microhardness values, $\varphi(1 - w_c^\alpha - w_c^\beta)$ is the mass fraction of amorphous PBT with a microhardness H_a, and $(1 - \varphi)$ is the mass fraction of the soft segments having a microhardness H_{soft}.

If we assume the value of 122 MPa for H_c^β, which was recently obtained from the study of the stress-induced polymorphic transition in homo-PBT by microhardness (53), then the calculations of H for both cases, before the stress-induced polymorphic transition ($H = 34$ MPa) and after this transition ($H = 24$ MPa), would lead to negative values of H_{soft}, -112 and -70 MPa, respectively, which are of course not physically acceptable.

For this reason one has to revise the deformation mechanism during microhardness determination commonly used for complex systems comprising components having T_g values above room temperature. For these purpose it is convenient to remember the structural peculiarity of the system under investigation. The fact that PBT crystallites are "floating" in a matrix of low viscosity has important consequences on the microhardness behavior. By floating in a liquid of low viscosity, the crystallites of PBT as well as the PBT amorphous phase cannot respond properly to the external stress to demonstrate in this way their inherent microhardness, but they are doing this mostly via the response of the liquid matrix.

A quite similar situation was observed (40) by studying the microhardness behavior of thermoplastic elastomers of the poly(ether ester carbonate) (PEEC) type comprising PBT as hard segments (60 wt %) and PTMO and aliphatic polycarbonate (PC) as soft segments. Scheme 2 shows the general formula that represents the PEEC terpolymers. For the synthesis of the terpolymers, a three-stage polycondensation method was used (15).

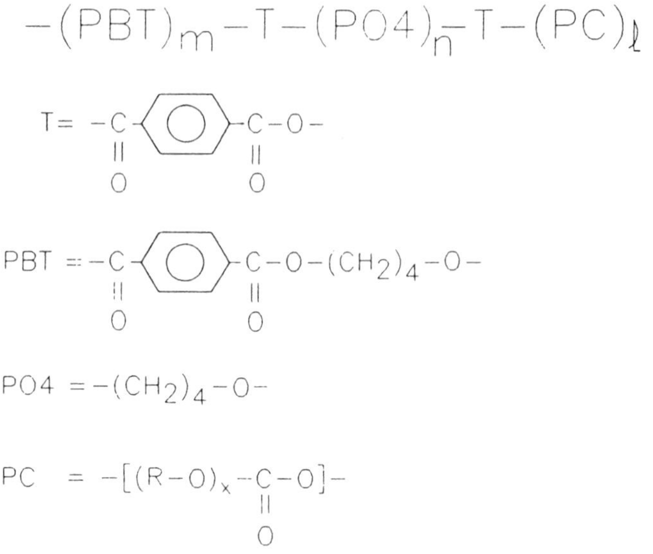

$$-(PBT)_m-T-(PO4)_n-T-(PC)_l$$

$$T= -C-\underset{O}{\overset{||}{\underset{}{}}}\!\!\!\!\left\langle\bigcirc\right\rangle\!\!-C-O-$$

$$PBT = -C-\left\langle\bigcirc\right\rangle-C-O-(CH_2)_4-O-$$

$$PO4 = -(CH_2)_4-O-$$

$$PC = -[(R-O)_x-C-O]-$$

Scheme 2

The obtained terpolymers were extruded from the reactor, pelletized, dried for 8 h at 70°C under reduced pressure ($\simeq$ hPa), and finally injection molded at $\simeq$50 MPa. From these 1-mm-thick plates, films were prepared by pressing the plates in a hot press at a temperature 5°C above the melting point. The films were removed and immediately placed in a cooled press to obtain 0.4-mm-thick films suitable for physical measurements.

The data obtained from calorimetric, dynamic mechanical, and x-ray analyses indicate the existence in the copolymers of three phases, two amorphous and one crystalline. The hard PBT segments give rise to the crystalline phase and to one of the amorphous phases; the soft segments contribute to the second amorphous phase. These three phases are characterized by their transition temperatures, T_g and T_m (66).

Table 1 shows the clear influence of PC content on the microhardness values. The dominating contribution of crystallinity in the mechanical properties of polymers is well-known (1,50). Therefore, to understand better the correlation between the chemical composition and the microhardness of the copolymers, one has first to examine the behavior of the crystalline phase. Figure 12 shows the nearly parallel dependence of microhardness and crystallinity on the PC content within the copolymers. H drops continuously with increasing PC content, almost in the same manner as crystallinity does. Such a parallel behavior of H and w_c shows that the effect of chemical composition on the microhardness is transferred via the crystallinity.

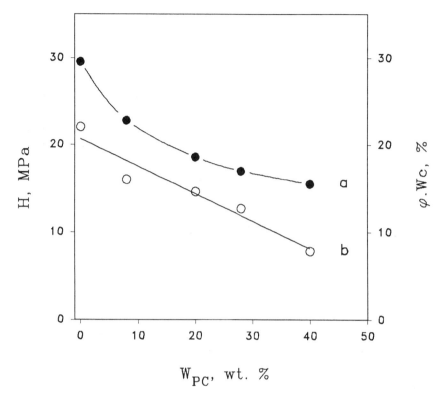

Fig. 12 Dependence of (a) microhardness H and (b) total degree of crystallinity φw_c (the sum of PBT plus PO4 and PC components) on the PC content of polyblock copolymers. (From Ref. 66.)

The plot of H vs. w_c for copolymers with various compositions offers the opportunity to derive, by means of extrapolation, the values of H for a completely amorphous copolymer H_a^{PEEC} and for a copolymer with a 60% fully crystallized PBT component. The extrapolation of the straight line to $w_c = 0$ and 100% leads to the following values: $H_a^{PEEC} \simeq 6.0$ MPa and $H_c^{PEEC} \simeq 106$ MPa (66). It is noteworthy that these values are significantly lower than those for the homopolymer PBT; the samples studied contain only 60 wt % of this polymer. The remaining molecular strains belong to a much softer material, which causes the drop in the H values.

The measured H values for the terpolymer studied (Table 1) are very low in comparison to the values known for common synthetic polymers, even for those in the amorphous state (1). What is more, H values are more than three times smaller than the ones for the same terpolymer calculated by means of the additive law:

$$H = \varphi H^{PBT} + (1 - \varphi)H^{\text{soft}} \tag{12}$$

where φ is the weight fraction of the hard PBT segments.

Taking into account that the soft-segment amorphous phase has T_g values between $-50°C$ and $0°C$, depending on the PC content (Table 1), one can expect that at room temperature such a phase will be a very soft material, much closer to a liquid than to a solid. For this reason, one may expect that $H^{\text{soft}} \simeq 0$ and, as a result, that H will be depressed with increasing values of φ according to the following simple expression:

$$H = \varphi H^{PBT} = \varphi [H_c^{PBT}w_c + H_a^{PBT} (1 - w_c)] \tag{13}$$

By applying the numerical values $\varphi = 0.6$, $H_c^{PBT} = 287$ MPa, and $H_a^{PBT} = 54$ MPa (7), one can derive the calculated values H_{cal} for the terpolymer depending on w_c (Table 1).

In order to explain such a hardness depression, it is convenient to consider the behavior of the harder component dispersed in the liquid component (with "zero") hardness). It seems reasonable here to assume that the total hardness of such a system will depend also on the viscosity of the soft component in which the particles of the harder component are "floating." The deviation of the H values from the H_{cal} ones actually may reflect the viscosity of the soft-segment phase, which also contributes to the resistance of the total system against the applied load.

The viscosity of the soft phase introduces two important effects. The first one is related to the aforementioned floating effect of the hard segments within the liquid matrix. Actually, the soft phase plays a plasticizing role, reducing strongly the H^{hard} value, depending on its viscosity. The lower the viscosity, that is, the lower the T_g^{soft} (T_{g1}), the stronger the plasticizing effect will be, that is, the larger the depression in H^{hard} will be.

Table 2 Characteristics of Thermoplastic Elastomers of PEE or PEEC Type Used for

Sample no.	Copolymer	Composition (wt %)	Treatment		Crystallinity (%)	Ref.
			Drawing λ	Anneal. (°C)		
1	PBT/PTMG	60/40	—	70	37.7	15
2	PBT/PTMG/PC	60/32/8	—	70	25.8	15
3	PBT/PTMG/PC	60/20/20	—	70	24.3	15
4	PBT/PEG	57/43	5	170	41	55
5	PBT/PEG	57/43	—	25	37	54
6	PBT/PEG	57/43	—	170	41	54

Source: Ref. 68.

Obviously in both cases of thermoplastic elastomers, the PEE (55) and the PEEC (66), the contribution of the soft-segment phases—PEO- or OTMO-based—to the overall mechanical resistance against the external stress takes place via their viscosity. In contrast to the other soft amorphous solids, the viscosity of these soft phases is not high enough to be characterized by means of microhardness (for which a residual indentation impression should be formed). For this reason such low-viscosity liquids exhibit a microhardness-depression effect. The latter is the higher the lower the viscosity of the liquidlike phase (component) is.

The question arises: How can one account for this microhardness-depression effect? As already demonstrated, the simple assumption that the soft segments have $H \simeq 0$ does not solve the problem. It is necessary to characterize the ability of the harder phase to displace in the soft matrix, which will depend on the viscosity of the matrix, i.e., the soft-segments phase in the present case. Since the T_g and viscosity are closely related, we looked for an analytical relationship between microhardness of amorphous polymers and their T_g. A fairly good linear relationship between H and T_g (K) in the T_g interval between 20 and 250°C was found (67):

$$H = 1.97T_g - 571 \quad \text{(Pa)} \tag{14}$$

Equation (14) offers us the opportunity to characterize the contribution of the soft-segments phase in the case of PEE to the overall microhardness of the system. For the quantitative evaluation of the microhardness-reduction effect, one has to replace H^{soft} in Eq. (12) by Eq. (14) using for T_g the glass transition temperatures of the soft-segments phase $Ts\backslash g$ (in K):

$$H = \varphi H^{PBT} + (1 - \varphi)(1.97T_g^s - 571) \tag{15}$$

The calculation of the H value for PEE and PEEC by means of Eq. (15) gives data that are in a very good agreement with the measured values H_{exp}, as shown in Table 2.

Verification of the Relationship Between T_g and H

T_g^s		Microhardness (MPa)			
			H_{cal}	H_{cal}	H_{cal}	$H_{\text{cal}} -$
(°C)	Ref.	H_{exp}	(eq 12)	H_{exp}	(Eq. 15)	H_{exp}
−56	15	29.6	85.7	56.1	29.7	0.1
−45	15	22.8	69.5	46.7	21.8	1.0
−32	15	18.6	59.9	41.3	22.0	3.4
−44.5	54	34.2	85.2	51.0	34.4	0.2
−41	54	30.7	79.9	49.2	31.6	0.9
−43	54	32.9	85.2	52.3	35.4	2.3

Summarizing, one can conclude that owing to the linear relationship between H and T_g in a rather broad temperature interval of T_g (-50 up to $250°C$) covering a number of commonly used commercial polymers, one can account for the contribution of soft, liquidlike components and/or phases (characterized by a negligible small microhardness) to the microhardness of the entire system. It seems obvious that the plastic deformation mechanism of such systems is different from that when all the components and/or phases are solid, i.e., with a T_g above room temperature.

VII. CONCLUSIONS

The microhardness of amorphous-condensation flexible copolymers (PET/PEN, PBT/PC) can be explained in terms of a simple additive parallel model of the two independent components. Physical aging induces a significant hardening of the materials. Upon annealing, the microhardness of the copolymers increases owing to the reinforcing contribution of the arising crystalline regions. Composition crystal thickness and surface free energy are shown to be important factors in determining the final hardness value.

The microhardness of coreactive blends of condensation polymers (PET/PC) follows likewise the predictions of the microhardness additivity law of independent components. The occurrence of one single T_g in the DSC curves indicates that chemical reactions take place during melt blending, resulting in formation of polycondensates with random sequential order. The parallel linear decrease of microhardness and T_g with increasing PET content is associated to the formation of new copolymer molecules enriched in the component with lower H and T_g values. In this case one is dealing with copolymers distinguished by the creation of new chemical bonds, but not anymore with blends, although one mixes two homopolycondensates at the very beginning.

The microhardness of blends of condensation polymers (PET/PEN) prepared by coprecipitation from solution followed by melt pressing and quenching shows deviation from the additivity law of the single components. The H deviation depends on the time for which the blend was melt pressed and can be associated to the changes occurring from the two-phase structure up to the PET-co-PEN obtained by transesterification of the two components.

The microhardness of random-condensation stiff copolymers (copolyesters of HBA and HNA) show a clear deviation from the hardness linear additivity law of single components. Such a deviation is related mainly to changes in the packing of the rigid molecules. In this case, the degree of crystallinity and crystal size do not add any significant contribution to the microhardness value. The slightly higher H values found in the annealed materials is due to an increase in the H_c value together with a concurrent decrease of the hardness of the noncrystalline regions.

The microhardness method can be applied successfully to study the stress-induced polymorphic transition in crystalline-condensation multiblock copolymers (PEE/PBT/PEO). A sharp drop in microhardness due to the $\alpha = \beta$ polymorphic transition is observed in a narrow interval at deformations $\varepsilon = 25-30\%$.

The study of the multiblock terpolymers of PEEC with PBT shows the dominating contribution of crystallinity to the microhardness value. The much lower measured microhardness in contrast to the one calculated by the additivity law has been discussed in terms of the plasticizing effect of the soft amorphous phase in agreement with structure details derived from x-ray scattering data.

This chapter has emphasized areas of polymer research that offer new possibility for applications of the microindentation method to measurements of the mechanical properties of condensation copolymers. It is specific in suggesting further microhardness morphology correlations of flexible and rigid amorphous and crystallizable copolymers. Of particular interest is the applicability of the technique to characterize polymorphic phase transitions in copolymers. It may be also expected that the microhardness technique will lead to future developments in the understanding and characterization of physical aging phenomena in glassy copolymers. The localized nature of the microindentation test will allow information to be obtained regarding the heterogeneity of these multiphase systems that is often not available with other techniques. Finally, it is expected that nanoindentation techniques will offer novel possibilities to study the elastic and plastic properties of the near-surface region of copolymers.

ACKNOWLEDGMENTS

Grateful acknowledgment is due to DGICYT (Grant PB94-0049), Spain, for the support of this investigation.

REFERENCES

1. FJ Baltá Calleja. Trends Polym. Sci., 2(12), (1994) 419.
2. D Chen. PhD dissertation, University of Hamburg (1990).
3. C Santa Cruz, FJ Baltá Calleja, HG Zachmann, D Chen. J. Mater. Sci. 27 (1992) 2161.
4. S Buchner, D Chen, R Gehrke, HG Zachmann. Molec. Cryst. Liq. Cryst. 155 (1988) 357.
5. FJ Baltá Calleja, HG Kilian. Colloid Polym. Sci. 266 (1988) 29.
6. FJ Baltá Calleja, HG Kilian. Colloid Polym. Sci. 263 (1985) 697.
7. C Santa Cruz, FJ Baltá Calleja, HG Zachmann, N Stribeck, T Asano. J. Polym. Sci. Polym. Phys. 29 (1991) 819.
8. FJ Baltá Calleja, C Santa Cruz, D Chen, HG Zachmann. Polymer 32 (1991) 2252.
9. FJ Baltá Calleja, C Santa Cruz, RK Bayer, HG Kilian. Colloid Polym. Sci. 268 (1990) 440.

10. L Giri, Z. Roslaniec, TA Ezquerra, FJ Baltá Calleja. J. Macromol. Sci. Phys. B36 (1997) 335.
11. Z Roslaniec, H Wojcikiewicz. Polimery 33 (1988) 360.
12. Z Roslaniec, TA Ezquerra, FJ Baltá Calleja. Colloid Polym. Sci. 273 (1995) 58.
13. FJ Baltá Calleja, HG Kilian. Colloid Polym. Sci. 263 (1985) 697.
14. FJ Baltá Calleja, HG Kilian. In: LA Kleintjens and PJ Lemstra, eds. Integration of Fundamental Polymer Science and Technology. Elsevier, London, 1986, p. 517.
15. FJ Baltá Calleja, L Giri, TA Ezquerra, S Fakirvo, Z Roslaniec. J. Macromol. Sci. Phys. B36 (1997) 655.
16. RS Porter, LH Wang. Polymer 33 (1992) 2019.
17. XY Chen, AW Birley. Br. Polym. J. 17 (1985) 326.
18. P Godard, JM Dekoninck, V Pevlesaver, J Devaux. J. Polym. Sci. Polym. Chem. Ed. 24 (1986) 3315.
19. LH Wang, M Lu, X Yang, RS Porter. J. Macromol. Sci. Phys. B29 (1990) 171.
20. LH Wang, Z Huang, T Hong, RS Porter. J. Macromol. Sci. Phys. B29 (1990) 155.
21. GvD Velden, G Kolfschoten-Smitsamans, A Veermans. Polym. Commun. 28 (1987) 169.
22. S Fakirov, M Sarkissova, Z Denchev. Macromol. Chem. Phys. 197 (1996) 2837.
23. W Zheng, Z Wang, Z Qi. Polym. Int. 34 (1994) 301.
24. FJ Baltá Calleja, L Giri, HG Zachmann. J. Mater. Sci. 32 (1997) 1117.
25. H Yoon, Y Feng, Y Quiu, CC Han. J. Polym. Sci. B: Polym. Phys. 32 (1994) 1485.
26. FJ Baltá Calleja, C Santa Cruz, T Asano. J. Polym. Sci. Polym. Phys. Ed. B31 (1993) 557.
27. J Martínez-Salazar, FJ Baltá Calleja. J. Mater. Sci. Lett. 4 (1985) 324.
28. E Andresen, HG Zachmann. Colloid Polym. Sci. 272 (1994) 1352.
29. F Ania, J Martínez-Salazar, FJ Baltá Calleja. J. Mater. Sci. 24 (1989) 2934.
30. A Flores, F Ania, FJ Baltá Calleja. Polymer 38 (1997) 5447.
31. KF Wissbrun, HN Yoon. Polymer 30 (1989) 2193.
32. H O'Neil. In: Hardness Measurements of Metals and Alloys. Chapman & Hall, London, 1967.
33. J Martínez-Salazar, J García Peña, FJ Baltá Calleja, HG Zachmann, D Chen. Polymer Commun. 26 (1985) 57.
34. Y Deslandes, E Alva Rosa, F Brisse, T Meneghini. J. Mater. Sci. 26 (1991) 2769.
35. K Tashiro, H Tadokoro. Crystalline polymers. In: Encyclopedia of Polymer Science and Engineering. Ed. Supplement. Wiley, New York, 1987, p. 187.
36. L Mandelkern. In: Crystallization of Polymers. McGraw-Hill, New York, 1964.
37. R Jakeways, IM Ward, MA Wilding, IH Hall, IJ Desborough, MG Pass. J. Polym. Sci. Polym. Phys. 13 (1975) 799.
38. M Yokouchi, Y Sakakibara, Y Chatani, H Tadokoro, T Tanaka, K Yoda. Macromolecules 9 (1976) 266.
39. K Tashiro, M Hitamatsu, M Kobayashi, H Tadokoro. Sen'i Gakkaishi 42 (1986) 659.
40. Y Takahashi, I Sumita, H Tadokoro. J. Polym. Sci. Polym. Phys. Ed. 11 (1973) 2113.
41. K Tashiro, H Tadokoro. Rep. Progr. Polym. Phys. Jpn. 21 (1978) 417.
42. K Miyasaka, K Ishikawa. J. Polym. Sci. A-2 6 (1968) 1317.
43. JWS Hearle, BM Chapman, GS Senior. Appl. Polym. Symp. 18 (1971) 775.
44. IH Hall, MG Pass. Polymer 17 (1976) 807.

45. IJ Desborough, IH Hall. Polymer 18 (1977) 825.
46. Z Mencik. J. Polym. Sci., Polym. Phys. Edn. 13 (1978) 2173.
47. K Tashiro, Y Nakai, M Kobayashi, H Tadokoro. Macromolecules 13 (1980) 137.
48. R Jakeways, T Smith, IM Ward, MA Wilding. J. Polym. Sci. Polym. Lett. Ed. 14 (1976) 41.
49. IM Ward, MA Wilding. Polymer 18 (1977) 327.
50. FJ Baltá Calleja, S Fakirov. Trends Polym. Sci. 5 (1997) 246.
51. J Martínez-Salazar, J García, FJ Baltá Calleja. Polym. Commun. 26 (1985) 57.
52. FJ Baltá Calleja, J Martínez-Salazar, T Asano. J. Mater. Sci. Lett. 7 (1988) 165.
53. S Fakirov, D Boneva, FJ Baltá Calleja, M Krumova, AA Apostolov. J. Mater. Sci. Lett. 17 (1988) 453.
54. S Fakirov, T Gogeva. Macromol. Chem. 191 (1990) 603.
55. AA Apostolov, D Boneva, FJ Baltá Calleja, M Krumova, S Fakirov. J. Macromol. Sci.-Phys. B37(4) (1988) 543.
56. S Fakirov, C Fakirov, EW Fischer, M Stamm. Polymer 32 (1991) 1173.
57. S Fakirov, C Fakirov, EW Fischer, M Stamm. Polymer 33 (1992) 3818.
58. AA Apostolov, S Fakirov. J. Macromol. Sci. Phys. B31 (1992) 329.
59. S Fakirov, C Fakirov, EW Fischer, M Stamm, AA Apostolov. Colloid Polym. Sci. 271 (1993) 881.
60. N Stribeck, AA Apostolov, HG Zachmann, C Fakirov, M Stamm, S Fakirov. Int. J. Polym. Materials 25 (1994) 185.
61. S Fakirov, Z Denchev, AA Apostolov, M Stamm, C Fakirov. Colloid Polym. Sci. 272 (1994) 1363.
62. N Stribeck, D Sapundjieva, Z Denchev, AA Apostolov, HG Zachmann, M Stamm, S Fakirov. Macromolecules 30 (1997) 1329.
63. K Nakamae, M Kameyama, M Yoshikawa, T Matsumoto. J. Polym. Sci. Polym. Phys. Ed. 20 (1982) 319.
64. J Roebuck, R Jakeways, IM Ward. Polymer 33 (1992) 227.
65. S Fakirov, AA Apostolov, P Boeseke, HG Zachmann. J. Macromol. Sci. Phys. B29 (1990) 379.
66. FJ Baltá Calleja, S Fakirov, Z Roslaniec, M Krumova, TA Ezquerra, DR Rueda. J. Macromol. Sci. Phys. B37 (1998) 219.
67. S Fakirov, FJ Baltá Calleja, M Krumova. J. Polym. Sci. Polym. Phys. Ed. 37 (1999) 1413.

8

Correlation Between Phase Behavior, Mechanical Properties, and Deformation Mechanisms in Weakly Segregated Block Copolymers

Roland Weidisch* and Goerg H. Michler
Martin Luther University Halle-Wittenberg, Halle, Germany

I. INTRODUCTION

The development of materials with improved property profiles is one of the important fields of materials science. The use of block copolymers aiming at specific mechanical properties opens up a wide field of possibilities due to different available highly ordered structures. It is well known that enhancement of toughness of polymers is possible in high-impact polymers via the incorporation of a dispersed elastomeric phase. This effect is caused by multiple crazing or multiple cavitation with shear yielding, which is macroscopically often shown by the phenomenon of stress-whitening (1,2). Whereas polymer blends show macrophase-separated structures, which often leads to a deterioration of mechanical properties due to the immiscibility of the components, in block copolymers microphase-separated morphologies at the typical size scale 10–100 nm are observed (3). Many authors report the existence of different ordered structures, including BCC spheres, hexagonally packed cylinders (HEX), ordered bicontinuous double diamonds

---
Current affiliation: Institute of Polymer Science and Engineering, University of Massachusetts at Amherst, Amherst, Massachusetts

(OBDDs), cubic bicontinous ("gyroid") structures, and lamellar and perforated lamellar structures (HPL) (4–7).

Most studies in the field of block copolymers are focused on morphology and phase behavior. In contrast, there are only few studies on the mechanical properties of block copolymers. Many authors have investigated block copolymers consisting of polystyrene (PS) and polybutadiene (PB) or polyisoprene (PI). However, it was shown that block copolymers can show improved mechanical properties. Poly(styrene-butadiene-styrene) triblock copolymers (SBS) are one example of thermoplastic elastomers (TPEs), which are well-known for their unique thermomechanical properties associated with a phase morphology of PS domains dispersed in a continuous rubbery PB matrix. In such TPEs the mechanical performances of vulcanized rubbers are combined with the straightforward processing of thermoplastics due to the physical network of the flexible chains (8). Special attention has recently been paid to poly(alkyl methacrylate)s as candidates for the outer blocks (9). In contrast to SBS triblock copolymers, styrene-butadiene (SB) diblock copolymers show only a small tensile strength due to the absence of the bridged midblock conformation that is generally observed in SBS triblock copolymers. The deformation behavior of the PS cylinders in SBS triblock copolymers at higher strains has been intensively investigated via various methods (10–16).

Russell et al. (17) recently found a microphase separation of poly(styrene-b-butyl methacrylate) (PS-b-PBMA) in small-angle x-ray scattering (SAXS) experiments that have shown the existence of both an upper critical order transition (UCOT) and a lower critical order transition (LCOT). In contrast to other intensively investigated block copolymers, PS-b-PBMA shows a weak segregation at higher molecular weights, which allows us to investigate the correlation between phase behavior, mechanical properties, and deformation mechanism in the molecular weight range of $M_n > 200$ kg/mol, where the mechanical properties do not show a molecular weight dependence.

The aim of this section is to report the latest results on the correlation existing between phase behavior, mechanical properties, and deformation mechanisms in weakly segregated block copolymers.

II. EXPERIMENTAL DETAILS

A. Sample Preparation

The samples investigated are model materials of several block copolymers and include:

> Poly(styrene-b-butyl methacrylate) (PS-b-PBMA) diblock copolymers
> Poly(butyl methacrylate-b-styrene-butylmethacrylate) (PBMA-b-PS-b-PBMA) triblock copolymers

Poly(methylmethacrylate-*b*-butylmethacrylate) (PMMA-*b*-PBMA) diblock copolymers

The detailed composition of the samples studied is characterized in Tables 1–3 (see pages 229, 234, and 245). Details of preparation are described in Ref. 18. All polymerizations were carried out in carefully flamed glass reactors with THF as the solvent at $-78°C$ under argon atmosphere using syringe techniques. After several cycles of degassing the monomer from calcium hydride, the monomer was introduced into the reactor by condensation under reduced pressure. The THF was condensed from oligomeric anions. For PS-*b*-PBMA and PMMA-*b*-PBMA diblock copolymers, the desired amount of initiator was added at once, and after 15 min the living polystyrene or methylmethacrylate anions were end-capped with diphenylethylene. Butyl methacrylate as the second monomer was added dropwise very slowly with a syringe. The living anions were terminated by adding methanol after another 30 min. Then the polymer was precipitated in a ⁷/₃ methanol/water mixture at $-30°C$ and washed and dried in a vacuum for several days. For PBMA-*b*-PS-*b*-PBMA triblock copolymers, 1,4-diphenyl-1,4-dilithium was used as bifunctional iniator. After dagassing, the styrene was introduced into the reactor together with naphthyllithium (in situ reaction of naphthyllithium to 1,4-diphyenyl-1,4-dilithium). Then the living polystyrene anions were end-capped with diphenyllithium, and butyl methacrylate was added dropwise as a second monomer, which is the same procedure as already described for diblock copolymers. In contrast to diblock copolymers, a bicunctional start of anionic polymerization occurs.

All samples were dissolved in toluene, and the solvent was allowed to evaporate slowly over 5–7 days at room temperature. Then the films were dried to constant weight in a vacuum oven at 120°C for 3 days.

B. Techniques

Size exclusion chromatography (SEC) measurements were carried out using a Knauer-GPC with an RI/Visco detector and a PSS linear column. The volume fraction of the diblock copolymers were estimated by ¹H-NMR.

Transmission electron microscopy (TEM) investigations were performed using ultrathin sections (50 nm) cut at room temperature using glass knives in an Ultramicrotome (Reichert). The polystyrene blocks were stained with RuO_4 vapor. Small-angle neutron scattering (SANS) measurements were performed at the KWS II small-angle scattering facility located at GKSS Geesthacht. All samples were melt-pressed into 1-mm-thick and 13-mm-diameter disks. The instrument configuration was $\lambda = 0.91$ nm and $\Delta\lambda/\lambda = 0.2$ due to the velocity selector and sample detector distance of 5.6 m. The scattering data were corrected for detector sensitivity, background scattering, used-sample thickness, and transmission and placed on an absolute basis using several standards. The SANS profiles at different temperatures are discussed as a function of the scattering vector $q = (4\pi/\lambda)$ $\sin(\theta)$, where 2θ is the scattering angle. The dynamic storage and loss shear moduli, G' and G'', were determined with a Rheometrics RMS 800 using the temperature step mode and a frequency of 1 Hz. Tensile tests were performed using a universal testing machine (Zwick 1425) at different strain rates and temperatures. For each temperature and strain rate, at least 10 samples were measured in order to avoid preparation effects. The sample dimensions of tensile specimens had a thickness of 0.5 mm and a total length of 50 mm. The toughness of the diblock copolymers was estimated as absorbed energy from the stress–strain curves.

To investigate the micromechanical deformation behavior, semithin sections with a thickness on the order of 0.5 μm were strained in a high-voltage electron microscope (HVEM, Jeol 1000) equipped with an in situ tensile device. The advantage of HVEM investigations is the possibility to use thicker sections for closer comparison with bulk materials.

III. MORPHOLOGY AND PHASE BEHAVIOR

A. PS-b-PBMA Diblock Copolymers

The order–disorder transition (ODT) of block copolymers has been intensively studied by many authors during recent years. The theory of Leibler (19) gives a description of this transition, which predicts that the ODT for a symmetric block copolymer belongs to a second-order phase transition. In contrast to this, for asymmetric compositions a first-order phase transition can be expected. For exact symmetric compositions the phase transition takes place by spinodal decomposition due to the second-order transition. However, the theory of Fredrickson and Binder (20) predicts that the ordering of block copolymer melts also occurs via nucleation and growth for nearly symmetric compositions due to a thermal-force-induced first-order phase transition. The ordering process of diblock copolymers has been intensively investigated by means of SAXS, SANS, rheology, and depolarized light scattering (21–29). It was shown that the ordering process proceeds via nucleation and growth with an incubation period (30). Recently, Sakamoto and Hashimoto (31) found, by TEM, a coexistence of ordered and disordered regions in a PS-b-PI diblock copolymer close to ODT.

The Leibler theory predicts that the Flory–Huggins segmental interaction parameter χ depends on temperature according to $\chi = A + B/T$, where A and B are constants. Furthermore, the theory predicts that the maximum of the reciprocal scattering intensity I^{*-1} and the square of width at half maximum $(\Delta q)^2$ change linearly with the reciprocal value of temperature. This behavior is expected from the mean field theory of Leibler, where thermal fluctuations are neglected. However, for different systems, a deviation from this behavior was found.

A discontinuous change of peak intensity and $(\Delta q)^2$ at ODT can be observed for many systems due to the thermal force effect. The product χN dictates the degree of segregation of A and B blocks. For a symmetrical diblock copolymer, the order–disorder transition is predicted by Leibler (17) in terms of a mean field model by:

$$(\chi N)_{ODT} = 10.495 \tag{1}$$

The theory of Fredrickson and Helfand (32) includes fluctuation effects that lead to a correction of the mean field theory on the order of $N^{-1/3}$. The prediction of composition fluctuations changes the mean field prediction of a second-order phase transition for symmetric diblock copolymers to a first-order transition. Furthermore, the theory of Fredrickson and Helfand predicts the order–disorder transition at higher χN values than obtained by mean field theory. In contrast to mean field theory, Fredrickson and Helfand predict direct transitions from disordered state to hexagonal and lamellar structures for asymmetric compositions. Leibler's theory was extended to triblock copolymers, star-block copolymers (33), and multigraft copolymers (34).

Most block copolymer systems show a transition to a disordered state as temperature is raised, which is denoted as the order–disorder transition or upper critical order transition (UCOT). One example is poly(styrene-b-isoprene) diblock copolymers, where the UCOT was intensively investigated by SAXS and rheology (35–38). In contrast to this, Russell and coworkers (17,39) reported the existence of both an upper (UCOT) and lower critical order transition (LCOT) in poly(styrene-b-butylmethacrylate) diblock copolymers. Hashimoto et al. (40) reported an LCOT in poly(vinyl methyl-ether-b-styrene) diblock copolymers (PS-b-PVME) by SANS. The temperature dependence of the interaction parameter χ for PS-b-PVME diblock copolymers was determined as $0.048 - 31/T$. This means that χ increases with increasing temperature, indicating a transition to an ordered state at high temperatures. In contrast to the classical enthalpically driven UCOT, the LCOT results suggesting a decrease in entropy (41). Systems showing a LCOT can be categorized either as those with strong interactions (e.g., dipole/dipole interactions) or as weakly interacting systems with molecular packing differences. One example for systems with strong interactions are PS-b-PVME diblock copolymers, where the loss of configurations is due to the preferred orientation of the interacting chemical groups. PS-b-PBMA diblock copolymers represent an example of weakly interacting systems, where molecular packing differences lead to differences in the free volume of the pure components that are reflected in their specific volumes and thermal expansion coefficients (41).

In Fig. 1 the scattering profiles for a dPS-b-PnBMA diblock copolymer with 90 kg/mol and 55 vol % PS are shown as a function of temperature. At lower temperatures, the sample exhibits a diffuse scattering maximum, with a wide full width at half-maximum, Δq. With increasing temperature, the peak intensity in-

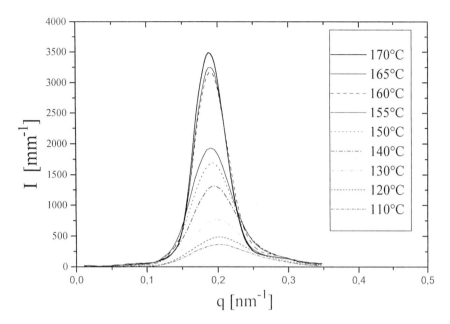

Fig. 1 Small-angle neutron scattering (SANS) profiles for a P(d-S-*b*-BMA) diblock copolymer with 55% PS and M_n = 90,000 g/mol at various temperatures as a function of the scattering vector *q*.

creases and sharpens. This means that the copolymer is approaching an ordering transition with increasing temperature (LCOT) (42).

As the disorder–order transition is approached, the composition profile becomes increasingly sinusoidal and the softening of the ordered state permits the stretched coils to get into their unperturbed conformation with decreasing temperature (43). Whereas for poly(styrene-*b*-isoprene) diblock copolymers a relatively large dependence of χ on temperature was found, for PS-*b*-PMMA diblock copolymers the dependence of χ is quite small as compared to other diblock copolymers (44). While for this block copolymer UCOT behavior can be observed, for dPS-*b*-PBMA, χ increases with increasing temperatures (42):

$$\chi = (0.0243 \pm 0.0004) - \frac{4.56 \pm 0.169}{T} \tag{2}$$

As shown in Fig. 2, χ rises with increasing temperature, indicating LCOT behavior. The temperature dependence of χ is small as compared to other block copolymers. It is clear from Eq. (2) that the entropic contribution to χ (first term in Eq. 2) is relatively large in contrast to that of the enthalpic contribution (second term in Eq. 2). As recently shown by Ruzette et al. (45) for PS-b-poly(alkyl methacrylate) diblock copolymers with long alkyl side-chain methacrylates, $n > 5$, the entropic and enthalpic contributions are comparable in size to our system. However, this systems show UCOT behavior. The small temperature dependence obtained for all systems is due to the large entropic contributions to χ. While PS-b-alkyl methacrylate diblock copolymers with long side-chain methacrylates, $n \geq 6$, show UCOT behavior, diblock copolymers with short side-chain methacrylates, $n < 5$, reveal LCOT behavior (45).

As χN increases to close to ODT, the phases are only weakly segregated where the microscopic density profile of the components is considered to vary sinusoidaly in space and the chains of the components are highly interpenetrating. It is shown by various authors (46) that between the weak segregation limit (WSL) and the strong segregation limit (SSL), an intermediate segregation regime (ISR) between $12.5 < \chi N < 95$ can be identified (crossover between ISR and SSL at $\chi N \approx 50$). The chains are stretched due to the coarsening of the density profile as χN

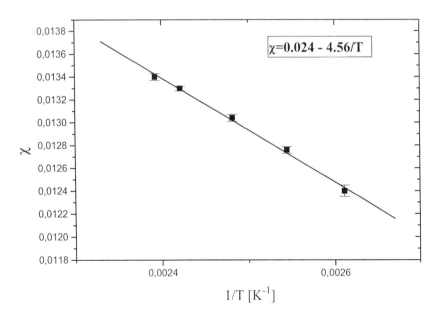

Fig. 2 Interaction parameter χ as a function of reciprocal temperature for a dPS-b-PnBMA diblock copolymer with $M_n = 90$ kg/mol and 55% PS.

is increased from the WSL (47). The interface is relatively broad, and the junction points are not completely localized in the interfacial region. For a symmetric dPS-b-PBMA diblock copolymer with 278 kg/mol, χN is about 29 (determined from Eq. 2). This means that this sample is intermediately segregated. In contrast to this, PS-b-PI diblock copolymers are strongly segregated at such high molecular weights due to a much larger interaction parameter between PS and PI (48).

For strongly segregated block copolymers, the theory of Helfand and Wasserman (49) assumes that the interface between A and B domains is narrow as compared to the domain size, which depends only on composition. Hagesawa et al. (50) observed in PS-b-PI diblock copolymers an interface width of about 2 nm, which supports the theoretical prediction. In contrast, for a symmetric dPS-b-PBMA diblock copolymer with 248 kg/mol, a large interface width of 8.4 nm at 151°C was determined by NR due to the intermediate segregation at this molecular weight, which confirms the theoretical prediction (42,51). The observed large interface width and low χ parameter for the investigated system is confirming the observed partial miscibility at $M_n > 200$ kg/mol and different compositions observed by dynamic mechanical analysis (DMA) (52,53). Figure 3 shows that for PS-b-PBMA diblock copolymers, only the glass transition temperature (T_g) of the PBMA block is shifted to higher temperatures, whereas the T_g of the PS block remains at approximately 100°C. The T_g of the PBMA block for a sample with a molecular weight of 212 kg/mol and 40% PS is shifted from 31°C for pure PBMA up to 43°C (52). This behavior was also found for other compositions. To explain these results we have to assume an asymmetric phase diagram, where asymmetric phase compositions are also expected. Therefore, an essentially pure polystyrene matrix phase and a PS/PBMA mixed phase may exist (52,54).

Block copolymers usually show a grain structure at a size scale of 1–10 μm as well as structures on the nanometer-scale. Such materials will exhibit isotropic properties in the case of the absence of macroscopic orientations. Therefore, block copolymers show a high transparency, which leads to interesting applications. In poly(styrene-b-isoprene) (PS-b-PI) diblock copolymers, the following morphologies were reported: BCC spheres, hexagonally packed cylinders, and lamellar structures (4,5). In the weak segregation limit, the perforated layers and the cubic bicontinuous structure ("gyroid") were also found (6,7).

For a PS-b-PBMA diblock copolymer with 39% PS, a bicontinuous structure was found by TEM, which, however, does not exist at compositions of 65–70% PS, where for PS-b-PI diblock copolymers the gyroid structure was observed. The existence of perforated (HPL) and modulated lamellar structure was described by Bates and coworkers for PS-b-PI diblock copolymers (6). In Fig. 4 one example of the morphology consisting of perforated lamellae of PS in PS-b-PBMA diblock copolymers with 40% PS is shown. It is obvious that parts of the sample are cut perpendicularly through the perforated lamellae and other parts are

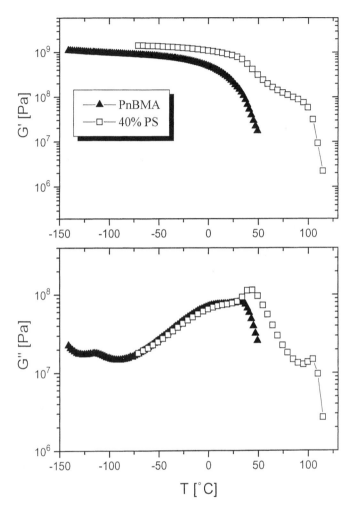

Fig. 3 Dependence of storage modulus (G') and loss modulus (G'') on the temperature for a PS-b-PBMA diblock copolymer (sample SBM40 Φ_{PS} = 0.40, M_n = 212 kg/mol) and pure PBMA (M_w = 363 kg/mol) measured at a frequency of 1 Hz.

cut in the plane of the lamellae. In both directions the holes in the stained PS layers are visible. Whereas for PS-b-PI diblock copolymers the complex structures exist on both sides of the phase diagram, for PS-b-PBMA diblock copolymers with 72% and 74% PS a coexistence of regions with lamellae and hexagonally packed cylinders (LAM/HEX) was observed (52). This structure cannot be ex-

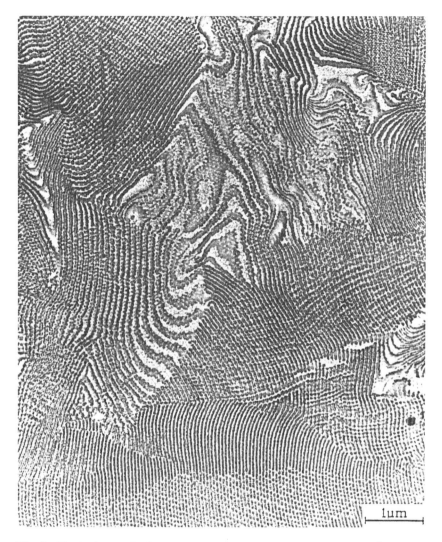

Fig. 4 TEM micrograph of a PS-*b*-PBMA diblock copolymer (SBM40 with $\Phi_{PS} = 0.40$, $M_n = 212$ kg/mol) with perforated lamellae of PS (stained with RuO_4) (52).

pected from Leibler's theory (17), which is based on equal Kuhn segment length and volume of the respective homopolymers. However, in the case of PS-*b*-PBMA diblock copolymers, the segment length and monomer volume of PS and PBMA are different. Furthermore, in the weak segregation limit (WSL) and intermediate segregation regime (ISR), the differences between the free energies for different structures is very small, which makes it difficult to achieve the equilib-

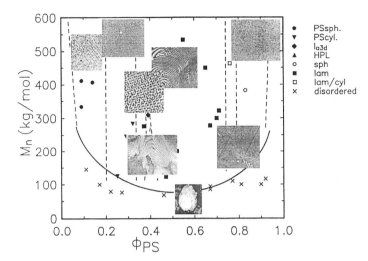

Fig. 5 Morphological phase diagram of the system PS-*b*-PBMA. The inserted pictures are typical for the morphologies found for the different marked areas. The dashed lines separate areas of different morphologies from each other. The abbreviations have the following meaning: PS_{sph} = PS spheres, PS_{cyl} = PS cylinders, I_{a3d} = gyroid structure, HPL = hexagonal perforated lamellae of PS, lam = lamellae, lam/cyl = coexistence of lamellae and cylinders, sph = PnBMA rich material in PS matrix, disordered = mixed phase (54).

rium morphology for samples with high molecular weight. A further increase of PS content, up to 76%, leads to a morphological change into hexagonally packed PBMA cylinders. For PS-*b*-PBMA diblock copolymers, HPL or bicontinuous structures were not observed for high PS contents, which indicates the existence of an asymmetric phase diagram for PS-*b*-PBMA diblock copolymers (Fig. 5) (52,54).

B. PBMA-*b*-PS-*b*-PBMA Triblock Copolymers and PMMA-*b*-PBMA Diblock Copolymers

For PBMA-*b*-PS-*b*-PBMA triblock copolymers, a partial miscibility is also found, where only the T_g of the PBMA block is shifted to higher temperatures. For a triblock copolymer with 37% PS and 275 kg/mol (sample Tri37), the Tg of the PBMA block is shifted up to 50°C, which is about 7°C higher than observed for PS-*b*-PBMA diblock copolymers at the same composition (55). These findings can be explained by an enhanced miscibility of triblock copolymers in comparison to that of diblock copolymers. While for a PS-*b*-PBMA diblock copolymer with 35% PS (sample SBM35) the value of χN is 27.9, for the sample Tri37 χN-eff, χN is only about 20. In order to compare block copolymer systems with dif-

ferent interaction parameters χ between the components, PS-b-PBMA were compared with PMMA-b-PBMA diblock copolymers (55). For PMMA-b-PBMA diblock copolymers the interaction parameter between PMMA and PBMA was determined as $\chi_{PMMA/PBMA} = 0.071$ at $T = 140°C$ (42,56), which is larger than
obtained for PS-b-PBMA diblock copolymers. A smaller interface width was observed for PMMA-b-PBMA (4 nm at $T = 140°C$) diblock copolymers as compared to that of PS-b-PBMA (7.7 nm at $T = 140°C$) diblock copolymers. This is
due to the larger interaction parameter between the components in PMMA-b-
PBMA diblock copolymers. In contrast to PS-b-PBMA diblock copolymers, for
PBMA-b-PS-b-PBMA and PMMA-b-PBMA diblock copolymers, HPL,
LAM/HEX, and bicontinuous structures were not observed. PMMA-b-PBMA diblock copolymers show spherical structures, hexagonally packed cylinders, and
lamellar structures, depending on composition. The reason for this observation
could be the decreased miscibility of PMMA-b-PBMA diblock copolymers as
compared to that of PS-b-PBMA diblock copolymers. It was shown that HPL and
bicontinuous structures exist near the order–disorder transition, where the microphases are only weakly or intermediately segregated. PBMA-b-PS-b-PBMA
triblock copolymers present spheres, cylinders, and lamellar structures in the investigated composition regime (55).

IV. MECHANICAL PROPERTIES

A. PS-b-PBMA Diblock Copolymers

In contrast to SBS triblock copolymers, SB diblock copolymers show only a small
tensile strength, due to the absence of the midblock conformation that is believed
to be important for TPEs. Since the tensile properties of TPEs arose mainly from
these midblock conformations, a heterophase morphology resulting from a strong
incompatibility of the components is necessary. In this case the interface width between the components is quite small. In contrast to TPEs, weakly segregated block
copolymers do not show a strong segregation forming a sharp interface between
the blocks. It is noteworthy that in the case of weakly segregated block copolymers, even diblock copolymers can lead to improved properties. For PS-b-PBMA
diblock copolymers with 76% PS, a tensile strength was found that is about 40%
higher than that of pure PS (Fig. 6) (57). In the composition range of 72%–76%
PS, not only was a higher tensile strength and a larger strain at break observed as
compared to pure PS.

Furthermore, Young's modulus at this composition, as indicated by the
slope of the stress–strain curves at small strain, shows a higher value than that for
pure PS. Young's modulus shows a maximum for hexagonal structures in the
composition range of 72–76% PS (Fig. 7) (58). For block copolymers with 83%
PS, Young's modulus decreases and shows almost the same value as that for pure

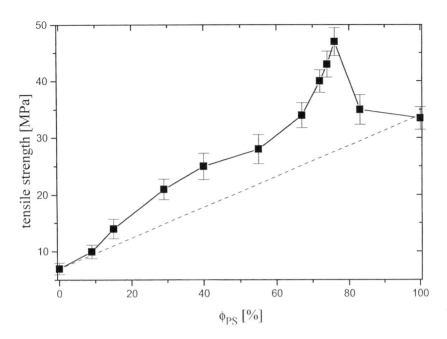

Fig. 6 Dependence of tensile strength on volume fraction of PS for PS-*b*-PBMA diblock copolymers and pure PS and PBMA at a strain rate of $\dot{\varepsilon} = 1.6 \times 10^{-4} s^{-1}$.

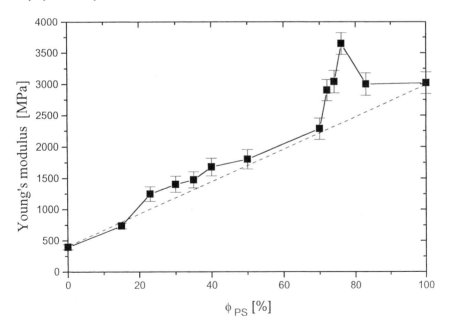

Fig. 7 Dependence of Young's modulus on the volume fraction of PS for PS-*b*-PBMA diblock copolymers.

PS. This means that for hexagonally packed cylinders at 76% PS, important properties such as tensile strength, stiffness (Young's modulus), and strain at break are significantly improved as compared to that of pure PS. In contrast to TPE's, PS-*b*-PBMA diblock copolymers also show thermoplastic properties at small PS content, due to the high glass transition temperature of the soft component PBMA, which is a rubbery component in the case of TPEs. While SBS triblock copolymers reveal optimum tensile properties at 20–35% PS, for PS-*b*-PBMA diblock copolymers a maximum of tensile strength and stiffness were observed at a much higher PS content, where a PS matrix is present. This finding arises from the differences of toughening and stiffening mechanisms between TPEs and the weakly segregated block copolymers, which will be discussed later.

Several authors have reported the existence of perforated lamellar and bicontinuous structures in block copolymer systems. It is worthy to investigate the correlation between these morphologies and mechanical properties, especially for bicontinuous structures, because improved mechanical properties may be expected due to their high storage modulus (59). In Fig. 8 the mechanical properties of PS-*b*-PBMA diblock copolymers in the composition range of 30–50% PS are shown. Diblock copolymers with 39% PS reveal a bicontinuous structure and

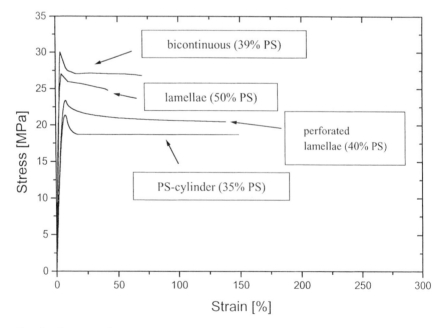

Fig. 8 Stress–strain curves for PS-*b*-PBMA diblock copolymers with different morphologies in the composition range of 35–50% PS at a strain rate of $\dot{\varepsilon} = 1.6 \times 10^{-4}\text{s}^{-1}$.

show not only a higher strain at break but also a much higher stiffness and tensile strength than a diblock copolymer with lamellar structure (50% PS). For hexagonally packed PS cylinders (sample SBM35—see Table 1), the strain at break is notably higher than observed for other morphologies. One may conclude that for PS-b-PBMA diblock copolymers with hexagonally packed PS cylinders (SBM 35), the presence of a PBMA matrix leads to a more homogeneous deformation. Perforated lamellar structures (40% PS) show intermediate mechanical properties. The perforated lamellar structure shows a higher strain at break but a lower stiffness and tensile strength. Our studies reveal that the perforated lamellar structure and hexagonally packed PS cylinder are more effective for an enhancement of toughness than lamellar structures. In contrast, the bicontinuous structure is very effective for an enhancement of tensile strength and stiffness (58).

It is important to describe the tensile properties at higher strain rates, which are more important for technical applications. As already found for other polymers, for PS-b-PBMA diblock copolymers, the tensile properties also show a strong dependence on strain rate. It was shown that the synergetic effects on tensile strength and absorbed energy also exist for high or strain rates (58). In contrast to the properties at lower strain rates, the maximum tensile strength was ob-

Table 1 Sample Characterization Data (58)

Sample	M_n [kg/mol][a] (M_w/M_n)	Φ_{PS} Block[b]	Morphology (TEM)	χN
PBMA	285.0 (1.03)	0	—	—
SBM 15	406.8 (1.05)	0.15	PS, spheres	38.3
SBM 25	277.0 (1.05)	0.23	PS, cylinder	27.0
SBM 30	278.0 (1.06)	0.30	PS, cylinder	27.5
SBM 35	270.0 (1.08)	0.35	PS, cylinder	27.2
SBM 39	254.0 (1.05)	0.39	Bicontinuous	25.9
SBM 40	212.1 (1.05)	0.40	Perforated lamellae	21.7
SBM 50	278.0 (1.07)	0.51	Lamellae	29.5
SBM 67	450.0 (1.05)	0.67	Lamellae	50.1
SBM 70	268.0 (1.05)	0.70	Lamellae	32.1
SBM 72	426.0 (1.04)	0.72	Lamellae/PBMA, cylinder	48.1
SBM 74	463.0 (1.03)	0.74	Lamellae/PBMA, cylinder	52.6
SBM 76	459.0 (1.09)	0.76	PBMA, cylinder	52.4
SBM 83	383.1 (1.04)	0.83	PBMA, spheres	44.6
PS	315.0 (1.02)	1	—	—

Molecular weight (M_n), volume fraction (Φ_{PS}), and polydispersity (M_w/M_n), χN values at 120°C (determined from Eq. 2) and morphology (TEM) for PS-b-PBMA diblock copolymers used in this study.
[a] Total molecular weight and polydispersity determined by size exclusion chromatography (SEC); values are based on the PS standards.
[b] Volume fraction of PS determined by ^{1}H-NMR.

served already at a PS content of 72% (sample SBM72; LAM/HEX) due to the premature failure of block copolymers with 76% PS (Fig. 9). This means that the maximum tensile strength is shifted to higher PBMA content as strain rates increase. Block copolymers with LAM/HEX (sample SBM72, SBM74) and HEX (sample SBM76) structures show improved mechanical properties at all measured strain rates as compared to those of PS.

The absorbed energy, estimated from the stress–strain curves, shows a maximum at 35% PS for hexagonally packed PS cylinders (58) (Fig. 10). A sharp decrease of absorbed energy occurs for perforated lamellae at 40% PS. Lamellar structures show a small absorbed energy (toughness) as compared to the morphologies in the composition range of 25–40% PS. The generally higher absorbed energy of samples with hexagonally packed PS cylinders (25–35% PS) as compared to that of perforated lamellar structures (HPL, 40% PS) may be attributed to a change of the deformation mechanism.

For SBS triblock copolymers with hexagonally packed PS cylinders, large deformations of the PS cylinders in the PB matrix are found, which are responsible for their enhanced strength. For diblock copolymers with LAM/HEX (72–74% PS) and HEX (76% PS) structures, the tensile strength also exceeds the value of pure PS for higher temperatures up to about 70°C. This means that LAM/HEX and HEX structures are also effective at higher temperatures due to different deformation mechanisms as compared to those of pure PS (58).

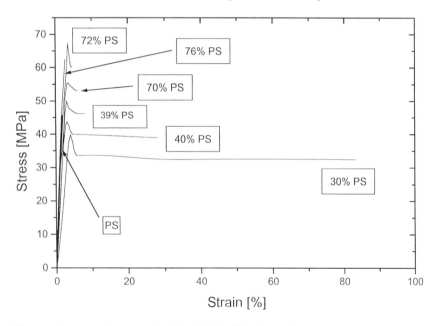

Fig. 9 Stress–strain curves for PS-*b*-PBMA diblock copolymers depending on volume fraction of PS at a strain rate of $\dot{\varepsilon} = 5.5 \times 10^{-2}\,\mathrm{s}^{-1}$.

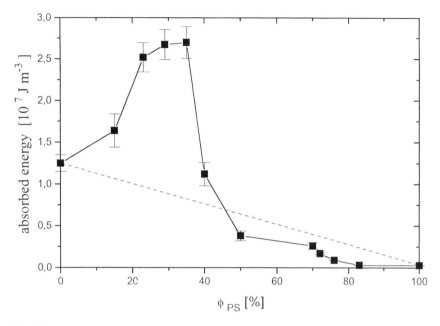

Fig. 10 Dependence of absorbed energy for PS-*b*-PBMA diblock copolymers on volume fraction of PS at a strain rate of $\dot{\varepsilon} = 5.5 \times 10^{-2}\text{s}^{-1}$.

The properties of TPEs arising from the midblock conformation due to the triblock architecture can lead to the formation of bridges between the PS blocks. The conditions for this behavior are strongly segregated components with narrow interfaces, and a distinct molecular weight of the polystyrene blocks. The lower limit of polystyrene block size is set by the incompatibility requirements. The upper limit of molecular weights is set by viscosity considerations, which affect both processability and efficiency of phase separation in the melt. This means that the styrene–diene ratio (morphology) controls the modulus of the TPE, where the polystyrene domains act as "fillers." Because of these conditions, the molecular weight of the PS blocks lies generally in the range of 10,000 to 15,000, while the polydiene molecular weight varies from 50,000 to 70,000 g/mol (3).

In weakly segregated block copolymers, the interfaces are broadened (42,58). The investigations of Bühler (60) have shown that a broadened interface in block copolymers can enhance the tensile strength, in accordance with the properties observed for PS-*b*-PBMA diblock copolymers. Usually one can assume that in the interfacial region a larger number of defects results in a low interfacial strength. For diblock copolymers, chain ends are localized preferentially in the middle of the respective components, and the chemically coupled chains in the interfacial region can enhance the strength as compared to blends. It is further shown that for block copolymers an increasing interface width leads to an decreasing interfacial energy. This is connected with a decreasing stress concentration at the interface, which is responsible for the increase of tensile strength. Using Eq. (2) it is possible to determine the χ-parameter of PS-b-PBMA at 120°C ($\chi = 0.0127 \pm 0.001$) at which the samples were annealed (42). Table 1 gives the χN value for different samples. As already discussed, PS-b-PBMA diblock copolymers are intermediately segregated. In contrast, for sample SBM76, which shows the largest value of tensile strength, the χN value is about 52. We can assume that samples with asymmetric compositions are also intermediately segregated because for those compositions the value of $(\chi N)_{\text{critical}}$ increases as compared to symmetrical compositions. This means that all investigated samples in this study are intermediately segregated. This is confirmed by the high measured value of the interface width (42,51). In contrast, block copolymers with molecular weights of about 130 kg/mol are weakly segregated or disordered. It was shown that the disordered state is combined with a deterioration of mechanical properties, reflecting the influence of phase behavior on tensile properties (61).

Also, Young's modulus can be influenced by the morphology. For TPEs the tensile modulus appears to be dependent mainly on the PS content and independent of the molecular weight of the center elastomeric block. The molecular weight of the rubbery center block is at least of 40,000 g/mol, whereas the entanglement molecular weight for PI is 7000. Hence, there are obviously a number of such chain entanglements between the PS blocks, and the "network chains" can really be considered as equivalent to the entanglement molecular weight value (3). The stiffening mechanisms in weakly segregated block copolymers are complex and cannot be described by different models for elastic constants of composites, including the models of Reuss, Voigt, and Takayanagi. These models cannot give a reliable description of the properties of our system with complex morphologies because they include only the elastic constants of the respective homopolymers and the volume fraction and the Poisson's constant but do not consider the specific morphologies of the polymers and their interface width (58).

B. PBMA-b-PS-b-PBMA Triblock Copolymers

For triblock copolymers also, tensile strength strongly increases with increasing polystyrene content (Fig. 11). Here, even triblock copolymers with a polystyrene content of 37% show a similar tensile strength as polystyrene. At $\phi_{PS} = 0.72$, a maximum tensile strength was found, which is also significantly higher than that of polystyrene (55,57). This result indicates that for triblock copolymers the synergetic effect of tensile strength occurs over a wider composition range than for diblock copolymers, due to the enhanced miscibility of triblock copolymers, as already discussed in Sec. III.B. It is surprising that triblock copolymers with a polystyrene middle block can show such high strength, because, for example, the tensile strength of PS-b-PB diblock copolymers or PB-b-PS-b-PB triblock copolymers with about 30% polybutadiene content is markedly smaller (3). The

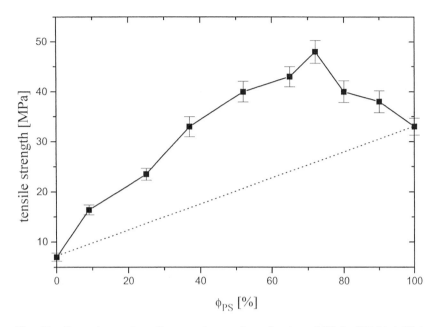

Fig. 11 Dependence of tensile strength on volume fraction of PS for PBMA-*b*-PS-*b*-PBMA triblock copolymers and pure PS and PBMA at a strain rate of $\dot{\varepsilon} = 1.6 \times 10^{-4} s^{-1}$.

strain at break strongly decreases with increasing PS content, which is generally smaller than that observed for PS-*b*-PBMA diblock copolymers. Even for triblock copolymers at a PBMA content of about 50%, a transition to brittle properties can be observed. It has been shown that Young's modulus for triblock copolymers is higher than that observed for diblock copolymers (55). For a triblock copolymer with 72% PS, Young's modulus markedly exceeds the value for pure PS of about 30%. The improved tensile strength of triblock copolymers as compared to that of diblock copolymers can be explained by the enhanced miscibility of triblock copolymers in comparison to that of diblock copolymers. As shown in Table 2, the χN_{eff} values are generally smaller in the case of triblock copolymers. This should be associated with an increased interface width, which is responsible for the improved tensile strength of triblock copolymers.

C. PMMA-*b*-PBMA Diblock Copolymers

In the case of PMMA-*b*-PBMA diblock copolymers, a maximum tensile strength is also observed that exceeds the value for pure PMMA. However, in contrast to di- and triblock copolymers of PS and PBMA, the tensile strength for a sample with 77% PMMA is only about 15% higher than that of pure PMMA (42,55) (Fig. 12). The strain at break decreases with increasing PS content. This was even ob-

Table 2   Sample Characterization Data

Sample	$10^{-3} \times M_n$ copolymer[a] (M_w/M_n)	Φ_{PS} Block[b]	Morphology (TEM)	χN_{eff}
Tri9	334.0 (1.07)	0.09	PS, spheres	17.2
Tri25	312.3 (1.12)	0.25	PS, cylinder	20
Tri37	275.1 (1.08)	0.37	Lamellae	20.1
Tri52	201.1 (1.10)	0.52	Lamellae	17.0
Tri65	305.3 (1.06)	0.65	Lamellae	29
Tri72	324.6 (1.09)	0.72	PBMA, cylinder	32.6
Tri80	325.5 (1.12)	0.80	PBMA, spheres	34.7
Tri90	299.2 (1.11)	0.90	PBMA, spheres	34.2

Molecular weight (M_n), volume fraction (Φ_{PS}), and polydispersity (M_w/M_n), and χN_{eff} values at 120°C (determined from Eq. 2) and morphology (TEM) for PBMA-*b*-PS-*b*-PBMA diblock copolymers used in this study.
[a] Size exclusion chromatography (SEC); values are based on the PS standards.
[b] Volume fraction of PS determined by ^{1}H-NMR.

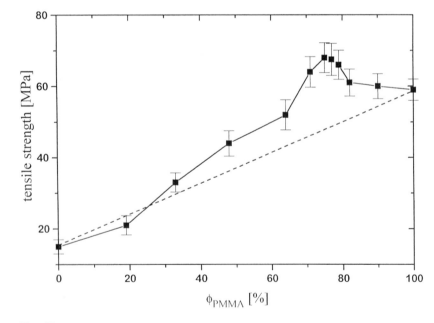

Fig. 12   Dependence of tensile strength on volume fraction of PMMA for PMMA-*b*-PBMA diblock copolymers and pure PMMA and PBMA at a strain rate of $\dot{\varepsilon} = 1.6 \times 10^{-4}\,s^{-1}$.

served for other block copolymers. The strain at break of PMMA-*b*-PBMA diblock copolymers is smaller than that of PS-*b*-PBMA diblock copolymers, which is attributed to the enhanced incompatibility of this block copolymer system. In contrast to di- and triblock copolymers of PS and PBMA, Young's modulus does not exceed the value of pure PMMA (55).

The reason for the observed different tensile properties of PS-*b*-PBMA and PMMA-*b*-PBMA diblock copolymers can be discussed on the basis of the different interaction parameters χ and interface width. It was already pointed out that the interface width for PS-*b*-PBMA diblock copolymers is larger than observed for PMMA-*b*-PBMA diblock copolymers. This leads us to the conclusion that a decreasing compatibility and interface width between the components are associated with a worsening of tensile properties.

V. DEFORMATION MECHANISMS

A. Influence of Morphology of PS-*b*-PBMA Diblock Copolymers

To investigate the influence of morphology on deformation mechanisms, PS-*b*-PBMA diblock copolymers with different compositions and $M_n > 200$ kg/mol were used (Table 1).

PS-*b*-PBMA diblock copolymers with PS contents lower than 39% PS reveal structures with a PBMA matrix. These samples show a homogeneous deformation of the PBMA matrix, which is connected to a small tensile strength. For block copolymers with 50% PS and with lamellar structures, a transition to a cavitation mechanism can be observed by HVEM investigations. This is shown in Fig. 13 for sample SBM55 with a lamellar structure (534 kg/mol, 55% PS). It is obvious in the HVEM micrograph that the light PBMA lamellae are largely deformed up to their breaking point. This means that cavitation occurs in the PBMA phase. In contrast to the crazing process in PS, the yielding and cavitation mechanism in this block copolymer is more delocalized and not concentrated in the form of crazes. In this sample a macroscopically stress-whitening was observed, which is attributed to the large-scale delamination of the lamellar structure.

For block copolymers with 67% PS, which also reveal a lamellar structure, a more localized cavitation mechanism, where the lamellae are stretched in the direction of applied stress, can be observed. Figure 14 shows narrow zones of largely deformed lamellae. These deformation zones propagate perpendicularly to the direction of the applied stress, which also was found for craze growth in homopolymers and blends. The crazes have thicknesses from less than 1 μm up to 5 μm. The thicker ones are the result of a coalescence of thinner deformation zones into thick crazes. From micrographs with lower magnifications it was possible to determine the length of these crazes, revealing lengths of up to more than 300 μm.

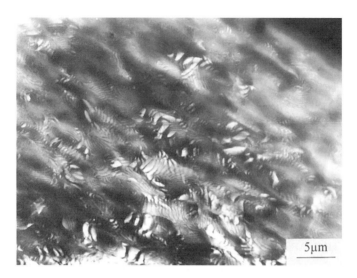

Fig. 13 HVEM micrograph of local deformation zones in sample SBM55 ($\Phi_{PS} = 0.55$, $M_n = 534$ kg/mol).

Fig. 14 Lower magnification of crazes in a PS-*b*-PBMA diblock copolymer with lamellar structure (sample SBM67, $\Phi_{PS} = 0.67$, $M_n = 450$ kg/mol).

When comparing with crazes observed in pure PS it is shown that PS-*b*-PBMA diblock copolymers with 67% PS (sample SBM67) exhibit fewer but much longer crazes.

As shown in Fig. 15, the internal structure of the crazes consists of highly extended craze fibrils of PS and PBMA. The PS fibrils appear as dark zones due to the higher thickness compared to the cavitated-light PBMA. The light PBMA lamellae can easily cavitate. Due to the microvoids, a stress concentration is built up at the PS lamellae that leads to a large plastic deformation of these PS lamellae. This is the same mechanism as described by Argon for PS-*b*-PB diblock copolymers. In contrast to PS-*b*-PB diblock copolymers, the existence of asymmetric phase compositions found by dynamic-mechanical analysis in PS-*b*-PBMA diblock copolymers has a pronounced influence on the level of craze initiation stress (52). From our results we conclude that an essentially pure PS matrix and a PS/PBMA mixed phase coexist. The mixed phase increases the craze initiation stress, σ_c, as compared to that of PB and can be responsible for the improved tensile strength. In previous studies (52) it was shown that the tensile strength of diblock copolymers with 67% PS shows almost the value of pure PS, which could be explained with high values of σ_c, as discussed later.

The PS fibrils in Fig. 15 arise from highly deformed PS lamellae, which is clearly visible in Fig. 16. The diameter of the PS fibrils is 15–25 nm, and the distance between the PS fibrils about 100–130 nm. Figure 17 illustrates the microstructure of crazes in pure PS with a molecular weight of 313 kg/mol. Typically, values for the thickness and distance between the fibrils are 2.5–10 nm and

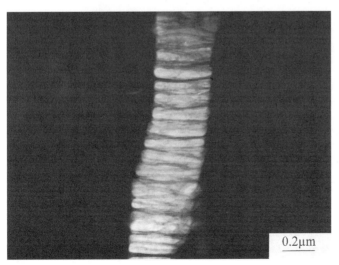

Fig. 15 HVEM micrograph of craze structure in a lamellar PS-*b*-PBMA diblock copolymer (SBM67 $\Phi_{PS} = 0.67$, $M_n = 450$ kg/mol).

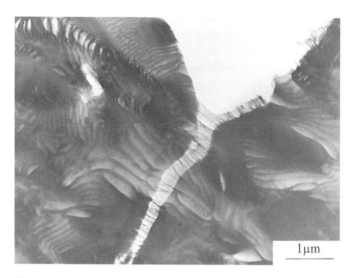

Fig. 16   HVEM micrograph of a craze in a PS-*b*-PBMA diblock copolymer with lamellar morphology (SBM67, Φ_{PS} = 0.67, M_n = 450 kg/mol).

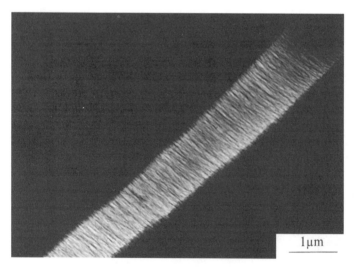

Fig. 17   HVEM micrograph of craze structure in pure PS with M_n = 313 kg/mol.

10–50 nm, respectively. These values agree with data observed by other authors using different methods (62–64). If we compare these values with the craze microstructure of sample SBM67 (67% PS) in Figs. 15 and 16, it is obvious that the thicknesses of the PS fibrils, as well as the distance between the fibrils, are much larger than that for unmodified PS. This suggests that the microstructure of crazes in lamellar PS-*b*-PBMA diblock copolymers arises from their microphase-separated morphology. This finding is in line with the results of Schwier et al. (65) that the nucleation of crazes in block copolymers occurs via cavitation. From the TEM analysis of the undeformed morphology of sample SBM67 (67% PS) we have obtained the thickness of the lamellae: The observed values of 80–90 nm for the PS lamellae are much larger than that observed for the thickness of the deformed PS lamellae, clearly indicating their large plastic deformation. From these data it was possible to determine values for the extension ratio λ_{PS}: $\lambda_{PS} = 5$–6, which is higher than values observed for pure PS ($\lambda_{PS} = 4.3$) (66). A similar result was observed by Creton et al. (66) for PS-*b*-PVP block copolymers, which can be attributed to the stretched chain conformation in block copolymers.

An influence of morphology on craze growth and propagation was observed for lamellar structures. In Fig. 18 a diversion of crazes depending on orientations

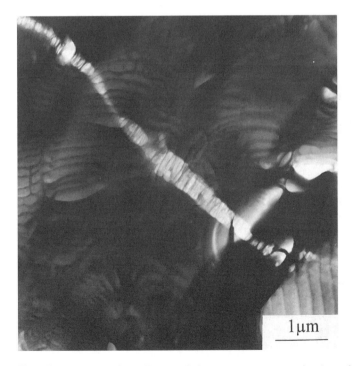

Fig. 18   Influence of lamellar morphology on craze propagation (sample SBM67: $\Phi_{PS} = 0.67$, $M_n = 450$ kg/mol).

in the morphology is shown. Crazes normally propagate perpendicular to the direction of the applied external stress. This also holds true for block copolymers with lamellar structures if the lamellae are in parallel orientation to the stress. If, however, the lamellae are tilted to the applied stress, the deformation follows these lamellae (Fig. 14). This means that craze propagation is influenced by the orientation of the lamellae and does not generally occur perpendicular to the external stress field. Furthermore, it was also found that stacks of lamellae were twisted into the direction of stress applied due to rotation mechanisms if the tilt angle between the lamellae and the external stress direction is relatively small. Another observed mechanism is the stopping of craze formation at stacks of lamellae (Fig. 19). This is connected with a rounded craze tip of the stopped craze (craze tip blunting). In pure PS, the craze tip is typically very sharp and the smallest resolvable voids near the tip are on the order of 20 nm (67). The broad craze tip shown in Fig. 19 can clearly be attributed to the influence of microphase-separated structures on the deformation mechanism. Stop and diversion of crazes, as well as the rounded craze tip, cause higher energy dissipation, resulting in a higher strain at break.

PS-b-PBMA diblock copolymers with PBMA cylinders (76% PS) show an increased yield stress (in contrast to PS-b-PB diblock copolymers). This is one reason for the improved tensile strength of PS-PBMA diblock copolymers as compared to that of pure PS (see Fig. 6). Figure 20 shows an HVEM micrograph of a craze in sample SBM 76 (76% PS). The cellular structure of the craze is quite

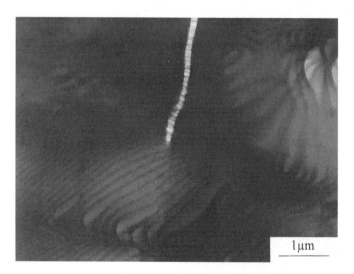

Fig. 19 Mechanism of craze stop in a lamellar PS-b-PBMA diblock copolymer (SBM67 $\Phi_{PS} = 0.67$, $M_n = 450$ kg/mol) causing improved strain at break due to energy dissipation. Also shown is the mature round craze tip, which is sharp in the case of pure PS.

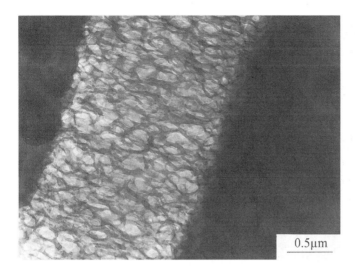

Fig. 20   HVEM micrograph of craze structure of sample SBM76 ($\Phi_{PS} = 0.76$, $M_n = 459$ kg/mol) with hexagonally packed PBMA cylinders.

similar to the cavitation mechanism in block copolymers a suggested by Schwier et al. (65). The light PBMA cylinders cavitate, followed by a large plastic deformation of the dark PS parts. The diameter of the cylinders is about 30–40 nm, which is much smaller than the thickness of the lamellae in the samples discussed before. To study the influence of morphology on craze propagation, ultrathin sections were cut from externally deformed samples. As illustrated in Fig. 21 the light PBMA cylinders are highly deformed in the narrow crazes, and the PS matrix shows a large plastic deformation. Mechanisms of diversion of crazes and craze stop were observed in sample SBM72 (72% PS) with a coexistence of lamellar and hexagonally packed cylinders (52,58). From Fig. 22 it is obvious that craze growth and propagation is strongly influenced by the direction of lamellar regions. For samples with spherical morphology at 83% PS (sample SBM83), the structure of the crazes is almost the same as observed for pure PS (Fig. 23). This means that a transition from cavitation to crazing occurs, in agreement with the results observed by Schwier et al. (68) in PS-*b*-PB diblock copolymers with about 5% PB.

B.   Craze Formation

Cavitation was observed in PS-*b*-PBMA diblock copolymers with spherical, lamellar, and hexagonal morphologies.

To discern a correlation between deformation mechanisms and mechanical properties it is necessary to consider the influence of composition on craze initia-

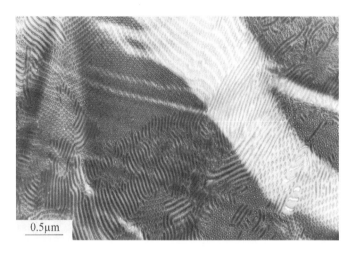

Fig. 21 TEM micrograph of craze structure in a PS-*b*-PBMA diblock copolymer with hexagonally packed PBMA cylinders (SBM76, Φ_{PS} = 0.76, M_n = 459 kg/mol); stained with RuO$_4$.

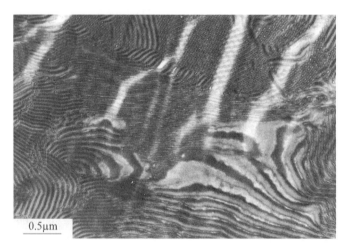

Fig. 22 TEM micrograph of craze structure and craze growth in sample SBM72, revealing a coexistence of lamellae and hexagonally packed PBMA cylinders (Φ_{PS} = 0.72, M_n = 426 kg/mol); stained with RuO$_4$.

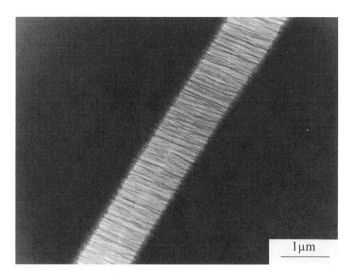

Fig. 23 HVEM micrograph of a craze structure of sample SBM83 with PBMA spheres ($\Phi_{PS} = 0.83$, $M_n = 383$ kg/mol). At this composition, a transition from cavitation to crazing can be observed.

tion stress σ_c. In Fig. 24, σ_c is plotted against the volume fraction of PS, obtained from the point of stress–strain curves where the curves deviate from their linear slope. Surprisingly, σ_c increases in the composition range 60–80% PS and exceeds the value for pure PS. Koltisko et al. (69) observed a decrease of craze initiation stress for SBS triblock copolymers with 20–30% PS, which is related to a decrease in yield stress. In contrast, for PS-*b*-PBMA diblock copolymers, maximum tensile strength and σ_c was found at 76% PS, which exceeds the value for PS of about 40% (58,71).

Creton et al. (66) found that the extension ratio of the fibrils in PS-*b*-PVP block copolymers was always greater when the lamellae were oriented parallel to the craze fibril direction than when the lamellae were oriented perpendicular to this direction. This observation was attributed to the stretched chain conformation in block copolymers normal to the interfaces of the lamellae. Furthermore, in both directions the values of the extension ratio λ were larger than that measured for the homopolymers PS and PVP. Also, for PS-*b*-PBMA diblock copolymers the observed values of λ_{PS} are higher than for pure PS. It was shown elsewhere (70) that a prestretching of PS leads to a marked increase in λ. This leads to the conclusion that a stretched chain conformation in block copolymers has a pronounced influence on the extension ratio of the fibrils.

It was further shown that the microphase-separated morphologies for different compositions have a pronounced influence on craze propagation. The observed mechanisms of diversion and craze stopping can be considered responsible

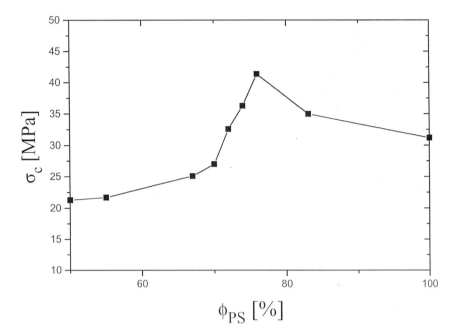

Fig. 24 Dependence of craze initiation stress σ_c on the volume fraction of PS for PS-*b*-PBMA diblock copolymers.

for the relatively high strain at break of about 40% for sample SBM67 (67% PS, lamellae). PS-*b*-PB diblock copolymers with lamellar structures are quite brittle, due to the low strength of the PB lamellae. PMMA-*b*-PBMA diblock copolymers with nearly symmetric composition reveal a strain at break of only 20% (for a symmetric PS-*b*-PBMA diblock copolymer the strain at break is about 50%). The reason for this difference is the higher miscibility of PS-*b*-PBMA diblock copolymers, which is connected to a σ_c increase and with the observed mechanisms of diversion and stopping of crazes. Recently, it was shown that the miscibility of PS-*b*-PBMA diblock copolymers is much larger than that of PMMA-*b*-PBMA, which is also connected to a larger interface width for PS-*b*-PBMA (51). As shown in Table 1, PS-*b*-PBMA diblock copolymers are only intermediately segregated. This provides an explanation for the partial miscibility, the interface width, and the observed high craze initiation stress.

VI. CONCLUSIONS

The present study clearly shows that the phase behavior, microphase-separated morphologies at the nanometer scale, interface formation, and chain architecture have a pronounced influence on the mechanical properties of block copolymers.

While the properties of TPSs arise mainly from the midblock conformation of the bridged rubbery block and strongly segregated morphologies, the synergetic effects on tensile properties of weakly segregated block copolymers are due to their weak segregation, asymmetric phase compositions, broadened interfaces, and different patterns of morphology. For PS-*b*-PBMA diblock copolymers, synergetic effects on tensile strength, absorbed energy, and stiffness were found. For PMMA-*b*-PBMA diblock copolymers, a deterioration of properties was found to be due to the decreased miscibility and interface width compared to those of PS-b-PBMA diblock copolymers (Table 3). In contrast, for PBMA-*b*-PS-b-PBMA triblock copolymers, tensile strength and stiffness are improved as compared to those of diblock copolymers, demonstrating the influence of chain architecture on tensile properties. For triblock copolymers, the enhanced miscibility between the phases is the factor responsible for the improved mechanical properties.

The large miscibility (small interaction parameter) of weakly segregated block copolymers investigated in this study, combined with a large interface width and asymmetric phase compositions, has been shown to be responsible for synergistic effects on tensile properties. For PS-*b*-PBMA diblock copolymers, a higher craze initiation stress was found, as compared to PS-*b*-PB diblock copolymers, which yields the improved tensile strength of this material.

The deformation mechanisms of PS-*b*-PBMA diblock copolymers were examined as an example of weakly segregated block copolymers. Here it was shown that all samples exhibit a cavitation mechanism that was earlier discussed for PS-*b*-PB diblock copolymers. For samples with lamellar structures at 67% PS, a tran-

Table 3 Sample Characterization Data

Sample	$10^{-3} \times M_n$ copolymer[a] (M_w/M_n)	Φ_{PMMA}[b]	Morphology (TEM)	χN
BMAMMA19	387.6 (1.09)	0.19	PMMA, spheres	207.3
BMAMMA33	426.3 (1.10)	0.33	PMMA, cylinder	242.7
BMAMMA48	429.7 (1.13)	0.48	Lamellae	258.1
BMAMMA64	277.4 (1.13)	0.64	Lamellae	175.9
BMAMMA71	212.5 (1.11)	0.71	Lamellae	137.9
BMAMMA75	214.0 (1.12)	0.75	PBMA, cylinder	140.7
BMAMMA77	226.7 (1.08)	0.77	PBMA, cylinder	161.3
BMAMMA79	214.2 (1.10)	0.79	PBMA, cylinder	141.5
BMAMMA80	202.6 (1.13)	0.80	PBMA, cylinder	135.3
BMAMMA90	226.3 (1.11)	0.90	PBMA, spheres	155.9
PMMA	205.0 (1.04)	—	—	—

Molecular weight (M_n), volume fraction (Φ_{PS}), and polydispersity (M_w/M_n), and χN values at 140°C ($\chi = 0.071$) for PMMA-*b*-PBMA diblock copolymers (55).
[a] Size exclusion chromatography (SEC); values are based on the PS standards.
[b] Volume fraction of PS determined by ^{1}H-NMR.

sition from a delocalized cavitation to more localized deformation zones and crazes was found. Furthermore, for lamellar structures, mechanisms of diversion and the stop of crazes were revealed, which are connected with relatively high strains at break. Samples with LAM/HEX structures reveal a combination of deformation mechanisms of pure HEX and LAM structure, which has a pronounced influence on tensile properties. For samples with PBMA spheres at a PS content of 83%, a transition from cavitation to crazing occurs, which explains the brittleness of these block copolymers.

The observed large draw ratios for PS fibrils clearly show that the stretched chain conformation in block copolymers has a marked influence on properties of block copolymers. The strong increase in craze initiation stress of PS-*b*-PBMA block copolymers with 60–80% PS is in contrast to the previous observation in SBS triblock copolymers, and explains the improved tensile strength of weakly segregated block copolymers. The deformation mechanisms presented provide a further explanation for the synergism on tensile strength in PS-*b*-PBMA diblock copolymers.

As qualitatively shown in Fig. 25, the tensile properties of weakly segregated block copolymers of major interest, such as tensile strength, stiffness, and toughness, are improved as compared to those of pure homopolymers. This is usually not observed for rubber-toughened polymers and polymer blends. However,

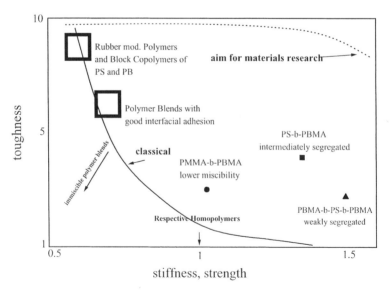

Fig. 25 Qualitative dependence of toughness on tensile strength and stiffness for different polymeric systems aiming at a property profile of weakly segregated block copolymers.

the toughness of PS-*b*-PBMA diblock copolymers is lower than that of rubber-toughened polymers and several polymer blends, especially at lower temperatures, due to the high T_g of the PBMA block. In contrast to TPEs, which show elastomeric behavior, the most interesting properties of weakly segregated block copolymers are their high tensile strength and stiffness, which are combined with a good toughness at higher temperatures. Weakly segregated block copolymers demonstrate the possibilities of creating polymers with improved properties based on materials with structures at the nanometer scale. This shows that the property profile of block copolymers enables one to create a new class of materials.

ACKNOWLEDGMENTS

The authors thank Mrs. E. Hörig (Halle) for the TEM investigations of some of the block copolymers. Roland Weidisch acknowledges financial support for Habilitation from Deutsche Forschungsgemeinschaft (DFG). It is our pleasure to thank Prof. M. Stamm (IPF Dresden), Prof. Stadler† (Bayreuth), Dr. V. Abetz (Bayreuth), Dr. D. W. Schubert (GKSS Geesthacht), Dr. H. Fischer (TU Eindhoven) and Prof. V. Alstädt (Hamburg-Harburg) for fruitful discussions and cooperations. We acknowledge the synthesis of the used block copolymers in the groups of Doz. Dr. M. Arnold (Halle, Germany), Doz. Dr. S. Höring (Halle, Germany) and Prof. R. Jerome (Liege, Belgium). We also thank Prof. Dr. U. Gösele, director of the Max Planck Institute of Microstructure Physics (Halle), for allowing the use of HVEM and Dipl.-Phys. M. Ensslen for performing deformation tests.

REFERENCES

1. CB Bucknall. Toughned Plastics. Appl. Sci. Publ. Ltd., London, 1977.
2. GH Michler. Kunststoffmikromechanik. Hanser Verlag, Munich, 1992.
3. G Holden. In: Legge, NR, Holden, G; Schroeder, HE, eds. Thermoplastic Elastomers. Hanser, Munich, 1987, 481–506.
4. T Hashimoto, K Yamasaki, S Koizumi, H Hasegawa. Macromolecules 26, (1994), 1562.
5. EL Thomas, DB Alward, DJ Kinning, DC Martin, DL Handlin, LJ Fetters. Macromolecules 20 (1987), 1651.
6. IW Hamley, KA Koppi, JH Rosedale, FS Bates, K Almdal, K Mortensen. Macromolecules 26 (1993), 5959.
7. DA Hadjuk, PE Harper, SM Gruner, CC Honeker, G Kim, EL Thomas, LJ Fetters. Macromolecules 27 (1994), 4063.
8. CP Rader. In: Kaplan, WL, ed. Modern Plastics: Encyclopedia. McGraw-Hill, New York, 1996, vol. 72.

9. JM Yu, P Dubois, R Jérôme. Macromolecules 29 (1996), 8362.
10. JA Odell, J Dlugosz, A Keller. J. Polym. Sci. Polym. Phys. Ed. 14 (1976), 847.
11. JA Odell, A Keller. Polym. Eng. Sci. 17 (1977), 544.
12. T Pakula, K Saijo, H Kawai, T Hashimoto. Macromolecules 18 (1985), 1295.
13. T Kurauchi, T Ohta. J. Mat. Sci. 22 (1984), 1699.
14. S Polizzi, P Bösecke, N Stribeck, G Zachmann, R Zietz, R Bordeinu. Polymer 31 (1990), 638.
15. J Sakamoto, S Sakurai, K Doi, S Nomura. Polymer 34 (1993), 4837.
16. S Sakurai, J Sakamoto, M Shibayama, S Nomura. Macromolecules 26 (1993), 3351.
17. TP Russell, TE Karis, Y Gallot, AM Mayes. Nature 368 (1994), 729.
18. M Arnold, S Hofmann, R Weidisch, A Neubauer, S Poser, GH Michler. Macromol. Chem. Phys. 199 (1998), 31.
19. L Leibler. Macromolecules 13 (1980), 1602.
20. GH Fredrickson, K Binder. J. Chem. Phys. 91 (1989), 7265.
21. B Stühn, R Mutter, T Albrecht. Europhys. Lett. 18 (1992), 427.
22. T Wolff, C Burger, W Ruland. Macromolecules 26 (1993), 1707.
23. T Hashimoto, T Ogawa, CD Han. I. Phys. Soc. Jpn. 63 (1994), 2206.
24. N Skamoto, T Hashimoto. Macromolecules 28 (1995), 6825.
25. FS Bates, JH Rosedale, GH Fredrickson. J. Chem. Phys. 92 (1990), 6255.
26. JH Rosedale, FS Bates, K Almdale, K Mortensen, D Wignall. Macromolecules 28 (1995), 1429.
27. JH Rosedale, FS Bates. Macromolecules 23 (1990), 2329.
28. G Floudas, T Pakula, EW Fischer, N Hadjichristidis, S Pispas. Acta Polymerica 45 (1994), 176.
29. G Floudas, G Fytas, N Hadjichristidis, M Pitsikalis. Macromolecules 28 (1995), 2359.
30. N Sakamoto, T Hashimoto. Macromolecules 31 (1998), 3292.
31. N Sakamoto, T Hashimoto. Macromolecules 31 (1998), 3815.
32. GH Fredrickson, E Helfand. Macromolecules 87 (1987), 697.
33. AM Mayes, M Olvera de la Cruz. J. Chem. Phys. 91 (1989), 7228.
34. AV Dobrynin, IY Erukhimovich. Macromolecules 26 (1993), 276.
35. FS Bates. Macromolecules 17 (1984), 2607.
36. K Almdal, JH Rosedale, FS Bates. Macromolecules 23 (1990), 4336.
37. DJ Pochan, SP Gido, S Pispas, et al. Macromolecules 29 (1996), 5091.
38. C Lee, SP Gido, Y Poulos, et al. Polymer 39 (1998), 4631.
39. TE Karis, TP Russell, Y Gallot, AM Mayes. Macromolecules 28 (1995), 1129.
40. T Hashimoto, H Hasegawa, T Hashimoto, H Katayama, M Kamigaito, M Sawamoto, M Imai. Macromolecules 30 (1997), 6819.
41. O Olabisi. Polymer–Polymer Miscibility. Academic Press, New York, 1978.
42. R Weidisch, M Stamm, DW Schubert, M Arnold, H Budde, S Höring. Macromolecules 32 (1999), 3405.
43. MD Foster, M Sikka, N Singh, FS Bates, SK Satija, CF Majkrzak. J. Chem. Phys. 96 (1992), 8605.
44. TP Russell, RP Hjelm, PA Seeger. Macromolecules 23 (1990), 890.
45. A Ruzette, P Banerjee, AM Mayes, M Pollard, TP Russell, R Jerome, T Slawecki, R Hjelm, P Thiyagarajan. Macromolecules 31 (1998), 8509.

46. MW Matsen, F Bates. Macromolecules 29 (1996), 1091.
47. J Melenkevitz, M Muthukumar. Macromolecules 24 (1991), 4199.
48. AK Khandpur, S Förster, FS Bates, IW Hamley, AJ Ryan, W Bras, K Almdal, K Mortensen. Macromolecules 26 (1995), 8796.
49. E Helfand, ZR Wasserman. Macromolecules 9 (1976), 897.
50. H Hasegawa, H Tanaka, K Yamasaki, T Hashimoto. Macromolecules 20 (1987), 1651.
51. DW Schubert, R Weidisch, M Stamm, GH Michler. Macromolecules 31 (1998), 3743.
52. R Weidisch, GH Michler, H Fischer, S Hofmann, M Arnold, M Stamm. Polymer 40 (1999), 1191.
53. R Weidisch, GH Michler, M Stamm, H Budde, S Höring. Macromolecules. In press.
54. H Fischer, R Weidisch, M Stamm, H Budde, S Horing. Polymer & Coll. Sci. In press.
55. R Weidisch, GH Michler, M Arnold. J. Mat. Sci. In press.
56. J Scherble, B Stark, B Stühn, J Kressler, DW Schubert, H Budde, S Höring, P Simon, M Stamm. Macromolecules 32 (1999), 1859.
57. R Weidisch, GH Michler, M Arnold, S Hofmann, M Stamm, R Jérôme. Macromolecules 30 (1997), 8078.
58. R Weidisch, M Stamm, GH Michler, H Fischer, R Jerome. Macromolecules 32 (1999), 742.
59. FS Bates, GH Fredrickson. Annual Rev. Mater. Sci. 26 (1996), 501.
60. F Bühler. Dissertation, Freiburg, 1984.
61. R Weidisch, GH Michler, M Arnold. Polymer 41 (2000), 2231.
62. GH Michler. Ultramicroscopy 15 (1984), 81.
63. P Beahan, M Bevis, D Hull. J. Mat. Sci. 8 (1973), 162.
64. AM Donald, T Chan, EJ Kramer. J. Mat. Sci. 16 (1981), 669.
65. CE Schwier, AS Argon, RE Cohen. Polymer 26 (1985), 1985.
66. C Creton, EJ Kramer, G Hadziioannou. Macromolecules 24 (1991), 1846.
67. SN Zhurkov, VS Kuksenko, AI Slutsker. In: Fracture. Chapman and Hall, New York, 1969.
68. CE Schwier, AS Argon, RE Cohen. Philos. Mag. A 52 (1985), 581.
69. B Koltisko, A Hiltner, E Baer. J. Polymer Sci. Polymer Phys. 24 (1986), 2167.
70. C Maestrini, EJ Kramer. Polymer 32 (1991), 609.
71. R Weidisch, M Enßlen, GH Michler, H Fischer. Macromolecules 32 (1999), 5375.

9
Ultrasonic Characterization of Block Copolymers

Franciszek Lembicz, Miroslawa El Fray, Jerzy Majszczyk, and Jerzy Slonecki
Technical University of Szczecin, Szczecin, Poland

I. INTRODUCTION

The ultrasonic method used a few times for the characterization of polymer materials is indicated in some literature reviews (1–4), studies (5–8), and reports on the progress in this technique (9). Taking into consideration the rapid development in the investigation of multiblock copoly[ester-*b*-ether]s and different multiblock copolymers using high-frequency techniques such as dielectric spectroscopy (10,11) and electron spin resonance (12,13), it is reasonable to supplement these results with ultrasonic tests.

Thermoplastic poly[ester-*b*-ether] elastomers consisting of hard and soft segments exhibit a tendency to aggregate in the form of hard and soft blocks (14–16). Noncrystallizable oligodiols are commonly applied as components of the soft phase and show a substantial flexibility (mobility) and a low glass transition temperature, and they are responsible for the elastic behavior of elastomers. The hard segments exhibit a higher glass transition temperature and cohesion temperature, and they are present in the elastomer matrix as crystalline forms that are able to create so-called physical crosslinks (15).

Thermal properties, such as the glass transition temperature and the melting point, depend on the concentration of hard and soft segments in polymer. The latter also had an influence on mechanical properties such as hardness, strain, and storage and loss moduli, and, consequently, on the parameters characteristic of the propagation of ultrasonic waves (6,16–19), namely, ultrasonic velocity (v) and attenuation (α).

The materials to be described in this chapter are three groups of elastomers. The soft segments of these three groups are composed of:

Poly(1,2-oxypropylene)diol (PO3) of molecular weight 1100 (series A)
Oligo(oxyethylene)diol (PO2) of molecular weight 1000 (series B)
Dimerized fatty acid (DFA) of molecular weight 570 (series C)

The hard-segment material was composed of poly(butylene terephthalate)(PBT). The elastomer structures can be thus characterized by some kind of crystalline heterogeneity dispersed in a relatively uniform polymer matrix in which the amorphous PBT is miscible with the soft segments. These polymers can be considered, as well, as helpful models in the analysis of polymer properties in which the soft-segment material shows different crystallization abilities.

II. EXPERIMENTAL PART

A. Materials

Multiblock copolymers of series A and B were obtained as described in Ref. 18 and references therein. The synthesis of polymers from series C was outlined elsewhere (20). Chemical structures of all studied polymers are shown in Fig. 1.

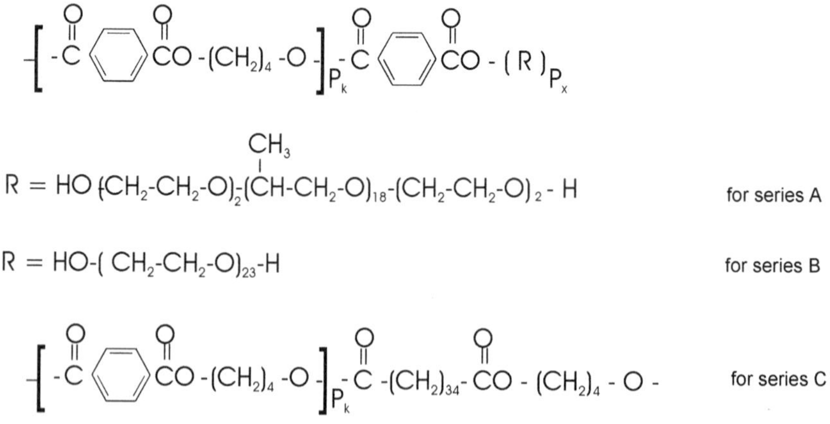

$$\left[-\overset{O}{\overset{\|}{C}}\!\!\bigcirc\!\!\overset{O}{\overset{\|}{C}}O\text{ -(CH}_2)_4\text{ -O -} \right]_{P_k}\!\!\overset{O}{\overset{\|}{C}}\!\!\bigcirc\!\!\overset{O}{\overset{\|}{C}}O\text{ - (R)}_{P_x}$$

$$R = HO\{CH_2\text{-}CH_2\text{-}O\}_2\text{-}\overset{CH_3}{(CH\text{-}CH_2\text{-}O)}_{18}\text{-}(CH_2\text{-}CH_2\text{-}O)_2\text{ - H} \qquad \text{for series A}$$

$$R = HO\text{-}(CH_2\text{-}CH_2\text{-}O)_{23}\text{-}H \qquad \text{for series B}$$

$$\left[-\overset{O}{\overset{\|}{C}}\!\!\bigcirc\!\!\overset{O}{\overset{\|}{C}}O\text{ -(CH}_2)_4\text{ -O -} \right]_{P_k}\!\!-\overset{O}{\overset{\|}{C}}\text{ -(CH}_2)_{34}\text{- }\overset{O}{\overset{\|}{C}}O\text{ - (CH}_2)_4\text{ - O -} \qquad \text{for series C}$$

P_k - degree of polycondensation of hard segments

P_x - degree of polymerization of oligoether

Fig. 1 Chemical structure of the investigated elastomers.

In addition, the nonelastomeric material, abbreviated as PO3-TTM (degree of polycondensation of the hard segments $D_h = 0$; i.e., equimolar concentration of the oligoether PO3 and terephthalate sequences TTM) for series A and the fully amorphous material DFA/BD (dimerized fatty acid/1,4-butanediol) have been synthesized.

B.   Method of Investigation

Samples for ultrasonic investigations were prepared by compression molding as 5-mm-thick disks, conditioned at 150°C. Ultrasonic parameters of velocity υ and attenuation α were investigated as a function of temperature using the pulse technique, heating at a rate of 0.5 K/min and at a frequency of 7 MHz. Two piezoelectric transducers of PZT ceramics were placed onto both sides of the sample and located in a chamber with temperature-controlled nitrogen atmosphere. In order to obtain good acoustic contact between the sample and the transducers, a thin layer of silicone oil was used. Short (3–5-µs) pulses containing HF oscillation were subsequently provided to one of transducers, whereas the other one worked as the ultrasonic wave receiver. These pulses were displayed after amplification at the CRO screen. The ultrasound wave velocity was calculated from the pulse transit time and sample thickness. The value of the attenuation factor was obtained by adjustment of the first pulse amplitude after passing through the sample to the "zero" level by means of a calibrated attenuator; then a constant reference value (estimated using a metal standard) was subtracted from these results. The uncertainity of the values were less than 0.5% and 5% for the velocity and attenuation, respectively.

The measurement of dynamic mechanical properties (dynamic mechanical thermal analysis, DMTA) was carried out at 35 Hz using a heating rate of 10 K/min.

III.   RESULTS AND DISCUSSION

A.   Ultrasonic Attenuation

Generally, the temperature dependence of attenuation $\alpha(T)$ exhibits a maximum at a certain temperature, called T_2 (Figs. 2–4), which can also be estimated from DMTA data for the temperature dependence of the loss modulus $E''(T)$ (5,17,19) (Table 1) and compared on the relaxation diagram (12). These maxima show different widths, depending on the type of the soft segments and the content of the hard segments. The parameter T_2 corresponds to the $\alpha + \beta$ relaxation of the soft phase (12,13); the α and β relaxations overlap at the frequency used in our experiments (see also Ref. 11). The mentioned relaxation process, commonly called the soft-segment glass transition (T_g), determines the elastomeric behavior of the material.

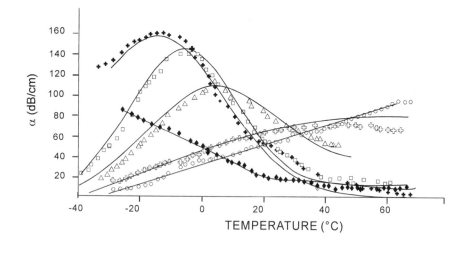

♦PO3;　○PBT;　✦PO3-TTM;　⊕A1100/80;　△A1100/40;　▫A1100/20;

Fig. 2　Temperature dependence of ultrasonic attenuation $\alpha(T)$ for the A series and samples PBT, PO3, and PO3-TTM. Single points: experimental data; solid lines: curves simulated from Eq. (1).

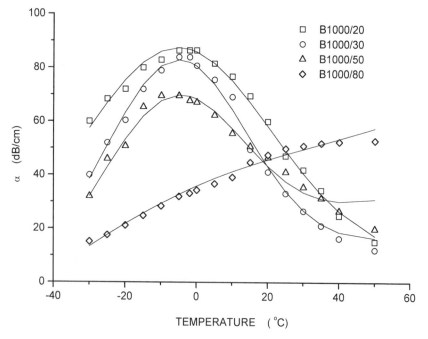

Fig. 3　Temperature dependence of ultrasonic attenuation $\alpha(T)$ for the B series. Single points: experimental data; solid lines: curves simulated from Eq. (1).

254

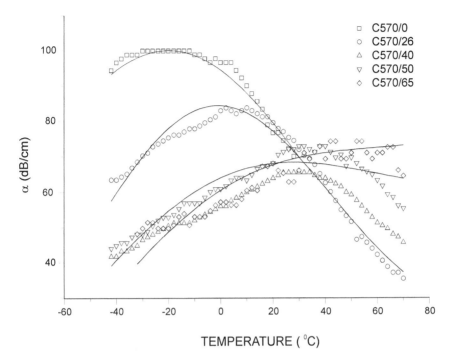

Fig. 4 Temperature dependence of ultrasonic attenuation $\alpha(T)$ for the C series and sample DFA/BD. Single points: experimental data; solid lines: curves simulated from Eq. (1).

Figure 2 illustrates the shift of $\alpha(T)$ as the polymer is changed from a viscous, nonelastomeric material PO3-TTM to the hard PBT. This results suggests the possibility of describing the elastomer properties as an addition of the properties of both components: the soft segments (i.e., PO3-TTM), represented here by a curve with a maximum ("bell" type), and the hard component (i.e., PBT), represented by the straight line:

$$\alpha(T) = CK_1 \exp\left[-\frac{(T - T_2)^2}{2\delta^2}\right] + K_2(aT + b) \tag{1}$$

where δ, K_1, and K_2 are the parameters listed in Table 1.

The PO3-TTM material was chosen as a model for the soft segment because the PO3 oligomer does not give an absorption curve with a maximum in the appropriate temperature range. It is also known that the soft phase in our block elastomers (series A) consists of some inclusions (here, the terephthalate sequences) that modify the PO3 properties (18,21).

In a computer simulation of the elastomeric series A, the constant C was determined by fitting the first term of Eq. (1) to the data for sample PO3-TTM, while

Table 1   Sample Description and Fitting Parameters of Experimental Data Used in Eq. (1)

Sample	W_h (%)	δ (°C)	T_1 (°C)	T_2 (°C) Max. α	Max E''	K_1	K_2
PBT	100		9				
PO3	0						
PO3-TTM	10		-6				
DFA/BD	0	59		-20		1.0	0.0
Series A							
A1100/20	20	18	5	-14	-50	0.9	0.15
A1100/30	30	21	13	-5		0.75	0.4
A1100/40	40	22	12	3	-40	0.6	0.4
A1100/50	50	22	15	4		0.6	0.6
A1100/60	60	24	22	12	-30	0.4	0.6
A1100/70	70	30	27	20		0.25	0.7
A1100/80	80	35		27	-20	0.2	0.7
A1100/90	90	36		25		0.2	0.8
Series B							
B1000/20	20	27	21	-6	-45	0.85	0.1
B1000/30	30	20	15	-6	-47	0.78	0.2
B1000/50	50	20	13	-7	-42	0.60	0.4
B1000/80	80	36	—	-3	-20	0.15	0.7
Series C							
C570/26	26	47	—	-5	-40	0.80	0.16
C570/50	50	62	—	-2	-9	0.52	0.40
C570/65	65	61	—	-2	-8	0.44	0.55

W_h: concentration of the hard segments; δ: parameter characteristic of the width of the bell-shaped curve; T_1: temperature at estimated break point of the straight line; T_2: temperature of the $\alpha(T)$ or $E''(T)$ maximum; K_1, K_2: coefficients estimated by comparison of the simulation with the experimental data using Eq. (1)

the a and b constants were assumed to be the same as those for the hard-phase PBT. The parameters δ, K_1, and K_2 were estimated by comparison with the simulation from the experimental data using Eq. (1). These values are listed in Table 1. One can see that the K_1 and K_2 values give a satisfactory characterization of the individual phases present in the elastomer, the sum of $K_1 + K_2$ being close to 1. The simulation curves are shown in Fig. 2 as solid lines. A good agreement between experimental and simulation curves is observed. However, it should be noted that it requires a variability of the bell curve width δ in the approximation process (Table 1). It can thus be concluded that a description of the elastomer properties cannot be made by a simple addition of the properties of the initial phases (the δ parameter cannot be fixed).

On the other hand, the good correlation between coefficient K_2 and hard-segment concentration W_h encouraged us to compare these results with those we obtained for other material (series B and C). The results obtained for series B (Fig. 3 and Table 1) (i.e., values $K_1, K_2, K_1 + K_2$) and the increase in the δ values shifting from B1000/30 to B1000/80 samples (increase in the degree of crystallinity of the hard segments at the expense of the amorphous soft phase) seem to confirm the foregoing conclusion. The relatively high value of δ for sample B1000/20 is probably caused by crystallization of the soft phase; this kind of broadening was also observed previously (17). Similar attempts to apply Eq. (1) to series C also give satisfactory results (Table 1). However, the simulation curves do not fit as well as the previous ones for series A and B (Fig. 4).

From the foregoing, the possibility of the approximate, "uniform" description of some properties of the block elastomers using Eq. (1) seems to be confirmed. In order to explain it qualitatively, let us remember some generally accepted rules. An increase in the hard-segments concentration can lead to: (a) stiffening of the amorphous sequences as a result of increasing concentration of terephthalate sequences and (b) an increase in the size of the crystalline regions. Both effects should be reflected in the relaxation-time distribution and, as a consequence, in the ultrasonic parameters. From this point of view, the first effect could be responsible for the shifting of the α maximum to higher temperatures, and the second one could be responsible for the broadening of the curves.

A more detailed discussion of the foregoing problem that considers the structural and relaxation effects seems to be difficult because of the comparative sizes of the crystalline aggregates and the ultrasound wavelength (for material from series C, the spherulites' sizes varied from 40 to 120 μm (22,23); typical sizes of spherulites for copoly[ester-b-ethers] are about 50–350 μm (24,25); and the length of ultrasonic waves in our measurements was 200–300 μm). In order to determine the relationship between the supramolecular structure and $\alpha(T)$, two further experiments were performed.

A special, two-layer sample of PO3-TTM/PBT with layer-width values of 2.15 mm and 3.15 mm, respectively, was prepared. Based on the density of the measured components and the $\alpha(T)$ data, it was estimated that the two-layer sample was composed of 50% hard segments. The results obtained for this sample are similar to those from series A, but with a lower hard-segment concentration, $W_h = 20$–30%.

The investigation of a sample, prepared as a mixture of grounded PBT in an amount guaranteed to be 50% of the PBT content in the PO3-TTM material, gave us a negative result that was caused by too high damping and dissipation, even for a very thin sample. This indicates that a simplified model of elastomer structure, where PBT (bulked of grounded) is present as a dispersion in liquid DFA does not allow one to reconstruct the properties of real elastomers. Hard and soft segments

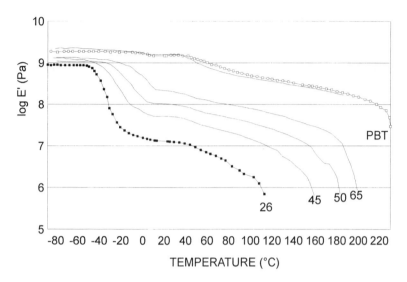

Fig. 5 Temperature dependence of storage modulus E' for series C.

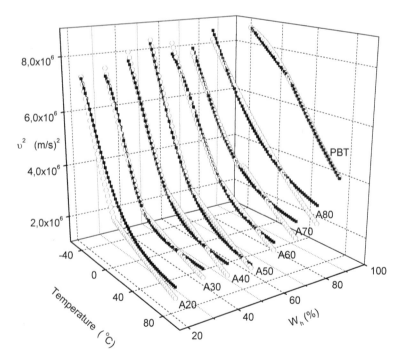

Fig. 6 Temperature dependence of ultrasonic velocity $v(T)$ for series A. Open circles: experimental data; solid squares: calculated from Eqs. (2) and (4).

in block elastomers are connected via chemical bonds; moreover, the soft-segments entanglements leads to a "stiffness" of the structure.

This is manifested in E' moduli changes (see Fig. 5), which in consequence leads to an increase in υ according to the change in PBT concentration. A comparison of the results for tested materials, which range from a liquid mixture of PO3-TTM, to elastomers, to the rigid PBT, can suggest that the elastomer is in the form of a "stressed liquid," with the degree of "stress" and the density of "crosslinking" dependent on the hard-segment concentration. This seems to be the main reason that the $\alpha(T)$ curves cannot be predicted without varying the δ parameter, but such a model can explain qualitatively the changes of attenuation as a function of PBT concentration.

B. Ultrasonic Velocity

The changes of the ultrasonic velocity as a function of temperature and hard-segments concentration take place in the following order: oligodiol (PO3), nonelastomeric material PO3-TTM, elastomers, and PBT (Figs. 6–8). These changes are

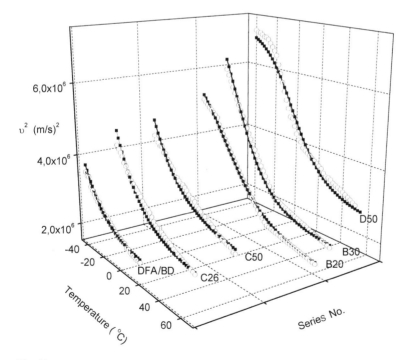

Fig. 7 Temperature dependence of ultrasonic velocity $v(T)$ for series B, C, and D. Open circles: from experimental tests; solid squares: simulated from Eqs. (2) and (4).

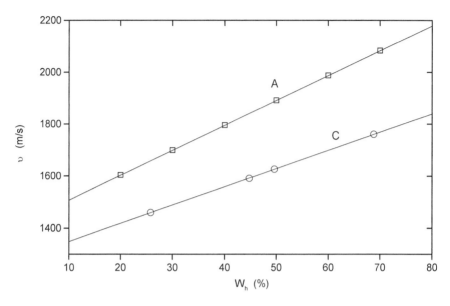

Fig. 8 Influence of the concentration of the hard phase (W_h) on ultrasonic velocity v at 22°C for series A and C.

a consequence of the increase in the "fast component" (PBT) concentration, as well as the chain stresses of the soft phase mentioned earlier.

From Figs. 6–8 one can see that all elastomers show different changes in the slopes of the curves of $v(T)$ vs. T. These changes are discrete for series A and B, while for series C the change has a continous character. The characteristic property of the investigated copolymers from series A and B is a large change in the slope of $v(T)$, which demonstrates a well-shaped maximum on corresponding $\alpha(T)$ curves.

An isothermal cross section of the plots from Figs. 6 and 7 gives the linear dependence between ultrasound velocity (e.g., v_{22} at 22°C) and hard-segment concentration W_h (Fig. 8), which can be expressed as follows:

$$v_{22} \ (m/s) = 9.6 \times W_h \ (\%) + 1412 \qquad \text{for series A and B}$$
$$v_{22} \ (m/s) = 7.0 \times W_h \ (\%) + 1280 \qquad \text{for series C}$$

A similar behavior was reported for more "chemically homogeneous" polythelene (3), where a linear dependence between degree of crystallinity and ultrasound velocity was observed. This allows us to suggest the use of the ultrasonic method for structural investigations.

The $v(T)$ curves were previously (17) characterized by a and b parameters of the appropriate straight lines and the temperature T_1 (break point) for elastomers of series A and B.

An attempt at a more general description of all investigated polymers and additionally of an elastomer based on poly(oxytetramethylene)diol (PO4) was made.

Hartmann (3) has proposed the following formula combining v and ω:

$$v^2 = v_r^2 + (v_u^2 - v_r^2) \times \frac{\omega^2 \tau^2}{1 + \omega^2 \tau^2} \qquad (2)$$

where ω and τ are the frequency and relaxation time, respectively, and v_u and v_r are the limiting sound velocities at high and low frequencies (i.e., at low and high temperatures), respectively. This equation is based on a similar equation for $E'(T)$ and $v^2 \sim E'$ at the same frequency, assuming that the system can be characterized by only one relaxation time.

However, the use of Eq. (2) to describe temperature-dependent data, for example, $v(T)$, requires the knowledge of the function $\tau = \tau(T)$ or the postulation of such a function. This was achieved by using the following well-known equation (with later modifications) (3):

$$\tau = \tau_0 \exp\left(\frac{E}{RT}\right) = \tau_0 \exp\left[\frac{E}{(R)(273 + t)}\right] \qquad (3)$$

where the temperature T is expressed in Celsius degrees.

The reasons for such attempt are as follows:

1. Examples presented in Refs. 2 and 26 suggest the possibility of describing ultrasound data by simple thermally activated processes (β or γ).
2. For the frequencies typical of ultrasounds (10^6–10^7 Hz), the low- and high-temperature relaxational processes merge for many polymers (12,27), and this effect was also observed in our investigations (28). This means in a first approximation, a similar run of the relaxations α and β in the relaxation diagram (the activation energies of the considered processes are equal in that region).
3. A similar value for activation energy (≈ 42 kJ/mol) was found for elastomers consisting of poly(ethylene oxide), poly(propylene oxide), poly(tetramethylene oxide) glycol as the soft segments and ≈ 28 kJ/mol for polymers consisting of 50 wt% of DFA in the soft segment (28,29).

The first attempts at fitting $v(T)$ were done by using Eq. (2) and (3). The probable reason for obtaining unsatisfactory results may be the simplifications made by transforming Eq. (3) or, for example, the existence of the relaxation-times distribution. The latter is favored, since the width of the E' "step" in Fig. 5 was variable while for a single relaxation time and for normalized amplitude $E'_v - E'_u$, this value should be constant.

The previously described changes of the curve widths δ are in support of such an assumption. This effect could be taken into account formally by a change of temperature scale: i.e., by introducing the coefficient s into Eq. (3):

$$\tau = \tau_0 \exp\left[\frac{E}{R(273 + st)}\right] \tag{4}$$

At this point our approach may look somewhat artificial, but the simplicity of Eq. (4) is in favor of it. It offers a better fitting of results and a monothonic variation of the coefficient s with the content of hard segments [larger diffusion of $\upsilon(T)$ curves]. The fitting was done by using Eqs. (2) and (4), the fitted parameters were υ_r, υ_u, τ_o, E, and s. The results of the fitting are presented in Figs. 6 and 7, respectively, and in Table 2.

While the mathematical description of experimental data might induce certain questions, there is no doubt that simulated curves describe the results satisfactorily. Figures 6 and 7 illustrate the glass transition process, which can be interpreted on the experimental data from $\upsilon(T)$ [in our earlier approach, $\alpha(T)$ was taken into account]. The results extracted from the $\upsilon(T)$ data are reported in Table 2. However, only some of them were analyzed and gave a satisfactory fit because of the "diffuse" character of some dependencies in series C.

Table 2 Fitting Parameters of Experimental Data Used in Eqs. (2 and 4)

Sample	W_h (%)	$\upsilon_r \times 10^{-3}$ (m/s)	$\upsilon_u \times 10^{-3}$ (m/s)	$\tau_0 \times 10^{-14}$ (m/s)	E (kJ/mol)	s
PBT	100	1.4	2.7	0.1	40	0.26
DFA/BD	0	1.4	2.50	11	25	0.68
Series A						
A1100/20	20	1.4	2.75	100	22	0.58
A1100/30	30	1.5	2.79	4	29	0.78
A1100/40	40	1.5	3.00	3	30	0.53
A1100/50	50	1.4	3.50	4	29	0.40
A1100/60	60	1.6	2.94	10	28	0.53
A1100/70	70	1.7	3.34	4	29	0.36
A1100/80	80	1.7	3.82	4	29	0.24
Series B						
B1000/20	20	1.3	2.48	10	27	0.55
B1000/30	30	1.4	2.52	10	27	0.68
Series C						
C570/26	26	1.3	3.27	10	25	0.54
C570/50	50	1.4	3.38	11	25	0.54
Series D[a]						
D1000/50	50	1.6	2.65	5	30	0.72

υ_r, υ_u: limiting sound velocities at high and low frequencies, respectively; τ_o: relaxation time; E, s: coefficients.
[a] Series represented by the polymer consisting 50 wt % of poly(oxytetramethylene) diol (PO4) of molecular weight 1000 g/mol as the soft-segments component.

In particular, the values of the parameters obtained independently, that is, solely in the fitting process, could be summarized as follows:

1. Values v_r and v_u are similar to those obtained from E', density ρ, and the Poisson coefficient μ (30).
2. The obtained value for $\tau_0 \cong 10^{-13}-10^{-14}$ is in fairly good agreement with values obtained by various methods for liquids and liquid polymers.
3. The E value seems to be constant and independent of sample content, although it is smaller for series C, and different for PBT. It corresponds qualitatively to the relation of appropriate activation energy of the β process of the soft segments of these elastomers.
4. The increase in the coefficient s, which is dependent on the content of the hard segments, correlates: (a) with the changes in the δ coefficient in the attenuation investigations; (b) with the width of transition in DMTA studies (Fig. 5); and (c) with the width of transition in dielectric studies (28).
5. A comparison of trends in the experimental and simulated curves in Figs. 6 and 7 with values of T_1 suggests that this temperature corresponds to the right edge of the "step" (high temperatures) on $E'(T)$ curves and also, consequently, to the more diffuse edge of the $v(T)$ drop, because of the dependence $v \sim \sqrt{E'}$.

IV. CONCLUSIONS

1. The ultrasonic velocity $v(T)$ and attenuation $\alpha(T)$ characteristics of copoly[ester-b-ether] elastomers and modified elastomers depend mainly on the glass transition of the soft phase of block copolymers with noncrystallizable soft phase. The ultrasonic technique is thus powerful for the determination of their properties.
2. The elastomer properties could not be considered a simple addition of properties of the initial components.
3. The ultrasonic attenuation can be described approximately by the following equation:

$$\alpha(T) = CK_1 \exp\left[-\frac{(T - T_2)^2}{2\delta^2}\right] + K_2(aT + b)$$

4. The temperature dependence of ultrasonic velocity can be described approximately by the following equation:

$$v^2 = v_r^2 + (v_u^2 - v_r^2) \times \frac{\omega^2\tau^2}{1 + \omega^2\tau^2}$$

together with the relationship between τ and T:

$$\tau = \tau_0 \exp\left[\frac{E}{R(273 + st)}\right]$$

5. The examples presented indicate that analysis of $\upsilon(T)$ curves can provide similar information [e.g., glass transition (T_g) or width of this transition] concerning the glass transition, as in the case of more convenient curves $\alpha(T)$.
6. The soft- or hard-segment concentration can be estimated on the basis of the α or υ parameters, which can be important from the application viewpoint.

The use of the foregoing method does not permit one to separate fully a viscoelastic effect from the dissipative one. Therefore, it would be necessary to perform ultrasonic investigations of these materials as a function of frequency for a physical understanding of the observed phenomena.

REFERENCES

1. AM Noth, RA Pethrick. Characterization of polymer solutions and melts by acoustic techniques. In: JV Davkins, ed. Developments in Polymer Characterisation—2. London: Applied Science, 1980, pp 183–206.
2. RA Pethrick. Ultrasonic Characterization of solid polymers. In: JV Davkins, ed. Developments in Polymer Characterisation—4. London: Applied Science, 1983, pp 177–209.
3. B Hartmann. Acoustic properties In: Encyclopedia of Polymer Science and Technology. Mark Bikales, Overberger, Menges. New York: Wiley, 1985, 1:131–160.
4. A Onabajo, Th Dormüller, G Fytas. J Polym Sci, Polym Phys 25:749–763, 1987.
5. J Tatibouet, L Piche. Polymer 32:3147–3151, 1991.
6. F Lembicz, J Slonecki. Kautsch Gummi Kunst 44:668–670, 1991.
7. K Matsuhige, N Hiramatsu, H Okabe. Adv Polym Sci 125:147–186, 1996.
8. A Sahnoune, F Massines, L Piche. J Polym Sci Polym Phys 34:341–348, 1996.
9. A. Meffezzoli, AM Luprano, G Montagna, L Nicolais. J Appl Polym Sci 67:823–831, 1998.
10. TA Ezquerra, Z Roslaniec, E Lopez-Cabarcos, FJ Balta-Calleja. Macromolecules 28:4516–4524, 1995.
11. Z Roslaniec, TA Ezquerra, FJ Balta-Calleja. Colloid Polym Sci 273:58–65, 1995.
12. F Lembicz, R Ukielski. Macromol Chem 186:1679–1683, 1985.
13. F Lembicz, J Slonecki. Polymer 30:1836–1838, 1989.
14. TW Sheridan. Copolyester thermoplastic elastomers. In: BM Walker, CP Rader, eds. Handbook of Thermoplastic Elastomers. New York: Van Nostrand Reinhold, 1988, pp 181–223.

15. S Abouzahr, GL Wilkes. Segmented copolymers with emphasis on segmented polyurethanes. In: MJ Folkes, ed. Processing, Structure and Properties of Block Copolymers. London: Elsevier Applied Science, 1985, pp 165–208.
16. J Slonecki. Sci. Papers of Technical University of Szczecin (Pol.) 479:25–135, 1992.
17. F Lembicz, J Majszczyk, J Slonecki, M El Fray. J Macromol Sci Phys B37(2):161–170, 1998.
18. J Slonecki. Acta Polym 42:655–660, 1991.
19. J Slonecki, M El Fray, K Pawlaczyk, W Bandera. Arch Mater Sci 17:149–161, 1996.
20. M El Fray, J Slonecki. Angew Makromol Chem 234:103–117, 1996.
21. M El Fray, J Slonecki. Polimery (Warsaw) 41:214–221, 1996.
22. M El Fray, J Slonecki. J Macromol Sci Phys B37(2):143–154, 1998.
23. M El Fray, J Slonecki, G Broza. Polimery (Warsaw) 42:35–39, 1997.
24. BP Sawille. Polarized light: theory and measurements. In: DA Hamsley, ed. Applied Polymer Light Microscopy. London: Elsevier Applied Science, 1989, pp 73–109.
25. UW Gedde. Polymer Physics. London: Chapman and Hall, 1995, pp 147–156.
26. DW Phillips, AM North, RA Pethrick. J Appl Polym Sci 21:1859–1867, 1977.
27. P Tormala. J Macromol Sci Rev Macromol Chem C17(2):297–357, 1979.
28. M El Fray. Influence of the soft segment structure on some properties of copoly(ester-ester)s. Unpublished results. Technical University of Szczecin, Szczecin (Poland), 1998.
29. J Slonecki. Unpublished results.
30. F Lembicz, M El Fray, V Altstädt, 7th EPF Symposium on Polymeric Materials, Szczecin, Polymers Friendly for the Environment, 20–24 Sept. 1988, pp 227–229.

10
Dielectric Relaxation of Block Copolymers

T. A. Ezquerra
Instituto de Estructura de la Materia, CSIC, Madrid, Spain

I. INTRODUCTION

The combination of chemically distinct species in a block copolymer notably influences the physical properties of the resulting material (1). In general, the phase behavior of block copolymers is controlled by two factors: (a) polymerization stoichiometry and (b) the Flory–Huggins interaction parameter (χ) of the two components that form the block copolymer (2). The first factor determines both the concentration of each copolymer (f) and the degree of polymerization (N). Depending on the values of the product χN and on the f parameter, block copolymers may suffer an order–disorder transition (ODT) as temperature is varied, from a disordered state, in which both components are mixed, to a locally segregated state, in which microdomains, rich in each component, appear (2,3).

Thermoplastic elastomers are block copolymers typically composed of "soft" segments, characterized by a low glass transition temperature, and "hard" segments having high strength and high glass and melting temperatures (1). The ability to crystallize the hard segments leads to the appearance of crystalline hard-segment-rich domains (1). The resulting semicrystalline copolymers exhibit combined properties of vulcanized rubbers and plastics. Both kinetics and phase-separation mechanisms control the final properties of the copolymers.

Dielectric spectroscopy is a technique that can provide useful information about the molecular dynamics of polymers (4,5). In particular cases, dielectric measurements can complement dynamic mechanical analysis, nuclear magnetic resonance, and quasi-elastic neutron scattering experiments (6). Indeed, dielectric measurements in block copolymers and thermoplastic elastomers have been used to elucidate the nature of the observed relaxations.

The aim of this chapter is to highlight some recent applications of dielectric spectroscopy to the study of relaxation behavior in block copolymers.

II. BRIEF DESCRIPTION OF DIELECTRIC SPECTROSCOPY

When an electric field is applied to a dielectric material, a displacement of the electric charge occurs that is characterized by the electric displacement vector $\vec{D}$. Due to the fact that the response of the material to an external excitation is not instantaneous, there appears a phase shift between $\vec{D}$ and the exciting field $\vec{E}$ that can be accounted for by introducing a complex constant of proportionality in the form $\vec{D} = \varepsilon^* \varepsilon_{vac} \vec{E}$, where $\varepsilon^* = \varepsilon' - i\varepsilon''$ is the complex dielectric permittivity of the material and ε_{vac} is the permittivity of the vacuum. The real part of ε^*, ε', corresponds to the dielectric constant and is associated with the energy stored in the material through polarization. The imaginary part, ε'', is related to the energy dissipated in the medium and therefore is frequently referred to as dielectric loss (7,8).

Dielectric spectroscopy is a technique developed to measure ε^* as a function of both the frequency of the exciting field and the temperature (7,8). A very schematic view about the principle of measurement is given in Fig. 1. When an alternating electric field $V = V_0 e^{i\omega t}$ of frequency $F = \omega/(2\pi)$ is applied to the system, a current $I = C (dV/d\omega)$ is produced, where $C = \varepsilon^* C_0$ is the complex capacity characterizing the sample, $C_0 = \varepsilon_{vac} (A/d)$ is the capacity of the empty capacitor, d is the thickness of the sample, and A is its surface area. By measuring the impedance of the sample, $Z = V/I = -1/\omega C$, it is possible to calculate ε^* as $\varepsilon^* = C/C_0$ (7,8). When dielectric spectroscopy is used to study molecular motion in polymeric materials, the frequency range of interest typically covers from 10^{-2} Hz to 10^9 Hz (4). Unfortunately, this broad frequency range cannot be covered by a single experimental setup. Within the $10^{-2} < F/\text{Hz} < 10^6$ range, ε^* measurements can be performed by using impedance or frequency response analyzers (Solartron Schlumberger, Hewlett-Packard 4192 A, etc.) and lock-in amplifiers (Standford, EG&G, etc). In this case, thin films with circular gold metallic electrodes in both free surfaces (typically 3–4 cm in diameter) are prepared and placed

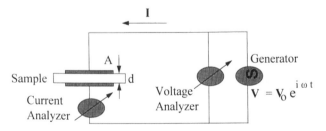

Fig. 1 Schematic view of the basic principle of dielectric measurements in frequency domain.

between two metallic electrodes, building up a capacitor. For frequencies between 10^6Hz and 10^9Hz, the just-described "sandwich geometry" is not valid (9) and reflectometer techniques are required. Here, ε^* can be obtained via reflection coefficient measurements (4). To obtain the temperature dependence of ε^*, the dielectric cell including sample and electrodes is introduced in a cryostat operating under controlled temperature conditions.

III. DIELECTRIC SPECTROSCOPY OF BLOCK COPOLYMERS

As in other polymeric systems, block copolymers may present a great variety of molecular motions that can be investigated by dielectric spectroscopy (DS), provided that dipolar groups are involved. In general, below the glass transition temperature (T_g), homopolymers present the occurrence of local motions, which give rise to subglass relaxations. Above T_g, segmental motions extended to several molecular units appear, provoking the appearance of a relaxation associated with the T_g. Block copolymers present a more complicated dynamic scenario, because both subglass and superglass relaxations associated with each block building up the block copolymer may appear.

To illustrate this, Fig. 2a shows ε'' values at 10^3 Hz as a function of temperature for 1,4-polybutadiene, 1,2-polybutadiene, and a 1,4-polybutadiene/1,2-polybutadiene block copolymer ($\Phi_{1,4} = 0.48$) (10). As one sees, 1,4-polybutadiene exhibits a well-defined peak at $T = -78°C$ and a shoulder at $T = -109°C$. Analogously, 1,2-polybutadiene presents a well-defined peak around 19°C and a weak and broad shoulder at $-26°C$. In both cases the most intense relaxations correspond to the segmental motions above T_g (α-relaxation), while the broad low-temperature shoulders are associated with local motions (β-relaxation) below T_g (10). The copolymer (BB7), which is in the ordered state, i.e., microphase separated, exhibits two clear maxima in ε'' around $-78°C$ and 19°C, which correspond to the individual α-relaxation of 1,4-polybutadiene and 1,2-polybutadiene blocks, respectively.

The sensitivity of DS for differentiating the molecular contribution of the individual blocks can be used to investigate the nature of the order-to-disorder transition (ODT) (10,11). Figure 2b shows ε'' values at 10^3 Hz as a function of temperature for three 1,4-polybutadiene/1,2-polybutadiene block copolymers with morphologies ranging from ordered to disordered structures as determined by small-angle neutron scattering (SANS) measurements (10). As already mentioned, the copolymer (BB7) is in the ordered state, i.e., microphase separated, exhibits the maxima in ε'' corresponding to the individual α-relaxation of 1,4-polybutadiene and 1,2-polybutadiene blocks, respectively. BB1 ($\Phi_{1/4} = 0.47$) and BB2 ($\Phi_{1/4} = 0.46$) samples, which are in the disordered state, i.e., mixed at the scale level explored by SANS, also exhibit the characteristic relaxations of the homopolymers, although a shift in temperature and a significant broadening is de-

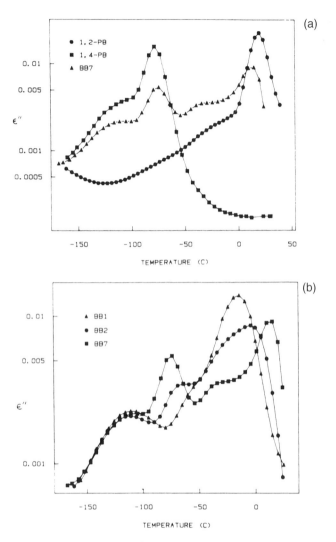

Fig. 2 (a) ε'' values at 10^3 Hz as a function of temperature for 1,4-polybutadiene, 1,2-polybutadiene, and a block copolymer with a 1,4/1,2 block ratio of 0.48 (BB7). (b) Similar representation for three 1,4-polybutadiene/1,2-polybutadiene block copolymers with a 1,4/1,2 block ratio of 0.47 (BB1), 0.46 (BB2), and 0.48 (BB7). BB1 and BB2 are in the disordered state, while BB7 is in the ordered state as revealed by SANS measurements. (Reprinted with permission from Ref. 10. © 1989 American Chemical Society.)

tectable. This study reveals that different local environments are present, even for samples that may look homogeneous at the length scales probed by scattering and rheological techniques (10).

Similar DS measurements have been performed in polystyrene-polyiso-prene block copolymers (11,12). In these kind of block copolymers the existence was also reported of two different segmental relaxations associated with each block, both in the disordered state and in the ordered state (11).

The foregoing results indicate that DS is an experimental technique capable of detecting composition fluctuations in block copolymers at a molecular level not accessible to other techniques.

IV. DIELECTRIC SPECTROSCOPY OF THERMOPLASTIC ELASTOMERS

Thermoplastic elastomers are a class of block copolymers in which one segment is in the rubbery state, imparting flexibility to the material, while the other is in either the glassy or the semicrystalline state, providing high elastic modulus, strength, and mechanical stability to the material. These polymers are of interest for a great variety of industrial processes, including extrusion, injection, and composite preparation. Special attention has been devoted to poly(esther-ether) (PEE) copolymers, in which the hard segment corresponds to poly(butylene terephthalate) (PBT) and the soft segment to poly(tetramethylene oxide) (PO4) (15). To such materials belong, for example, the commercial Hytrel (Du Pont), Arnitel (DSM), and Elitel (Elana). A general scheme of these systems is given in Fig. 3a. The typical molecular weight of the tetramethylene oxide moiety is 1000 g/mol.

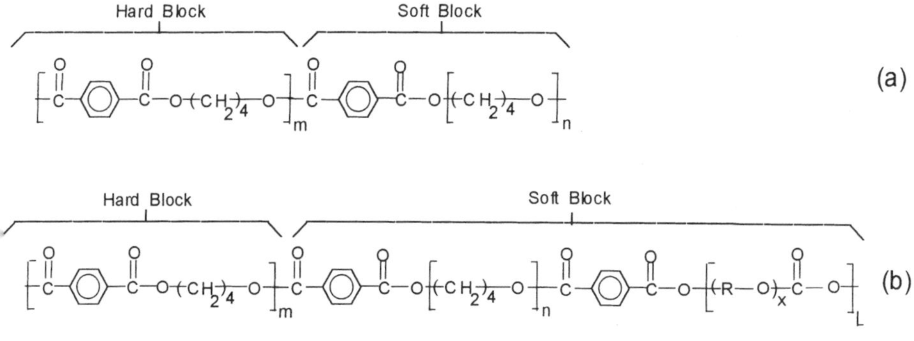

(a)

(b)

Fig. 3 (a) Scheme of a poly(ester-ether) (PEE) thermoplastic elastomer based on PBT (hard block) and polytetramethylene oxide (PO4) (soft segment). (b) Scheme of a poly(ester-ether-carbonate) (PEEC) terpolymer based on PBT PO4 and aliphatic poly(carbonate) (PC). $l \approx 10, \chi \approx 1.12, R = C_6H_{12}$ isomers. The given data are average values derived from stoichiometric ratio and NMR analysis.

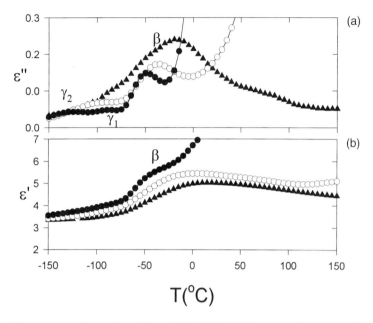

Fig. 4 (a) (ε'') and (b) ε' for a 50/50 PBT-PO4 block copolymer as a function of temperature. (●) 10 Hz, (○) 10^3 Hz, and (▲) 10^5 Hz.

The dielectric behavior of these block copolymers has been the subject of several investigations (17–19). In order to illustrate the general relaxation behavior, Fig. 4 shows isochronal plots of ε'' and ε' as a function of temperature at different frequencies for PBT-PO4 copolymer with a PBT weight percent of 50%, rendering a polycondensation degree of $m \approx 5$. These measurements were taken with a Novocontrol dielectric spectrometer (20). The most intense relaxation ($T = -50°C$, 10 Hz) appears as a broad maximum in the dielectric-loss measurements (Fig. 4a) and as a concurrent step in the dielectric-constant measurements (Fig. 4b). This process corresponds to large-scale motions appearing above the glass transition temperature in the amorphous phase of the copolymer (17,18). This relaxation is currently referred to in the literature as β. The β relaxation depends on hard-segment concentration and, in accordance with the Fox–Flory expression, shifts towards higher temperatures as PBT content increases (18).

At lower temperatures, PBT-PO4 block copolymers exhibit two relaxations clearly detectable at 0.1 Hz (γ_1 and γ_2, in order of decreasing temperature). These processes correspond to local molecular motions, which are due to the—COO ester group and to the ether group present in the PO4 moiety (15,18). Due to the higher rigidity of the ester group as compared with that of the ether group, the local relaxation of the ester is expected to appear at lower frequencies, i.e., higher temperatures (18), giving rise to γ_1. At lower temperatures, γ_2 can be attributed to the ether group present in the soft segment. As frequency increases, the maximum loss peaks for both relaxations shift toward higher temperatures and merge together at 10 Hz.

The large increase of ε'' at higher temperatures for low frequencies is characteristic of semicrystalline polymers and is due to dc conductivity (7,8). Worth mentioning is the appearance of a small shoulder in ε'', detectable at 10^5 Hz in the temperature region where the β relaxation of pure semicrystalline PBT appears (18). This relaxation is strongly affected by the dc conductivity contribution and therefore is highly dependent on the preparation conditions. Although it could be tentatively attributed to amorphous PBT segments within segregated PBT semicrystalline domains, a clear assignment is under discussion.

V. DIELECTRIC SPECTROSCOPY OF MULTIBLOCK COPOLYMERS

Due to the fact that molecular architecture controls the phase behavior and properties of block copolymers, a great synthesis effort is devoted to the preparation of well-defined multiblock copolymers. As far as DS measurements are concerned, multilock copolymers based on polyisoprene (PI) have been intensively studied. Polyisoprene possesses a component of its net dipole moment parallel to the chain backbone, giving rise to a block end-to-end relaxation (normal mode) in addition to the segmental relaxation (13) (Fig. 5). The PI block acts as a dielectric label sen-

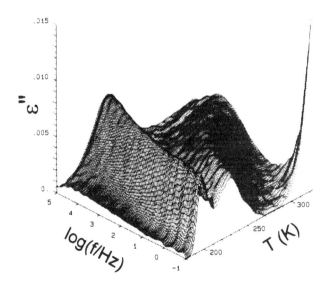

Fig. 5 Frequency and temperature dependence of ε'' for an SIB star terpolymer, showing, as temperature increases, first the segmental and second the normal relaxation processes characteristic of the polyisoprene block. (Reprinted with permission from Ref. 21. © 1996 American Chemical Society.)

sitive to different phenomena. For example, the normal mode of PI in ordered PS-PI copolymers reflects the confinement of the PI blocks within the PI microphase-separated regions (11,14).

In polystyrene-polyisoprene-polystyrene (SIS) triblock copolymers, besides the expected contributions from segmental motions of polyisoprene- and polystyrene-rich regions, a process associated with PI chain dynamics (normal mode) due to a certain mobility of the PI-PS junction point at the interface was observed (12). This finding indicates that DS can be used as a dynamic probe of the interface in ordered copolymers.

The dynamic behavior of star terpolymers has also been studied via DS (21). Polystyrene-polyisoprene-polybutadiene (SIB) and polystyrene-(polysoprene)$_2$ (SI$_2$) star terpolymers present a morphology consisting of PS cylinders embedded in a PI or PI/PB matrix, respectively. The polysioprene block in these terpolymers exhibits the characteristic segmental and normal relaxation processes, as shown in Fig. 5. Here, the study of the PI dynamics allowed the characterization of the effect on the local friction caused by the replacement of polybutadiene by polyisoprene blocks.

Based on PBT, a great variety of thermoplastic elastomers, including multiblock terpolymers, can be generated. The modification of the nature of the soft and hard segments has a tremendous effect on the relaxation behavior (15,22).

Figure 3b shows a general formula representing a poly(ester-ether-carbonate) (PEEC) terpolymer (16,22). The dielectric loss, ε'', values for three PEEC terpolymers with different PBT/PO4/PC ratios are shown in Fig. 6. The dielectric behavior of 60/40/0 is similar to that presented in Fig. 4a and reveals the existence of a broad β relaxation appearing at $\approx 0°C$ for 10 kHz. The β process is accompanied by a lower-temperature γ relaxation, which appears as a shoulder at $\approx -90°C$ and 1 kHz. As frequency is increased, the maximum loss peak for both relaxations shifts toward higher temperatures. The large increase of ε'' at higher temperatures ($\approx 100-110°$) for low frequencies is characteristic of semicrystalline polymers and is due to Maxwell–Wagner–Sillars polarization, which appears due to the existence of interfaces between crystalline and amorphous domains (8). The 60/0/40 copolymer (Fig. 6c) also presents two relaxation process, β and γ, in order of decreasing temperatures. In this case the β and the γ relaxations appear at higher temperatures than those for the 60/40/0 copolymer. Both relaxation maxima are clearly separated from each other, even at lower frequencies (1, 10, and 100 kHz). The effect of varying the amount of PC and PO4 segment content in the dielectric relaxation of the terpolymers is illustrated in Fig. 6b for the 60/20/20 terpolymer. The maxima in ε'' corresponding to both the β and the γ processes shift toward higher temperatures as the number of PC segments increases and of PO4 segments decreases. The transition temperatures for the β process at 1 kHz are represented in Fig. 7 as a function of the PC hard-segment content. For comparison, the glass transition temperatures measured by DSC experiments are also

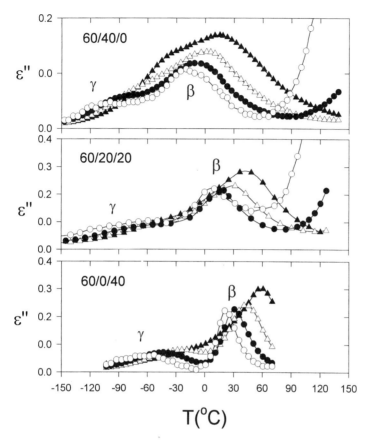

Fig. 6 ε'' data for different terpolymers as a function of temperature. (○) 10^3 Hz, (●) 10^4Hz, (△) 10^5Hz, and (▲) 10^6Hz.

shown (22). Both transition temperatures increase with PC content. This behavior suggests that the amorphous phase in the terpolymers partially consists of a mixture of PO4, PC soft segments, and PBT hard segments, giving rise to a single T_g value (18,22,23).

In PBT-PO4 copolymers, the γ process is due to the merging of two local mode motions: the ether groups present in the soft-segment units (γ_2) and the—COO moiety of the PBT segments in the amorphous phase (γ_1). For PBT-PC, the γ process is due mainly to the contribution of the—COO groups from amorphous PBT and of the—OCOO groups from the soft PC segments. The present results reveal that the γ relaxation in PEEC terpolymers, as measured by dielectric spectroscopy, detects a change from an ester-ether–dominated γ process at low PC concentrations to an ester-carbonate γ process at higher PC soft-segment content.

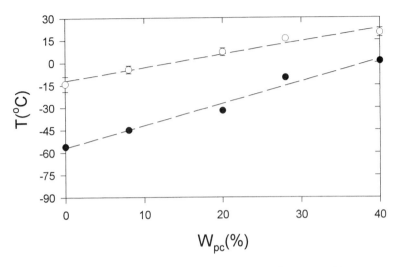

Fig. 7 Glass transition temperature measured by DSC (●) and transition temperature for the β relaxation process at 1 kHz (○) as a function of the weight percent of PC soft segments for PEEC terpolymers.

VI. BLOCK CRYSTALLIZATION AS REVEALED BY DIELECTRIC SPECTROSCOPY

The use of impedance analyzers capable of measuring over several decades of frequencies during measuring times of less than minutes allows one to apply dielectric spectroscopy to a great variety of kinetic processes (5,25,26). In particular, DS has been applied to follow the PBT block crystallization of a PBT-PC (60/40) copolymer (Fig. 3b) in real time (27).

Figure 8 shows the real-time evolution of the β relaxation during an isothermal crystallization process at $T_c = 31 \pm 0.5°C$ for the investigated block copolymer. As crystallization time increases, a reduction in the intensity of the β relaxation process is observed. In contrast, the frequency of the maximum loss remains almost constant. Figure 9 shows the variation of the logarithm of the maximum loss frequency, F_{max}, and of the relative variation of the dielectric loss value at the maximum as a function of the crystallization time. If one considers that $(2\pi F_{max})^{-1}$ is an average relaxation time, the observed constancy of F_{max} suggests that the overall chain mobility in the remaining amorphous phase is almost unaffected by the crystallization process.

Since ε''_{max} is related to the density of dipoles involved in the relaxation process (7), the decrease in ε'' with t can be associated with the progressive reduction of the amorphous phase as hard segments crystallize. This reduction is larger than

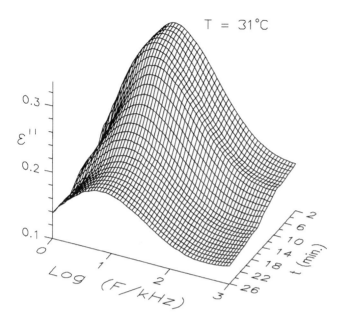

Fig. 8 Real-time evolution of the β relaxation dielectric loss, ε'', values versus frequency and crystallization time during an isothermal crystallization process at $T = 31 \pm 0.5°C$.

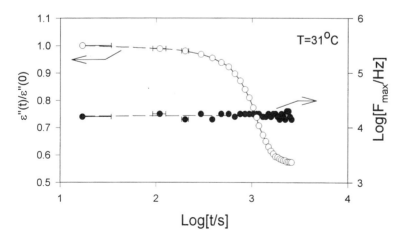

Fig. 9 Variation of the frequency of maximum loss values, F_{max} (●), and relative dielectric loss, $[\varepsilon''_{max}(t)/\varepsilon''_{max}(0)]$ (○), as a function of crystallization time (t) for the isothermal crystallization at $T = 31°C$.

the decrease of amorphous material as derived from DSC or x-ray crystallinity (27). From this result, the existence has been inferred of a tightly bound amorphous phase, which appears provoked by the influence of the crystal lamellae. The observed invariance of F_{max} for the PBT-PC (60/40) block copolymer during the isothermal treatment (Fig. 8) can be understood by considering the morphology of this type of block copolymer. Typically, thermoplastic elastomers based on PBT can build up chain-folded lamellae of the crystallizable hard segments (23,28) embedded on an amorphous phase consisting of a mixture of soft and uncrystallized hard segments. On a larger scale, crystalline lamellae arrange themselves on spherulitic structures (29). The relation of PBT crystallized material to the overall spherulite volume is smaller than for typical spherulites of homopolymers, due to the existence of the soft-segment moieties. The amorphous phase, consisting of a mixture of both soft and hard components, is expected to be less constrained by the crystalline lamellae than when they arrange within more compact spherulitic superstructures such as those formed by homopolymers.

VII. CONCLUSIONS

Dielectric spectroscopy has been shown to be a powerful technique when dealing with the molecular dynamics of polymeric materials. As far as block copolymers are concerned, DS is useful for studying different aspects of interest, including the nature of the relaxation behavior, the order-to-disorder transition, and the dynamics at the interphase region in ordered block copolymers. Dielectric spectroscopy can also be used to monitor structural changes due to block crystallization. The combination of dielectric measurements over a broad frequency range with precise structural characterization opens up new possibilities for the study of the structure–dynamics relationships in block copolymers.

REFERENCES

1. NR Lesse, G Holden, HE Schroeder, eds. Thermoplastic Elastomers: A Comprehensive Review. Munich: Hanser, 1987.
2. FS Bates, GH Fredrickson. Block copolymer thermodynamics: theory and experiment. Annu. Rev. Phys. Chem. 41:525–557, 1990.
3. G Floudas, T Pakula, EW Fischer, N Hadjichristidis, S Pispas. Ordering kinetics in a symmetric diblock copolymer. Acta Polymer 45:176–181, 1994.
4. F Kremer, D Boese, G Meier, EW Fischer. Broadband dielectric spectroscopy on the molecular dynamics in polymer model-systems. Prog. Colloid Polym. Sci. 80:129–139, 1989.
5. G Williams. Dielectric properties of polymers. In: RW Cahn, P Haasen, EJ Kramer,

eds. Materials Science and Technology. Vol 12: Structure and Properties of Polymers. Weinheim: VCH, 1993.

6. J Colmenero, A Alegria, JM Alberdi, F Alvarez, B Frick. Dynamics of the α relaxation of a glass-forming polymeric system. Phys. Rev. B 44:7321–7329, 1991.

7. P Hedvig. Dielectric Spectroscopy of Polymers. Bristol: Adam Hilger, 1997.

8. AR Blythe. Electrical Properties of Polymers. Cambridge: Cambridge University Press, 1979.

9. RP Feynman, RB Leighton, M Sands. The Feynman Lectures on Physics. Vol. II. New York: Addison-Wesley, 1989, pp. 23–1.

10. X Quan, GE Johnson, EW Anderson, FS Bates. Block copolymers near the microphase separation transition. 4. Dielectric spectroscopy. Macromolecules 22:2451–2456, 1989.

11. K Karatasos, SH Anastasiadis, AN Semenov, G Fytas, M Pitsikalis, N Hadjichristidis. Composition fluctuation effects on dielectric normal-mode relaxation in diblock copolymers. Macromolecules 27:3543–3552, 1994.

12. I Alig, G Floudas, A Avgeropoulos, N Hadjichristidis. Junction point fluctuations in microphase separated polystyrene-polyisoprene-polystyrene triblock copolymer melts. A dielectric and rheological investigation. Macromolecules 30:5004–5011, 1997.

13. D Boese, F. Kremer. Molecular dynamics in bulk *cis*-polyisoprene as studied by dielectric spectroscopy. Macromolecules 23:829–835, 1990.

14. M Yao, H Watanabe, K Adachi, T Kotaka. Dielectric relaxation behavior of styrene-isoprene diblock copolymers: bulk systems. Macromolecules 24:2955–2962, 1991.

15. JL Castles, MA Vallance, JM McKenna, SL Cooper. Thermal and mechanical properties of short-segment block copolyesters and copolyetheresters. J. Polym. Sci. Polym. Phys. 23:2119–2147, 1985.

16. Z Roslaniec, H Wojcikiewicz. Synteza i charakterystyka elastomerów kopoli(tereftalanowo-weglanowych). Polimery 33:360–363, 1988.

17. AM North, RA Pethrick, AD Wilson. Dielectric properties of phase separated polymer solids: 2. Butanediol terephthalate-poly(tetramethylene oxide terephthalate) copolymers. Polymer 19:923–930, 1978.

18. J Runt, L Du, LM Martynowicz, DM Brezny, M Mayo. Dielectric properties and cocrystallization of mixtures of poly(butylene terephthalate) and poly(ester-ether) segmented block copolymers. Macromolecules 22:3908–3913, 1989.

19. KP Gallagher, X Zhang, JP Runt, G Huynh-ba, JS Lin. Miscibility and cocrystallization in homopolymer-segmented block copolymer blends. Macromolecules 26:588–596, 1993.

20. A Szymczyk. Synthesis and properties of poly(ester-ether) ionic elastomers. PhD dissertation, Technical University of Szczecin, Szczecin, Poland, 1999.

21. G Floudas, N Hadjichristidis, H. Iatrou, T Pakula. Microphase separation in model 3-miktoarm star co- and terpolymers. 2. Dynamics. Macromolecules 29:3139–3146, 1996.

22. Z Roslaniec, TA Ezquerra, FJ Baltá-Calleja. Dielectric relaxation of poly(ester-ether-carbonate) multiblock terpolymers. Colloid Polym. Sci. 273:58–65, 1995.

23. S Fakirov, A Apostolov, P Boeseke, HG Zachmann. Structure of segmented

poly(ether-ester)s as revealed by synchrotron radiation. J. Macromol. Sci. Phys. B29(4):379–395, 1990.

24. G Williams. Molecular Aspects of Multiple dielectric relaxation processes in solid polymers. Adv. Polym. Sci. 33:59–92, 1979.

25. AK Nass, JC Seferis. Analysis of the dielectric response of thermosets during isothermal and nonisothermal cure. Polym. Eng. Sci. 29:315–324, 1989.

26. TA Ezquerra, J. Majszczyk, FJ Baltá-Calleja, E López-Cabarcos, KH Gardner, BS Hsiao. Molecular dynamics of the α-relaxation during crystallization of a glassy polymer: a real-time dielectric spectroscopy study. Phys. Rev. B 50:6023–6031, 1994.

27. TA Ezquerra, Z Roslaniec, E López-Cabarcos, FJ Baltá-Calleja. Phase separation and crystallization phenomena in a poly(ester-carbonate) block copolymer: a real-time dielectric spectroscopy and x-ray scattering study. Macromolecules 28:4516–4524, 1995.

28. S Fakirov, C Fakirov, EW Fischer, M Stamm. Deformation behavior of poly(ether ester) thermoplastic elastomes as revealed by small-angle x-ray scattering. Polymer 32:1173–1180, 1991.

29. RW Seymour, JR Overton, LS Corley. Morphological characterization of polyester-based elastoplastics. Macromolecules 8:331–335, 1975.

11
Polymer Blends with Block Copolymers

G. P. Hellmann
Deutsches Kunststoff-Institut, Darmstadt, Germany

I. INTRODUCTION

In discussions on polymer blends with block copolymers, everybody inevitably seems to have the schematic picture in Fig. 1 in mind: a monolayer of diblock copolymer chains $\alpha b\beta$ covers the interface between two phases of polymers A and B. The blocks α and β of the copolymer penetrate these phases selectively, entangling with the A and B chains. This structure is suggested by models (1–5) and confirmed by many observations on blends and composites A/B/$\alpha b\beta$, where (a) the chain pairs A-B and the block pairs α-β are incompatible, which is the rule, and (b) the chain–block pairs A-α and B-β are compatible, which is often (but not always) the case if, chemically, A = α and B = β, and sometimes even if A $\neq$ α and B $\neq$ β.

Two decades ago, when the high R&D costs for chemically new thermoplastics became prohibitive, polymer blends moved into the focus of attention (11–13). Blending is cheap and leads to new thermoplastics, many properties of which vary monotonously with the blend composition and can thus be predicted from those of the components. Blends would probably dominate today's market of thermoplastics if it were not for a serious flaw: due to the incompatibility of their components, most blends feature coarse phase morphologies with weak interfaces between the phases. These interfaces break easily under stress; consequently the blends are brittle.

Block copolymer monolayers such as in Fig. 1 remedy this brittleness. The $\alpha b\beta$ chains (a) lower the thermodynamic tension of the A–B interface (1–5,14,15), thus permitting finer phase dispersion (6–10,16), and (b) tie the phases together,

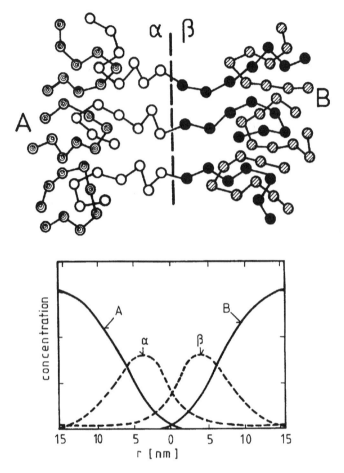

Fig. 1   Compatibilized blend A/B/αbβ: monolayer of a diblock copolymer αbβ at the interface between the phases of two polymers A and B with chain–block entanglements A-α and B-β; concentration profiles of the components are shown schematically.

entangling with the chains of both. Therefore, the inherent weakness of the interface is more or less removed (17–19); i.e., the block copolymer *compatibilizes* the blend. Much R&D effort in industry has been and will further be devoted to commercially viable compatibilized blends (20).

Blends with block copolymers A/B/αbβ pose scientifically challenging problems on all levels, from the molecular to the macroscopic scale. The compatibility, the interfaces, the morphologies, and the properties of blends A/B/αbβ of all kinds of compositions and copolymer architectures have been investigated.

Before these topics are addressed, it is necessary to define the *class of block copolymers.* Symmetric diblock copolymers with long blocks such as indicated in Fig. 1 are excellent representatives of this class. But (a) diblock copolymers with two short blocks or with one long and one short block, (b) triblock, (c) multiblock, and (d) graft copolymers have a blocky architecture, too, as have terblock and blockgraft copolymers, which consist of three monomer units (Fig. 2). It is thus hard to distinguish the block copolymers from the others.

But a copolymer demonstrates its blockiness definitely when its blocks segregate from one another. In the context of this chapter, therefore, one can define *good block copolymers for blends* as follows: They must be able to behave as in Fig. 1, proving their blockiness by segregating in the interfaces of blends. Using this definition, all copolymers in Fig. 2 but the multiblock and the random copolymers can be block copolymers. Theoreticians prefer to deal with diblock copolymers, but triblock and graft copolymers are, in fact, much more important in industry (20). Graft copolymers $\alpha g \beta$ may have the biggest potential.

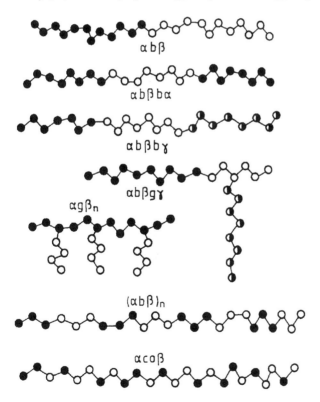

Fig. 2 Types of copolymers: diblock ($\alpha b \beta$), triblock ($\alpha b \beta b \alpha$), terblock ($\alpha b \beta b \gamma$), graft ($\alpha g \beta_n$), blockgraft ($\alpha b \beta g \gamma$), multiblock (($\alpha b \beta)_n$), random ($\alpha co \beta$).

To keep the notation simple, all blocky copolymers will in this chapter be referred to as block copolymers $\alpha b\beta$, where general effects are discussed. The special architecture will be specified where necessary.

After early studies on the phase behavior of blends $A/B/\alpha b\beta$ (21–23), models were developed for the thermodynamics and the phase separation behavior of compatible blends and for the miscibility gaps of incompatible blends $A/B/\alpha b\beta$ (1,2). A key issue is that block copolymers $\alpha b\beta$ and blends A/B demix in a different manner (Fig. 3):

Copolymers $\alpha b\beta$ form well-ordered, thermodynamically stable structures of fine *microphases* (24–28) on the scale of polymer chain coils (Fig. 3a).

Blends A/B form comparatively ill-defined, unstable structures of *macrophases* (29). These are born on a submicroscopic scale (30–33) but coarsen in the melt monotonously (34), a process that can be stopped only

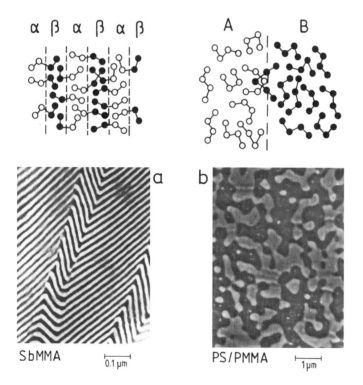

Fig. 3 Characteristic phase structures of block copolymers $\alpha b\beta$ and binary blends A/B: (a) microstructure of a symmetric diblock copolymer SbMMA, (b) cocontinuous macromorphology of a symmetric blend PS/PMMA (S: styrene; MMA: methylmethacrylate).

by cooling in the solid state. Competitive solid blends feature morphologies on the micrometer scale (Fig. 3b).

The tendencies toward macrophase and microphase separation compete with each other when blends with a block copolymer $A/B/\alpha b\beta$ demix (35–38). One compromise is the macrophase-monolayer structure in Fig. 1. But the copolymer can also form multilayers, micelles, or even macrophases of its own. The pivotal point is always: How well can the copolymer $\alpha b\beta$ transfer the thermodynamic stability of its microstructure to the blend $A/B/\alpha b\beta$?

The mechanically weak spots of most two-phase blends are the interfaces. They are extremely thin, at a thickness on the scale of block coil dimensions just like the copolymer microphases in Fig. 3a. New techniques using thin films (39–41) were designed to study these interfaces. Their concentration profile and the chain conformations in them are current issues. Of outstanding importance is the chain entanglement network across interfaces. It is poor and weak in simple blends A/B, so block copolymers $\alpha b\beta$ are needed to restore it. Unfortunately, it is difficult to measure entanglements directly. Their effect shows only indirectly, in the mechanical response (17–19). Details about the entanglement network in interfaces can be better extracted from computer simulations (42,43) than from experiments.

While the blend compatibility and the interface structure lend themselves to accurate investigation and exact model interpretation, more complex yet important characteristics of incompatible blends $A/B/\alpha b\beta$ so far cannot be described and predicted in similar detail. Especially the phase morphology and many properties, above all the toughness, of blends depend on too many factors: (a) the properties of the components, (b) the blend thermodynamics, (c) the architecture of the copolymer, (d) the melt rheology, and in particular (e) the method of blend preparation.

The interpretation of all blend morphologies, including those of blends $A/B/\alpha b\beta$, is problematic because of their nonequilibrium character. The phase patterns result from phase dispersion and coalescence, where kinetics and thermodynamics are often inseparably intermingled. This disorder weakens the power of experimental and theoretical techniques that have excelled in the investigation of block copolymers, such as electron and force microscopy, x-ray and neutron scattering, relaxation, fluorescence, and NMR spectroscopy as well as thermodynamic modeling and computer simulation.

In industry, incompatible blends $A/B/\alpha b\beta$ are prepared by very different techniques. Some blends (not many) are prepared by simple *melt blending,* where stress fields control the morphology as well as the thermodynamic (30,44–47) and rheological (48–52) incompatibility of the phases. More $A/B/\alpha b\beta$ blends, or, rather, $A/B/\alpha g\beta$ blends with graft copolymers, are produced in processes involving "in situ" reactions. Very many blends are prepared via *reactive processing*

(53,54), where the copolymer is created during blending via chain coupling at the interfaces. Some blends are produced in *in situ processes* (55–58), which involve simultaneous polymerization, grafting, phase separation, and structure formation. This coupled process yields directly blends $A/B/\alpha g\beta$ with the desired morphology, ready for sale. Tailor-made phase structures can be designed in situ that owe their sometimes-fascinating superstructures entirely to nonequilibrium effects (55–61).

The mechanical performance is the ultimate test of any compatibilized blend. The copolymer $\alpha b\beta$ must toughen the blend A/B (17–20,62,63). Yet the goals are often modest. If A and B are two thermoplastics, real synergism is rare. It is already satisfactory if the blend $A/B/\alpha b\beta$ interpolates more or less linearly in strength between A and B.

The expectations are much higher if B is an elastomer. Competitive thermoplastic–elastomeric blends $A/B/\alpha g\beta$, normally referred to as *rubber-modified thermoplastics* (62,63), are always much tougher than thermoplastic A. The toughening mechanisms are still in doubt, despite decades of research. No class of blends $A/B/\alpha b\beta$ is as complex as these morphologically and micromechanically finely tuned high-impact thermoplastics.

Evidentally, blends with block copolymers have many facets. While some basic effects are already understood in detail, many complex phenomena are still an enigma.

This chapter deals with $A/B/\alpha b\beta$ blends (6–10), which consist of two homopolymers (or random copolymers) and a block (or graft) copolymer. Selected aspects will be addressed, to demonstrate the complexity of these blends. No comprehensive report is intended.

II. THERMODYNAMICS OF BLENDS WITH BLOCK COPOLYMERS

Attractive objects for thermodynamic modeling are the compatible and the weakly incompatible blends $A/B/\alpha b\beta$. Both are rare. But only they permit an insight into the molecular origins of the structures and the dynamics in these complex systems.

A. Compatibility

Basic phenomena are reflected by the random-phase approximation (RPA) model (30,44–47). More advanced theories and simulations are treated in Chap. 3 of this book. The RPA model describes blends or solutions of any number of components, made of homopolymers and block copolymers of any chain architecture, on the basis of single-chain conformations and interactions. In the regime of compatibility, it yields the pattern of the concentration fluctuations and the position of the critical point and, in the regime of incompatibility, the miscibility gaps. Of particular interest are the ternary-phase diagrams of blends A/B/$\alpha b\beta$ (64).

In homogeneous systems, the fluctuation structure is characterized by a matrix of structure factors S_{ij} (i,j: components) that are determined, via matrix inversion, by the secondary derivatives M_{ij} of the locally resolved (wave vector q) Gibbs energy of mixing ΔG (35,36):

$$[S] = [M]^{-1} \qquad M_{ij}(q) = \frac{\partial^2 \Delta G(q)}{\partial \phi_i(q) \partial \phi_j(q)} = \frac{1}{F_{ij}(q)} + \varepsilon_{ij} \qquad (1)$$

where the energy terms ε_{ij} describe the interactions and the form factors F_{ij} describe the chain or block conformations, which, in turn, depend on the length and the stiffness of the chains or blocks. The structure factors of blends A/B (S_{AB}) and block copolymers $\alpha b\beta$ ($S_{\alpha\beta}$) are given by

$$\frac{1}{S_{AB}} = \frac{1}{F_A} + \frac{1}{F_B} - 2\chi_{AB}$$

$$\frac{1}{S_{\alpha\beta}} = \frac{F_{\alpha\alpha} + F_{\beta\beta} + 2F_{\alpha\beta}}{F_{\alpha\alpha}F_{\beta\beta} - F_{\alpha\beta}^2} - 2\chi_{\alpha\beta} \qquad \chi_{ij} = \varepsilon_{ii} + \varepsilon_{jj} - 2\varepsilon_{ij} \qquad (2)$$

where χ_{ij} is Flory's interaction parameter. The structure of these systems is controlled basically by the interaction parameter and the chain and block lengths. Two important conclusions for A/B/$\alpha b\beta$ blends emerge from Eq. (2):

The critical points derived from S_{AB} and $S_{\alpha\beta}$ are similar at equal chain and block lengths. This means that adding a copolymer $\alpha b\beta$ to a blend A/B does not markedly improve the compatibility. The copolymer $\alpha b\beta$ may concentrate in the interfaces between the A and B phases, in the blends A/B/$\alpha b\beta$, but it will not noticeably induce mixing of A and B.

The structure factors S_{AB} and $S_{\alpha\beta}$ are very different in character (Fig. 4a). Blends A/B/$\alpha b\beta$ are described by a decay function with a maximum at $q^* = 0$, which indicates macroscopic fluctuations. On the contrary, block copolymers $\alpha b\beta$ are described by a peak with a maximum at a characteristic wave vector $q^* > 0$, which indicates submicroscopic fluctuations on the scale of the block coil diameters. These structure factors predict that phase separation will lead to macrophases, in blends A/B, and to microphases, in block copolymers $\alpha b\beta$, which is confirmed by the morphologies in Fig. 3. For the ternary blends A/B/$\alpha b\beta$, the model predicts that they demix primarily either into macrophases or into microphases, depending on the composition and the chain–block length ratios. The morphology of blends A/B/$\alpha b\beta$ should thus be either blendlike (A/B$_{\alpha b\beta}$) or copolymerlike ($\alpha b\beta_{A/B}$). In the A/B$_{\alpha b\beta}$ structure, the copolymer forms microphases inside the A and B macrophases. In the $\alpha b\beta_{A/B}$ structure, which is favored by attractive block–chain interactions and high

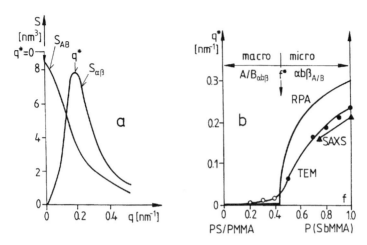

Fig. 4 Macrophases and microphases in blends A/B/$\alpha b\beta$: (a) RPA structure factors S_{AB} (Eq. 2) of a blend A/B and $S_{\alpha\beta}$ of a diblock copolymer $\alpha b\beta$ (corresponding to the morphologies in Fig. 3 but calculated for homogeneous solutions); (b) maximum wave vector q^* of $S_{AB\alpha\beta}$ for blends PS/PMMA/SbMMA varying in copolymer content f, macro–microphase transition at f^*; RPA model calculations, small-angle x-ray scattering (SAXS) and transmission electron microscopy (TEM). (From Ref. 35.)

block–chain length ratios, the microstructure of the copolymer $\alpha b\beta$ is swollen by A and B, which corresponds to the desired monolayer structure in Fig. 1. The dominance of the $\alpha b\beta_{A/B}$ structure is a measure of the compatibilizer efficiency. It increases, at constant overall molecular weight, with the blockiness of the copolymers, usually from graft over triblock to diblock copolymers.

Experimentally, the compatibility of blends A/B/$\alpha b\beta$ has been studied intensively (6–13,20). Particularly sensitive tools to study miscibility and phase separation are scattering techniques (37,38,65–68) as well as fluorescence spectroscopy (69,70). Since blends with a critical point in the melt are difficult to design, phase separation has been studied in demixing solutions, too.

A macrophase–microphase transition (37,38) is shown in Fig. 4b, in blends PS/PMMA/SbMMA of polystyrene, polymethylmethacrylate, and a styrene-MMA diblock copolymer (35). The blends demix into A/B$_{\alpha b\beta}$ macrostructures ($q^* = 0$) at low copolymer content and into $\alpha b\beta_{A/B}$ microstructures ($q^* > 0$) at high copolymer content. However, this is not yet the full picture: according to electron microscopy, blends A/B/$\alpha b\beta$ actually demix in two steps. The initially formed macrophase (A/B$_{\alpha b\beta}$) and microphase ($\alpha b\beta_{A/B}$) structures turn eventually into characteristically different macro–micro or micro–macro hybrid structures (35,36).

The kinetics of phase separation complicate the situation further. Macrophases of blends A/B are at birth submicroscopically fine, almost as fine as microphases of block copolymers $\alpha b\beta$ (30–33). The difference emerges only later: macrophases coarsen (34), microphases do not. Blends A/B/$\alpha b\beta$ coarsen often just like blends A/B, especially at low $\alpha b\beta$ contents, but more slowly (67,68). But at higher $\alpha b\beta$ contents, A/B/$\alpha b\beta$ morphologies seem sometimes totally controlled by the copolymer microphase structure (37). These morphologies may be thermodynamically stable, which would amount to a perfect compatibilizer effect of the copolymer $\alpha b\beta$. It is not absolutely certain yet if such a perfect compatibilizer exists. It is always hard to rule out that the A/B/$\alpha b\beta$ blend morphologies are partly kinetically stabilized.

B. Interfaces

The interfaces in A/B/$\alpha b\beta$ systems are special insofar as free chains A and B mix with more or less confined blocks α and β (Fig. 1). Antagonistic effects must be balanced: a monolayer of a suitably adapted $\alpha b\beta$ copolymer is energetically favored, since it diminishes adverse A-B contacts, but a strict monolayer geometry is entropically disfavored, due to chain localization and coil stretching (1–5).

Blend interfaces are extremely thin, commonly ≤ 30 nm thick. They are best measured in simple geometries with highly resolving techniques. True two-phase blends are inadequate objects because their interfaces are isotropically arranged and irregularly curved. Only scattering can yield some information on the interface thickness (71). Therefore, experiments were designed for model systems with macroscopic interfaces. The interface tension is measured on systems with a drop (pendant or spinning drop (14,15), rheology (72,73)) or a fiber (fiber retraction (74)) of a polymer A that is immersed in a polymer B. The interface structure is measured in thin-film composites of a thin top layer A on a flat substrate B with a straight interface that can be probed with neutron reflectivity and nuclear reaction analysis (39–41). The interface toughness is measured in cantilever tests on welded composites, again with a straight interface (17–19).

Even models (1–5) and simulations (42,43) are focused on straight interfaces. Both have revealed many details on interfaces that are less obvious from experiments. Simulations elucidated the narrowing of interfaces in blends of increasing incompatibility and elucidated the block coil stretching in the interfaces with increasing block length. Decreasing chain–block compatibility in blends A/B/$\alpha b\beta$ and (especially if, chemically, A $= \alpha$ and B $= \beta$) decreasing block–chain length ratio leads in the interface to a wet–dry brush transition. The copolymer monolayer is a wet brush if its blocks α and β mix readily with the free chains A and B, which is the desired situation, and it is a dry brush if blocks and chains do not interpenetrate markedly (1,5,37,75). For details on models and simulations, see Chap. 6 of this book.

Unfortunately, results obtained on macroscopic interfaces cannot be transferred straightforwardly to true blends A/B/αbβ. Macroscopic interfaces have a constant area that remains constant when the copolymer αbβ is added. The copolymer can be analyzed in low- or high-interface concentrations, up to a critical micelle concentration (CMC), where the interface is saturated with αbβ layer so further αbβ starts forming micelles in the phases A and B. In true blends A/B/αbβ, on the contrary, the interface area is not constant but increases as αbβ is added, so a CMC normally does not exist (Sec. III.A).

Despite these problems of translation, some conclusions drawn from thin-film composites help a lot in the design of blends A/B/αbβ: (a) long-chained copolymers with equally attractive A–α and B–β interactions are best at concentrating in the interface and improving its strength; (b) the compatibilizing effect of these copolymers is particularly pronounced in the very narrow interfaces of strongly incompatible blends A/B, where the αbβ block coils are stretched into the A and B phases and entangle well with their chains (1). The interface is narrow, in these blends, and thus fairly well represented by the monolayer structure in Fig. 1.

It is informative to compare the behavior of block and random copolymers at interfaces. Copolymers αcoβ are quite interface active if they are equally incompatible with A and B but less so than A and B with each other. Such copolymers exist often if, chemically, A = α and B = β. However, αcoβ chains are not blocky and cannot form monolayers. Instead, a random copolymer αbβ forms thicker interphases. The chains at the αcoβ–A and αcoβ–B interfaces of these interphases seem to entangle in a special pattern of interchain stitching (76,77). This ties the phases together less well than an αbβ monolayer, but αcoβ interphases, too, can improve the toughness of blends quite efficiently (78).

III. MORPHOLOGY

Morphologies that are born in homogeneous blends (Sec. II.A), by phase separation, can be modeled satisfactorily. But practically all blends of incompatible polymers are prepared by techniques where they never run through a homogeneous stage. The components cannot really be mixed but only dispersed.

Solutions or latices may be coprecipitated, in the laboratory. But in a factory, on a grand scale, blends A/B/αbβ are produced by melt blending, reactive blending, or in situ processes. Whatever the technique, the blends usually feature morphologies where the phases are rather irregularly dispersed, often on the scale of micrometers (the cocontinuous phase pattern in Fig. 3b is uncommonly regular; most melt blended morphologies are strongly perturbed).

Melt blending is treated in this section, reactive and in situ processes in Sec. IV.

Melt blending of polymers in kneaders or, predominantly, twin-screw extruders eventually always lead to a steady-state morphology, where the dispersing effect of the mechanical stresses and the coarsening effect of the interface tension are balanced (48–52). For binary two-phase blends A/B forming domain-matrix morphologies under shear, this balance of mechanical energy input and thermodynamic interface energy leads to a coarseness d of morphology of

$$d = \text{We}(R)\,\frac{\sigma}{\eta\dot{\gamma}} \tag{3}$$

where η is the matrix viscosity, $\dot{\gamma}$ is the deformation gradient, and σ is the interface tension. The capillary (or Weber) number We characterizes the effect of the A–B viscosity ratio ($R = \eta_B/\eta_A$, where A forms the matrix). Blends with $R \neq 1$ are rheologically incompatible and yield coarse morphologies, in particular at $R > 1$, where the domains of the dispersed phase B are more viscous than the matrix phase A. Fine morphologies result only if the blend components are thermodynamically and rheologically weakly incompatible, i.e., at $\sigma \to 0$ and $R \cong 1$.

Blend morphologies are often interpreted and predicted on the basis of Eq. (3), which is, however, limited to dilute blends A/B, i.e., to isolated drops of B that are immersed in A. Complex phase structures such as the cocontinuous morphologies of binary two-phase blends (Fig. 2a) and the three-phase morphologies of ternary blends so far are beyond modeling.

A. Blends with Block Copolymers

All competitive blends $A/B/\alpha b\beta$ are quasi-two-phase systems, since the copolymer forms interface layers and not a third phase. Therefore, Eq. (3) is a fair guideline for the interpretation of blends $A/B/\alpha b\beta$. The copolymer lowers primarily the interface tension σ (14,15) and lowers, therefore, the morphology coarseness d of the blend (16).

This effect is observed in Fig. 5, in a blend PS/PVC/SbMMA, which stands for many other blends. The symmetric diblock copolymer SbMMA is well adjusted to the blend PS/PVC because polyvinylchloride (PVC) and PMMA are compatible. As shown in Fig. 5c, the blend morphology is refined monotonously, from the micrometer scale for the binary blend PS/PVC to the nanometer scale for the pure SbMMA. The curve predicts the coarseness assuming that the homopolymers PS and PVC simply swell the microphase structure of the copolymer SbMMA. It fits the data fairly well, which indicates that all SbMMA chains are in the interfaces, forming a monolayer such as in Fig. 1. This characterizes a good compatibilizer.

Another, more demanding test is to verify whether the blend morphology is thermodynamically stable. Indeed, the fine morphology in Fig. 5b was almost stable in the melt. It coarsened only slightly upon annealing. This characterizes an al-

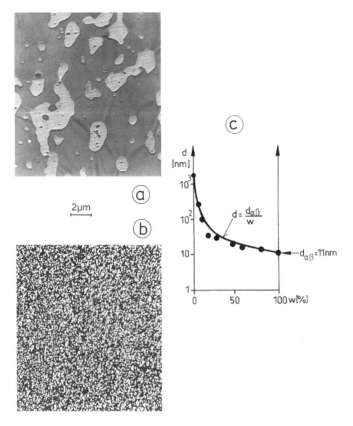

Fig. 5 Morphologies of melt-mixed blends without and with a symmetric diblock copolymer: (a) PS/PVC$_{1/1}$, (b) PS/PVC + SbMMA$_{10wt\%}$, (c) coarseness d of the blend morphology as a function of the copolymer weight content w; $d_{\alpha\beta}$: coarseness of the pure copolymer SbMMA.

most perfect compatibilizer, which transfers the equilibrium character of its own microstructure largely to the blend A/B/$\alpha b\beta$. Similarly attractive interactions in the pairs A–α and B–β are the recipe.

According to the CMC concept (Sec. II.B), adding copolymer $\alpha b\beta$ to a blend A/B should be useless above the CMC. But Fig. 5c proves otherwise. The blend morphology becomes finer as more copolymer is added. The interface area is steadily increased, preventing a CMC: good compatibilizers $\alpha b\beta$ always create, in blends A/B/$\alpha b\beta$, the interface they need (some less good compatibilizers do form micelles, but they do so at all concentrations, not only above a CMC).

Various factors can impair the copolymer efficiency. Copolymers $\alpha b \beta$:

With too short blocks are weakly interface active and form interface multilayers or a macrophase of their own

With too long blocks are thermodynamically good but shift themselves very slowly into the interfaces

With asymmetric block–chain interactions form micelles in the preferred phase

Unfortunately, block copolymer interface monolayers are invisibly thin, even in electron micrographs. In Sec. III.B, copolymers are introduced that can actually be seen.

B. Blends with Other Copolymers

At a given total chain length, diblock copolymers are the best compatibilizers. But triblock ($\alpha b \beta \beta b \alpha$) and graft ($\alpha g \beta$) copolymers (79–81) do well, too, provided that the single blocks and grafts are sufficiently long. The mechanism of compatibilization is the same.

But random copolymers $\alpha co \beta$ behave differently, forming interphases instead of interface monolayers. The interfaces between the phases of the blend PS/PMMA in Fig. 6 are covered by thick interphases of a random copolymer ScoMMA. A high copolymer content is shown, to emphasize the point. But extremely thin interphases are obtained, at lower ScoMMA contents, not much thicker than a monolayer. It is not yet known just how thin an $\alpha co \beta$ interphase can be without becoming unstable.

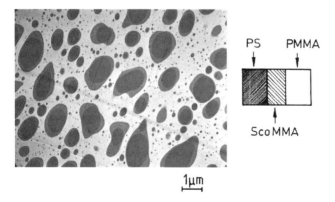

Fig. 6 Morphology of a blend with a random copolymer: PS/PMMA$_{1/3}$ + (S$_{68}co$MMA$_{22}$)$_{20wt\%}$, melt-blended and annealed: interphases of the random copolymer (despite the asymmetric weight composition of S$_{68}co$MMA$_{22}$, this copolymer interacts equally with PS and PMMA). (From Refs. 117–119.)

Random copolymers are a real competition for block copolymers, because they are cheap and can be prepared with more monomers by more techniques. But they are inferior. Blend morphologies A/B/αcoβ are less refined and remain instable.

Other alternatives to block copolymers of two monomer units are terblock (αbγbβ) (82–84) and blockgraft copolymers (αbγgβ) (85), which consist of three monomer units (Fig. 3). In blends, they can form interface monolayers just like diblock copolymers (86–88), as in the blend PS/PVC/SbBgMMA shown in Fig. 7 (88). The copolymer SbBgMMA was prepared by grafting MMA from a styrene-butadiene diblock copolymer SbB. The copolymer monolayer in Fig. 7 can be observed, as a row of dots, because the butadiene center block was selectively stained.

Terblock copolymers αbγbβ yield the most fascinating microphase structures of all copolymers (82–84). The combination of three blocks permits the design of copolymers with novel properties. Blockgraft copolymers αbγgβ (85) are less well ordered, especially due to the polydispersity of the center block, but they are particularly suited as compatibilizers for blends (88). Radical grafting on one diblock copolymer yields, at low cost, a whole family of blockgraft copolymers with homopolymer or random copolymer grafts.

The center butadiene block B of the copolymer SbBgMMA in Fig. 7 is short. Hence, the blockgraft copolymer is practically a diblock copolymer SbMMA, however, synthesized partly radically, not anionically. Indeed, SbMMA (Fig. 5) and SbBgMMA behave similarly in blends at similar block lengths.

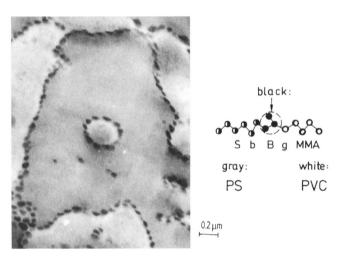

Fig. 7 Morphology of a blend PS/PVC$_{1/1}$ + SbBgMMA$_{20wt\%}$, melt-blended and annealed: the blockgraft copolymer SbBgMMA forms interface monolayers, of which only the short butadiene center block is observed (black dots). (From Ref. 88.)

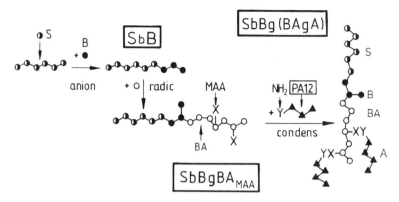

Fig. 8 Synthesis of the blockgraftgraft copolymer S*b*B*g*(BA*g*A): anionic polymerization for the styrene block and the diblock copolymer S*b*B, radical graft copolymerization with butyacrylate and some maleic anhydride for the blockgraft copolymer S*b*B*g*BA$_{MAA}$, melt grafting with nylon 12 at the MAA functions for the product.

Reactive functions can easily be incorporated into the grafts of $\alpha b\beta g\gamma$ block-graft copolymers. The synthesis of a reactive blockgraft copolymer is indicated in Fig. 8: (a) an S*b*B diblock copolymer is prepared anionically, and (b) functionalized butylacrylate grafts BA$_{MAA}$ with reactive maleic anhydride units are attached radically. This reactive copolymer S*b*B*g*BA$_{MAA}$ can (c) couple with polyamides, at the MAA functions, which leads to a blockgraftgraft copolymer S*b*B*g*(BA*g*A). This copolymer may look unappropriately complex, but it will be demonstrated in Sec. IV.A that it has very special properties. Each block is necessary.

IV. IN SITU BLENDS WITH GRAFT COPOLYMERS

Melt blending of two polymers A and B and a block copolymer $\alpha b\beta$ is as attractive as it is inexpensive. However, suitably designed copolymers $\alpha b\beta$ are needed, and these are usually not inexpensive. Long-chain diblock and triblock copolymers must be synthesized by costly anionic polymerization (89,90) processes. Industries are reluctant to produce these copolymers just for use as compatibilizers in blends. Moreover, anionic polymerization is restricted to a few monomers and is thus irrelevant for many blends. In future, perhaps, more and cheaper block copolymers can be synthesized by sleeping radical polymerization (91,92).

But so far, graft copolymers $\alpha g\beta$ are the real alternative (93,94). Grafting is less expensive and more versatile, and the β grafts can easily be functionalized by random copolymerization. But the most attractive aspect is that graft copolymers can often be generated in situ, in blends. The two main routes towards in situ

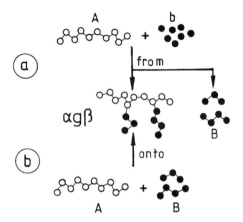

Fig. 9 Strategies for grafting: (a) grafting-from in polymer–monomer solutions A/b, (b) grafting-onto in reactive polymer–polymer melts A/B.

blends A/B/$\alpha g\beta$ are indicated in Fig. 9 (notice that all in situ blends have in common that, chemically, A = α and B = β).

> *Grafting-from* (Fig. 9a): In a polymer/monomer solution A/b, the monomer b is polymerized, which leads to free B chains and to B grafts that grow *from* A chains, whereby the copolymer $\alpha g\beta$ is created.
>
> *Grafting-onto* (Fig. 9b): While two polymers A and B carrying complementary reactive functions are melt blended, the functions are coupled, whereby the copolymer $\alpha g\beta$ is created by grafting B chains *onto* A chains.

Complete grafting is usually not aimed at and, in fact, is often difficult. Most products of these in situ reactions are not pure graft copolymers $\alpha g\beta$ but blends A/B/$\alpha g\beta$. Both grafting-onto and grafting-from are established in industry and attract much interest. Some current issues are addressed next.

A. Grafting-Onto

The technology of reactive processing, or *reactive extrusion (REX)* (53,54), is steadily gaining importance because, in REX processes, thermoplastics are chemically modified in extruders, kneaders, or injection molding machines. These are more widespread in industry than are chemical reactors: polymer chemistry is shifted from the producers to the compounders.

Grafting-onto is a REX process (63,95). As indicated in Fig. 10, typically a polymer A_X carrying functions X along its chains is blended in the melt with a polymer B_Y carrying a reactive end group Y. When X and Y couple, graft copolymer chains $\alpha g\beta$ with an A backbone and several or sometimes many β grafts result. Some functions X and Y are shown in Fig. 10. All react via X-Y addition, which is sufficiently fast at melt temperatures. Many polymers can serve as β grafts because they carry naturally reactive end groups Y. The backbones α are less easy. Normally, commercial polymers do not carry reactive functions X along their chains. They must be modified accordingly.

Frequently, unpolar polyolefines are coupled with polar polycondensates, especially with polyamides (20). These polymers contribute extremely different, often complementary property profiles that are worth combining in blends. Polyolefines are strong at low temperatures, and polyamides are strong at high temperatures; polyolefines keep off water, and polyamides keep off oxygen. All kinds of combinations of functionalized polyolefines and polycondensates have been tested, some of them, such as PP_{MAA}/PA of maleinated polypropylene and polyamides, are on the market.

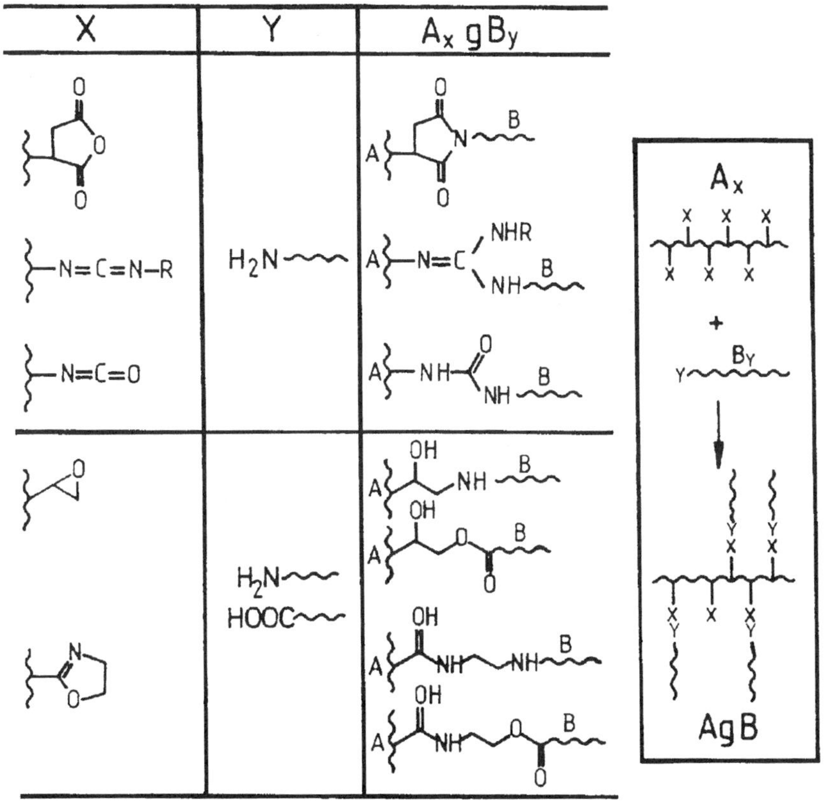

Fig. 10 X and Y functions for reactive blends A_X/B_Y that couple, during REX, by grafting-onto.

These blends are evidence that the principle of creating graft copolymers in situ in REX processes is a success. But they show also problems of grafting-onto:

The polyolefines must be functionalized, which often calls for more effort than the actual grafting-onto process. Functions (Fig. 10) are either (a) copolymerized into the chains or (b) attached to the polymer chains via radical reactions. The former method affords the synthesis of new, modified polymers; the latter suffers from side reactions: the commonly used peroxide initiators tend to crosslink PE and to degrade PP (96,97).

The processing temperatures of some polycondensates are beyond the stability limit of the polyolefines.

The graft reaction always proceeds in two-phase blend melts. Polyolefines and polycondensates, above all polyamides, are thermodynamically extremely incompatible and sometimes also rheologically incompatible. The phase dispersion often is initially appallingly bad and improves only as the grafting-onto proceeds.

Due to this incompatibility, the kinetics of grafting-onto are extraordinary. They are dominated by the facts that the grafting-onto (a) proceeds exclusively in the interfaces of a two-phase morphology and (b) leads to an interface-active copolymer that (c) keeps changing its composition. As a consequence, grafting-onto in blend melts is not at all a random process.

Convenient model blends for kinetic investigations are PS_{MAA}/PA blends of maleinated polystyrene (PS_{MAA}) and polyamides (PA) (98–100). The stages of the grafting-onto process are indicated in Fig. 11 (98). Polymer chains A_X and B_Y react at the interface, coupling to graft copolymer chains $\alpha g\beta$ (Fig. 11a). The relevant point, kinetically, is that these $\alpha g\beta$ chains tend to stay in the interface, as a compatibilizer should. But this hinders further grafting: The $\alpha g\beta$ chains separate the A_X and B_Y phases and prevent A–B contacts, which amounts to an autoinhibition of the grafting.

Nonetheless, the grafting goes on. While new chains A_X and B_Y cannot readily couple to new copolymer chains $\alpha g\beta$ anymore, the chains $\alpha g\beta$ already sitting in the interface keep taking up grafts and become overgrafted (Fig. 11b). Therefore, grafting-onto in two-phase blends often deviates in two respects from homogeneous kinetics. The grafting is:

Incomplete: Fewer chains B_Y are grafted than expected from the reactivity of the functions X and Y.

Nonrandom: Some A_X chains become overgrafted, while others remain undergrafted or even ungrafted.

Both effects depend sensitively on the blend composition and the density of the X functions on the A_X chains. A high X density favors nonrandom grafting. Chains $\alpha g\beta$ can become so overgrafted that they lose their interface activity, leaving the interface to float off into the B_Y phase, which reactivates the grafting (Fig. 11c).

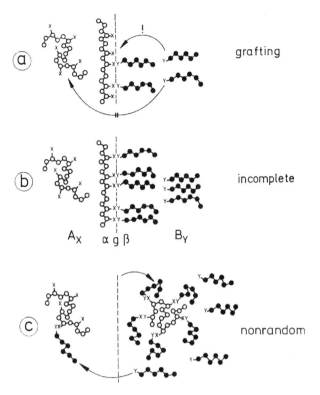

Fig. 11 Stages of grafting-onto in a blend interface: (a) initial grafting of chains A_X and B_Y in the interface; (b) overgrafted chains $\alpha g\beta$ hinder further grafting, which remains incomplete; (c) overgrafted chains $\alpha g\beta$ float off the interface, permitting further grafting, which leads to an inhomogeneous, nonrandom overgrafted–undergrafted product.

These effects are illustrated in Fig. 12. A PS_{MAA}/PA blend of a heavily maleinated PS and PA12 is shown in an intermediate state and in the final state of grafting-onto (98). In this system, ungrafted PS_{MAA} would already have disappeared in the early stage in a homogeneous system; and in the late stage, one pure graft copolymer PSgPA would have been formed. But the reality is different.

The electron micrograph in Fig. 12a shows much ungrafted PS_{MAA} in the early stage. Moreover, as specified in Fig. 12c, the graft copolymer PSgPA forms (a) interphases between the PS_{MAA} and the PA phase but also (b) micelles inside the PA phase. It appears that two copolymers PSgPA coexist, one overgrafted and one undergrafted, created by the mechanism in Fig. 12c. Indeed, these two copolymers PSgPA coexist in the end product, too. Two phases are observed in Fig. 12b instead of one, neither of them due to the original polymers PS_{MAA} and PA. These are the phases of overgrafted and undergrafted PSgPA, proving nonrandom grafting.

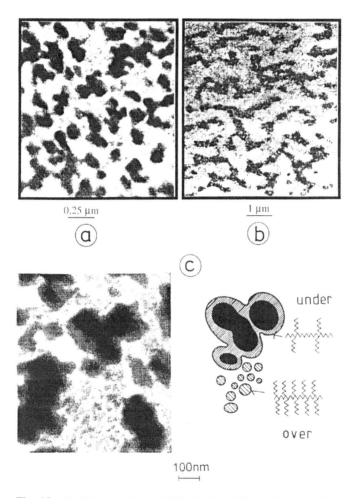

Fig. 12   Grafting-onto during REX of a blend PS_{MAA}/PA: (a) early stage with PS_{MAA} domains (black) in a PA matrix (white), compatibilized by a graft copolymer PSgPA (gray) created in situ; (b) late stage with a cocontinuous structure of two graft copolymers PSgPA (dark and light gray); (c) magnification of (a) distinguishing undergrafted PSgPA forming interphases and overgrafted PSgPA forming micelles in the PA matrix.

In Fig. 12b, all PA is grafted. This complete conversion suggests that grafting-onto may be a promising route not only toward stabilized blends A/B/$\alpha g \beta$ but also toward novel copolymers $\alpha g \beta$. However, a high degree of grafting, which is essential, seems to be limited to special blend compositions and polymer chain lengths. Frequently, grafting remains incomplete, so the end products feature blend morphologies, as in Fig. 12.

Other reports on grafting-onto (20,53,54) confirm this level of complexity. Definite conclusions on the kinetics and the optimization of grafting-onto certainly cannot yet be drawn. Too many variables are involved that have not been studied systematically. But the morphology of the blend PS/PA/SgA in Fig. 12 may turn out to be quite representative (101). The main features of the morphology are:

The structure is fine, finer by one or two orders than in PS/PA blends without the copolymer. The copolymer PSgPA is thus an efficient compatibilizer.

Grafting is incomplete and nonrandom, as discussed earlier.

The copolymer concentrates in the interfaces but does not form monolayers as predicted by Fig. 1 for block copolymers. Instead, it forms rather thick interphases, more like a random copolymer (Fig. 6). The PA grafts are apparently not long enough to let the graft copolymer PSgPA behave like a good block copolymer. This is a built-in flaw of the grafting-onto technique in blends A/B/αgβ: the grafts β cannot be made longer at will because they stem from the polymer B.

Grafting-onto can also be used to compatibilize rod–coil blends of a flexible-coil polymer and a stiff-rod polymer. Two decades ago, the concept of molecular reinforcement was advanced (102–105): Stiff polymer chains, dispersed in a matrix of flexible chains, can reinforce this matrix like fibers of molecular dimensions. But the concept failed, due to rod–coil incompatibility (106–108). As demonstrated in Fig. 13, grafting-onto helps. Stiff-rod chains of a polyester PES, grafted with flexible-coil chains of PS (Fig. 13b), form cylindrical micelles in a

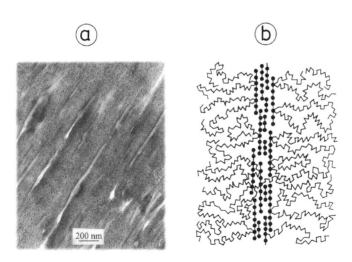

Fig. 13 Micellar reinforcement in PS: (a) submicroscopic fibers of a rod–coil graft copolymer PESgPS in the PS matrix; (b) cylindrical micelles of the copolymer PESgPS, schematically (the stiff PES backbones are much shorter than the flexible PS grafts).

PS matrix. The submicroscopic high-modulus micelles of the copolymer PESgPS, shaped like fibers (Fig. 13a), double the modulus of PS (109).

Sometimes, A/B blends are compatibilized via a technique mixed from simple melt blending and REX. Some block copolymers $\alpha b \beta$ are not, as such, compatibilizers but can be turned into compatibilizers by in situ grafting. PP/PA6 blends were compatibilized without functionalizing the PP by adding a maleinated SEBS triblock copolymer. This copolymer is fairly compatible with PP and reacts with PA6 by grafting (110,111).

In Fig. 14, a PS/PA12 blend containing the blockgraft copolymer SbBg-BA$_{MAA}$ introduced in Sec. III.B is shown. The copolymer was converted in situ into a compatibilizer because PA12 chains were grafted onto the BA$_{MAA}$ grafts, so the blockgraftgraft copolymer SbBg(BAgA) with butylacrylate-amide grafts BAgA was formed (Fig. 8) (113). Under stress, the resulting blend PS/PA12/ SbBg(BAgA) yielded profuse multicrazing, as shown in Fig. 14b. It was tough, while the original PS/PA12 blend is awfully brittle.

The complexity of the copolymer SbBg(BAgA) is justified: it forms interface layers (Fig. 14a) that are *elastomeric*. As indicated in Fig. 14c, the copolymer chains are anchored into both the PS phase and the PA phase by their end blocks. The interface itself consists, therefore, of anchored elastomeric blocks BgBA (of which gBA is much longer). These elastomeric interface monolayers cavitate under stress and induce controlled multicrazing, which toughens the blend (Fig. 14b). This blockgraftgraft copolymer is thus not only a compatibilizer for the blend PS/PA but also an impact modifier, a rare combination. Many blends need two or three additives to become mechanically competitive, i.e., one compatibilizer plus one or two impact modifiers.

Grafting-onto is involved in some thermoplastic–elastomeric blends A/B/$\alpha g \beta$, too. These systems are better known as *rubber-modified thermoplastics* (62,63). The elastomer is always the minor component. It is dispersed in domains and acts in the thermoplastic matrix as an impact modifier, by activating shear yielding or multicrazing under stress.

The elastomeric domains in these rubber-modified thermoplastics have an architecture that must be adjusted to the brittleness of the thermoplastic matrix. The types of architecture are (exemplified by commercially important thermoplastics):

> *Micelles:* Supertough nylons (113) are prepared from PA/EP blends with functionalized ethylene-propylene (EP) rubbers, by grafting-onto. The graft copolymer EP-PA forms micelles.
>
> *Superstructures:* High-impact PS (HIPS) (62) is prepared by polymerization under conditions of grafting-from in demixing solutions of PB and styrene. The elastomer PB forms microphase superstructures, the so-called salami domains (Sec. IV.B).

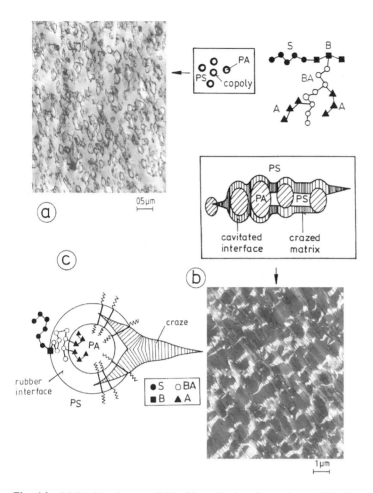

Fig. 14 PS/PA blend compatibilized by a blockgraft copolymer SbBg BA$_{MAA}$, which reacts in situ to a blockgraftgraft copolymer SbBg(BAgA): (a) copolymer monolayers around PA domains; (b) controlled cavitation of the interfaces and multicrazing in the PS matrix; (c) craze entering the elastomeric interface monolayer of anchored elastomeric BgBA grafts.

Latex particles: ABS (114) has a SAN matrix, which is impact modified by adding elastomeric-thermoplastic core-shell latex particles butadiene-SAN. The submicroscopic particles are stabilized by crosslinking.

The ABS type is different, but the types of the supertough nylons and HIPS are, in fact, A/B/$\alpha g \beta$ blends.

B.   Grafting-From

Impact-modified thermoplastics prepared by grafting-from (Fig. 9a) are particularly fascinating. High-impact PS, which is commonly viewed as PS modified with PB but is, in fact, a PS/PB/SgB blend, was developed decades ago. Other high-impact thermoplastics were later prepared analogously. High-impact PS is famous for its salami domains (Fig. 15), which are unique superstructures of graft copolymer microphases. These elastomeric-thermoplastic domains toughen the brittle PS matrix by stress whitening. Surprisingly, the discussion on the salami morphologies themselves (55–58) and on the mechanism of multicrazing (62) has not ended yet. The following models may find acceptance.

Grafting-from is quite an indirect method of blend preparation (Fig. 9a). It starts from a homogeneous polymer–monomer solution A/b, where A is an elastomer. Some stages of the polymerization of b are indicated in Fig. 16. Early, when some of the thermoplastic polymer B is formed, the solution A/B/b demixes. Later, when more B than A is formed, the two-phase solution undergoes matrix inversion: the thermoplastic B takes over the matrix. The polymerization of b is accompanied by grafting (disregarded in Fig. 16). The polymerizing system is thus in all intermediate stages and in the end a blend $A/B/\alpha g\beta$ (where, chemically, $A = \alpha$, $B = \beta$).

The salami domains (Fig. 15), made of the elastomer A but heavily filled with subdomains of the thermoplastic B, are born in the moment of matrix inver-

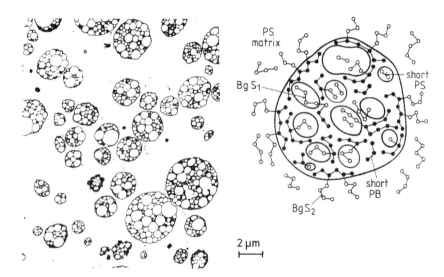

Fig. 15   Salami domain morphology of HIPS: elastomeric PB domains, heavily filled with PS subdomains, stabilized by graft copolymers BgS_1 with one styrene graft and BgS_2 with two styrene grafts.

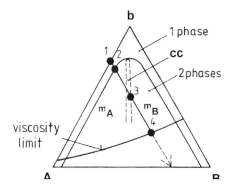

Fig. 16 Polymerization of a monomer b in a polymer–monomer solution A/b, indicated by the arrow: miscibility gap with sectors m_A, m_B (matrix A or B) and cc (cocontinuous); points 1 (initial homogeneous solution), 2 (phase separation), 3 (matrix inversion), 4 (viscosity limit from which point on stirring is inefficient); grafting is disregarded.

sion and are later, under quiescent conditions, largely preserved and eventually stabilized by crosslinking. This means that the polymerization leads directly to products with the desired morphology. Grafting-from is, therefore, said to yield *products by process.*

The key issue of grafting-from is: Why does this process yield complex, big salami superstructures, instead of simple, fine morphologies as in A/B/$\alpha b\beta$ melt-mixed blends (Fig. 5b)?

A recent model (59) proposes that the natural polydispersity of the graft copolymer $\alpha g\beta$ accounts for the salami structure. At the stage of the matrix inversion, most $\alpha g\beta$ chains carry one ($\alpha g\beta_1$) and some two ($\alpha g\beta_2$) grafts β. These copolymers behave differently in blends. Since the β grafts are shorter than the main chain α, $\alpha g\beta_1$ is dominated by A while $\alpha g\beta_2$ is compositionally more balanced between A and B. Therefore, as indicated schematically in Fig. 15, it is straightforward to conclude that the copolymer $\alpha g\beta_2$ stabilizes the salami domains, covering their surface as a compatibilizer, while the copolymer $\alpha g\beta_1$ forms micelles inside the domains, providing the internal substructure.

A discussion on the mechanism of the impact resistance of HIPS seemed, until recently, unnecessary. It was known that, under stress, myriads of crazes open up in the PS matrix, at and between the salami domains. It was generally accepted that these many tiny crazes absorb most of the impact energy, thus toughening the PS matrix (62). However, new model studies suggest (115,116) that the primary energy-absorbing effect is actually the cavitation of the rubber lamellae on the surface (as in Fig. 14) and in the interior of the salami domains. The salami substructure of the elastomeric HIPS domains intensifies the cavitation, which, in turn, initiates the crazing, which is thus a secondary effect.

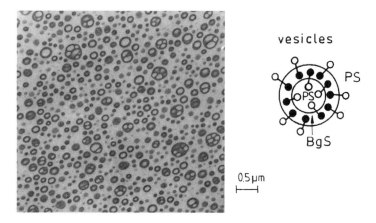

Fig. 17   Late stage of grafting-from for HIPS: elastomeric vesicles of PB.

The salami structure is the most important morphology of grafting-from products but by no means the only one (55–58). Depending on a balance of the chemistry of polymerization and grafting, the physics of incompatibility and interdiffusion, and the process technology, in particular the irrigation, very different morphologies, ranging from rubbers filled by thermoplastic domains to thermoplastics with elastomeric micelles, are generated before and after the matrix inversion (60,61). Late in the process, the salami domains run through a stage of vesicles, as shown in Fig. 17, which serves to toughen optically transparent HIPS derivatives.

As many variables are involved in grafting-from as are in grafting-onto. Both in situ methods have still a great potential, and their analysis is a challenge now and will be in the future. More complex systems may widen the scope. Promising systems with three or more types of chains and blocks, some of which were mentioned, are just beginning to attract attention.

V. CONCLUSIONS

Predictions are difficult in the world of polymer materials. Not long ago, expensive polymers with ultrahigh moduli or high-temperature resistance were expected to conquer the market, which they did not. Instead, standard polymers are dominating more than ever. Evidently, polymer blends are suited to increasing the versatility, and thus the attractiveness, of this class of cheap, standard thermoplastics. One may dare to predict, therefore, that blends have a future.

Blocky copolymers will become indispensable as compatibilizers for these blends if they can be prepared at low cost, either prior to blending or in situ, during the blending. The foregoing discussion was meant to show that well-adjusted simple diblock copolymers are certainly excellent compatibilizers but that other blocky copolymers bear more potential, because they can be prepared by cheaper techniques or because they can be provided with reactive functions that enable them to couple in situ in various ways with the blend components.

The ultimate goal of blend design should be to prepare the blends as *products by process* (Sec. IV.B), i.e., in one single step that can involve the polymerization of some components, the coupling of all components, and the creation of the final phase morphology. But specialists will know what this means: almost every blend will call for a new, special process.

ACKNOWLEDGMENTS

The author thanks, regarding his own studies, the financial support of Deutscher Bundesminister für Wirtschaft through the Arbeitsgemeinschaft Industrieller Forschungsvereinigungen (AiF, Nr. 6699, 8869, 9264, 9511, 10516, 11929).

REFERENCES

1. L Leibler. Makromol. Chem. Macromol. Symp. 16:1, 1988.
2. J Noolandi, KM Hong. Macromolecules 14:727, 736, 1229, 1981; 15:482, 1982.
3. TA Vilgis, J Noolandi. Makromol. Chem. 16:225, 1988; 23:2941, 1990.
4. KR Shull, EJ Kramer. Macromolecules 23:4769, 1990.
5. J Noolandi. Macromol. Chem. Theory Simul. 1:295, 1992.
6. JW Barlow, DR Paul. Polym.Eng.Sci. 24:525, 1984.
7. RJ Roe, D Rigby. Adv. Polym. Sci. 82:103, 1987.
8. DR Paul. Compatibilisation of polymer blends by styrene-hydrogenated-polybutadiene copolymers. In: NR Legge, G Holden, HE Schroeder, eds. Thermoplastic Elastomers. Hanser, Munich, 1987, Chap. 12.6.
9. P Teyssie, R Fayt, R Jerome. Makromol. Chem. Macromol. Symp. 16:41, 1988.
10. AK Kandpur, P Guegan, FS Bates, CW Macosko. Polymer 1999, in press.
11. DR Paul, S Newman, eds. Polymer Blends. Academic Press, New York, 1978.
12. O Olabisi, LM Robeson, MT Shaw. Polymer-Polymer Miscibility. Academic Press, New York, 1979.
13. LA Utracki. Polymer Alloys and Blends—Thermodynamics and Rheology. Hanser, Munich, 1989.
14. S Wu. Polymer Interface and Adhesion. Marcel Dekker, New York, 1982.
15. SH Anastasiadis, I Gancarz, JT Koberstein. Macromolecules 22:1449, 1989.
16. R Fayt, R Jerome, P Teyssie. J. Polym. Sci. Polym. Lett. 19:79, 1981; J. Polym. Sci. Phys. 20:2209, 1982; J. Polym. Sci. Lett. 24:25, 1986; Makromol. Chem. 182:837, 1986; Polym. Eng. Sci. 27:328, 1987; J. Polym. Sci. Lett. 27:481, 1989; J. Polym. Sci. Phys. 27:775, 1989; Polym. Eng. Sci. 30:937, 1990.

17. HR Brown. Ann. Rev. Mater. Sci. 21:463, 1991.
18. C Creton, EJ Kramer, CY Hui, HR Brown. Macromolecules 25:3075, 1992.
19. C Creton, HR Brown, VR Deline. Macromolecules 27:1774, 1994.
20. S Datta, DJ Lohse. Polymeric Compatibilizers—Use and Benefit in Polymer Blends. Hanser, Munich, 1996.
21. G Riess, J Kohler, C Tournut, A Banderet. Makromol. Chem. 101:58, 1967.
22. T Inoue, T Soen, T Hashimoto, H Kawai. Macromolecules 3:87, 1970
23. G Riess, Y Jolivet. Adv. Chem. Ser. 142:243, 1975.
24. MJ Folkes, A Keller. The morphology of regular block copolymers. In: RN Haward, ed. The Physics of Glassy Polymers. Applied Science, London, 1973, Chap. 10.
25. FS Bates, GH Fredrickson. Annu. Rev. Phys. Chem. 41:525, 1990.
26. S Sakurai. Trends Polym. Sci. 3:90, 1995.
27. L Leibler. Macromolecules 13:1602, 1980.
28. ST Milner, TA Witten, ME Cates. Macromolecules 21:2619, 1988.
29. LN Andradi, GP Hellmann. Polymer 34:925, 1993.
30. JW Cahn, JE Hilliard. J. Chem. Phys. 28:258, 1958.
31. K Binder. J. Chem. Phys. 79:6387, 1983.
32. GR Strobl. Macromolecules 18:558, 1985.
33. FS Bates, W Maurer, TP Lodge, MF Schulz, MW Matsen, K Almdal, K Mortensen. Phys. Rev. Lett. 75:4429, 1995.
34. ED Siggia. Phys. Rev. A20:595, 1979.
35. B Löwenhaupt, GP Hellmann. Colloid Polym. Sci. 268:885, 1990; 272:121, 1994; Polymer 32:1965, 1991.
36. B Löwenhaupt, A Steurer, GP Hellmann, Y Gallot. Macromolecules 27:908, 1994.
37. H Tanaka, H Hasegawa, T Hashimoto. Macromolecules 23:4378, 1990, 24:240, 1991.
38. DJ Kinning, KI Winey, EL Thomas. Macromolecules 21:3502, 1988.
39. TP Russell. Mater. Sci. Rep. 5:171, 1990.
40. M Stamm. Adv. Polym. Sci. 100:357, 1992.
41. R Schnell, M Stamm, F Rauch. Macromol. Chem. Phys. 1999, in press.
42. A Werner, F Schmid, K Binder, M Müller. Macromolecules 29:8241, 1996.
43. K Matyjaszewski, D Greszta, T Pakula. ACS Polym. Prepr. 38:707, 1997.
44. PG de Gennes. Scaling Concepts in Polymer Physics. Cornell University Press, Ithaca, 1978.
45. L Leibler. Macromolecules 15:1283, 1982.
46. H Benoit, W Wu, M Benmouna, B Mozer, B Bauer, A Lapp. Macromolecules 18:986, 1985.
47. K Mori, T Tanaka, H Hashimoto. Macromolecules 20:381, 1987.
48. FD Rumscheidt, SG Mason. J. Colloid Sci. 16:210, 1961.
49. HP Grace. Chem. Eng. Commun. 14:225, 1982.
50. JJ Elmendorp, RJ Maalke. Polym. Eng. Sci. 25:1041, 1985.
51. JJ Elmendorp, AK vander Vegt. Polym. Eng. Sci. 26:1332, 1986.
52. LN Andradi, GP Hellmann. Polym. Eng. Sci. 35:693, 1995.
53. M Xanthos. Polym. Eng. Sci. 28:1392, 1988; Reactive Extrusion—Principles and Practice. Hanser Munich, 1992.
54. A Grefenstein. Reaktive Extrusion und Aufbereitung. Hanser, Munich, 1996.

55. A Echte. Angew. Makromol. Chem. 58/59:175, 1977; Adv. Chem. Ser. 222:15, 1989.
56. H Keskkula. Plast. Rub. Mater. Appl. 16:56, 1979.
57. A Echte, F Haaf, J Hambrecht. Angew. Makromol. Chem. 93:372, 1981.
58. F Haaf, H Breuer, A Echte, BJ Schmidt, J Stabenow. J. Sci. Ind. Res. 40:659, 1981.
59. M Fischer, GP Hellmann. Macromolecules 29:2498, 1996.
60. A Echte, H Gausepohl, H Lütje. Angew. Makromol. Chem. 90:95, 1980.
61. GP Hellmann, M Walter, M Dietz, A Steurer. Macromol. Symp. 112:175, 1996.
62. CB Bucknall. Toughened Plastics. Applied Science, London, 1977.
63. AA Collyer. Rubber Toughened Engineering Plastics. Chapman and Hall, London, 1994.
64. M Banaszak, MD Whitmore. Macromolecules 25:249, 1992.
65. GD Wignall. Enc. Polym. Sci. Eng. 10:112, 1987.
66. EH Hellmann, GP Hellmann. Macromol. Chem. Phys. 198:329, 1997.
67. RJ Roe, CM Kuo. Macromolecules 23:4635, 1990.
68. T Hashimoto, T Izumitani. Macromolecules 26:3631, 1993.
69. R Gelles, CW Frank. Macromolecules 15:747, 1982.
70. MD Major, JM Tokelson, AM Brearley. Macromolecules 23:1711, 1990.
71. P Perrin, RE Prud'homme. Macromolecules 27:1852, 1994.
72. JF Palierne. Rheol. Acta 29:204, 1990.
73. D Graebling, R Muller. Coll. Surf. 55:89, 1991.
74. CJ Carriere, A Cohen, CB Arends. J. Rheol. 33:681, 1989.
75. AN Semenov. Macromolecules 25:4967, 1992.
76. AC Balazs, MT DeMeuse. Macromolecules 22:4260, 1989.
77. Y Lyatskaya, D Gersappe, NA Gross, AC Balazs. J. Phys. Chem. 100:1449, 1996.
78. HR Brown, K Char, VR Deline. Macromolecules 26:4155, 4164, 1993.
79. M Olivera de la Cruz, IC Sanchez. Macromolecules 19:2501, 1986.
80. N Hadjichristidis, H Iatrou, SK Behal, JJ Chludzinski, MM Disko, RT Garner, KS Liang, DJ Lohse, RT Milner. Macromolecules 26:5812, 1993.
81. ST Milner. Macromolecules 27:2333, 1994.
82. U Krappe, R Stadler, IG Voigt-Martin. Macromolecules 28:4458, 1995.
83. U Breiner, U Krappe, V Abetz, R Stadler. Polym. Bull. 40:219, 1998.
84. U Breiner, U Krappe, EL Thomas, R Stadler. Macromolecules 31:135, 1998.
85. M Fischer, GP Hellmann. Polymer 37:4547, 1996.
86. G Riess, M Schlienger, S Marti. J. Macromol. Sci.-Phys. B17:355, 1980.
87. C Auschra, R Stadler. Macromolecules 26:6364, 1993.
88. D Braun, M Fischer, GP Hellmann. Polymer 37:3871, 1996.
89. A Noshay, JE McGrath. Block Copolymers—Overview and Critical Survey. Academic Press, New York, 1977.
90. HL Hsieh, RP Quirk. Anionic Polymerisation. Marcel Dekker, New York, 1996.
91. CJ Hawker. Acc. Chem. Res. 30:373, 1997.
92. D Colombani. Progr. Polym. Sci. 22:1649, 1997.
93. C Price, R Singleton, D Woods. Polymer 15:117, 1974.
94. M Rabeony, DG Pfeiffer, WD Dozier, MY Lin. Macromolecules 26:3676, 1993.
95. NC Liu, WE Baker. Adv. Polym. Technol. 11:249, 1992.
96. SH Ryu, CG Gogos, M Xanthos. Adv. Polym. Technol. 11:121, 1992.

97.   VJ Triacca, PE Gloor, S Zhu, AN Hrymak, AE Hamielec. P. Eng. Sci. 33:445, 1993.
98.   A Steurer, GP Hellmann. Polym. Adv. Technol. 9:297, 1998.
99.   U Sundararaj, CW Macosko. Macromolecules 28:2647, 1995.
100. K Dedecker, G Groeninckx. Polymer 39:4985, 5001, 1998.
101. S Yukioka, N Higashida, T Inoue. Polymer 35:1182, 1994.
102. T Takanayagi, T Ogata, M Morikowa, T Kai. J. Macromol. Sci. Phys. B17:591, 1980.
103. CS Wang, IJ Goldfarb, TE Helminiak. Polymer 29:825, 1988.
104. M Ballauff. Angew. Chem. 101:261, 1989.
105. J Wendling, JH Wendorff. Macromol. Theory Simul. 5:381, 1996.
106. PJ Flory. Macromolecules 11 (1978) 1138; Adv. Polym. Sci. 59:2, 1984.
107. BY Shin, IJ Chung. Polym. Eng. Sci. 30:13, 1990.
108. M Ballauff. Polym. Adv. Technol. 1:109, 1990.
109. V Wilhelm, GP Hellmann. Polymer 1999, in press.
110. VJ Triacca, S Ziaee, JW Barlow, H Keskkula, DR Paul. Polymer 32:1401, 1991.
111. R Mülhaupt, J Rösch. Makromol. Chem. Rapid Commun. 14:503, 1993; Kunststoffe 84:1153, 1994.
112. D Braun, M Dietz, S Richter, GP Hellmann. Macromol. Chem. Phys. 1999, in press.
113. BN Epstein. U.S. Pat. 4.172.859, 4.174.358, Du Pont, 1979.
114. DM Kulich, PD Kelley, JE Pace. Ind. Polym. Sci. Eng. 1:388, 1985.
115. CB Bucknall, AM Karpodinis, XC Zhang. J. Mater. Sci. 29:3377, 1994.
116. A Lazzeri, CB Bucknall. Polymer 36:2895, 1995.
117. PR Kohl, AM Seifert, GP Hellmann. J. Polym. Sci., Phys. 28:1309, 1990.
118. D Braun, D Yu, R Kohl, X Gao, LN Andradi, E Manger, GP Hellmann. J. Polym. Sci. Phys. 30:577, 1992.
119. KJ Winey, ML Berba, ME Galvin. Macromolecules 29:2868, 1996.

12
Two Multiblock Copolymer Blends

Zbigniew Roslaniec
Technical University of Szczecin, Szczecin, Poland

I. INTRODUCTION

The possibility of obtaining a homogeneous (one-phase) structure of blends of two or more polymers is limited because most macromolecular components are immiscible. According to the classical Flory–Huggins–Scott theory, the high molecular weights and the differences in chemical structures of the polymers are the deciding factors (1). A system is said to be compatible when one of its components is a diblock (AB) or a triblock (ABA) copolymer in which one of the blocks has the same structure as the other component (homopolymer A or B). Many experimental and theoretical studies (2–4) have been devoted to the solution of this problem (see also: Chapters 11 and 18).

A specific microphase structure is shown by block copolymers, which have two kinds of segments in the chain that vary distinctly in chemical structure: flexible (soft) and rigid (hard) segments (5). Such polymer materials, called *thermoplastic elastomers*, exhibit a unique combination of strength, flexibility, and processability.

In a multiblock copolymer $(AB)_n$ the polymer chain is formed by alternately repeating flexible and rigid segments. Such a structure occurs, for example, in poly(ether-ester) (6), poly(ester-urethane), poly(ether-urethane) (7), poly(ether-amide) (8–9), and poly(carbonate-ester) (10).

A literature survey shows that one can gather reliable information about the correlation between the physical properties and phase structure of multiblock copolymers by studying mixtures of those copolymers with homopolymers or random copolymers with physical properties similar to those of one of the segments, such as polarity and crystallizability. The cocrystallization of copoly(ether-ester) with poly(butylene terephthalate) (PBT) (11) and copoly(ether-amide) with

311

polyamide (12) has been investigated, among others. More information on the physical properties of multiblock copolymer/homopolymer blends is included in Chapter 13. The aim of the present chapter is to report on the phase structure and the physical properties of poly(ether-ester) multiblock copolymers and poly(ester-urethane), poly(carbonate-urethane) or poly(ether-urethane) multiblock copolymers blends.

II. THEORETICAL ASPECTS: INTERMOLECULAR INTERACTIONS AND PHASE STRUCTURE

Central to phenomena of miscibility or immiscibility of polymers is the free energy of mixing (ΔG_M), which must be nonpositive for mixing to occur:

$$\Delta G_M = \Delta H_M - T \Delta S_M \tag{1}$$

where ΔH_M and ΔS_M are the enthalpy and entropy changes of mixing, respectively, and T is the absolute temperature (1,4,13).

The heat of mixing (enthalpy term) can be expressed as a function of the Flory–Hugins interaction coefficient χ_{12} (Eq. 2) or as the difference of Hildebrandt solubility parameters (δ_i) (Eq. 3):

$$\Delta H_M = RT n_1 \varphi_2 \chi_{12} \tag{2}$$
$$\Delta H_M = V(\delta_1 - \delta_2)^2 \varphi_1 \varphi_2 \tag{3}$$

where n_1 is the mole number of component 1 and φ_1, φ_2 are the volume contents of components 1 and 2 in the blends.

It is known that for the miscibility criterion, the term $T \Delta S_M$ becomes more favorable for higher temperatures and for low-molecular-weight polymer components (14). The enthalpy term is favorable for low Flory–Hugins interaction coefficient or low difference of Hildebrands solubility parameters. The change of free energy of mixing can be negative for strong specific intermolecular interactions (for example: hydrogen bond, donor–acceptor, or ionic interaction) (1,15).

The interaction coefficient χ_{12} depends not only on physical factors related to the contacting fragments of the polymer chains, but it is also related to changes in the free volume of the system (1,13–14,16–17).

If the phase separation concerns the fragments of macromolecules of various chemical structures, as happens in block copolymers, one should recognize the occurrence of an interphase and the specific function of chemical bonds between blocks in the formation of the phase structure. In such a case, the heat of mixing that appears under specific circumstances, such as an enthalpy of forming rigid-segment aggregates (hard domains), diminishes in the same amount as the fraction of stable interphase (18):

$$\Delta H_M = V(\delta_1 - \delta_2)^2 \varphi_1 \varphi_2 (1 - f) \tag{4}$$

where f is the volume fraction of the interphase.

The change in entropy is influenced by: (a) the components related to the location of chemical bonds between the blocks (segments) in the interphase layer (ΔS_1), (b) the replacement of rigid segments (A) into the forming domain (ΔS_A), and (c) the removal of flexible segments (B) outside (ΔS_B):

$$\Delta S_M = \Delta S_1 + \Delta S_A + \Delta S_B \tag{5}$$

These values depend on the number and length of the rigid segments, the dimensions and morphology of the forming domains, and the distances between them (19–20). Figure 1 shows the influence of rigid-segment contents on the change of system energy and morphology of the domains in block copolymers.

The total change of free energy causing the destruction of the domain structure of the polymer system (block copolymer) in the molten state (ΔG_T) consists of a change in the mixing free energy of macromolecules (segments A) forming microphase (ΔG_M) and the energy of elimination of other macromolecules (segments B) from the domains (ΔG_2):

$$\Delta G_T = \Delta G_M - \Delta G_2 \tag{6}$$

In an easy phase microseparation, i.e., at $\Delta G_M > 0$, a spontaneous solution of domains in the matrix does not occur. Instead, one can expect a homogeneous system formed under the influence of external forces, for example, under an adequately high shear stress gradient (τ) (see Chapter 14).

The preceding characteristics suggest, for the most important critical conditions of polymer miscibility, that the occurrence in the molten state is often the one

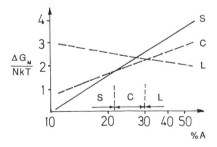

Fig. 1 Influence of rigid-segment content (A) in di- and triblock copolymers on the change of the free energy of the system (ΔG_M) and morphology of the domains. The shape of the domains: S-spherical, C-cylindrical, L-lamellar. (From Ref. 23.)

of a heterogeneous material, its phase structure depending on (21–22):

> Intermolecular interactions in the system (cohesion force, interfacial tension, solubility parameters, diffusion of lattices, etc.)
> Composition (position on the phase diagram)
> Rheological properties of components and the whole system

In practice, the most interesting properties of that type for polymer materials result precisely from their multimicrophase structure, the limited miscibility (or immiscibility) being a condition of advantageous features. There is a strict relationship between the mechanism of domain formation and the diffusion coefficient of the polymer components (23–25). If the blend component crystallizes or forms another type of ordering (e.g., hydrogen bonds), then the phase structure is more complicated. The phenomenon of phase separation interferes with solid–liquid phase transformation (Fig. 2) (26). There are also possible micro- and macrophase-separation processes, particularly in the case when one of the components is a block copolymer (27–30).

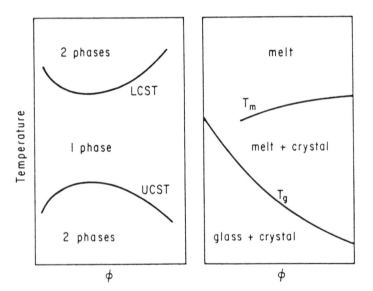

Fig. 2 Scheme of phase transitions in the polymer system: (left) liquid–liquid and (right) solid–liquid. LCST and UCST denote, respectively, lower and upper critical solubility temperatures, T_g is the glass transition temperature, and T_m is the melting point. (From Ref. 26.)

Among immiscible or partially miscible polymer systems, the systems of elastothermoplastic properties (blends and block copolymers) are of great importance. Apart from the practical aspects, the investigation of such systems may contribute to relevant new information relating to polymer miscibility theory.

III. PHYSICAL PROPERTIES OF POLY(ETHER-ESTER) (PEE)/POLYURETHANE (PUR) MULTIBLOCK COPOLYMER BLENDS

A. Chemical Structure and Solubility Parameters

The physical properties of block copolymers are related, among others, to the intermolecular factors of stabilization of their microphase structure (22). In blends of two block copolymers, the possible interactions comprise four segments of different chemical structure, which are located preferentially in two different macromolecules. Poly(ether-ester) (PEE) (Formula 1) and poly(ester-urethane) (PUA) (Formula 2a) block copolymers have been used as examples. They show a similar capability of phase separation resulting from a similar multiblock structure of $(AB)_n$ type, but they essentially differ in the nature of the intermolecular interaction within the hard phases conditioning this separation. The domains of polyester (PBT) rigid segments have semicrystalline structure, while polyurethane (PU) segments have pseudocrystalline structure resulting from intermolecular hydrogen bonds (6–7). In both cases the flexible segments form a similar liquid (elastic) microphase (matrix) in which the hard domains are dispersed. The phase equilibrium achieved for each copolymer can be disturbed by the interaction of segments originating from another copolymer. The character of these interactions is easier to investigate by changing the chemical structure or the fraction of flexible segment in one of the blend components. Apart from PUA (31–32), copoly(carbonate-urethane) (PUC) (32–33) and copoly(ether-urethane) (PUE) (34) (Formulas 2b and 2c) were used.

Formula 1

Copoly(tetramethyleneoxide-*block*-butylene terephthalate); copoly(polyoxyte-tramethylene terephthalate-*run*-tetramethylene terephthalate)

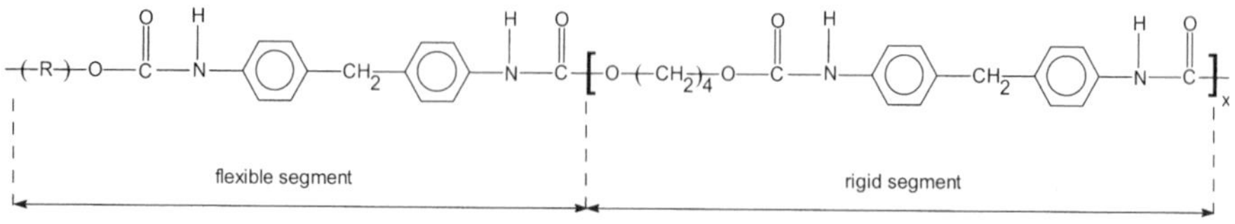

where:

a)

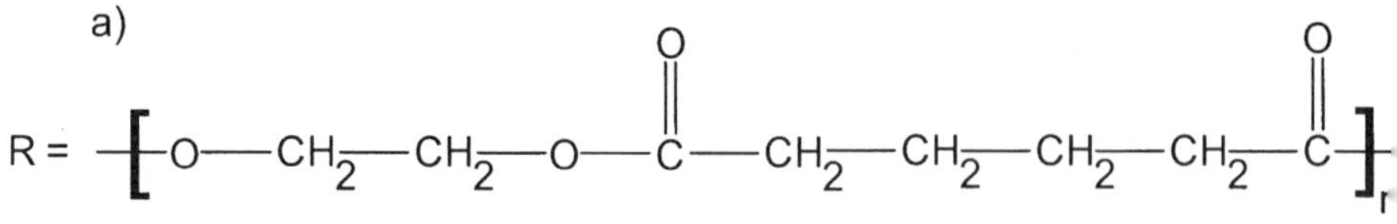

$$R = \left[O - CH_2 - CH_2 - O - \overset{\overset{\displaystyle O}{\|}}{C} - CH_2 - CH_2 - CH_2 - CH_2 - \overset{\overset{\displaystyle O}{\|}}{C} \right]_r$$

Copoly(ethylene adipate-*block*-tetramethylene 4,4′-methylene-diphenylene-di-urethane)

b)

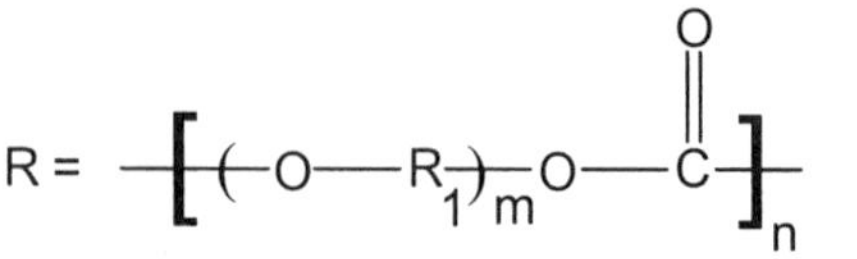

$$R = \left[(O - R_1)_m O - \overset{\overset{\displaystyle O}{\|}}{C} \right]_n \quad ; \quad R_1 = alkyl\,(C_6 - C_8)$$

$$m > 1$$

Copoly(oligooxyalkylene carbonate-*run*-alkylene carbonate)-*block*-tetramethy-lene 4,4′-methylene-diphenylene-diurethane)

c)

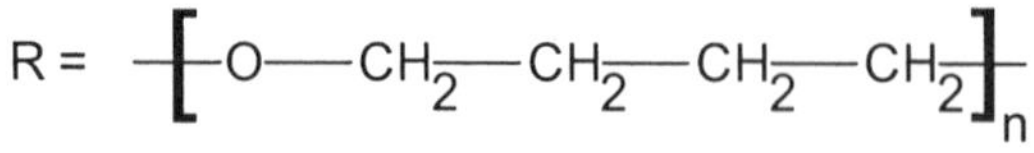

$$R = \left[O - CH_2 - CH_2 - CH_2 - CH_2 \right]_n$$

Copoly(oxytetramethylene-*block*-tetramethylene 4,4′-methylene-diphenylene-diurathane)

Formula 2

The important feature of the system is the possibility (or its lack) of satisfy-ing the conditions of the component's miscibility, which can be estimated on the basis of the square of the difference of solubility parameters $\Delta\delta^2$ from Eq. (3) (see also Chapter 18).

Table 1 contains the physical properties of polymer segments, including sol-ubility parameters calculated from the so-called molar attractions (35–36) or taken

Table 1 Chemical Structure and Physical Properties of Polymers Used as Segments in PEE and PUR

Type of polymer forming-segment	Denotation	Average polymerization degree n (x)	Density d (g/cm³)	Molar volume Vu (cm³/mol)	Solubility parameter δ (MPa$^{1/2}$)					Glass Transition temperature T_g (°C)	Melting point T_m (°C)
					δ_d^c	δ_p^c	δ_h^c		δ^d		
Polyether	PO4	~14 and 28	0.98^a 1.18^b	73.5	16.0	5.4	8.1	18.7	17.5 [40] 16.6 [41]	−92.8 [40] −85 [42]	35 [35] 33–40 [43]
Polyester	PBT	~4 – 5	1.31^b 1.396 [44] 1.405 [45]	168.1	18.6	10.3	9.8	23.5	21.6 [107] 23.1 [108]	37–53 and −25 [45]	227 [46]
Polyester (aliph)	PEA	~12	1.18$^{(1)}$	145.6	16.5	11.9	10.5	22.9	19.4 [35]	−40 [35]	47 [35]
Polyether-carbonate	POC	—	1.08	107$^{(3)}$	15.6	10.4	12.9	22.8	—	—	55–60 [47]
Polyurethane	PU	~2.5 – 4.5	1.25^a 1.69^b	274.4$^{(3)}$	18.2	6.6	9.1	21.4	27.0 [39]	110 [40,48]	~241^e [49] ~206^f [50]

^{a} For amorphous structure [35], [51].
^{b} For semicrystalline structure [52], [45], [51], [44].
^{c} Values calculated according to Van Krevelen [35] and Barton's Handbook [36].
^{d} Values taken from other publications.
^{e} For x = 6.
^{f} For x = 3.

from experimental data in the literature. Rather considerable differences between δ values from different sources are observed. In particular, in the case of polyurethane, the data should be treated as approximate. Chapiro et al. (37) have determined the solubility parameter of PUE containing approximately 75% flexible segments, obtaining the value $\delta = 19.6$ MPa$^{1/2}$. For PUA with approximately 79% of polyester segments, Piglowski and Laskawski (38) found a solubility parameter $\delta = 22$ MPa$^{1/2}$. Indirectly estimated values of solubility parameters for urethane homopolymer are those given by Camberlin and Pascault (39).

Even a superficial comparison of the squared difference of the solubility parameters for particular pairs of segments suggest that PBT/PEA and PBT/POC systems are characterized by distinctly smaller $\Delta\delta^2$ values, thus yielding:

PU/PO4	PBT/PO4	PBT/UP	PBT/PEA PBT/POC
$\Delta\delta^2 \cong 69$ MPa $>$	$\Delta\delta^2 \cong 23$ MPa $>$	$\Delta\delta^2 \cong 12$ MPa $>$	$\Delta\delta^2 \cong 0.4$ MPa $\Delta\delta^2 \cong 0.5$ MPa

For particular segments one can expect significant differences in the interaction resulting from thermodynamic factors. Apart from that, technological parameters such as viscosity in the molten state, mixing conditions, and processing parameters may also have a determined effect. All systems discussed in this chapter were prepared in the same way, by a two-stage homogenization in the molten state using a Brabender mixer and a laboratory extruder. Detailed descriptions for preparing the mixtures and samples are given in the subject papers (31–32). All block copolymers used for comparative investigations are presented in Table 2.

B. Glass Transition Temperature

A miscible polymer system (homogeneous phase) shows a single glass transition temperature (T_g), the value of which is usually found between the glass transition temperature of individual components. In the simplest case, the additive dependence of glass transition temperature on blend composition is observed:

$$T_g = w_1 T_{g1} + w_2 T_{g2} \tag{7}$$

where:

w_1, w_2 = weight fractions of polymer components in the blends

$T_{g1}, T_{g2},$ = glass transition temperature of the individual components

As result of intermolecular interactions, market deviations from linearity of T_g plots against composition for blends are observed (1,52). For approximate results, Flory's and Fox's equation (Eq. 8) (12,53) or Gordon–Taylor's equation

Table 2 Chemical Structure and Physical Properties of Components of Multiblock Copolymers Blends

Type of polymer	Denotations used in the text	Molecular weight of flexible segment M_n	Polycondensation degree of rigid segments x^a	Density d (g/cm³)	Viscosity $[\eta]$ (dl/g)	Other information
Copoly (etherester)		1000	20.5	1.24	1.12[c]	Elitel 4420 Commercial products from
	PEE	1000	7.7	1.20	1.16	Elitel 4440 Elana Co., Torun,
	PEE1	1000	5.1	1.16	1.43	Elitel 4450 Poland
		1000	4.2	1.13	1.52	Elitel 4455
Copoly (ethercarbonate-urethane)	PUC	2000	2.9[b]	1.15	1.20[d]	Desmopan 786, Bayer AG
Copoly (ester-urethane)	PUA	2000	2.5	1.20	0.59[d]	Research Laboratory products
	PUA1	2000	3.0	1.21	0.99	Jelur 85A SO Commercial products
		2000	4.0	1.23	0.81	Jelur 90A SO from Jelchem Co.,
		2000	4.5	1.2	0.76	Jelur 95A SO Jelenia Gora, Poland
Copoly (ether-urethane)	PUE	2000	3.0	1.12	0.73	Research Laboratory products from
	PUE1	2000	3.5	1.13	0.67	Jelchem Co., Jelenia Gora, Poland
		2000	4.5	1.16	0.92	

[a] Determined on the basis of stoichiometric calculations.
[b] Value estimated basing on IR examinations and literature data (Ref. 86).
[c] Phenol–trichloroethylene 1:1 vol., 30°C.
[d] DMF, 25°C.

(55) (Eq. 9) is frequently applied:

$$\frac{1}{T_g} = \frac{w_1}{T_{g1}} + \frac{w_2}{T_{g2}} \tag{8}$$

$$T_g = \frac{w_1 T_{g1} + kw_2 T_{g2}}{w_1 + kw_2} \tag{9}$$

The coefficient k is defined as the difference between expansion coefficients in the glassy and in the elastic state, $\Delta\alpha$ (15), or by the difference in specific heat at the glass transition temperature, ΔC_p (15,54): $k = \rho_1 \Delta\alpha_2/\rho_2 \Delta\alpha_1$; $k = \Delta C_{p2}/\Delta C_{p1}$, where ρ is the density and indices 1 and 2 refer to the initial components of the mixture. Couchman (54) describes a general dependence of T_g on the composition of such systems, based on a thermodynamic model of glass transition transformation:

$$\ln T_g = \frac{w_1 \Delta C_{p1} \ln T_{g1} + w_2 \Delta C_{p2} \ln T_{g2}}{w_1 \Delta C_{p1} + w_2 \Delta C_{p2}} \tag{10}$$

If $\Delta C_{p1} \cong \Delta C_{p2}$ and T_{g1}/T_{g2} is close to zero or $\Delta C_{p2}/\Delta C_{p1} = k$, then Eq. (10) transforms into one of the earlier-mentioned equations (52).

Blends of PEE with PUA, PUC, and PEE are characterized by two temperature ranges of glass transition transformation, from approximately $-80°C$ to approximately $20°C$ (T_{g1}; low-temperature range) and from approximately $+30°C$ to approximately $+80°C$ (T_{g2}). It is self-evident that the low-temperature range refers to the soft phase. The dependence of T_g on flexible segment contents over this range and in such types of copolymers has been pointed out earlier (56–59). Figures 3 and 4 present DSC plots in the low-temperature range for PEE/PUA, PEE/PUC, and PEE/PUE systems. The content of flexible segments in initial copolymers was, respectively, 50%, 72%, 65%, and 69%. It is noteworthy that there are two inflection points on the DSC plots for the PEE/PUA mixture, between T_1 and T_2 or T_1' and T_2' temperatures, respectively, but there is only one inflection point on the plots for the PEE/PUC and PEE/PUE blends (Table 3, Figs. 3 and 4). This means that the systems exhibit, respectively, one or two glass transition temperatures T_{g1}. This is confirmed by dynamic-mechanical characteristics (Figs. 5 and 6) where the temperature of β relaxation (T_β) determined from the maximum of the loss modulus (E'') is interpreted as the glass transition temperature of the soft phase (60–61). The dependence of the glass transition temperature T_{g1} and the relaxation temperature β on blend composition is shown in Fig. 7.

The presence of one or two T_{g1} values is related to the chemical structure of the segments. PEE1/PUA1 blends composed of copolymers containing, respectively, 55% and 69% of flexible segments are also characterized by two T_{g1} values, located close to the T_g of the initial components (Fig. 7b).

In the second range of glass transition, almost all of the blends show a double glass transition temperature of values close to T_{g2} of the initial components: $T_{g2,PEE} = 46°C$; $T_{g2,PUA} = 63°C$; $T_{g2,PUC} = 65°C$; $T_{g2,PUE} = 76°C$; $T_{g2,PUA1} = 74°C$ (32,34).

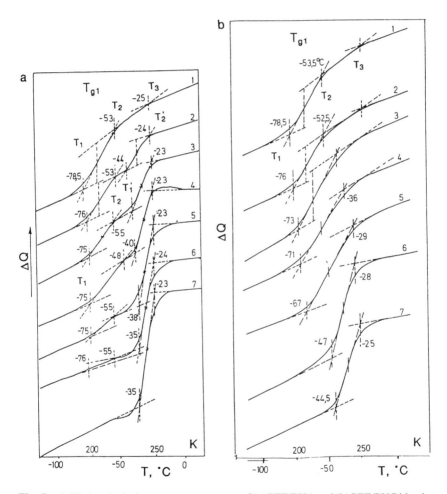

Fig. 3 DSC plots in the low-temperature range of (a) PEE/PUA and (b) PEE/PUC blends. T_{g1}: glass transition temperature. Blend composition (PEE/PUR wt%): 1 = 100/0, 2 = 90/10, 3 = 70/30, 4 = 50/50, 5 = 30/70, 6 = 10/90, 7 = 0/100; heating rate: 10°C/min.

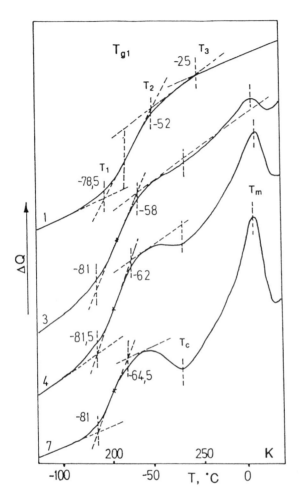

Fig. 4 DSC plots in the low-temperature range of PEE/PUE blends. T_c, T_m, respectively, crystallization and melting point temperatures. Blend composition (PEE/PUR wt%): 1 = 100/0, 2 = 90/10, 3 = 70/30, 4 = 50/50, 5 = 30/70, 6 = 10/90, 7 = 0/100; heating rate: 10°C/min.

The single glass transition temperature of the mixture in the low-temperature range can be interpreted as the result of:

The existence of only one soft phase.

The overlapping of ΔC_p effects on the DSC plots (and E''_{max} peaks on DMTA plots too), also due to low differences in the glass transition tem-

Table 3 Glass Transition Temperatures from DSC and TMDA Scans

PEE/PUR (wt %)	PEE/PUA			PEE/PUC			PEE/PUE			PEE/PUA1		
	T_{g1} (°C)	T_{g2} (°C)	T_β (°C)	T_{g1} (°C)	T_{g2} (°C)	T_β (°C)	T_{g1} (°C)	T_{g2} (°C)	T_β (°C)	T_{g1} (°C)	T_{g2} (°C)	T_β (°C)
100/0	−67	46	−62	−67	46	−62	−67	46	−62	−68	46	−65
90/10	−65, −35	47	−61	−66	36, 47	−62	−68	50	−63	−68, −33	48	−61
70/30	−64, −32	37, 47	−60, −27	−62	37, 47, 65	−58	−70	51	−67	−65, −31	50	−57, −31
50/50	−62, −30	37, 47, 62	−60, −25	−53	49, 64	−45	−72	37, 53, 77	−68	−64, −29	52, 74	−58, −27
30/70	−64, −30	38, 48, 62	−60, −27	−44	47, 65	−31.5	−72	37, 53, 77	−67	−65, −30	52, 74	−55, −25
10/90	−65, −28	36, 63	−25	−37	41, 65	−30	−73	38, 54, 77	−69	−65, −27	51, 75	−24
0/100	−28	36, 63	−25	−35	65	−28	−72.5	38, 76	−73	−27	74	−20

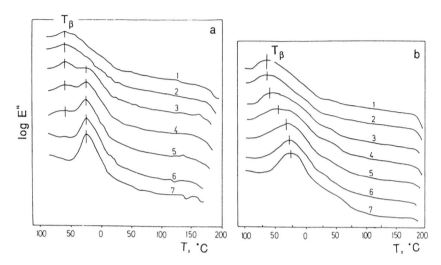

Fig. 5 Dynamic loss modulus (E″) of (a) PEE/PUA and (b) PEE/PUC blends as a function of temperature: T_{β}-β relaxation temperature. Blend composition (PEE/PUR wt%): 1 = 100/0, 2 = 90/10, 3 = 70/30, 4 = 50/50, 5 = 30/70, 6 = 10/90, 7 = 0/100; (Rheovibron Viscoelastometer, DDV: IIc, Toyo Baldwin, 35 Hz).

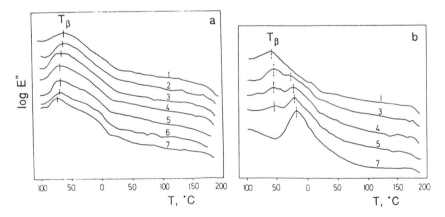

Fig. 6 Dynamic loss modulus (E″) of (a) PEE/PUE and (b) PEE1/PUA1 blends as a function of temperature. Blend composition (PEE/PUR wt%): 1 = 100/0, 2 = 90/10, 3 = 70/30, 4 = 50/50, 5 = 30/70, 6 = 10/90, 7 = 0/100; (35 Hz).

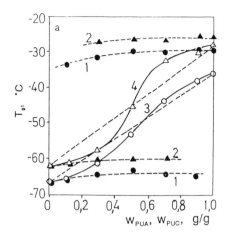

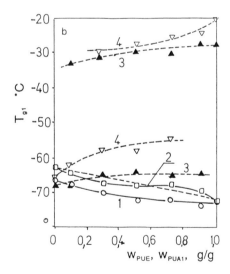

Fig. 7 Dependence of the glass transition temperature in the low-temperature range (T_{g1}) on the blends of (a) PEE/PUA (1, 2) and PEE/PUC (5,4) and (b) PEE/PUA (1, 2) and PEE1/PUA1(3, 4); plots 1, 3 = DSC data, plots 2, 4 = DMTA data ($T_{g1} = T_\beta$).

peratures of the components; MacKnight et al. (62) showed that such effect is not observed if $\Delta T_g > 30°C$.
The small size of the soft microphases.

The difference in the glass transition temperatures of PEE and PUA is indeed too small ($\Delta T_{g1,PEE/PUE} = 6-9°C$), while for the PEE/PUC system the difference is higher than 30°C. The effect of overlapping is also limited if the measurements are carried out at a low heating rate and low frequencies. The aforementioned results on dynamic studies were obtained applying 35-Hz frequency. However, PEE/PUC blends were also examined at 3.5 Hz (32). In such circumstances one β relaxation temperature is also obtained (see Fig. 8). The differences in glass transition temperatures determined in the second cycle of heating (T_{g1}) and while cooling (Fig. 9) do not exceed 4°C.

The plot of the glass transition temperature of the soft phase in the PEE/PAC blend shows positive–negative deviation from the additivity behavior of the single components, and it cannot be approximated with any of the earlier-cited equations. The latter will be discussed later from the point of view of the quantitative estimation of intermolecular interactions.

C. Degree of Phase Separation

The specific character of block copolymers is their ability for microphase separation, on which the elastic properties of such materials depend. For multiblock elastomers there is a strict correlation between the degree of phase separation and the length and fraction of particular segments and the difference of their solubility parameters (39–41,43,46–47).

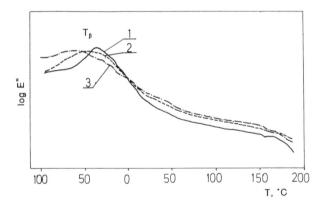

Fig. 8 Loss modulus (E'') as a function of temperature (T) for PEE/PUC blends: 70/30 (1), 50/50 (2), and 30/70 (3) determined at 3.5 Hz.

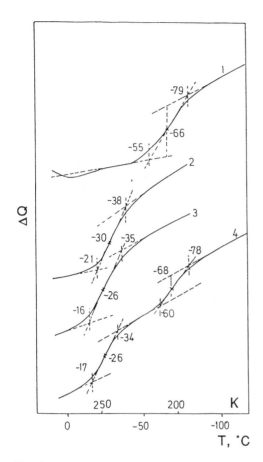

Fig. 9 DSC plots for copolymers and their blends during cooling: 1 = PEE, 2 = PUA, 3 = PUA1, 4 = PEE1/PUA1 50/50. Cooling rate = 10°C/min.

The degree of soft-phase separation (SR_S) may be expressed by the product of the flexible-segment fraction in a soft phase (w_{SS}) and the soft-phase fraction in the material w_{sp} in relation to the flexible-segment contents in the polymer (w_S) (63):

$$SR_S = w_{SS} \cdot \frac{w_{SP}}{w_S} \qquad (11)$$

The change in specific heat of the soft phase observed on the DSC diagram at the glass transition temperature (ΔC_p^{obs}) is the sum of the ΔC_p fractions originating from the flexible segment (ΔC_p^S) and the rigid segment (ΔC_p^H) forming the soft phase:

$$\frac{\Delta C_p^{\text{obs}}}{w_{SP}} = w_{SS} \Delta C_p^S + w_{HS} \Delta C_p^H \tag{12}$$

Introducing the Eq. (12) into Eq. (11), the following equation is obtained:

$$SR_S = \frac{w_{SS}(\Delta C_p^{\text{obs}}/w_S)}{w_{SS} \Delta C_p^S + w_{HS} \Delta C_p^H} \tag{13}$$

The fraction of specific heat of the rigid segment in the soft phase is much lower than the fraction of the flexible segment $\Delta C_p^H < \Delta C_p^S$. The expression $(1 - w_{SS}) \Delta C_p^H$ can be disregarded as very small, and then (41):

$$SR_S = \left(\frac{\Delta C_p^{\text{obs}}}{w_S}\right)\left(\frac{1}{\Delta C_p^S}\right) \tag{14}$$

Therefore, the degree of soft-microphase separation in block copolymers may be estimated on the basis of the ΔC_p^{obs} (in the low-temperature range) and the change in specific heat of the polymer constituting the flexible segment (ΔC_p^S), read from DSC scans of these materials. Typical SR_S values for urethane multi-block elastomers and poly(ether-amide) elastomers are listed in Table 4 (39,41). For binary systems one can expect a reciprocal influence on the degree of their phase separation.

In the course of various investigations (31,34) it has been stated that for poly(ether-ester) copolymers and their blends the determination of the phase separation degree by this method does not give explicit results. The calculated SR_S

Table 4 Degree of Phase Separation (SR_S) in Polyurethane and Poly(ether-amide) Multiblock Elastomers

Rigid segment	Flexible segment	ΔC_p^S (J/g · K)	$\Delta C_p^{\text{obs}}/w_s$ (J/g · K)	SR_s
MDI-BD	Polybutylene 1,2 (90%), $M_n = 2280$	0.430	0.356	84
(3:2)	Polybutadiene $-1,2$ (90%), $M_n = 2280$	0.566	0.426	75
	PO4; $M_n = 2400$	0.815	0.520	63
PA-12	PO4; $M_n = 2000$; $w_s = 77\%$	0.72	0.21	66[a]
	PO4; $M_n = 2000$; $w_s = 70\%$	0.72	0.20	63[a]
	PO4; $M_n = 2000$; $w_s = 62\%$	0.72	0.22	55[a]
	PO4; $M_n = 2000$; $w_s = 50\%$	0.72	0.17	50[a]
	PO4; $M_n = 2000$; $w_s = 33\%$	0.72	0.13	37[a]

[a] Value considering a crystallized part of the flexible segments (from Ref. 41):
$SR_s^a = SR_s + (\Delta H_m^{\text{obs}} - \Delta H_c^{\text{obs}}) / \Delta H_m^s$; ΔH_m^s $_{PO4} = 120$ J/g
Source: Refs. 39 and 41.

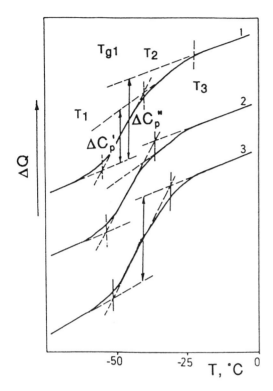

Fig. 10 Method of determination of changes of specific heat in the low-temperature range for PEE and the blends: 1 = PEE, 2 = PEE/PUC 90/10, 3 = PEE/PUC 70/30.

values were higher than could be expected on the basis of earlier publications. The detailed analysis of results reveals that on the DSC plots of sample containing a prevailing amount of poly(ether-ester) block copolymer at the temperature denoted as T_3 (Fig. 10), an additional inflexion point, related to the glass transition transformation of the totally amorphous interphase (PBT) is observed: $T_g = 248$ K ($-25°C$), $\Delta C_{p}^{a}, PBT = 0.48$ J/g·K. Using the Camberlin and Pascault method (39–40), it is possible to read for copoly(ether-ester) that the change in specific heat of the soft phase is $\Delta C_p^s = \Delta C_p'$ (between T_1 and T_2, Fig. 10) or $\Delta C_p^s = \Delta C_p''$ (between T_1 and T_3). For the blends, only the $\Delta C_p''$ value is accessible. In Table 5, data of ΔC_p^{obs} for the investigated copolymers and their blends in the low-temperature range are given. Assuming that the difference originates exclusively from the amorphous PBT phase, it is possible to estimate its fraction in the copolymer, $w_p^{obs} = (\Delta C_p'' - \Delta C_p')/\Delta C_{p}^{\alpha}, PBT$, as 0.25 and 0.27 for PEE and PEE1, respectively. In such a case it should be assumed that this part of PBT does not enter into the composition of the soft phase. In this case, value changes in specific heat in the

Table 5 Specific Heat Change ΔC_p^{obs} of the Blend of Two Multiblock Copolymers in the Low-Temperature Range

Blend composition (wt %)	ΔC_p^{obs} (J/g · K)			
	PEE/PUA	PEE/PUC	PEE/PUE	PEE1/PUA1
100/0	0.172; (0.279)[a]	0.172; (0.279)[a]	0.172; (0.279[a])	0.209; (0.322[a])
90/10	0.218; 0.180	0.318; (0.210[a]; 0.219[b])	0.230	0.260; 0.078
70/30	0.141; 0.252	0.258; (0.230[a]; 0.208[b])	0.230	0.170; 0.132
50/50[c]	0.170; 0.329	0.305	0.238	0.119; 0.194
30/70	0.062; 0.441	0.503	0.247	0.072; 0.244
10/90	0.016; 0.491	0.523	0.272	0.019; 0.330
0/100	0.544	0.484	0.297	0.400

[a] For the glass transition range between T_1 and T_3.
[b] After substraction of the correction related to the change in specific heat of totally amorphous PBT (interphase).
[c] For PEE/PUA1 50/50 blend: respectively 0.119 and 0.203 J/g · K; for PEE/PUE1 50/50 blend: 0.244 J/g · K; for PEE1/PUC 50/50 blend: 0.309 J/g · K; for PEE1/PUE1 50/50 blend: 0.557 J/g · K.

low-temperature range (ΔC_p^{obs}) should be interpreted as the sum of components: $\Delta C_p^{obs} = \Delta C_{p1} + w_{PEE} \cdot w_{PBT}^{a} \cdot \Delta C_p^{a}$, PBT, where ΔC_{p1} refers solely to the homogenous soft phase composed mainly of flexible segments. Such an assumption allows one to interpret the fact that the PEE/PUC 30/70 and 10/90 blends show ΔC_p^{obs} values greater than those for PUC (Table 5). For a soft phase originating from PUC, larger ΔC_p^{obs} values are observed than would result from their additivity.

Thus, calculating the degree of microphase separation of blends containing PEE, the corrected ΔC_p^{cor} value should be utilized:

$$\Delta C_p^{cor} = \Delta C_p^{obs} - w_{PEE} \cdot w_{PBT}^{a} \cdot \Delta C_{p, PBT}^{a} \tag{15}$$

where ΔC_p^{obs} is the change in specific heat in the low-temperature range, including the effects of the amorphous PBT interphase, $\Delta C_{p,PBT}^{a}$ is the change in specific heat of amorphous PBT at the glass transition temperature, w^a is the fraction of amorphous PBT in copoly(ether-ester), and w_{PEE} is the PEE fraction in the blend. Equation 15 would then have the form:

$$SR_S^{cor} = \frac{\Delta C_p^{obs} - w_{PBT}^{a} \cdot w_{PEE} \cdot \Delta C_{p, PBT}^{a}}{w_S \, \Delta C_{p, PBT}^{s}} \tag{16}$$

The results of calculations of the degree of phase separation for the copolymers and their blends are shown in Table 6. It turns out that the degree of phase separation of various multiblock copolymers depends on their composition (see Fig. 11).

Table 6 Degree of Phase Separation SR_s of Multiblock Copolymers and Their Blends

Polymers		Degree of separation of the soft phase				
PEE	w_s; wt %	20	40	50	55	
	SR_S^{cor}; %	21	35	48	52	
PUE	w_s; wt %	60	65	69	65[b]	
	$SR_S^{cor\,a}$; %	56	61	62	82[b]	
PUA	w_s; wt %	60	65	69	72	
	SR_S; %	60	63	67	78	
	Blends					
PEE/PUA	w_{PUA}; wt %	10	30	50	70	90
	$SR_{S,PEE}$; %	67	56	94	57	44
PEE1/PUA1	w_{PUA1}; wt %	10	30	50	70	90
	$SR_{S,PEE1}$; %	72	61	60	60	48
PEE/PUC	w_{PUC}; wt %	10	30	50	70	90
	SR_S^{cor}; %	58	52	60	—	—
PEE/PUE	w_{PUE}; wt %	10	30	50	70	90
	$SR_S^{cor;a}$ %	62	59	56	55	56

w_s = content of flexible segments in the copolymer.

[a] Value calculated considering the crystalline part of PO4 as in Table 4: $\Delta H_{m,PO4}^S = 120$ J/g (from Ref. 41); $\Delta C_{p,PO4}^S = 0.72$ J/g (from Ref. 41); $\Delta C_{p,PEA}^S = 0.83$ J/g (from author's investigations).

[b] Value for PUC estimated on the basis of IR spectrophotometric examinations $\Delta C_{p,POC}^s$ assumed as for PEA.

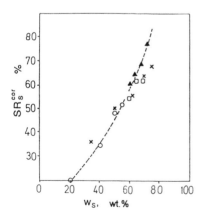

Fig. 11 Influence of flexible-segment contents in block copolymers (w_s) on the degree of soft-phase separation (SR_s). (For copoly(ether-esters) and copolyurethanes, SR_S^{cor} from author's investigations; for copoly(ether-amides), from Ref. 41.)

Generally the degree of phase separation of the soft phase in the blends is greater than that of the initial copolymer. This effect may be due to:

A change of flexible-segment contents in the soft phase as result of inter-molecular interaction with another component

An apparent increase of the soft-microphase fraction resulting from dis-persing in it the determined amount of another copolymer

The SR_S values of PEE/PUE and PEE/PUA blends do not require correc-tion because their glass transition transformation is beyond the T_g range of amor-phous PBT. However, in the case of the PEE/PUE blend, the crystalline part of PO4 should be considered.

The source of error in the estimation of the degree of phase separation by this method may be the assumption of a constant fraction of amorphous PBT in-terphase in PEE regardless of blend composition, ignoring the fraction of rigid segment in the soft phase. On the basis of DSC examinations, unfortunately, it is impossible to determine the fractions of flexible (w_{SS}) and rigid (w_{HS}) segments in the soft phase. The location of amorphous PBT phase ($T_g = -25°C$) in the phase structure of PEE requires further examination, applying other methods.

D. Melting Behavior

In copolymers or in polymer blends with one crystallizing component, a depres-sion of the melting point is observed (64–66). Häslinn et al. (67) investigated the influence of the type and molar volume of end groups on the depression of the PBT melting point. Furthermore, Schmidt and Dröscher (68) found that the de-pression of the melting point of block ether-ester copolymers depends on the frac-tion of flexible segments and not on their polymerization degree or the type of structure (triblock or multiblock). The influence of urethane groups on the de-pression of T_m for linear polyoxytetramethylene was investigated by Bill et al. (69). By this method Castles et al. (60) determined the influence of intermolecu-lar interaction of polyoxytetramethylene segments on the crystallization of poly(butylene isophthalate) phase.

Table 7 shows the results of DSC analysis for the melting range of PEE/PUR blend from the second heating run in the heating-cooling-heating cy-cle, at a rate 10°C/min. The crystallization conditions were the same for all sam-ples. The influence of these conditions on the fusion point and heat of melting of PEE is well known (70). In copoly(ether-esters) and blends of them that are rich in PEE/PUA and PEE/PUC, the main endotherm of melting (T_8, Figs. 12 and 13) is observed in the range 191–187°C. Two smaller endotherms related to condi-tions of crystallization (annealing peak) (T_7) and recrystallization of PBT form from α to β (T_9) (71–72) respectively over the ranges 155–154°C and 204–202°C. In the case of PUA, only one wide endotherm with maximum temperature over-

Table 7 Results from DSC Analysis of Multiblock Copolymer Blends

Composition of the blend		T_m^a (°C)	T_{mk}^b (°C)	ΔH_m (J/g)	T_c (°C)	ΔH_c (J/g)
PEE/PUA	100/0	155, <u>191</u>, 204	212	29.5	154	30.2
	90/10	<u>189</u>, 204	212	26.3	151	24.3
	70/30	<u>187</u>, 204	211	19.7	152	20.6
	50/50	<u>186</u>, 204	211	13.1	148	12.2
	30/70	<u>186</u>, 201	207	9.0	153	7.7
	10/90	160, <u>187</u>, 201	200	6.1	147	3.0
	0/100	<u>185</u>	197	5.5	105, 140	6.4
PEE/PUC	90/10	154, <u>189</u>, 203	212	27.1	155	26.7
	70/30	105, 154, <u>188</u>, 202	209	25.2	121, 154	23.9
	50/50	114, <u>187</u>, 202	207	21.4	119, 152	20.9
	30/70	122, <u>178</u>, <u>188</u>,	207	15.9	101, 153	16.8
	10/90	124, <u>178</u>, <u>188</u>,	220	12.2	103, 153	14.2
	0/100	126, <u>175</u>, <u>189</u>,	220	8.9	101	9.3
PEE/PUE	90/10	148, <u>185</u>, 200	207	26.7	149	24.9
	70/30	132, <u>177</u>, 196	208	19.3	135	18.6
	50/50	<u>150</u>	178	8.8	98	11.3
	30/70	<u>162</u>	192	9.8	108	10.4
	10/90	<u>184</u>, <u>206</u>, <u>225</u>	230	8.6	147	9.1
	0/100	<u>188</u>, <u>209</u>, 226	231	8.2	114, 134	10.1
PEE1/PUA1	100/0	150, <u>186</u>, 200	208	26.9	150	27.0
	90/10	146, <u>183</u>, 198	206	25.1	144	25.8
	70/30	142, <u>180</u>, 196	203	18.0	141	18.0
	50/50	<u>168</u>	192	15.9	121	14.0
	30/70	<u>162</u>, 176, 187	210	15.1	140, 101	9.2
	10/90	<u>141</u>, 170, 180, 209	218	11.5	156, 96	5.6
	0/100	165, <u>186</u>, <u>200</u>, 218	224	9.8	88	11.4
PEE/PUA1	50/50	154, <u>188</u>	209	22.2	152, 77	21.0
PEE1/PUE1	50/50	<u>167</u>	182	14.3	117	13.2
PEE/PUE1	50/50	<u>179</u>, <u>197</u>, 211	190	18.3	139	14.1

^{a} Main endothermal effect underlined.
^{b} T_{mk} = temperature of end-melting range.

lapping T_8 is observed. For PUC, three maxima (T_{10}, T_{11}, T_{12}) are observed. They occur due to disordering of the pseudocrystalline structure of hard polyurethane and dispersion of the polymerization degree (7).

Thermograms of the blends in both cases result from overlapping of thermograms of the initial components. The temperature maxima of the main endotherm of PEE, PUA, and PUC (T_8, T_{11}, T_{12}), assumed as the melting point (T_m) of copolymers and blends, are similar. For PEE/PUA and PEE/PUA blends

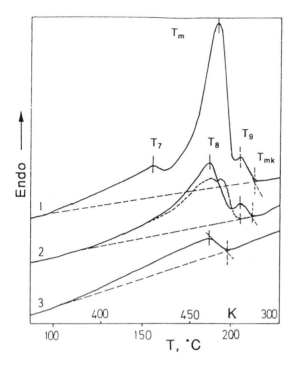

Fig. 12   DSC plots for PEE (1), PEE/PUA 50:50 blend (2), and PUA (3) in the range of melting; dashed line = first heating plots, solid line = second heating; 10°C/min.

(Fig. 14), an explicit melting point depression has been observed. It is not clear yet whether the lowering of temperature results from the interaction of the rigid or of the flexible segments.

Assuming that the enthalpy of melting of totally crystalline PBT is 145 J/g (45), it is possible to calculate the approximate degree of crystallinity for the PEE hard phase: $w_c = \Delta H_m / 145 \cdot w_H$). The approximation results from the assumption of equal fractions of hard phase and hard segments in copoly(ether-esters) ($w_H = w_{HP}$). Figure 15 shows the dependence of the degree of crystallinity of the PEE hard phase on the volume fraction of polyurethane segments, assuming a hypothetical mixture of hard segments from both copolymers.

For most copolymer blends and for PEE, the investigations that are the basis of this work, the ratio of crystallization temperature to melting point T_c/T_m is approximately 0.92; for PUC, PUE, and PUA1 it is close 0.82, i.e., the universal value given by Van Krevelen (35). According to Runt and Gallagher (73) the difference between the melting point and the crystallization temperature of the polymer is related directly to the nucleation conditions and the structure of the crys-

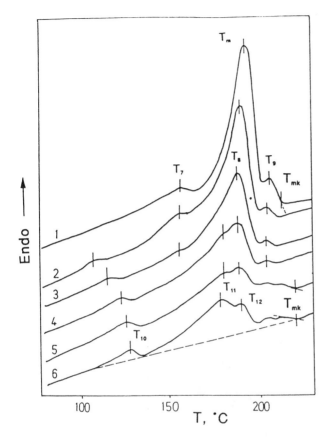

Fig. 13 DSC plots for PEE/PUC blends: 1 = PEE, 2 = PEE/PUC 70/30, 3 = 50/50, 4 = 30/70, 5 = 10/90, 6 = PUC. (From Ref. 32, with permission from Elsevier Science.)

tallites. The crystallization temperatures are also of great technological importance (see Sec. V).

E. Ordering in the Hard Phase

More information on the supermolecular structure of polymer systems, including
block copolymers, can be obtained from x-ray scattering studies (24,74–78) (see
also Chap. 7). Among others, this refers to the crystalline-phase fraction, the shape
and dimensions of domains, and the thickness of the interface.

Analysis of wide-angle x-ray (WAXS) diffraction patterns allows the separation of the crystalline and amorphous phases occurring in the samples investigated. In the method used by Hindeleh and Johnson (79–80), a corrected and nor-

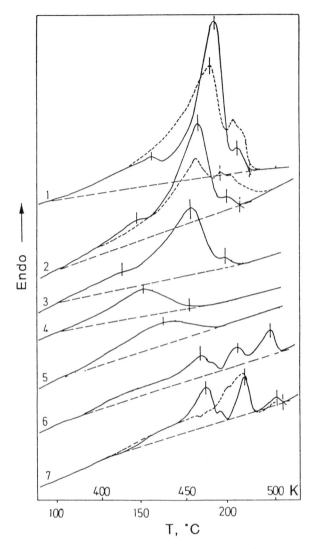

Fig. 14 DSC plots for PEE/PUE blends: 1 = PEE, 2 = PEE/PUE 90/10, 3 = 70/30, 4 = 50/50, 5 = 30/70, 6 = 10/90, 7 = PUE.

malized diffraction pattern is resolved into individual peaks and a polynomial background. An experimental x-ray diffraction pattern is approximated by a theoretical function Y_c of the form:

$$Y_c = \sum_{i=1}^{n} Q_i + B \tag{17}$$

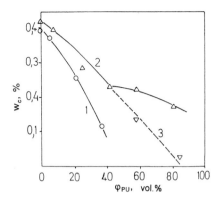

Fig. 15 Dependence of crystallization degree of PEE hard phase (w_c) on volume fraction of polyurethane segments (φ_{PU}) in the hypothetical mixture of hard segments for (1) PEE/PUA and PEE1/PUA1 (2) before and (3) after subtracting the fraction of endothermic effects from PU.

where n is the number of crystalline peaks. Each peak is represented by a "peak" function Q_i, which is a combination of Gaussian and Cauchy profiles:

$$Q_i = f_i A_i \exp\left\{-\ln 2 \left[2\,\frac{(x - P_i)}{w_i}\right]^2\right\} + \frac{\{A_i \cdot (1 - f_i)\}}{\{1 + [2\,(x - P_1)/w_i]^2\}} \qquad (18)$$

The peaks are defined by the following four parameters: the profile function parameter f_i, peak height A_i, peak width at half height w_i, and peak position P_i; x is the scattering angle 2Θ. An amorphous background B has a polynomial form. All the parameters of the "peak" functions Q_i and background function B are found by minimization of the sum of squares. The degree of crystallinity can be calculated as the ratio of the scattering area under the resolved crystalline peaks to the total area from the normalized x-ray scattering curve. Calculations are usually performed between two arbitrarily chosen scattering angles (81).

Small-angle x-ray scattering (SAXS) may originate from the diffraction of crystalline structures having long spacings (10–20 nm) or from the heterogeneity in the electron density related to the interfaces of two amorphous phases (domain structure), slits, micropores, and the like (82–84).

From the scattering intensity curve, an invariant quantity (invariant) Q can be derived as the integrated intensity of diffracted radiation, I, over the whole scattering angles Θ:

$$Q = \int_0^\infty h^2 I(h)dh = 2\Pi^2 I_T \langle \Delta\rho_e^2 \rangle \qquad (19)$$

where $h = 4\Pi/\lambda \sin \theta$, λ is the wavelength of x-ray radiation, I_T is Thomson's factor, and $\langle \Delta\rho_e^2 \rangle$ is the average square of electron density fluctuation.

Information as to the shape and spatial distribution of domains of volume fractions φ and $(1 - \varphi)$ and densities ρ_1 and ρ_2 are given by the correlation function. The correlation between the intensity of radiation scattered by the system (SAXS) and the correlation function is formed by direct and inverse Fourier transformation. Vonk and Kortleve have shown (85) that the one-dimensional correlation function for an isotropic, lamellar structure can be calculated from the SAXS intensity distribution using the equation:

$$\gamma_1(x) = \frac{\displaystyle\int_0^{\infty} h^2 I(h) \cos(h\ 2\pi hr)\ dh}{\displaystyle\int_0^{\infty} h^2 I(h) dh} \tag{20}$$

where $h = 4\pi \sin v/\lambda$ and $I(h)$ is the intensity scattered at small angles.

The x-ray scattering intensity curve for the aforementioned structural model (44,86) can be determined applying the general scattering function:

$$Y = I(h) \cdot h^2 \tag{21}$$

or by using the Vonk–Kortleve function (Eq. 20) (85–86).

The scattering curves of amorphous and/or semicrystalline polymer blends can be analyzed, and the contributions arising from each individual component can be separated (87–88). Let us comment on some results of the investigations of PEE/PUA and PEE1/PUA1 blends obtained by the preceding methods.

Figures 16a and b illustrate the WAXS diffractograms of PBT and PEE showing the appearance of two rather complex diffraction maxima over the angle range $2\Theta \sim 14^0 – 18^0$ and $2\Theta \sim 22^0 – 26^0$. They are related to the crystal β and α forms of PBT, respectively (45,72,89). The first maximum can be considered as a superposition of the (010) reflections for the β and α forms, respectively, the latter as a superposition of (100) reflections for both α and β phases. The broad maximum observed in the range $2\Theta \sim 26^0 – 33^0$ should be considered as the superposition of the (101) reflections of the phases β and α. The diffraction patterns of copoly(ether-urethanes) (Fig. 16) are characterized by a single broad maximum ($2\Theta \sim 12^0 – 28^0$), which can be assumed to be the effect of overlapping of the peak related to polyurethane amorphous (pseudocrystalline) hard phase (1) and an amorphous "halo," originating from the soft phase (2), similar to that of copoly(ester-urethanes) (Fig. 16d). Asymmetry of interferencial maximum for PUR is related to the hardly detectable ordering of the hard phase (faintly formed) pseudocrystalline structure of polyurethane domains.

Figure 17 illustrates the WAXS diffractograms of the PEE/PUE and PEE1/PUA1 blends. These results show that in the blends containing more than 70% PUA, no crystalline phase of PBT appears. The effect may be ascribed to an interpenetration of lamellar structures. Low amounts of PEE in the blends are also

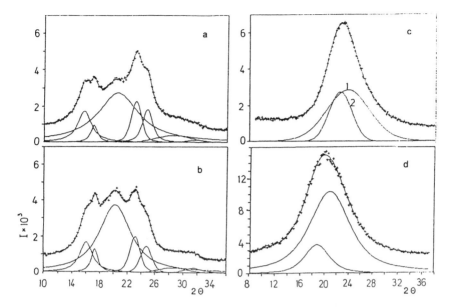

Fig. 16 WAXS diffractograms of (a) PBT homopolymer and block copolymers: (b) PEE, (c) PUE, and (d) PUA; $++++$ experimental data, — approximation according to Hindeleh and Johnson; diffractometer TUR M62, Cu Kα, $U = 30$kV, $I = 25$ mA, step 0.2, $t = 20$ s.

not detectable, but they influence the k_u value (see Fig. 18). The diffractograms of all PEE1/PUA1 blends show the peaks originating from PBT, as a result of a much greater dispersion. The fractions of high-ordered structures in the foregoing blends in relation to the contents of particular hard segments are listed in Table 8. The sum fraction of ordered structures k_s (or $k_\alpha + k_\beta$) is the equivalent of crystallinity degree w_c obtained from microcalorimetric examinations.

The diffraction maximum occurring at small scattering angles (Fig. 19) reveals the existence of a quasi-periodic lamellar structure. In PEE/PUA blends and copolyurethanes, the occurrence of such a structure has been confirmed after correction of the collimation error using Vonk's method (87,90) and after analysis of data employing a one-dimensional scattering function (Eqs. 20–21) (Fig. 20). All the investigated systems show a distinct maximum of the scattering function, the angular position of which is given in Table 9. Values of the long period L of each sample were determined from the position of the first maximum of the correlation function $\gamma(r)$ (85). The amorphous layer thickness l_a was determined by Ruland's method (91). Table 9 also shows the squared average of electron density fluctuation $\langle \Delta \rho_e^2 \rangle$ calculated from the following equation:

$$\langle \Delta \rho_e^2 \rangle_t = \varphi_c (1 - \varphi_c)(\rho_{ec} - \rho_{ea})^2 \tag{22}$$

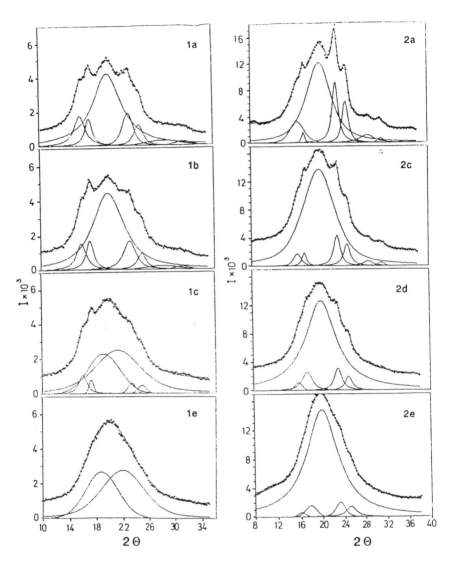

Fig. 17   WAXS diffractograms of (1) PEE/PUE and (2) PEE1/PUA1 blends. Copolymer weight fraction PEE/PUR: a = 90/10, b = 70/30, c = 50/50, d = 30/70, e = 10/90 wt %; ++++ experimental points, — approximation by correlation function.

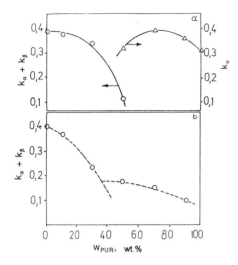

Fig. 18 Contents of crystalline structure of PBT ($k_\alpha + k_\beta$) and pseudocrystalline structure of PU (k_u) converted to content of hard segments versus urethane copolymer content (w_{PUR}) in (a) PEE/PUA and (b) PEE1/PUA1 blends.

Table 8 Results from WAXS Analysis for PPE/PUA and PEE1/PUA1 Blends

Polymer		k_s (%)	k_α (%)	k_β (%)	k_u (%)
PBT		44	16	28	—
Elitel 4420		42	16	26	—
Elitel 4440		38	17	21	—
PEE		39	17	22	—
PEE/PUE	90/10	38	18	20	—
	70/30	34	18	16	—
	50/50	44	4	7	33
	30/70	40	—	—	40
	10/90	36	—	—	36
PUE		33	—	—	33
PEE1		40	16	24	—
PEE1/PUA1	90/10	37	14.4	22.6	—
	70/30	23	9.2	13.8	—
	50/50	18	8.4	9.6	—
	30/70	15.6	8.8	9.8	—
	10/90	10	5.9	4.1	—
PUA1		18	—	—	18

k_s = general fraction of ordered structures; k_α = weight fraction of α PBT crystalline form; k_β = weight fraction of β PBT crystalline form; k_u = weight fraction of polyurethane pseudocrystalline structure; all values converted to hard-segment content.

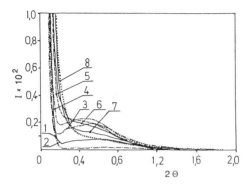

Fig. 19 SAXS diffractograms of PEE/PUA blends: 1 = PBT homopolymer, 2 = Elitel 4420, 3 = PEE/PUE 100/0, 4 = 90/10, 5 = 70/30, 6 = 50/50, 7 = 30/70, 8 = PUE 100 (%wt). Rigaku–Denki diffractometer, Cu K_α, $U = 32.5$; step 0.01–0.05; $t = 100$ s.

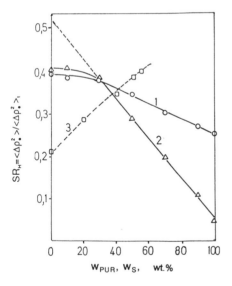

Fig. 20 Degree of hard-phase separation (SR_H, wt %) calculated from SAXS results vs. contents of PUR (w_{PUR}) in (1) PEE/PUE and (2) PEE1/PUA1 blends and (3) contents of flexible segments (w_s, wt %) in PEE.

Table 9 Results from SAXS Analysis for PEE/PUA and PEE1/PUA1 Blends

Polymer		$\langle \Delta\rho_e^2 \rangle \times 10^3$ (mol $\cdot$ el^2/cm^6)	$2\theta_{m1}$	$2\theta_{m2}$	L (nm)	l_a (nm)
Homopolymer PBT		0.745	0.68	0.90	9.8	2.0
Elitel 4420		2.08	0.67	0.86	10.2	1.9
Elitel 4440		3.30	0.59	0.85	11.4	2.1
PEE		3.56	0.54	0.75	12	2.1
PEE/PUE	90/10	3.51	0.46	0.71	11.8	2.5
	70/30	3.48	—	0.72	11.6	2.2
	50/50	3.09	—	0.71	11.8	2.1
	30/70	2.75	—	0.78	10.5	1.8
	10/90	2.42	—	0.77	10.3	1.5
PUE		2.31	—	0.83	10	1.9
PEE1		3.57	0.58	0.72	12	1.9
PEE1/PUA1	90/10	3.64	0.28	0.60	23.2	1.6
	70/30	3.34	0.30	0.64	22	1.5
	50/50	2.54	0.25	0.60	21	1
	30/70	1.75	0.35	0.50	21	0.9
	10/90	0.94	0.40	0.45	23.5	1
PUA1		0.40	—	—	—	—

$\langle \Delta\rho_e^2 \rangle$: average squared electron density fluctuation; $2\theta_{m1}$: maximum angular position on curve I (2θ); $2\theta_{m2}$: maximum angular position for one-dimensional scattering function $y = I(h) h^2$; L: long period; l_a: thickness of amorphous layer.

Generally, as the copolyurethane content in the blends increases, $\langle \Delta\rho_e^2 \rangle$ decreases. It is observed that an increase of PEE content in PEE/PUE blends does not involve essential changes of long period L. In the case of PEE1/PUA1 blends, the long period is twice as large as that for PEE1 and does not depend on blend composition. This may be a result of different phase-separation phenomena. Bandara et al. (44,92) relate the change in long period in copoly(ester-ethers) to the crystallization conditions, specifically to the $T_m - T_c$ difference. In blends such a correlation has not been observed. Yet, by analogy (44,93–94), a lamellar model for their structure can be assumed.

The average square of electron density fluctuation $\langle \Delta\rho_e^2 \rangle$ can be an approximate measure of the degree of separation of the hard (crystalline) phase (44,95). If $\langle \Delta\rho_e^2 \rangle$, experimentally determined for an actual sample, is related to the $\langle \Delta\rho_e^2 \rangle$ calculated value for the case when two phases—crystalline PBT phase ($w_{PBT} = w_{PBTc}$) and totally amorphous PO4 phase—exist, then the SR$_H$ = $\langle \Delta\rho_e^2 \rangle / \langle \Delta\rho_e^2 \rangle_t$ changes depend on the fraction of crystalline structure in the system. According to Eq. (22) and on the basis of the data presented in Sec. III.c, we can write:

$$\langle \Delta \rho_e^2 \rangle_t = \left(\frac{\langle \Delta \rho_e^2 \rangle_{PBT}}{\langle \Delta \rho_e^2 \rangle_{PO4}} \right) \cdot \varphi_{PBT} \, (1 - \varphi)_{PBT} \tag{23}$$

$$\varphi_{PBT} = w_H \cdot \frac{\rho_{PEE}}{\rho_{PBT}} \tag{24}$$

PBT_c and $PO4_\alpha$ electron densities, respectively, 0.736 mole el/cm³ and 0.543 mole el/cm³ and densities $\rho_{PBT_c} = 1.4$ g/cm³, and $\rho_{PO4} = 0.98$ g/cm³ were assumed (96–97) for the calculation. According to Bronschlege and Bonart (96), the total electron density of amorphous PBT has been assumed as 0.674 mole × el/cm³. Results are shown in Fig. 20.

Figure 20, plot 3, illustrates, for the copoly(ether-ester)s, the obtained correlation between the degree of hard-phase separation, SR_H, and flexible-segment content. These calculations are obviously simplified. In fact, the blends of two multiblock copolymers may be expected to be composed of two hard phases, each one containing (in relation to x-ray scattering conditions) a part of greater ordering density (semicrystalline or pseudocrystalline), one or two soft phases rich in flexible segments, and an interphase. In this case SR_H appears to reflect more the degree of formation of the ordered semicrystalline and pseudocrystalline phase, including the interphase. On the basis of plots 1 and 2 (Fig. 20), a model can be assumed in which flexible segments from copoly(ether-ester)s and the whole copolyurethane component (PUR) form a common matrix.

F. Mechanical Properties

The mechanical properties of polymer blends depend mainly on their composition and phase structure and on those components' interaction. The lack of miscibility causes heterogeneity of the blend and lowers mechanical properties such as tensile strength and strain of break. The additivity and synergism always mean the occurrence of compatibility (among others as a result of specific intermolecular interaction) (52,98–101).

Systems of incompatible polymers usually show a significant lowering of mechanical properties at tension. An earlier example is given by the blends of copoly(ether-ester) and polyethylene (102). In the case of the investigated two multiblock copolymer blends, the lowering of tensile strength has been observed for PEE/PUA and PEE1/PUA1 blends of 50–90% PUR content (Fig. 21, plots 1 and 4), while the decrease in elongation at tension for all blends is negligible (Fig. 22).

In summary, the investigated blends of two multiblock elastomers show mechanical properties specific for compatible systems and are thermoplastic elastomers.

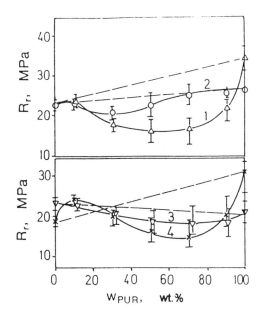

Fig. 21 Tensile strength (R_r) of (1) PEE/PUA, (2) PEE/PAC, (3) PEE/PUE, and (4) PEE1/PUA1 blends vs. PUR fraction (w_{PUR}).

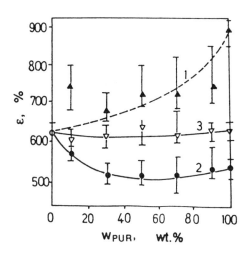

Fig. 22 Dependence of elongation at break (ε_r) on PUR contents (w_{PUR}) in (1) PEE/PUA, (2) PEE/PUC, and (3) PEE/PUE blends.

IV. INTERMOLECULAR INTERACTIONS

Several authors have pointed out that for the quantitative estimation of inte molecular interactions, a glass transition temperature may be applied (64,103–105). Kwei
(103) suggests the adoption of Eq. (25), where coefficients k and q reflect the values of deviations from the additive change of T_g as a function of blend composition:

$$T_g = \frac{w_1 T_{g1} + k w_2 T_{g2}}{w_1 + k w_2} + q w_1 w_2 \tag{25}$$

where T_{g1} and T_{g2} are the glass transition temperatures of components and w_1 and
w_2 the corresponding weight fractions of the components. The expression $q w_1 w_2$
has been interpreted as measure of hydrogen bond participation in intermolecular
interactions.

Schneider (15,104–105) worked out a versatile dependence:

$$\frac{T_g - T_{g1}}{T_{g2} - T_{g1}} = (1 + K_1) w_{2c} - (K_1 + K_2) w_{2c}^2 + K_2 w_{2c}^3 \tag{26}$$

where $w_{2c} = K w_2/(w_1 + K w_2)$; $w_1 + w_2 = 1$; $K = T_{g1}/T_{g2}$; and K_1 and K_2 are the
respective interaction coefficients taking into account the contact between hete-
rochains and the influence of the immediate surrounding on this interaction, respectively. For components having similar chemical structure, the coefficients K_1
and K_2 are close to zero. For the polymer blends with a specific intermolecular interaction (e.g., donor–acceptor), K_1 is larger than zero (15,106).

The preceding dependencies have been applied in this work for the estimation of interactions in PEE/PUC blends showing a single glass transition temperature in the low-temperature range (Sec. III.B). The results of the calculations carried out according to Schneider's equation are presented in Table 10 and Fig. 23.
According to assumptions presented previously, $K = \Delta C_{p2}/\Delta C_{p1}$. The calculated

Table 10   Calculation of Intermolecular Interaction Coefficients from Schneider's Equation for the PEE/PUC Blend

T_g (°C)	w_1	w_2	Y	w_{1c}	w_{2c}
−37	0.1	0.9	0.99	0.94	0.06
−44	0.3	0.7	0.885	0.81	0.19
−53	0.5	0.5	0.67	0.65	0.35
−62	0.7	0.3	0.35	0.44	0.56
−66	0.9	0.1	0.18	0.16	0.84

$Y = (T_g - T_{g1})/(T_{g2} - T_{g1}) \cdot 1/w_{2c}$; $w_{2c} = (K w_2/(w_1 + K w_2)$
$w_{1c} + w_{2c} = 1$; $K = \Delta C_{p2}/\Delta C_{p1}$; $\Delta C_{p1} = 0.262$ J/g· K; $\Delta C_{p2} = 0.484$ J/g· K; $T_{g1} = T_{gPEE} = -67°C$;
$T_{g2} = T_{gPUE} = -35°C$; $K_1 = -1.01$; $K_2 = 0.11$.

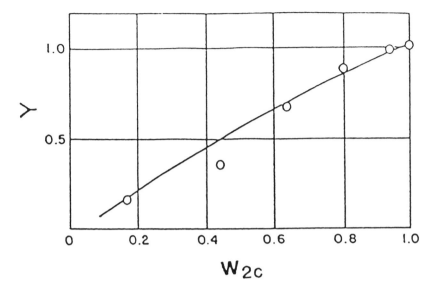

Fig. 23 Correlation between experimentally determined ($\bigcirc$) and calculated according to Eq. (22) (——) T_g values for PEE/PUC blends. (From Ref. 33, with permission from Elsevier Science.)

coefficient K_1 is smaller than zero, meaning that in the system investigated there are no specific interactions. The constant K_2 is small. These constants offer approximate information on the strength of intermolecular interactions but do not say much about their nature.

Equation (25) is equivalent to Couchman's (Eq. 10) and Gordon's and Taylor's (Eq. 9) equations, where $q = 0$ and coefficient K is also defined as a ratio of changes in the specific heat of the components at the glass transition temperature. Within such a context it appears interesting to write Eq. (25) in the form:

$$T_g = w_{1c}T_{g1} + w_{2c}T_{g2} + qw_{1c}w_{2c} \tag{27}$$

where w_{1c} and w_{2c} are the corrected contents of the components as in Eq. (26).

The differences between the experimental and calculated T_g values, using Eq. (27), (for $q = -25$) are not larger than 3°C (Fig. 24).

The line expressing the dependence of T_{gs} on the PEE/PUC blend composition (see also Fig. 7) allows one to conclude that with increasing PEE content, the soft phase of PUC grows richer in one more polyether flexible segment. When PEE is predominant, the opposite phenomenon is observed. The system stabilizing force may be the so-called physical networks; i.e., hard microphase domains between which the chains of flexible segments forming soft microphase are

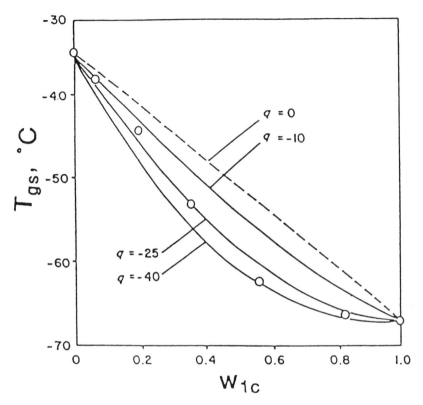

Fig. 24 Glass transition temperature of soft phase (T_{gs}) versus corrected PEE content (w_{1c}) in PEE/PUC blend: (○) experimental data; (—) calculation according to Eq. (27). (From Ref. 33, with permission from Elsevier Science.)

"fixed." Such systems may be defined as "interpenetrating polymer physical networks." The scheme of such a structure is shown in Fig. 25.

V. APPLICATIONS

Polymer systems, as a result of two-block copolymers mixing, have many specific physical properties, which depend on the properties of the individual components. Their unique characteristic makes them suitable for applications like:

    Hydraulic hose and fire hose
    Electrical-wire coating with heating resistance
    Damping elements used in wagons, ship construction, and vibro-insulation
    CVJ boots

During the extrusion and cooling of hoses and films (Fig. 26) made of copoly(ether-ester)s, local shrinkage can occur after crystallization. Similarly,

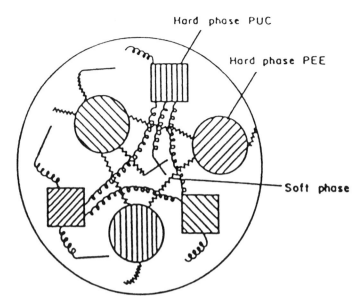

Fig. 25 Scheme showing the formation of interpenetrating polymer physical networks (IPPN) in the system of two multiblock copolymers. (From Ref. 33, with permission from Elsevier Science.)

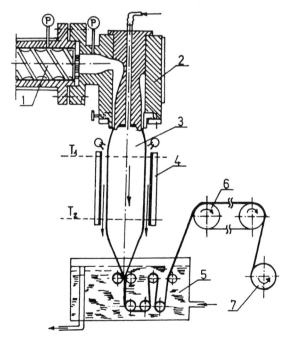

Fig. 26 Extrusion of elastomeric foils in order to receive fire hoses. 1: extruder; 2: die; 3: "bubble"; 4: air jet and water cooler; 5: bath with cooling water; 6: pull rolls and conveyor; 7: wind-up roll.

during the winding of polyurethane foils, the sticking of foil walls is possible. This is because of overcooling effects. If the features of semicrystalline polyester segments prevail in the system and, at the same time the crystallization temperature drops considerably, then the PEE/PUA extrusion of foils runs without disruption (109–110).

Aging and thermal degradation during copoly(ether-ester) and poly(ester-urethane) processing produce opposite symptoms, such as brittleness and cracking as well as softening and tackiness, respectively. These symptoms neutralize each other in the blend. Additionally, the PUA with phenol and amine antioxidants is used as a PEE costabilizer. Costabilizing effects can be used in the manufacturing of cables (111–113). In an efficient way the PEE/PUA blends are useful in polyurethane recycling. Investigations on energy absorption show that these polymer systems present a high damping factor as well as very good mechanical properties during high load compression (Fig. 27). Regarding mechanical properties of PEEs, absorption energy and damping factor increase with flexible-segment concentration increases. However, a PEE/PUA mixture shows the most optimal value of relative and permanent set (see Table 11) (114–115). The data

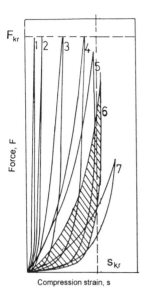

Fig. 27 Features of multiblock elastomers during the compression process. The wagon bumper model set. Dependence of the force on a particular jump of the set, which was measured using poly(ether-ester) elastomers with the following amounts of soft phase: 1—20%, 2—40%, 3—50%, 4—55%. The blends of PEE/PUA: 5—90/10, 6—70/30, 7—PUA. F_{kr} and s_{kr} approximate the values of the critical force and maximum jump as compared to the real conditions of bumper work.

Table 11 Comparison of Selected Results for Different Multiblock Elastomers Used to Produce Damping Elements

Static testing cycle	Parameters	Unit of measure	Elitel 4455	Ditel[a] 15/M-1	Elastollan 1180/m
First preload	Absorption Energy, W_e	kJ	16.6	13	5.2
	Damping, d	J/J	0.8	0.79	0.65
	Jump, s	mm	54[b]	45	54
	Final force, F_k	kN	730	835	230
Fifth working load with preliminary clip	Absorption Energy, W_e	kJ	6.9	8.2	3.5
	Damping, d	J/J	0.62	0.63	0.54
	Jump, s	mm	33[b]	36	38
	Final force, F_k	kN	645	700	210
	Compression set ε_s	%	2.5	1.7	1.6
Test in temperature −30°C	Absorption energy, W_e	kJ	9.1	11	8.8
	Damping, d	J/J	0.7	0.75	0.81
	Jump, s	mm	29	31	37
	Final force, F_k	kN	850	1035	630
Testing load after the series examinations made in low temperatures. Room temperature	Absorption Energy, W_e	kJ	8.8	11.6	3.5
	Damping, d	J/J	0.65	0.7	0.54
	Jump, s	mm	34	35	37
	Final force, F_k	kN	720	915	235

[a] PEE/PUE blend.
[b] Relative yield is 45% and 28%, respectively.
The model set consists of six rings at a real scale. One ring volume is about 375.5 cm^3.

shown in Table 11 refer to real conditions of bumper work. These polymer system properties make them suitable for vibro-insulation applications, especially of heavy and rotating machine foundations.

VI. CONCLUSIONS

Phase structure and physical and utility properties of poly(ether-ester) multiblock copolymer blends with poly(ester-urethane), poly(carbonate-urethane), and poly(ether-urethane) multiblock copolymers have been discussed. Two-multiblock copolymer systems develop a multiphase structure as a result of micro- and macrophase separation. The influence of the chemical structure and location of the segments within the chains on (a) the degree of phase separation, (b) the content of ordered areas, and (c) phase transition temperatures has been examined. The correlation between content and the degree of phase separation of flexible segments in multiblock copolymers has been discussed.

It is shown that the stabilization of the phase structure of two-multiblock copolymer blends is possible, owing to a double contribution of (a) intermolecular interactions between segments from distinct block copolymers and (b) microphase separation of multiblock components. On the basis of the Kwei and Schneider equations, the coefficients of intermolecular interactions in the blends have been estimated. In these systems the occurrence of interpenetrating physical networks (IPPNs) is also possible. These blends of two types of multiblock copolymers show elastomeric properties as well. These multiphase-structure systems and, especially, their stabilization make them suitable for good mechanical properties and many technological applications.

REFERENCES

1. Walsh DJ, Rostanmi S: Adv. Polym. Sci. 1985, 70:119–169.
2. Koning C, Van Duin M, Pagnoulle C, Jérôme R: Prog. Polym. Sci. 1998, 23:707–757.
3. Chun SB, Han CD: Macromolecules 1999, 33:4030–4042.
4. Löwenhaupt B, Steurer A, Hellman GP, Gallat Y: Macromolecules 1994, 27:908–916.
5. Folkes MJ: Processing, Structure and Properties of Block Copolymers. Elsevier Applied Science, London, 1985.
6. Stevenson JC, Cooper SL: Macromolecules 1988, 21, 1309; J. Polym. Sci., Polym. Phys. Edn 1988, 26:953.
7. Bonart R: Polymer 1979, 20:1389.
8. Alberola N: J. Appl. Polym. Sci. 1988, 36:797–804.
9. Faruque HS, Lacabanne C: Jap. J. Appl. Phys. 1986, 25:473.

10. Roslaniec Z, Ezquerra T, Balta Calleja FJ: Colloid Polym. Sci. 1995, 273:58–65.
11. Runt J, Lei Du, Martynowicz LM, Brezny DM, Mayo M: Macromolecules 1989, 22:3908.
12. Chung IZ, Kuo DL, Tsai HB: J. Polym. Sci. Polym. Phys. 1994, 32:1149–1156.
13. Utracki L: Polymer Alloys and Blends. Hanser, Munich, 1989.
14. Kammer HW, Inoue T, Ougizawa T: Polymer 1989, 30:888–892.
15. Schneider HA: Polymer 1989, 30:771–779.
16. Patterson DD, Robard A: Macromolecules 1987, 11:690–695.
17. Patterson DD: Polym. Eng. Sci. 1982, 22, No 2, 103–112.
18. Leary DF, Wiliams MC: J. Polym. Sci. Polym. Phys. 1973, 11:345–358.
19. Xie Y, Liu H: Makromol. Chem. Rapid. Com. 1989, 10:115–120.
20. Spontak RJ, Williams MC: Makromolecules 1988, 21:1377–1387.
21. Menges G: Makromol. Chem. Macromol. Symp. 1989, 23:13–35.
22. Lohmar J, Meyer K: Makromol Chem. Macromol. Symp. 1988, 16:161–173.
23. Meier DJ: NATO ASI Ser., Ser. E 1985, 89:173–194.
24. Kinning DJ, Thomas EL, Fetters LJ: J. Chem. Phys. 1989, 90:5806–5824.
25. He X., Herz J., Guenet JM: Macromolecules 1987, 20:2003–2009.
26. Paul DR, Barlow JW: J. Macromol. Sci. Rev. Macromol. Chem. 1980, C-18:109–168.
27. Izhizu K: Polym. Plast. Technol. Eng. 1989, 28:601–630.
28. Löwenhaupt B, Hellman GP: Polymer 1991, 32:1065–1076.
29. David DJ: Adv. Polym. Technol. 1996, 15:315–326.
30. Lipatov YS, Nesterov AE: Thermodynamics of Polymer Blends. Technomic, Lancaster, PA, 1997.
31. Roslaniec Z: Sbornik VSChT (Praha) 1989, S-19:213–223.
32. Roslaniec Z: Polymer 1993, 34:1249.
33. Roslaniec Z: Polymer 1993, 34:359.
34. Roslaniec Z: Miscibility and Melting in Polyetherester Block Copolymer-Polyurethane Elastomer Blends. Conference Internationale du Caoutchouc "IRT-90", Paris, 11–16 June 1990.
35. Van Krevelen DW, Hoftyzer PJ: Properties of Polymers, Their Numerical Estimation and Production Correlation with Chemical Structure. 3rd ed. Elsevier Science, Amsterdam, 1990.
36. Barton AFM: Handbook of Solubility Parameters and Other Cohesion Parameters. CRC Press, Boca Raton, FL, 1985.
37. Chapiro A, Lamothe M, Doan T: Eur. Polym. J. 1978, 14:647–680.
38. Piglowski J, Laskawski W: Angew. Makromol. Chem. 1979, 82:157–170.
39. Camberlin Y, Pascault JP: J. Polym. Sci. Polym. Phys. 1984, 22:1835–1844.
40. Camberlin Y, Pascault JP: J. Polym. Sci. Polym. Chem. 1983, 21:415–423.
41. Xie M, Camberlin Y: Makromol. Chem. 1986, 187, 383–400.
42. Brandrup J, Immergut EH: Polymer Handbook. Interscience, Wiley, New York, 1989.
43. Du Pont Glycols—Terathane Polyetherglycol: Techn. Inf., Du Pont Co. 1982.
44. Bandara U., Dröscher M: Colloid. Polym. Sci. 1983, 261:26–39.
45. Cheng SZD, Pan R, Wunderlich B: Macromol. Chem. 1989, 189:2443–2458.
46. Nichols ME, Robertson RE: J. Polym. Sci. Polym. Phys. 1992, 30:755–768.

47. Hamb MH, Gorman SB: Weathering Stability of Polycarbonate Urethanes Compared with Polyester and Polyurethane Elastomer Blends. Conference Internationale du Caoutchouc "IRT-90", Paris, 11–16 June 1990.
48. Ma EC: Rubb. World 1989, No. 3, 30–35.
49. Dobrowski SA, Goodman J, Johnson AF: Makromol. Chem. Rapid Commun. 1986, 7:273–279.
50. Eisenbach CD, Baumgartner M, Gunter C: In: Lal J., Mark JE: Advances in Elastomers and Rubber Elasticity. Plenum Press, New York, 1989, p. 51.
51. Phillips RA, Stevenson JC, Nagarajan MR, Cooper SL: J. Macromol. Sci. Phys. 1988, B-27:245–274.
52. Kalfoglou NK: J. Appl. Polym. Sci. 1981, 26:823–831.
53. Li C, Yu X, Speckhard TA, Cooper SL: J. Polym. Sci. Polym. Phys. 1988, 26:315–337.
54. Couchman PR: Macromolecules 1987, 20:1712–1717.
55. Gordon M, Taylor JS: J. Appl. Chem. 1953, 2:493–496.
56. Paik Sung CS, Hu CB, Wu CS: Macromolecules 1980, 13:111–116.
57. Slonecki J: Acta Polymerica 1991, 42:655–660.
58. Slonecki J: Polymer 1990, 31:1464–1466.
59. Degatre C, Camberlin Y, Pillot C, Pascault JP: Angew. Makromol. Chem. 1978, 72: 11–30.
60. Castles JL, Vallance MA, Mckenna JM, Cooper SL: J. Polym. Sci. Polym. Phys. 1985, 23:2119–2147.
61. Seymour RN, Cooper SL: Macromolecules 1973, 6:48–53.
62. MacKnight WJ, Karasz FE, Frud JR: In: Paul DR, Newman S., eds.: Polymer Blends. Vol. 1. Academic Press, New York, 1979.
63. Wagener KB, Matayabas JC: Polym. Prep. 1989, 30:243–244.
64. Chee KK: Polym. Eng. Sci. 1989, 29:609–613.
65. Nishi T, Wang TT: Macromolecules 1975, 8:909–915.
66. Bohdanecki M, Simek L, Petrik S: Polymer Com. 1990, 31:137–139.
67. Hässlin HW, Dröscher M, Wegner G: Makromol. Chem. 1980, 181:301–323.
68. Schmidt FG, Dröscher M: Makromol. Chem. 1983, 184:2669–2677.
69. Bill R, Dröscher M, Wegner G: Makromol. Chem. 1978, 179:2993–2996.
70. Yeh JT, Runt J: J. Polym. Sci. Polym. Phys. 1989, 27:1543–1550.
71. Stevenson JC, Cooper SL: J. Polym. Sci. Polym. Phys. 1988, 26:9953–9966.
72. Davidson IS, Manuel AJ, Ward IM: Polymer 1983, 24:30–36.
73. Runt J, Gallagher KP: Polymer Com. 1991, 32:180–182.
74. Striberg N, Fakirov S, Sapoundjieva D: Macromolecules 1999, 32:3368–3378.
75. Pilatti F, Toselli M, re A, Bottino FA, Pollicino A, Recca A: Macromolecules 1990, 23:348–350.
76. Yoon SCh, Ratner BD: Macromolecules 1988, 21:2392–2400.
77. Hedrick JL, Haidar B, Russell TP, Hofer DC: Macromolecules 1988, 21:1967–1977.
78. Owens JN, Gancarz IS, Koberstein JT, Russell TP: Macromolecules 1989, 22: 3388–3394.
79. Hindeleh AM, Johnson D: J. Phys. 1971, 4:259.
80. Hindeleh AM, Johnson DJ: Polymer 1978, 19:27–32.

81. Wlochowicz A. (Ed.): Sci. Biul. Lodz Technical University No. 16; TU Press, Bielsko-Biala, Poland, 1993.
82. Ruland W: Macromolecules 1987, 20:87–93.
83. Rabiej S: J. Appl. Polym. Sci. 1991, 27:947.
84. Rabiej S, Wlochowicz A: J. Appl. Polym. Sci. 1992, 46:1205.
85. Vonk CG, Kortleve G: Kolloid Z.Z. Polymer 1967, 220:19–24.
86. Kortleve G, Vonk CG: Kolloid Z.Z. Polymer 1968, 225:124–131.
87. Rabiej S: Europ. Polym. J. 1993, 29:625.
88. Rabiej S, Wlochowicz A: Angew. Macromol. Chem. 1990, 175:81.
89. Striberg N., Sapoundjieva D, Denchev Z, Apostolov AA, Zachmann HG, Stamm M, Fakirov S: Macromolecules 1997, 30:1329–1339.
90. Baltá Calleja FJ, Vonk CG: X-Ray Scattering of Synthetic Polymers. Elsevier, Amsterdam, pp. 241–304, 1989.
91. Santa Cruz C, Stribeck N, Zachmann HG, Baltá Calleja FJ: Macromolecules 1991, 24:5980–5990.
92. Zhu Li-Lan, Wegner G, Bandara U: Makromol. Chem. 1981, 182:3693–3651.
93. Bandara U, Dröscher M: Angew. Makromol. Chem. 1982, 107:1–23.
94. Fakirov S, Gogeva T: Makromol. Chem. 1990, 191:603–615.
95. Bonart R, Müller J: J. Macromol. Sci. Phys. 1974, B-10:177–189.
96. Bornschleggl E, Bonart R: Coll. Polym. Sci. 1980, 258:315–331.
97. Bonart R: Angew. Makromol. Chem. 1977, 58/59:259–297.
98. Kalfoglou NJ: J. Polym. Sci. Polym. Phys. 1982, 20:1259–1267.
99. Margaritis AG, Kallitsis JK, Kalfoglou NK: Polymer 1987, 28:2122–2129.
100. Kerner EH: Proc. Phys. Soc. 1956, 69-B:808:813.
101. Borggreve RJM, Gaymans RJ: Polymer 1988, 29:1441–1446.
102. Roslaniec Z, Wojcikiewicz H: Physical Properties of Polyetherester Copolymer—Polyethylene Blends. International Conf. "Plastko 1987," Gottwaldov, Czech Republic, 7–9 April 1987, 58–61.
103. Kwei TK: J. Polym. Sci., Polym. Lett. 1984, 22:307–313.
104. Schneider HA: Chemische Ind. 1989, 10:116–118.
105. Brekner MJ, Schneider HA, Cantow HJ: Makromol. Chem. 1988, 189:2085–2097.
106. Brekner MJ, Schneider HA, Cantow HJ: Polymer 1988, 29:78–85.
107. Kallitsis JK, Sotiropoulou DD, Kalfoglou NK: Polymer 1988, 29:1233–1239.
108. Pukanszky B, Tudos F: International Conf. Miscibility and Mechanical Properties of Polymer Blends. Prague 17–20 July 1989, SL9
109. PL Pat. 16 00 20 (1988, 1993).
110. PL Pat. 16 47 64 (1990, 1994).
111. PL Pat. 142 846 (1985, 1988).
112. PL Pat. 15 78 84 (1987, 1992).
113. Roslaniec Z., Wojcikiewicz H: Polimery (Warsaw) 1986, 31:204–206.
114. PL Pat. 16 10 24 (1989, 1993).
115. PL Pat. 16 41 53 (1988, 1994).

13
Multiblock Copolymer/Homopolymer Blends

Nikos K. Kalfoglou, Ioannis K. Kallitsis, and Efstathia G. Koulouri
University of Patra, Patra, Greece

I. INTRODUCTION

Blends of multiblock copolymers with different homopolymers are reviewed in this chapter. More specifically, poly(ether ester) (PEE) of different compositions (soft versus hard segments) and thermoplastic polyurethanes (TPUs) (polyether urethanes, PU-eth, and polyester urethanes, PU-est) are the two multiblock copolymers studied the most. Limited work was carried out with other copolymers, such as polyester amides and polyether amides; this is summarized and presented in Sec. III.c.

Since PEE and TPU contain ester or amide bonds in addition to intermolecular interactions, interchange reactions may take place during blending with polyester or polyamide homopolymers. The origin of physical or reactive compatibilization between blend components has been identified for each case and relevant characterization techniques noted.

This review does not cover special classes of blends, e.g., interpenetrating networks, blends obtained via crosslinking reactions, and filled systems. Prior to the references listing, some general considerations related to structure-compatibility principles for these copolymer blends are summarized.

II. STRUCTURE–COMPATIBILITY CONSIDERATIONS

The composite chemical structure of the multiblock copolymers considered in this chapter gives rise to several factors that may determine compatibility. This term is used in a general context to include thermodynamic miscibility (1) and "mechanical compatibility" as shown, e.g., by a polymer alloy (2,3). These determining factors may be characterized as physicochemical and purely chemical. Both factors are affected by processing parameters and blending method—melt-blending vs. dissolving in a common solvent.

These copolymers usually comprise hard, crystallizable blocks interspersed within an amorphous phase of similar and/or different chemical structure. The hard/soft-phase ratio depends not only on chemical composition as determined by the copolymer synthesis, but also on thermal history during blend preparation and subsequent annealing at ambient or elevated temperature.

In addition to the intermolecular interactions considered later, solid-state phase composition of the copolymers influences compatibility, since the crystalline phase during solidification may reject the "foreign" polymer because of crystallographic requirements. This is not the case for the amorphous-phase component, which retains its liquidlike disorder. Thus in many semicrystalline polymer blends, only the amorphous phases are involved in the compatibilized blend. However, in some exceptional cases compatibilization is enhanced by the presence of a crystalline block in the copolymer that is structurally related to the crystalline homopolymer with which it is mixed. This leads to isomorphism; i.e., both crystallographically similar components cocrystallize (4). The blending mode is also important, since not only does melt-blending lead to phase homogenization, but, due to high temperatures, interchange reactions may take place. In the case of solution blending, depending on the kind and degree of solvent polarity, selective solvation ($\Delta\chi$ effect) may take place (5). Thus solvent evaporation could lead to a phase-separated blend. This complication and practical processing considerations discouraged solution blending in the majority of blends studied in the past.

The chemical features are all important in determining compatibility. These give rise to intermolecular interactions and/or chemical reactions between blend partners. Intermolecular interactions are of various types, depending on the specific groups involved (6), e.g., electron donor–acceptor, dipole–dipole, π-electron interactions. A practical generalization that may provide guidance for compatibility prediction is the principle of *complementary dissimilarity* (7). Though this includes various types of molecular interactions, as will be described later, by definition it does not include the athermal thermodynamic mixing between chemically similar segments present in both blend partners.

On purely chemical reactivity considerations and depending on mixing conditions, specific groups in the polymeric chain of the blend components may come into play, leading to reactive compatibilization (8,9), e.g., ester, or ester–amide interchange reactions. These are thermally activated organic reactions, and their extent depends, in addition to concentration, group reactivity, etc., on temperature and duration of mixing, which is usually carried out in the melt.

Reactive compatibilization with or without an added third component (compatibilizer and/or catalyst) significantly enlarged the number of compatible polymer blends and stimulated the involvement of the organic polymer chemist in the field of polyblends.

In the following paragraphs the contributing role of the foregoing factors to the compatibilization of the specific multiblock copolymers with various homopolymers will be pointed out.

A. Poly(ether-ester) (PEE) Blends

A typical PEE may be considered as being formed by randomly joining soft poly(tetramethylene) ether glycol terephthalate (PTMEG-T)$_m$ segments:

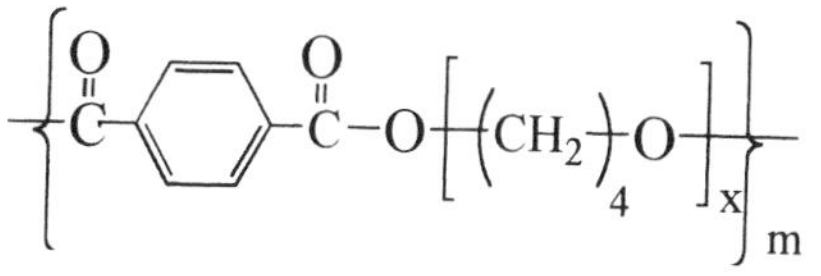

Structure 1

and hard poly(tetramethylene glycol terephthalate) (TMGT)$_n$ segments:

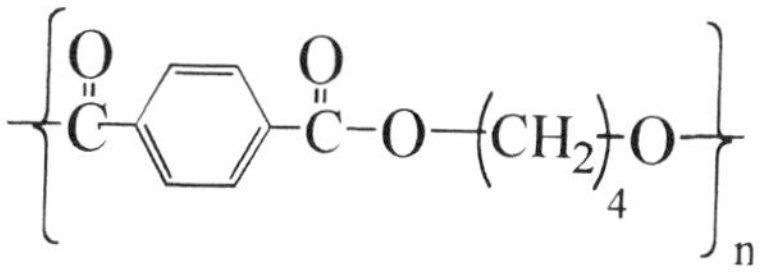

Structure 2

One important structural difference compared to PU is the absence of strong $C{=}O\cdots H$ intramolecular interactions characterizing the latter. Thus a less ordered crystalline domain structure results; therefore, a higher participation to mixing is anticipated for the hard (TMGT)$_n$ segments compared to the PU hard blocks.

Compatibility may arise from the electron donor capability of the carbonyl and ether groups, leading to donor–acceptor intermolecular interactions with other "complementary dissimilar" polymers. On this principle, poly(vinylchroride) (PVC) and chlorinated polymers (CPE) having an acidic hydrogen are expected to give compatible blends (10–19). These studies showed that the amorphous components are involved mainly in the compatibilization process (15,19). The crystalline phase may also contribute to compatibilization, since depending on the content of hard (TMGT)$_n$ segments in PEE, partial or complete cocrystallization was reported in PEE/poly(butylene terephthalate) (PBT) blends (20–22).

Though transesterification reactions during melt-mixing of PEE/Polyester blends are possible, related studies on PEE/PBT (23) and PEE/polycarbonate (PC) (24) do not support such interchange reactions. Partial miscibility in PEE/poly(ethylene terephthalate) (PET) blends (25) was attributed to π-electron interactions between the aromatic rings of (TMGT)$_n$ and PET, similar to PET/PBT blends (26). Such an interaction could also account for the miscibility origin of PEE/Polyarylate blends (27).

Interchange ester-amide reactions of melt-mixed PEE/polyamide 6 (PA 6) (28) and PEE with various aliphatic PA blends were also reported (29). These reactions required high temperatures and adequate mixing time. Thus no such reactions were reported in solution-prepared PEE/PA 6 blends (30).

B. Polyurethane (PU) Blends

Polyurethane consists of a hard block, e.g.,

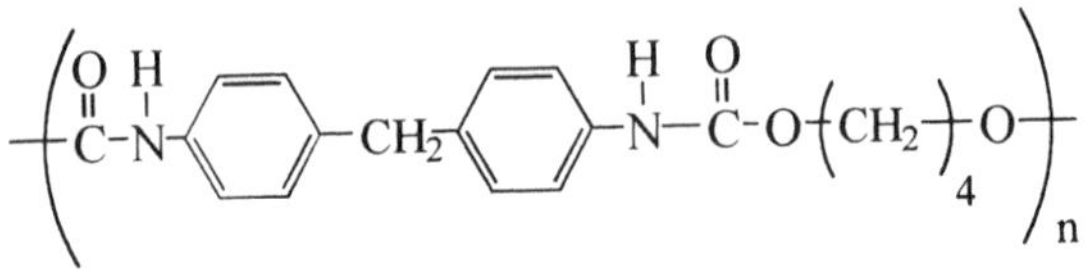

Structure 3

and a soft block of a polyether, e.g., polytetramethylenoxide (PTMO):

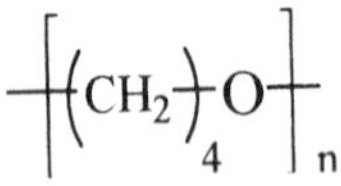

Structure 4

or a polyester, e.g., poly(ε-caprolactone) (PCL):

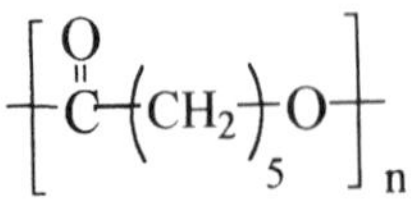

Structure 5

Based on the electron-donating capability of the carbonyl and ether groups of the soft segments (1,2), these may interact with proton-donating polymers via donor–acceptor interaction. Thus, compatible blends were reported for the pair PU/PVC (31,32), the latter providing the acidic hydrogen at the α-position. Unexpectedly, the higher acidic strength of the OH of poly(vinylphenol) (PVPh) did not yield a miscible PU/PVPh blend (33). However, TPU/phenoxy miscibility was reported (34) originating from the carbonyl group of the PCL segment with the acidic hydroxyl of phenoxy. Dipole–dipole interactions are not sufficiently strong to yield compatible blends as was the case for chlorinated PVC (CPVC) and poly(vinylidenechloride) (PVDC) with polyesters (35). Thus, blends of PU/CPVC were reported (36) to be partially miscible and PU/poly(vinylidene fluoride) (PVDF) incompatible at all compositions (37). It is also reasonable to assert that the C=O and —NH— groups of the hard blocks are less available for interactions, since after solidification their proximity leads to a tight hydrogen-bonded crystalline phase. This could lead to partial miscibility if blend composition is not corrected for the immiscible crystalline phase. Consider, e.g., partial miscibility of TPU/PVC depending on PU structure (31,38–41). In such cases miscibility prediction afforded by practical criteria, e.g., T_{gb}-composition, will be improved if only the amorphous components are considered (42). This approach is not always successful, since amorphous/crystalline phase separation in multiblock copolymers is incomplete. In addition, annealing may lead to phase separation because of the reduction of intermolecular interaction and/or increase of the crystalline/amorphous-phase ratio. Hydrogen bonding was invoked to explain miscibility of a PU-eth/PA 6 blend (43,44). Stronger interactions are involved in the blend of sulfonated PS and a PU containing a tertiary amine group (45). Partial miscibility due to ester interchange reactions was also reported for TPU/PC blends (46). In certain studies the effect of solvent on PU blend miscibility was also reported (31).

Structural features of other classes of multiblock copolymers, e.g., polyester amides, polyether amides, combine the basic chemical building blocks of PEE and PU. Therefore, their compatibility behavior may be explained along similar lines.

III. SURVEY OF BLENDS

A. Blends of Poly(ether-ester) (PEE)

1. PEE/PVC Blends

Among PEE blends, those with PVC have been most extensively studied, using a variety of techniques. Nishi et al. in an early study have reported (10) on the compatibility of PEE/PVC using thermal and dynamic mechanical analysis, thermal expansivity (TMA) and NMR. Blends obtained from solution casting or quenched from the melt have been shown to be compatible, while annealing resulted in phase separation.

The effect of heat treatment in the preceding system has also been studied by Nishi (11). He reported that the impact strength at room temperature may be improved using the appropriate heat treatment. The mechanism of energy dissipation appeared to consist of both crazing and shear flow.

Hourston and Hughes, in two publications, have reported on the miscibility behavior of PEE/PVC (12,13). In the first of these, blends have been examined using DSC, DMA, and sonic velocity measurements. Dynamic mechanical analysis has shown that blends containing 25–50% PEE are completely miscible, using as a criterion the single glass transition observed (12). Annealing performed on the same blends was also reported using the same techniques (13). It was concluded that annealing increases the heterogeneity of blends with 45–75% PEE. In a study related with the dynamic viscoelestic behavior of various PVC blends (14), the PEE/PVC blend at an intermediate composition was shown to be compatible, using dynamic shear oscillation measurements.

Kwak and Nakajima (15) examined the morphology of PEE/PVC blends using SEM. In addition, application of high-resolution pulsed-FT ^{13}C-NMR spectroscopy capable of magic-angle spinning and cross-polarization techniques permitted the analysis of existing domains in the solid state, in the range of a few to a few hundreds of angstroms. They concluded that after mixing, microcrystalline domains of PEE and PVC in the size range of 14–24 Å existed in a molecularly mixed amorphous-phase matrix.

In a closely related work (16), Kwak and coworkers further quantified previous findings (15) and supplemented PEE/PVC blend characterization using DMA in the frequency-sweep mode. This interpretation of micro domain structure, has been criticized by Guo because of the complexity of the morphology of semicrystalline polymer blend components involved (17).

Ellis and Barry have reported (18) on the synergistic effects on the melt index and Izod impact properties of PEE/PVC blends using melt index, capillary rheometry, and DMA. Immiscible blends have been identified; based on the DSC results, it was suggested that phase separation was the cause of the synergism in the flow and impact properties.

Related to these studies, PEE/CPE blends have also been examined. Kallitsis et al. (19) reported on the compatibility of CPE containing 48 and 25 wt % Cl (CPE48 and CPE25) with PEE, in the complete composition range. To examine the effect of PEE structure on compatibility, a 50/50 PBT/CPE48 blend was also studied. The techniques used were DMA, DSC, and tensile testing. The DMA results indicated miscibility of PEE/CPE48 at all compositions and allowed the determination of the LCST (low critical solution temperature). CPE25 was partially miscible, while the crystalline component of PBT was immiscible with CPE48. Miscibility was attributed to polyether and polyester groups in the amorphous phase interacting with proton-donating CPE. Miscibility was rationalized using the copolymer–copolymer miscibility scheme. Data on fire retardancy were also reported.

2. PEE/Polyester Blends

Since the hard segments of PEE consist of PBT blocks, it is reasonable that blends of PEE with the PBT homopolymer were studied. Thus, Gallagher and coworkers examined (20) the crystallization and phase behavior of PBT/PEE blends using DSC, SAXS, and dielectric techniques on solution-cast films melt-pressed and quenched to 25°C. It was reported that compatibility was dependent on the hard-segment content of PEE and based on the β-relaxation shift of the crystalline phase, it was concluded that at high contents of hard block $(TMGT)_n$ units, ≥ 75 wt %, miscibility sets in. Evidence from DSC and SAXS supports the proposition that miscibility is due to cocrystallization. At other PEE compositions, the amorphous phases were reported to be partially miscible for 58 wt % $\leq (TMGT)_n \leq 75$ wt %, or immiscible for 40 wt % $\leq (TMGT)_n \leq 51$ wt %. Results on miscible blends were successfully interpreted in terms of the Flory–Huggins mean field approach and solubility parameter theory to correctly estimate the $(\chi_{12})_{cr}$ of the blend. T_{mb} depression was also determined and found in agreement with available theory.

The cocrystallization behavior of PEE/PBT blends was also examined on extruded samples drawn and annealed as well as undrawn samples, all prepared in vacuum (21). The absence of complete cocrystallization of homo-PBT and the PBT block from PEE in the blend was observed. This lack of complete cocrystallization and miscibility was attributed to the insufficient length of PBT hard segments in PEE required for the formation of a lamellar thickness, typical for the crystallites of homo-PBT.

Apostolov et al. reported (22) on melt-blended PBT and PEE that were zone-drawn and zone-annealed at various stresses (10–50 MPa), at temperatures of 160 and 190°C. Their structure and mechanical properties were compared to those of the same blend but cold-drawn and isothermally annealed. The samples were characterized by DSC, WAXS, SAXS, and tensile mechanical properties. The studies showed that the structural features created in the zone-drawn and annealed materials resulted in higher values of Young's modulus and tensile strength compared to those receiving the simple isothermal treatment.

Sauer et al. (23) applied thermally stimulated currents (TSCS) and ac dielectric spectroscopy to study PBT and its blends with PEE. The high selectivity of TSCS and ac dielectric techniques as compared to DSC was useful for the study of these blends where the elastomer phase was present at only 20 wt %. The combined techniques covered a frequency range of $\approx 5 \times 10^{-3}$–10^5 Hz, allowing detailed characterization of the amorphous relaxations in these phase-separated blends. Two separate glass transitions for the PBT/PEE blends were observed, unchanged from the T_g's of the pure components, indicating that they are immiscible and that ester exchange during processing was minimal.

The compatibility of melt-mixed blends of PET with two types of PEE was studied over the complete composition range by Skafidas and Kalfoglou (25). The PEE used had a low (PEE-s) or high (PEE-h) content of hard butylene glycol terephthalate (4GT) sequences. The techniques applied were DMA, DSC, tensile testing, and optical microscopy. The DMA results indicated partial miscibility, higher in PEE-h/PET than in PEE-s/PET blends. This was attributed to weak intermolecular π-electron interactions, leading to the miscibility of the segregated $(4GT)_n$ domains in PEE with PET, analogous to the PET/PBT blend (26). Partial miscibility led to good tensile properties for both blend series, and for the phase-separated PEE/PET blend the Kerner mechanics model was successfully applied to obtain information on phase morphology. Gaztelumendi and Nazabal studied (24) the compatibility behavior of melt-mixed blends of PEE/PC over the complete composition range. Experimental evidence excluded the possibility of transesterification reactions during blend preparation. The DSC and DMA results on the T_{gb}-composition variation indicated miscibility in the melt state but phase separation when blends were not quenched. The latter, as well as the asymmetry of the tan δ spectra, is attributed to the reorganization of the $(GT)_n$ hard-block component, leading to crystallinity increase and reduction of the amorphous-phase content of PEE. Though both UCST (upper critical solution temperature) and LCST were reported, no experimental evidence for the latter was presented.

Blends of PEE/polyarylate have also been examined (27), using DSC and DMA. The DSC results revealed T_m depression, while DMA showed a single T_g in all compositions examined, implying complete miscibility over the entire composition range.

3. PEE/Polyamide Blends

Kulak et al. reported (28) on melt-mixed blends of polyamide 6 with PEE. Using DMA, DSC, and tensile testing it was concluded that interchange reactions occurring during the melt-mixing procedure resulted in blends with very good mechanical properties.

In a closely related work, Koulouri et al. (29) investigated the mechanical properties of binary melt-mixed blends of various polyamides (PA 11; 6,6; 6,10; 6,12) with PEE. All blends prepared had very good mechanical properties and in some cases even better than those of the respective pure PA. Property improvement was attributed to interchange reactions between the constituents. Two of these systems (PA 11/PEE and PA 6,12/PEE) were selected for further study in terms of morphology, viscoelastic, and thermal properties. These blends showed two distinct but mutually converging T_g's and a low extracted amount of PEE during extraction experiments, supporting the view that a very good interfacial grafting had been achieved in these two-phase systems. This proposition was further supported by results on cryofractured and selectively etched blends (Fig. 1), where the existence of grafted PEE was established by FTIR (Fig. 2) as well as by micro-Raman spectroscopy (Fig. 3).

Blends of PA 6 and PEE were also investigated earlier by Gattiglia et al. (30), but there were no indications of interchange reactions during the mixing procedure used. These blends prepared by dissolving in a common solvent were examined using thermal analysis, the conclusion drawn was that the components were not miscible.

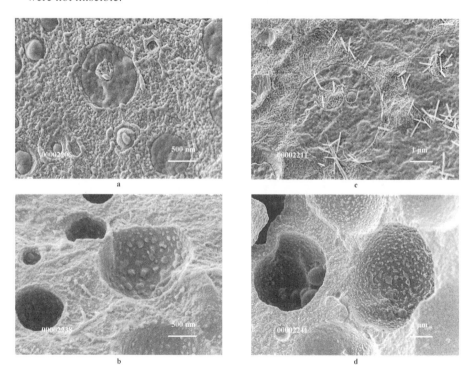

Fig. 1 Scanning electron micrographs of the cryofractured nylon 11/PEE blend: 75/25 (a) before and (b) after; 50/50 (c) before and (d) after the extraction of PEE with chloroform at room temperature. (Reprinted from EG Koulouri, EC Scourlis, JK Kallitsis. Characterization of melt-mixed blends of poly(ether-ester) with various polyamides. Polymers 40:4887–4896, 1999, with permission from Elsevier Science.)

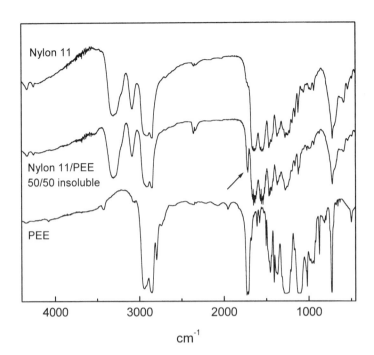

Nylon 11

Nylon 11/PEE
50/50 insoluble

PEE

4000   3000   2000   1000

cm⁻¹

Fig. 2 FTIR spectrum of the nylon 11/PEE 50/50 blend, after the extraction of PEE with chloroform at room temperature, in comparison with the pure polymers. (Reprinted from EG Koulouri, EC Scourlis, JK Kallitsis. Characterization of melt-mixed blends of poly(ether-ester) with various polyamides. Polymers 40:4887–4896, 1999, with permission from Elsevier Science.)

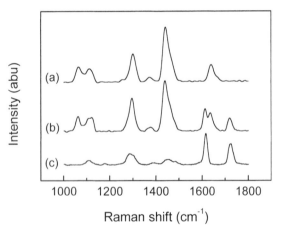

Intensity (abu)

(a)

(b)

(c)

1000   1200   1400   1600   1800

Raman shift (cm⁻¹)

Fig. 3 Micro-Raman spectrum of the nylon 11/PEE 50/50 cryofractured and selectively etched surface (b); in comparison with the spectra of the pure components; nylon 11 (a) and PEE (c). (Reprinted from EG Koulouri, EC Scourlis, JK Kallitsis. Characterization of melt-mixed blends of poly(ether-ester) with various polyamides. Polymers 40:4887–4896, 1999, with permission from Elsevier Science.)

B. Blends of Thermoplastic Polyurethanes (TPUs)

1. TPU/PVC Blends

Wang and Cooper studied the morphology and the properties of PVC-PU blends (31). Blends of a segmented PU-eth with PVC were examined utilizing DSC, DMA, tensile testing, IR, and infrared dichroism experiments. This thermodynamically incompatible system was made kinetically compatible by precipitation from tetrahydrofuran (THF) solutions. THF–dioxane solution casting and melt processing produced an incompatible system. The compatible PU/PVC system contained a well-mixed PVC/polyester matrix phase as evidenced by T_g shifts, orientation characteristics, and infrared peak position changes. The aromatic urethane segments, which exhibited microphase separation in the pure PU, were not solubilized by blending with PVC by any of sample preparation methods used in this study.

Kalfoglou reported (32) on the composition dependence of small- and large-deformation behavior of elastomeric PU-eth/PVC blends. Ultimate tensile properties were found to vary regularly with composition (see Fig. 4), and the signifi-

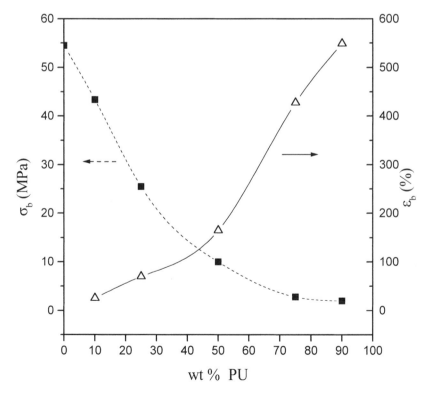

Fig. 4 Composition dependence of ultimate properties of PET/PU blends: (■) ultimate strength, σ_b, and (△) ultimate elongation ϵ_b.

cant shift toward each other of the main component relaxations support the view of a partly miscible blend. This is also corroborated by phase-contrast microscopy. Partial miscibility may be traced to the carbonyl/acidic α-hydrogen interaction of the PU/PVC amorphous phases, respectively.

The miscibility of TPUs with PVC was studied by Kim and coworkers (38). When TPUs were prepared with hydroxy-terminated poly(butylene adipate) (PBA) as the soft segment, they were miscible with PVC, judging from the single T_g that was observed. This result was independent of the degree of neutralization of the carboxylic acid groups of the dimethylolpropionic acid (DMPA) unit in TPU by triethylamine (TEA). But after neutralization, increased cohesion of hard segments, due to the strong Coulombic forces of the ionic sites in the TPU ionomer, restricted intimate segmental mixing, leading to incompatibility. When hydroxy-terminated poly(propylene glycol) was used as the soft segment in TPU, its blends with PVC showed two separate T_g's, of TPU and PVC, irrespective of neutralization.

The miscibility of linear segmented PU, poly(hexyleneadipateurethane), poly(butyleneadipateurethane), poly(ethyleneadipateurethane) (PHAU, PBAU, PEAU) based on three polyesters (PHA, PBA, PEA), MDI, and butanediol with PVC was also the subject of two studies by Zhu and his coworkers (39,40). Using the techniques of wide-angle x-ray diffraction (WAXD), DSC, DMA, solubility tests, and phase-contrast microscopy, they concluded that the blends of PHAU/PVC and PBAU/PVC were compatible while the blends of PEAU/PVC were not. The adhesive strengths of these blends were quite different from one another, in the order of PBAU ≥ PHAU ≫ PEAU. The influence of crystallinity and compatibility on adhesion was discussed, and an interpretation of adhesion for the PU/PVC system based on diffusion theory was proposed.

The miscibility behavior of TPU elastomers with chlorine-containing polymers were studied by Ahn et al. (41). Four kinds of TPUs, containing 20, 40, 60, and 80 wt % of PCL as a soft segment, were used for blending. In this study PVC and CPE with a chlorine content of 40% were used. The results of optical clarity and thermal characterization suggest that PVC was miscible with TPUs containing 40, 60, 80 wt % PCL segment, whereas CPE was immiscible with all the TPUs studied. Tensile measurements showed that all blends with TPU containing 60 wt % PCL segment were mechanically compatible.

Garcia studied (36) PU/CPVC blends over the complete composition range using DSC, DMA, and electron microscopy. On the basis of a single though broad composition-dependent T_g, it was concluded that blends were partially miscible.

Baranov and coworkers reported (47) on the dynamic viscoelastic characteristics of TPU/PVC blends at various compositions using the DMA technique. A mutual shift of T_g and shear storage modulus G' in blends indicated partial miscibility.

The kinetics of thermal degradation of PVC blends with TPU in the form of powders and films was investigated by Kolesov et al. (48b). Kinetic equations were proposed for dehydrochlorination of stabilized and nonstabilized blends. Effective rate constants of PVC dehydrochlorination catalyzed by urethane groups and their complexes with hydrogen chloride, and the rate constants of decomposition and formation of these complexes, were calculated.

More recently, Samios et al. reported (48a) on the compatibility behavior of melt-mixed blends of PET/TPU blends over the complete composition range. The techniques applied were tensile testing, DMA, optical microscopy, SEM, and high-resolution ^{1}H-NMR. Mechanical properties indicated blend compatibility typical of a polymeric alloy. Morphology examination revealed good component dispersion and strong interfacial adhesion. ^{1}H-NMR of blends showed the formation during melt-mixing of a PET-PU copolymer by ester-amide interchange reactions whose extent depended on the PET/PU ratio; see scheme 1. Reaction path a is favored at low and path b at high PET/PU ratios. Figures 5(a), (b), and (c) give

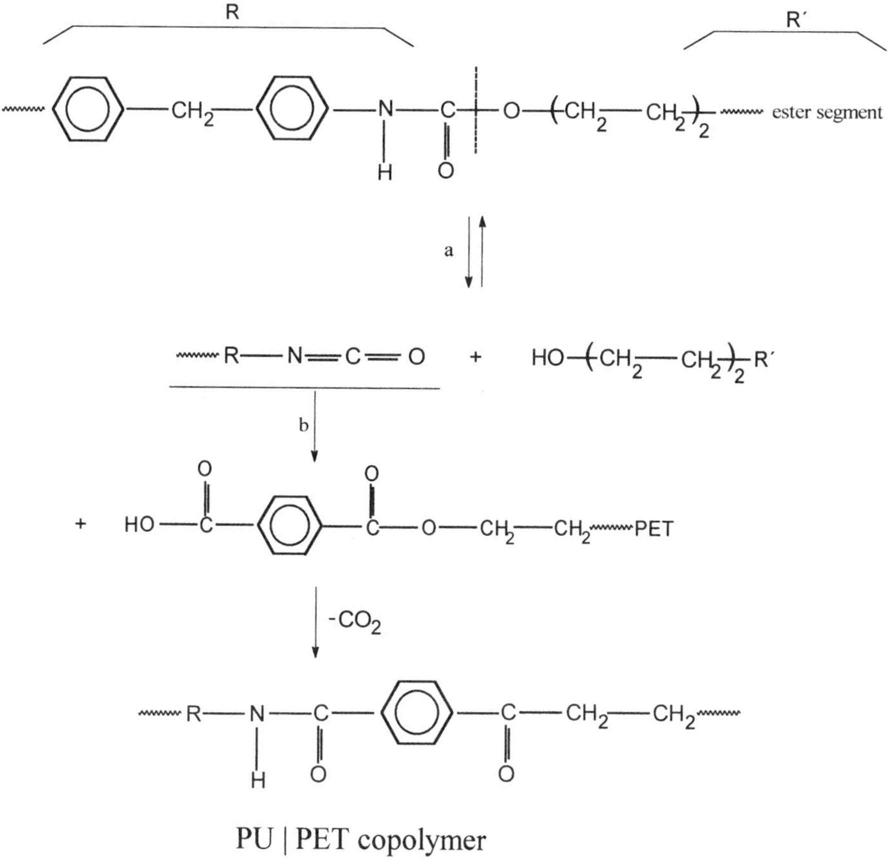

PU | PET copolymer

Scheme 1   (From Ref 48a.)

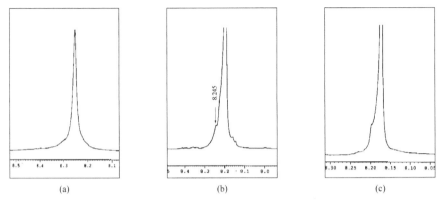

Fig. 5 ¹H-NMR spectrum of the aromatic H of PET: (a) pure PET; of PET/PU blends: (b) 2/1; (c) 1/2. (From Ref 48a.)

the ¹H-NMR spectra of the aromatic H of pure PET and of the same in blends at the 2/1, 1/2 PET/PU ratios, respectively. The small peak appearing at 8.245 ppm in Figs. 5(b) and (c) is caused by the change of the aromatic H environment of PET brought about by the in situ formation of the PET/PU copolymer.

2. TPU/Polyester Blends

Ahn et al. (46) reported on the miscibility of TPU/PC blends. TPUs contained varying amounts of the 1,4-butanediol hard segment covering a 24–100 wt % range, at five compositions. Blends at the 50/50 composition prepared by solvent casting were annealed between 230 and 270°C for 10 min. Convergence of the T_g's determined with DSC and supplementary data obtained with TGA support the view that interchange reactions with PC are responsible for the partial miscibility observed.

In another study PU elastomers based on 4,4′-diphenylmethane diisocyanate and aliphatic polycarbonate diols and/or polytetramethylene glycols were found to be miscible with PC (49b). Incorporation of 0–10% PU elastomers during melt-mixing decreases the glass transition temperature of PC (by 0–30°C), facilitates its processing, and improves its resistance to hydrothermal aging. No phase separation in blends was detected by means of the absorption of visible light and DSC measurements, while DMA and static bending measurements showed a slight increase in the glassy state modulus. These were ascribed to the strong interaction between the constituents, probably due to the formation of hydrogen bonds.

3. TPU/Polyether Blends

TPU/polyacetal (POM) blends have also been investigated by Kumar and coworkers (50) with respect to their mechanical and impact properties. The inclusion of up to 30% TPU resulted in effective toughening of POM, whereas additional amounts led to decrease of this property. As expected, the yield stress of POM decreases continuously with further addition of TPU elastomer. Stress relaxation measurements in simple extension were carried out for POM and its blends with 10, 20, and 30% TPU at a constant temperature (30°C). The rate of the relaxation modulus decrease was found to be a nonlinear function of time. All the stress relaxation curves for POM and TPU/POM blends could be satisfactorily superimposed, and smooth master curves were constructed.

In another work of Kumar and coworkers (51), the effect of TPU elastomer on the melting point and the percentage crystallinity of POM was studied by DSC. Wide-angle x-ray diffraction (WAXD) scans of POM, TPU, and their blends indicated that the crystalline structure of POM remained unaffected after the addition of amorphous TPU. The influence of defects, e.g., holes and notches, on the ultimate tensile strength was also examined. The resistance to crack initiation, the resistance to steady crack propagation, and the resistance to crack growth at maximum load were estimated. The TPU/POM blends displayed higher crack resistance values than pure POM. The hysteresis energy of blends was determined and was found to increase with TPU content.

Chang and Yang investigated (52) POM toughening with TPU elastomer in terms of rheological, mechanical, and morphological properties. POM can be effectively toughened by blending with TPU elastomer, and the improvement on toughness was found most significant with TPU content varying from 20 to 30 wt %. POM fractured in a ductile mode under an extremely low deformation rate. The rate at which the ductile–brittle transition occurred increased with the increase of the TPU elastomer content. The partial miscibility observed between POM and TPU was probably due to the hydrogen bond formation, which was also responsible for the TPU elastomer being such an efficient toughening agent for POM.

4. TPU/Polyamide Blends

Seo and Ha investigated (43) the phase behavior of a 6-block (PU-eth)-PA 6/PA 6 binary and a ternary blend where a PU was also included. They reported miscibility in the binary blend in the complete composition range due to hydrogen bonding between the urethane ($C=O$ group) and the amide groups. This was based on T_g variation and T_m depression determined by DSC. The same specific interaction was shown to affect compatibilization of the PU/PA 6 blend using PU-eth as a compatibilizer. Experimental evidence for this was obtained by DSC, DMA, and SEM.

Haponiuk investigated (44) melt-mixed blends obtained from PA 6 and PU-est or PU-eth by means of DMTA. Changes in composition did not influence the T_g of the amorphous fraction of the PA, but, also, no distinct transition for the separated PU soft segment was detected. Therefore, the blends seemed to be multiphase systems, where the elastomeric PU phase was dispersed in a continuous PA phase. Better miscibility of PA with PU-est than with PU-eth was observed. Based on changes in the β relaxation region of the PA, this was attributed to hydrogen bonding in the two-component common amorphous phase in the blends of PA with the PU-est.

5. TPU/Polyolefin Blends

Poetschke and coworkers investigated (53) the blend morphology, tensile properties, and rheology of melt-mixed PU-est and PU-eth with six polyolefins (PO). No blend component interactions were detected using DSC and DMA. The tensile strength shows a negative deviation from linearity, with a minimum at 40–60 wt % PO. In general, blends with PU-est show a stronger decrease than PU-eth in tensile strength. Blends with polypropylene (PP) show a smaller drop in properties compared to blends with PE.

Wallheinke and coworkers reported (54) on the coalescence behavior and its kinetics of the immiscible blends of TPU with PO. Two different types of measurements were used to observe coalescence in quiescent melt at the processing temperature. Coalescence was observed in situ using a light microscope or by static laser light scattering. It was observed that the higher-viscosity ratio of TPU/PE blend led to a more pronounced coarsening of the morphology of 80/20 blends than the corresponding TPU/PP. It was shown that the process of shape evolution during coalescence is one mechanism that takes place in quiescent melt. Another mechanism that was directly observed is a "domino effect," in which one coalescence event causes the next one.

In order to improve the dispersity, stability, and properties of TPU/PE blends, Poetschke and coworkers (55) used polyethylene grafted with maleic anhydride (PE-g-MA). With PE-g-MA as a blend component, the particle size was reduced dramatically compared to PE. Coalescence was reduced significantly, and the particle size increase with composition was less pronounced compared to blends with the unmodified PE.

6. TPU/Polystyrene Blends

Theocaris and Kefalas reported (56) on mechanical properties of blends of polystyrene (PS) with 5, 10, 20, and 40% TPU. Impact properties were compared with those of pure PS and commercial types of high-impact polystyrene (HIPS). Other techniques applied were DSC, mechanical spectroscopy, and rheology. It was found that the softer PU conglomerates embedded inside the PS matrix, increased the toughness of the blend (as expected from the addition of the softer particulate), and also expanded the glassy region of the blends by shifting their T_g's to higher temperatures. A theory based on the interaction of phases was proposed to explain this phenomenon.

Kefalas and Milios studied (57) blends of PS containing different amounts of a TPU elastomer that they synthesized. Blends were tested in uniaxial tension and compared with a commercial HIPS and a blend of PU/HIPS. Blends of PS with TPU as a dispersed phase showed improved mechanical properties and impact strength by ca. 30% of the energy to fracture of commercial HIPS. The materials produced confirmed the general rule that incorporation of an elastomeric phase into a glassy matrix enhances toughness.

Yang et al. reported (45) on the miscibility of sulfonated PS (SPS) with a PU containing a tertiary amine group (NPU). The blend cannot be considered truly miscible, since both main T_g relaxations of components are evident though converging to each other. However, strong interactions resulting from proton transfer from the sulfonic acid to the tertiary amine leads to synergism in tensile strength and a maximum in density at the 50/50 blend composition, i.e., around the stoichiometric ratio of sulfonic acid to tertiary amine. Various structural parameters examined affecting properties were the sulfonation level in SPS and the amount of chain extender in NPU.

7. TPU/Others

Wen and coworkers described (58) the surface modification of a PU-eth capped with poly(dimethylsiloxane) end groups [PDMS-(PU-eth)-PDMS], which finds applications as a biomaterial, and its blends with phenoxy (99 wt %). Using x-ray photoelectron spectroscopy (XPS) they investigated PDMS distribution in the film surfaces of air/polymer and glass substrate/polymer interfaces for the pure modified PU-eth and its blend. Parameters examined were annealing temperature and the presence of water.

A study on blends of a TPU with PVDF was described by Yue and Chian (37). This investigation focused on the effects that PVDF has on the mechanical behavior and morphological characteristics of the blends. Basic thermodynamic and structural considerations that were applied to predict blend miscibility revealed that the addition of PVDF disrupted the intermolecular chain interactions in TPU. As a result, a lowering of the mechanical properties of the blends was observed, with the intermediate composition (50/50) having the lowest mechanical strength. This was ascribed to the formation of layered structures caused by the immiscibility of the polymers. It was concluded that TPU/PVDF blends were incompatible at all compositions.

Landry et al. reported (33) on the effect of polymer structure on the blend miscibility of PU, with various polymers having potential hydrogen-bonding capability and among them PVPh. Solution-prepared blends of PU/PVPh showed gross phase separation, with two physically separated phases forming upon casting. Incompatibility of this blend was also reported by Seefried et al. (34).

Hill et al. reported (59) on melt-mixed blends of commercial TPU with low levels of a commercial PDMS fluid. The resultant modified PU showed improvement in wear resistance of up to 25 wt %, compared to virgin PU, with an optimal PDMS concentration of 1.5–2.0 wt %, beyond which the properties diminish rapidly. The mechanical properties of these blends were more significantly enhanced, by up to 40% for tensile strength and 50% for elongation to break. Surface studies on these blends were also reported by means of XPS, contact angles, and coefficient of friction.

In a subsequent article of Bremner et al. (60), a model was presented that accounted for the observed relationship between PDMS content and the properties of the blends. It was proposed that the addition of PDMS facilitated an improvement in packing efficiency (antiplasticization) in the polyurethane soft domain, leading to improved material performance. Beyond an optimum PDMS concentration of 1.5–2.0 wt %, phase separation of PDMS became significant and plasticization set in, leading to deterioration of mechanical properties. This model has been rigorously tested and shown to be realistic.

C. Miscellaneous Blends

Recently, in blends of multiblock copolymers with different homopolymers, a different approach that has been attempted utilized a multiblock copolymer containing polysulphone (PSU)/polybutadiene (PB), PC/PDMS, PSU/PDMS, or PTMO/PB (61). The homopolymer, which was immiscible with each of the copolymer constituent blocks taken separately, was either PS, polymethylmethacrylate (PMMA), or PVC. The blends were prepared by film casting from solutions of the components in a common solvent. The critical content of multiblock copolymer in the homopolymer matrix for transparency was found to be about 5 wt %, but the upper limit could reach as high as 20 wt %. It was found that the physical basis of the blend transparency was not thermodynamic miscibility of the components, but separation of the multiblock copolymer in the homopolymer matrix in the form of microphases (up to 0.1-μm diameter). These transparent blends were microheterogeneous systems with a morphology similar to that of the parent phase-separated block copolymers used.

Investigations have also been carried out in binary blends of multiblock copolymers containing PSU segments and a liquid crystalline polyester (LCP) [PSU-poly(ethylene terephthalate-co-oxybenzoate), (BCP)] with PSU (62). In these solution-cast blends, miscibility between both PSU phases was observed. Assuming a segment molecular weight of PSU of $\geq$8,400 g/mol, more than twice its entanglement molecular weight, compatibilization was proposed to take place in the amorphous phase of PSU, in agreement with the theoretical prediction of Noolandi (63).

IV. CONCLUSIONS

The blockiness of the main chain of PEE and PU has been usefully exploited in preparing modified self-reinforced thermoplastic materials covering a wide spectrum of mechanical properties.

Modification by blending with a suitable homopolymer was attained by utilizing the chemical features of these copolymers via:

1. The chemical and structural similarity of the block with the homopolymer, leading to athermal mixing and/or cocrystallization at the block level
2. The "complementary dissimilarity" of functional groups residing in the blend partners, leading to thermodynamic miscibility of the copolymer block with the homopolymer or to a polymeric alloy with good interphase adhesion
3. The reactive blending at suitable melt-mixing conditions of time and temperature, e.g., transesterification, ester-amide interchange reactions, anhydride-acidic hydrogen coupling reactions

Blends of these multiblock copolymers are inherently phase-separated. As a result, unless a suitable homopolymer is chosen, its impact modification can be readily achieved using a ductile-type PEE or PU as an additive.

Recently, optical clarity of the homopolymer matrix has also been attained by monitoring the amount and the size of the copolymer dispersed phase. In the past, characterization techniques usefully employed for all these blends included mechanical and electromagnetic spectroscopy, investigation of ultimate mechanical properties, as well as morphological studies using optical and electronic microscopy on fractured and/or etched specimens.

ABBREVIATIONS

BCP	poly(ethylene terephthalate-*co*-oxybenzoate)/polysulfone multiblock copolymers
CPE	chlorinated polyethylene
CPVC	chlorinated poly(vinylchloride)
DMA	dynamical mechanical analysis
DMPA	dimethylolpropionic acid
DMTA	dynamical mechanical thermal analysis

DSC	differential scanning calorimetry
FTIR	Fourier transform infrared
HIPS	high-impact polystyrene
LCP	liquid crystalline polyester
LCST	low critical solution temperature
NMR	nuclear magnetic resonance
NPU	polyurethane containing a tertiary amine group
PA	polyamide
PA 6	polyamide-6
PB	polybutadiene
PBA	poly(butylene adipate)
PBAU	poly(butyleneadipateurethane)
PBT	poly(butylene terephthalate)
PC	polycarbonate
PCL	poly(ε-caprolactone)
PDMS	poly(dimethylsiloxane)
PE	polyethylene
PEAU	poly(ethyleneadipateurethane)
PEE	poly(ether ester)
PE-g-MA	polyethylene grafted with maleic anhydride
PET	poly(ethylene terephthalate)
PHAU	poly(hexyleneadipateurethane)
PMMA	poly(methylmethacrylate)
PO	polyolefin
POM	poly(oxymethylene)
PP	polypropylene
PS	polystyrene
PSU	polysulphone
$(PTMEG-T)_m$	poly(tetramethylene) ether glycol terephthalate
PTMO	polytetramethylenoxide
PU	polyurethane
PU-est	polyester urethanes
PU-eth	polyether urethanes
PVC	poly(vinylchroride)
PVDC	poly(vinylidenechloride)
PVDF	poly(vinylidene fluoride)
PVPh	poly(vinylphenol)
SAXS	small-angle x-ray scattering
SEM	scanning electron microscopy
SPS	sulfonated polystyrene
TEA	triethylamine
TGA	thermogravimetric analysis

THF	tetrahydrofuran
TMA	thermal expansivity analysis
$(TMGT)_n$	poly(tetramethylene glycol terephthalate)
TPU	thermoplastic polyurethane
TSCS	thermally stimulated currents spectroscopy
UCST	upper critical solution temperature
XPS	x-ray photoelectron spectroscopy
WAXD	wide-angle x-ray diffraction
WAXS	wide-angle x-ray scattering

REFERENCES

1. O Olabisi, LM Robeson, MT Shaw. Polymer–Polymer Miscibility. New York: Academic Press, 1979, pp 1–9.
2. O Olabisi, LM Robeson, MT Shaw. Polymer–Polymer Miscibility. New York: Academic Press, 1979, pp 339–348.
3. LA Utracki. Polymer Blends and Alloys. Munich: Hanser, 1989, pp 1–27.
4. B Wunderlich. Macromolecular Physics. Vol. 1. New York: Academic Press, 1973, pp 147–161.
5. C Hugelin, A Dondos. Makromol Chem 126:206–216, 1969.
6. O Olabisi, LM Robeson, MT Shaw. Polymer–Polymer Miscibility. New York: Academic Press, 1979, pp 26–31, 206–211.
7. O Olabisi. Macromolecules 8:316–322, 1975.
8. M Lambla. Reactive processing of thermoplastic polymers. In: G Allen, S Aggarwal, S Russo, eds. Comprehensive Polymer Science. Suppl. 1. Oxford: Pergamon Press, 1992, pp 619–642.
9. M Xanthos, SS Dagli. Polym Eng Sci 31:929–935, 1991.
10. T Nishi, TK Kwei, TT Wang. J Appl Phys 46:4157–4165, 1975.
11. T Nishi, TK Kwei. J Appl Polym Sci 20:1331–1337, 1976.
12. DJ Hourston, ID Hughes. J Appl Polym Sci 21:3093–3103, 1977.
13. DJ Hourston, ID Hughes. Polymer 20:823–826, 1979.
14. SY Kwak. Polymer J 26:491–497, 1994.
15. SY Kwak, N Nakajima. Macromolecules 29:3521–3524, 1996.
16. SY Kwak, JJ Kim, UY Kim. Macromolecules 29:3560–3564, 1996.
17. M Guo. Macromolecules 30:1234–1235, 1997.
18. CL Ellis, CMF Barry. J Vinyl Addit Technol 2:326–329, 1996.
19. JK Kallitsis, DD Sotiropoulou, NK Kalfoglou. Polymer 29:1233–1239, 1988.
20. KP Gallagher, X Zhang, JP Runt, G Huynh-ba, JS Lin. Macromolecules 26:588–596, 1993.
21. AA Apostolov, S Fakirov, B Sezen, I Bahar, A Kloczkowski. Polymer 35:5247–5255, 1994.
22. AA Apostolov, M Evstatiev, S Fakirov, A Kloczkowski, JE Mark. J Appl Polym Sci 59:1667–1675, 1996.

23. BB Sauer, P Avakian, GM Cohen. Polymer 33:2666–2671, 1992.
24. M Gaztelumendi, J Nazabal. J Polym Sci Part B Polym Phys 33:603–610, 1995.
25. SS Skafidas, NK Kalfoglou. Polymer 38:1057–1064, 1997.
26. M Kimura, R Porter, G Salee. J Polym Sci Polym Phys Ed 21:367–378, 1983.
27. SM Hong, BC Kim. Polym Eng Sci 34:1605–1612, 1994.
28. WP Kulak, EG Koulouri, JK Kallitsis, Z Roslaniec. Some properties of polyamide 6 copoly(ether-ester) blends. Symposium on Composites and Polymer Blends, Szczecin, Poland, June 1997.
29. EG Koulouri, EC Scourlis, JK Kallitsis. Polymer 40:4887–4896, 1999.
30. E Gattiglia, E Pedemonte, A Turturro. Thermal and morphological analysis of Poly(ε-caprolactam)-poly(ether-ester) mixtures. In: LA Kleintgens, PJ Lemstra, eds. Integration of Fundamental Polymer Science and Technology. Essex: Elsevier, 1986, pp 148–151.
31. CB Wang, SL Cooper. J Appl Polym Sci 26:2989–3006, 1981.
32. NK Kalfoglou. J Appl Polym Sci 26:823–831, 1981.
33. MR Landry, DJ Massa, JT Lantry, DM Teegarden, RH Colby, TE Long, PM Henrichs. J Appl Polym Sci 54:991–1011, 1994.
34. CGJr Seefried, JV Koleske, FE Critchfield. Polym Eng Sci 16:771–776, 1976.
35. RE Prud'homme. Polym Eng Sci 22:90–95, 1982.
36. D Garcia. Polym Prepr Am Chem Soc Div Polym Chem 27:259–260, 1986.
37. MZ Yue, KS Chian. J Appl Polym Sci 60:597–603, 1996.
38. SJ Kim, BK Kim, HM Jeong. J Appl Polym Sci 51:2187–2190, 1994.
39. YQ Zhu, YJ Huang, ZG Chi, HJ Shen. Eur Polym J 30:1493–1500, 1994.
40. YQ Zhu, YJ Huang, ZG Chi. J Appl Polym Sci 56:1371–1379, 1995.
41. TO Ahn, KT Han, HM Jeong, SW Lee. Polymer Int 29:115–120, 1992.
42. Z Roslaniec. Polymer 34:359–361, 1993.
43. SW Seo, WS Ha. J Appl Polym Sci 54:1997–2011, 1994.
44. JT Haponiuk. J Thermal Anal 43:91–101, 1995.
45. W Yang, J Shen, S-H Zhu, C-M Chan. J Appl Polym Sci 67:2035–2045, 1998.
46. TO Ahn, S Jung, J Lee, HM Jeong. J Appl Polym Sci 64:2363–2369, 1997.
47. AO Baranov, VV Nizhegorodov, II Perepechko, MI Knunyants, EV Prut. Vysokomolekularnye Soedineniya Seriya A 34:66–76, 1992.
48. (a) CK Samios, KG Gravalos, NK Kalfoglou. Eur Polym J (In press, May, 2000.) (b) SV Kolesov, IV Meboilova, AM Steklova, SV Vladychina, KS Minsker. Polymer Science USSR 31:476–480, 1989.
49. L Fambri, A Penati, J Kolarik. Polymer 38:835–843, 1997.
50. G Kumar, MR Arindam, NR Neelakantan, N Subramanian. J Appl Polym Sci 50:2209–2216, 1993.
51. G Kumar, NR Neelakantan, N Subramanian. J Mater Sci 30:1480–1486, 1995.
52. F-C Chang, M-Y Yang. Polym Eng Sci 30:543–552, 1990.
53. P Poetschke, K Wallheinke, H Fritsche, H Stutz. J Appl Polym Sci 42:749–761, 1991.
54. K Wallheinke, P Poetschke, CW Macosko, H Stutz. Coalescence in blends of thermoplastic polyurethane and polyolefins. Technical Papers, Regional Technical Conference—Society of Plastics Engineers, 1997. Soc Plast Eng, Brookfield, CT, pp 509–544.

55. P Poetschke, K Wallheinke, H Stutz. Blends of thermoplastic polyurethanes and maleicanhydride grafted polyethylene. Technical Papers, Regional Technical Conference-Society of Plastics Engineers, 1997. Soc Plast Eng, Brookfield, CT, pp 476–508.

56. PS Theocaris, V Kefalas. J Appl Polym Sci 42:3059–3063, 1991.

57. V Kefalas, J Milios. Mechanical properties of polystyrene-polyurethane blends. Polymeric Materials Science and Engineering, Proceedings of the ACS Division of Polymeric Materials Science and Engineering, Washington, DC, 1990. Vol. 63, pp 195–199.

58. J Wen, G Somorjai, F Lim, R Ward. Macromolecules 30:7206–7213, 1997.

59. DJT Hill, MI Killeen, JH O'Donnell, PJ Pomery, D St John, AK Whittaker. J Appl Polym Sci 61:1757–1766, 1996.

60. T Bremner, DJT Hill, MI Killeen, JH O'Donnell, PJ Pomery, DSt John, AK Whittaker. J Appl Polym Sci 65:939–950, 1997.

61. VS Papkov, GG Nikiforova, VG Nikol'sky, IA Krasotkina, ES Obolonkova. Polymer 39:631–640, 1998.

62. L Häußler, D Pospiech, K Eckstein, A Janke, R Vogel. J Appl Polym Sci 66:2293–2309, 1997.

63. J Noolandi. Macromol Chem Theory Simul 1:295–298, 1992.

14
Structure and Melt Rheology of Block Copolymers

Jørgen Lyngaae-Jørgensen
Technical University of Denmark, Lyngby, Denmark

I. INTRODUCTION

The properties of polymer materials are determined by the molecular structure, the microstructure (morphology etc.), and the macrostructure (geometrical form) of the materials (1).

Block copolymers constitute a set of molecular structures that open special possibilities for tailor-making polymer materials. A nonexhaustive survey of basic structural factors can be represented schematically as follows:

1. Molecular structure of block copolymers

   Repetition units in the blocks
   Sequence-length distribution of the blocks
   Tacticity—stereospecific structure of block
   Number of blocks per polymer molecule
   Distribution of the number of blocks per polymer molecule
   Molecular weight distribution
   Branch structures
   Deviating repetition units in blocks etc.

2. Microstructure

   Multiphase morphology
   Crystallinity

Orientation effects
Compounding ingredients etc.

3. Macrostructure

Geometrical form

The variability of the molecular structure and microstructure of block copolymer–based materials is enormous. This presentation is limited to pure block copolymers.

Most systematic investigations (on the rheology) of block copolymer melts deal with diblock and triblock copolymers with very well-defined molecular structure. Commercially, the production of thermoplastic elastomers represents the largest use of block copolymers (2,3). Some of the commercial products are multiblock materials with large variation in, e.g., the distribution of block sequence length.

Block copolymers in the disordered (mixed) state show properties those of analogeous to homopolymer melt properties (1,4,5); that is, master curve principles can be constructed. Newtonian liquid–like behavior at low shear rate and the Cox–Merx rule (6) are applicable.

The same is not so for microphase-separated block copolymers. The block-length ratio controls the microdomain structure, e.g., spheres (S), cylinders (C), lamellae (L), or ordered bicontinuous double-diamond structures (B) (4,5) in block copolymers with well-defined uniform molecular structure.

When two-phase-structure materials are deformed, especially in the melt state, large changes in the microstructure are usually observed. These structural transformations depend on time and deformation. Consequently, two-phase (or multiphase) high-molecular-weight block copolymer melts are complex time-dependent viscoelastic materials. Thus studies of block copolymers melts should preferentially encompass the possibility of following structural changes during flow, that is, in situ structure measurements, and reporting transitional measurements. It is important to clearly distinguish steady-state measurements from transitional studies.

II. RHEOLOGICAL MEASURING TECHNIQUES

These are characterization methods used to determine the structure and rheological properties of block copolymers in the melt state. The rheological functions discussed in this chapter are limited to viscosity, normal stress differences measured in steady-shear flow, the complex viscosity and complex modulus, and the real and imaginary components in oscillatory-shear flow (dynamic mechanical spectroscopy = DMS) and the viscosity measured in uniaxial extensional flow. Comprehensive treatments on measuring techniques may be found in many textbooks on rheology (7,8).

In order to follow structure development during flow experimentally, a number of techniques combining rheological measurement and in situ determination of multiphase structure have been developed. A general review is given in Ref. 9. The most applied techniques for measuring on block copolymers are neutron scattering, small-angle x-ray scattering (SAXS), light-scattering, and birefringence. Transmission electron microscopy (TEM) and scanning electron microscopy (SEM) represent very useful supplementary techniques on shock-cooled samples. Master curve analyses can be used as diagnostic tools (1).

III. HYPOTHESES

In the following, a postulate concerning steady-state structures during melt flow will be formulated. Such hypotheses may serve as a basis for the type of expectations we might have concerning structure changes during flow, structure transitions, etc. These hypotheses should preferentially give guidance in connection with the development of new block copolymers.

Based partly on the work of Harry Hull (10,11), a (modified) Gibbs free energy is well defined for steady-state flow at constant pressure and temperature. Relative to a standard state, this (molar) change in free energy is written ΔG_{ss}. Without flow, the equivalent term is ΔG_{melt}.

Obviously one may write

$$\Delta G_{ss} = \Delta G_{melt} + \Delta G^* \tag{1}$$

where ΔG^* is simply the difference $\Delta G_{ss} - \Delta G_{melt}$. A stable steady-state structure corresponds to a structure where ΔG_{ss} is a minimum. A relatively large number of expressions have been published for ΔG_{melt} (15–21). No general consensus exists as to the evaluation of ΔG_{ss} or ΔG^*. Under a number of simplifying assumptions, ΔG^* may be evaluated. This author has evaluated the term by postulating that it represents an elastic contribution (1).

A necessary condition for flow is that part of a block (at least in an ABA(s), or multiblock copolymer) mix with B as the continuous phase. Yield stress is a measure of a solid/liquid transition, as stressed especially by M. Williams and coworkers (12,13). Yield stresses in block copolymers have recently been predicted by Doi et al. (14). In their case, this includes not only triblock copolymers with spherical microstructure.

The expression given in Ref. 1 for a critical shear stress (τ_{cr}) for a transition from a two-phase to a monomolecular melt is a reasonable approximation for this yield stress, even though a monomolecular melt is never established because oriented two phase microstructures have been documented to represent steady-state structures with the lowest ΔG_{ss} (Eq. 1.).

For strictly regular block copolymers the stresses corresponding to the so-called "gel destruction," or solid-gel/liquid transition, is approximately the same as τ_{cr} because mixing of the first molecules in pure B is the critical step. A derivation of $\tau_{gel.destr.}$ (τ_{cr}) and a comparison with experimental data will be given in a separate paper.

The limiting form for τ_{cr} for high-molecular-weight immiscible block copolymers is:

$$\tau_{cr} = k \left(\frac{\rho R}{M_c} \right)^{1/2} \frac{| \delta_A - \delta_B |}{H} (| T - \text{ODT} |)^{1/2} \tag{2}$$

where k is a constant (0.03 in cgs. units), M_c is the molecular weight between entanglements in the mixed sample, ρ is melt density, R is the gas constant, $H \equiv \overline{M_w}/\overline{M_n}$ is the ratio between the average molecular weight by weight ($\overline{M_w}$) and by number ($\overline{M_n}$), respectively, δ_A and δ_B are the solubility parameters for the A and B blocks, respectively, and ODT is the order–disorder temperature.

In general the estimated difference in ΔG_{melt} between the different ordered states found for such ordered systems is often relatively small. This means that order–order transitions (OOTs) in highly ordered systems are probable even in systems with large interaction parameters. The ΔG^* terms for a given microstructure are increasing functions of stress at constant temperature, pressure, and deformation rate. Consequently, one should expect order–order transitions or eventual order–disorder transitions at critical shear stresses.

In the other limit contrary to these well-ordered block copolymers—namely, less ordered multiblock systems, mainly systems with crystalline short blocks, the so-called hard segments—are of interest. Actually transitions dring flow for these systems resemble transitions in PVC systems or other systems with nearly atactic structure in the chains, but where, e.g., syndiotactic sequences crystallize.

For such systems, Eq. (1) leads to a prediction for melting of the last crystallite of the form (9,22,23)

$$T_{dyn} = \frac{T_m}{2} \left[1 + \sqrt{1 - \frac{4\tau_{cr}^2}{T_m Q}} \right] \tag{3}$$

That is, a stress larger than τ_{cr} is necessary to melt the last crystallite at temperature T_{dyn} in simple shear flow. Here,

$$Q = \frac{2c^4 R \Delta H_u}{a^2 \rho^2 M_A H^2 M_C} \tag{4}$$

a is a constant, ρ is the polymer density, T_m is the static melting temperature, and T_{dyn} is the melting temperature at constant shear rate in simple shear flow. M_c is twice the average molecular weight between entanglements (M_e). In the calculations we use an estimate that is approximately two times M_e, namely, the molecular weight where the exponent in the equation $\eta_o = KM_W^a$ changes from ~1 to ~3.5. $H = \overline{M}_W/\overline{M}_n$ is the ratio between weight- and number-average molecular weights and c is polymer concentration.

The general behavior of a material with a low degree of crystallinity, i.e., short crystalline sequences in the polymer molecules, is shown in Fig. 1 (plots of log τ against $1/T$). The principal sketch shown on Fig. 1 shows three areas demarked by the melting curve (the melt fracture curve and the so-called gel destruction curve, respectively). In area A, a melt state where the single polymer molecules constitute the "flow units," a monomolecular melt state exists. In this area normal melt flow behavior is expected and observed. In area B, stable crys-

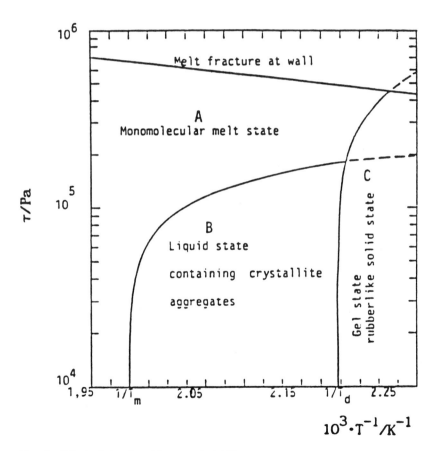

Fig. 1 Principal sketch and log τ against $1/T$.

tallite aggregates consisting primarily of a number of polymer molecules held together by one (or more) crystalline nuclei exist in the melt. The flow properties in this regime will be expected to reflect two effects: (1) that the crystallite aggregates will behave as a branched system and (2) that the number of crystallite aggregates may depend on the measuring conditions and the prehistory. In region C, that is, at temperatures below the so-called gel destruction temperature, T_d, the material will be in a state corresponding to a crosslinked rubber; that is, the material will be a "solid" in contrast with the two melt states in A and B.

Since the fringed-micelle concept in materials with a low degree of crystallinity is probably a reasonable description of the crystalline state and since the size distribution of crystals with varying sequence lengths is very broad, the (macro) structural state will reflect the prehistory. If the material has not been exposed to temperatures higher than T_d, then total fusion will not have taken place. In this chapter the degree of fusion is defined by considering an interface between two originally completely separated surfaces (a healing experiment). A 100% fusion exists when the original interfacial area cannot be distinguished. A reasonable conceptual definition could be: degree of fusion = degree of crystallization at the interface/degree of crystallization in the material.

It is quite obvious that an original structure, whether it is particular or not, cannot be extinguished in the fully fused sense below T_d. Above T_d, complete fusion is a matter of time. However, since diffusion times of large branched structures are very long, interfaces will disappear slowly. For plasticized samples, both T_m and T_d are suppressed to lower temperatures, and even the gelled system may be melted at high-enough shear stresses.

When the crystallizable sequences become more regular, $1/T_m$ moves toward $1/T_d$. The influence of the interaction parameter [$\chi \propto (\delta_A - \delta_B)^2$] is to increase the stresses necessary in order to induce a gel-destruction transition or a complete melting. If the gel-destruction stress becomes larger than the melt-fracture stress, the material cannot be processed in the liquid state in shear flow.

IV. EXPERIMENTAL INVESTIGATION OF BLOCK COPOLYMERS

The rheological behavior of noncompatible AB, ABA, $A_n B_n$ block copolymers is considerably different from that of homopolymers as well as random copolymers. The block copolymers are reported to have higher melt viscosities (at low rates of deformation) than random copolymers having the same composition and weight-average molecular weight. It seems to be generally observed that block copolymers, and sometimes also random copolymers, do not approach a constant viscosity at low shear rates in simple steady-state shear flow. Abrupt changes of slopes in the flow curves of block copolymers have been observed by many research groups.

Figure 2 depicts possible shear viscosity–shear rate curves for given three-block copolymer samples measured at steady state. The full drawn curve represents the normally observed curve type for a monomolecular melt. Deviations may be observed at high shear rates/high shear stresses, e.g., caused by melt fracture. The dashed lines represent possible curves for block copolymers. Curves A and B show non-Newtonian behavior (structure breakdown) at low shear rates and near coincidence with the viscosity curve for monomolecular melts at high shear rates. Curves C and D deviate from the viscosity–shear rate curve for a monomolecular melt at all measurable shear rates.

The complex melt flow behavior can be attributed to the existence of the two-phase domain structure and domain structure changes as a function of the variables that determine the flow. An examination of the literature, which is the subject of the following sections, will of course reflect the fact that most accessi-

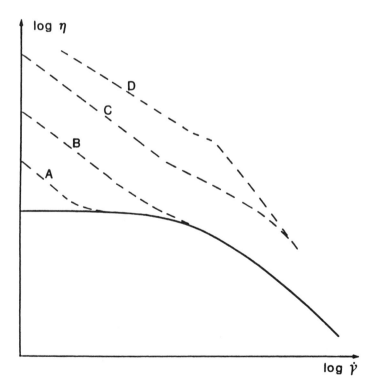

Fig. 2 Possible steady-state shear viscosity–shear rate curves for (three)-block copolymers.

ble data are reported for SBS types. The literature is biased in this respect, so generalizations must be made with caution.

A. Diblock Copolymers

1. Oscillatory Flow

A considerable number of papers have been dedicated to in situ studies of structure formation in oscillatory shear flow (DMS). If a lamellar diblock copolymer melt is subjected to an oscillatory shear field at large amplitude (LAOS), an ordered laminar structure tends to form. In 1992, Koppi et al. (24), using the large-amplitude oscillatory shear (LAOS) technique, documented that a shift between two structures oriented differently relative to the applied flow field was observed. Table 1 gives a review of the type of block copolymers studied and the observation of shifts between parallel lamellar ($=$) and perpendicular ($\perp$) orientation.

Figure 3 shows a schematic of DMS morphology data for diblock copolymers. The normal to the lamellar plane is parallel to the gradient direction in the parallel orientation but parallel to the neutral direction in the perpendicular orientation in simple shear flow.

In situ SANS measurement during reciprocating shear flow is reported to induce ordering in the perpendicular arrangement at temperatures above ODT, as seen in Fig. 3 (32,33) for lamellar-ordered diblock copolymers. Cylindrical-forming diblock copolymers (the fraction of PEP in a PEP-b-PEE block copolymer molecule $f_{PEP} = 0.7$, PEP-b-PEE) show parallel orientation in reciprocating shear flow at temperatures above ODT (34). Observations for PS-b-PB are analogous to observations for PS-b-PI. An important factor is the time scale for transition from an original lamellar structure, which often contains many kinds of dislocations, here called structural imperfections, to a final "ordered structure." A number of studies throw light on this question (5,35,36).

2. Hypothesis for the Steady-State Structure of Lamellar Diblock Copolymers

Can a simple hypothesis, such as sketched in Sec. III, be used to rationalize the observed behavior? Actually, Eq. (1) may be written $\Delta G_{ss} = \Delta G_{melt} + f(\tau)$ for shear flow between parallel plates with constant apparent overall shear rate $\dot{\gamma} \equiv V/h$, where V is the rate of the moving plate and h is the plate separation. $f(\tau)$ is an increasing function of shear stress: τ. Neglecting surface effects, ΔG_{melt} is the same for a parallel structure and a perpendicular structure. Thus, according to Eq. (1), a steady-state structure corresponds to a structure where the shear stress is a minimum.

Consider a perfect lamellar structure. The frictional resistance of the A block in an A-b-B diblock copolymer where one end is fixed in an (infinitely) thin interface layer is represented by a viscosity index term η_A such that the total fric-

Table 1 Structures Observed in Diblock Copolymers After Oscillatory Shear

Diblock copolymer	$\bar{M}_n$ (kg/mol)	Structure L, C, S[b]	Angular velocity (rad/s)	Temperature (°C)	Stable structure	In situ method	Refs.
PS-b-PI	18.4–56.5	L	Low frequencies	130–150	=	SAXS	25[a]
—	—	L	Middle	—	⊥	SAXS	25[a]
—	—	L	High	—	=	SAXS	25[a]
PEP-b-PEE		L	Lowest		=	SANS	24, 26
		L	Highest		⊥	SANS	24, 26
PS-b-PI		L	Low	High*	⊥	SAXS	27–29
		L	High	low	=	SAXS	27–29
PS-b-PI	19.9	L	0.1–10	123	⊥	SAXS	30
—		L	0.1–1	113	⊥	SAXS	30
—		L	10	113	=	SAXS	30
—		L	10	103	=	SAXS	30
—		L	1–10	93	=	SAXS	30
PS-b-PI	60	C	1	$T < 155$	=	SAXS, TEM	31
	60	S	1	$155 < T < 200$		SAXS, TEM	31

[a] Observations for PS-b-PB are reported analogous to observations for PS-b-PI.
[b] L, lamellar microstructure; C, cylindrical microstructure; S, spherical microstructure.

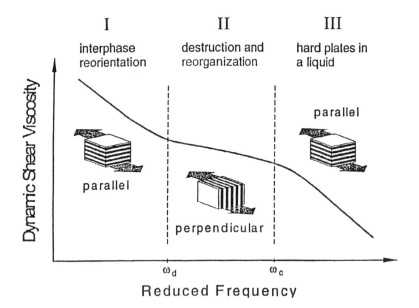

Reduced Frequency

Fig. 3 Double logarithmic plot of viscosity against shear rate for PS-*b*-PI diblock copolymer, plus principal sketch. (Reproduced from Ref. 25, by courtesy of the publisher, Wiley-VCH.)

tional resistance at steady state scales with the number of fixed ends. In the same way, a term η_B is defined for the B block friction.

In the case of a constant rate for the upper plate we have

$$\dot{\gamma}_{\text{Total}} = \frac{V}{h} = \dot{\gamma}_{T,\perp} = \dot{\gamma}_{T,=}$$

where $\dot{\gamma}_T$ represents the total apparent shear rate, $\dot{\gamma}_{T,\perp}$ in the perpendicular structure, $\dot{\gamma}_{T=}$ in the parallel structure. Assuming that stress is unbroken through interfaces, the perpendicular lamellar structure gives:

$$\dot{\gamma}_{T,\perp} = \dot{\gamma}_{A,\perp} = \dot{\gamma}_{B,\perp} \tag{5}$$

and

$$\tau_{T,\perp} = \varphi_A \tau_{A,\perp} + \varphi_B \tau_{B,\perp}$$
$$\tau_{T,\perp} = [\varphi_A \eta_{A,\perp}(\dot{\gamma}_T) + \varphi_B \eta_{B,\perp}(\dot{\gamma}_T)] \, \dot{\gamma}_T \tag{6}$$

and the parallel lamellar structure gives:

$$\tau_T = \tau_{T,=} = \tau_{A,=} = \tau_{B,=} \tag{7}$$
$$\dot{\gamma}_{T,} = \varphi_A \dot{\gamma}_{A,=} + \varphi_B \dot{\gamma}_{B,=} \tag{8}$$

Since

$$\dot{\gamma}_{T,=} \equiv \dot{\gamma}_{T,\perp} = \dot{\gamma}_T \tag{9}$$

the ratio between the shear stress in perpendicular lamellar $\tau_{T,\perp}$ and in parallel lamellar is:

$$\frac{\tau_{T,\perp}}{\tau_{T,=}} = \varphi_A^2 \frac{\eta_{A,\perp}(\dot{\gamma}_T)}{\eta_{A,=}(\dot{\gamma}_{A,=})} + \varphi_B^2 \frac{\eta_{B,\perp}(\dot{\gamma}_T)}{\eta_{B,=}(\dot{\gamma}_{B,=})}$$
$$+ \varphi_A \varphi_B \frac{\eta_{A,\perp}(\dot{\gamma}_T)}{\eta_{B,=}(\dot{\gamma}_{B,=})} + \varphi_A \varphi_B \frac{\eta_{B,\perp}(\dot{\gamma}_T)}{\eta_{A,=}(\dot{\gamma}_{A,=})} \tag{10}$$

If the viscosity indexes are all constant (independent of shear rate) at steady state, then Eq. (10) predicts that $\tau_{T,=} < \tau_{T,\perp}$ except when $\eta_A = \eta_B$, where $\tau_{T,=} = \tau_{T,\perp}$. That means that the parallel laminar structure is predicted to be the stable structure. Normally constant viscosity may be observed at low shear rate.

It is assumed that the viscosity indexes contributed to the A blocks and B blocks have the same shear-rate dependence independent of whether perpendicular or parallel structure is considered.

$$\eta_{A,\perp}(\dot{\gamma}) = \eta_{A,=}(\dot{\gamma}) \qquad \text{and} \qquad \eta_{B,\perp}(\dot{\gamma}) = \eta_{B,=}(\dot{\gamma})$$

If $\eta_A = \eta_B$ and the shear-rate dependence of η_A and η_B is the same, then $\tau_{T,\perp} = \tau_{T,=}$ and no preference exists.

The frictional properties of blocks A and B are probably different. The frictional properties depend heavenly on the difference between the measuring temperature T and the glass transition temperature and entanglement structure.

If we consider monodisperse samples of homopolymers, a principal sketch of the relative viscosity shear rate curves (double logarithmic) is as shown in Fig. 4 for the case $\eta_{A,O} > \eta_{B,O}$, where η_A and η_B are now the frictional viscosity indexes of blocks A and B, respectively. For $\eta_{B(\dot{\gamma})} = $ constant $= \eta_B$, we have $\tau_{T,\perp} < \tau_{T,=}$ for

$$3 > \frac{\eta_A(\dot{\gamma}_T)}{\eta_{A,=}(\dot{\gamma}_{A,=})} + \frac{\eta_B(\dot{\gamma}_T)}{\eta_{A,=}(\dot{\gamma}_{A,=})} + \frac{\eta_A(\dot{\gamma}_T)}{\eta_B}$$

which is fulfilled in an interval around the point P, where the two curves cross.

The parallel lamellar structure is the stable one for all large viscosity ratios. Consequently for systems with constant viscosity indexes or where the frictional properties of A and B blocks are the same, the parallel lamellar structure should be the only stable structure, whereas greater differences in frictional properties of the blocks will give a prediction with parallel lamellars at low shear rate perpendicular at medium shear rates and eventually change to parallel lamellars at high shear rates. Thus we expect that the condition $T_{g,A} \gg T_{g,B}$ will give the last behavior whereas $T_{g,A} \simeq T_{g,B}$ may tend to give stable parallel lamellars. If we com-

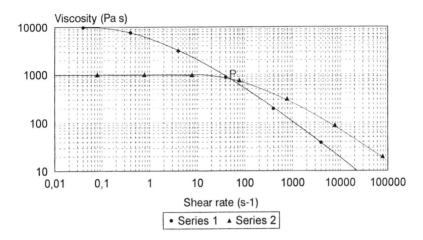

Fig. 4 Relative position of two monodisperse polymer samples. The samples depicted have the same reduced plot of log η/η_o against log $\lambda\gamma$, where η and η_o is viscosity and zero-shear viscosity, respectively, λ is a time constant proportional to $\eta_o M_e$, M_e is the molecular weight between entanglements, and γ is the shear rate.

pare these qualitative predictions with data from Table 1, they seem to fit reasonably well.

3. Steady-Shear Flow

Steady-shear measurements are encumbered with problems, e.g., the existence of secondary flows, that may make the formation of perfectly ordered lamellars difficult. Few investigations have reported on diblock copolymers (37–40). Nguen et al. (39) compare the flow behavior of a diblock copolymer of polyisoprene and polybutadiene, PI-b-PB, with 30% PB and a total molecular weight of 3.5×10^5 in the temperature range 25–90°C. The flow behavior of the block copolymer is compared with the behavior of homopolymers of PB and PI and blends of these homopolymers. The PI-b-PB block copolymer exhibited a Newtonian behavior at low shear rate. It is reported that the zero-shear viscosity of the block copolymer is approximately twice the zero-shear viscosity of either homopolymer for samples with approximately the same total molecular weight by weight. The mixing rule

$$\eta_0 = (w_1\eta_{0,1}^{1/\alpha} + w_2\eta_{0,2}^{1/\alpha})^\alpha \tag{11}$$

is reported to be followed for blends of samples of the same type of homopolymer (41); w_1 and w_2 are weight fractions of components with viscosities $\eta_{0,1}$ and $\eta_{0,2}$, respectively, and α is the exponent in the molecular weight dependence of η_0. Equation (11) did not apply to the investigated blends of PI and PB nor to the diblock copolymer. The flow energy of activation for the block copolymer is found to be between the energy of activation for the homopolymers.

Lyngaae-Jørgensen et al. (38) studied a diblock copolymer of polystyrene and poly(methyl methacryalte) with 25% PMMA and a total molecular weight by weight of 91,000. Transitions were observed in the flow curves, in qualitative agreement with those reported for S-B-S block copolymers, but the transitions observed at different temperatures did not take place at constant-shear stress.

Steady-shear flows (parallel plates) introduce imperfections in highly ordered PS-b-PI (lamellar microstructure); the highly oriented structures are obtained by means of large-amplitude oscillatory shearing (LAOS) (40).

B. Triblock Copolymers

For triblock coplymers and multiblock copolymers Eq. (1) points to the importance of morphological symmetry and especially the possibility of forming flow surfaces with low frictional resistance to mutual movements of the blocks. Morphology should influence the rheological properties and vice versa (reverse flow should influence morphology).

The best investigated types of block copolymer are those of styrene and butadiene or styrene and polyisoprene, and here again the three-block SBS or SIS type has received much attention. This is an obvious consequence of the fact that these materials were the first "large"-scale block copolymers commercially available.

It is also for an SBS block copolymer that the most comprehensive investigation of different melt flow properties has been performed. In particular, the IUPAC investigation reported in Ref. 42 should be mentioned. The data reported in the IUPAC Report (42) represent cooperative data obtained by 14 major research laboratories. These extensive studies have been performed on Cariflex TR-1102, an SBS type manufactured by Shell. The block lengths in molecular weight units are 11,000–56,000–11,000 and the styrene content is about 28% by weight ($\overline{M}_w / \overline{M}_n = 1.2$). Data for this sample are reported in Refs. 42 and 44.

C. Dynamic Measurements

The majority of published experimental data on block copolymers originate in dynamic measurements.

1. Influence of Frequency

Arnold and Meier (45) investigated a number of SBS samples with end block sizes in the range 10,000–14,000 mol^{-1} and midblock sizes in the range 50,000–70,000 g-mol^{-1}. The dynamic viscosity data exhibited two "regions," a high-frequency region, where the data corresponded to typical behavior for monomolecular melts, and a low-frequency region, where Newtonian behavior could not be established, the viscosity increasing continuously with decreasing frequency.

Qualitatively, these results are in agreement with dynamic data reported in the majority of other papers (30,31,40,42–64). Complementary observations for an SBS 11,000–56,000–11,000 show that the ratio between complex viscosity data and steady-shear viscosity at identical values of shear rate, $\dot{\gamma}$, and frequency, ω, is much higher than 1 for $\dot{\gamma} < 1 \text{ s}^{-1}$ (42). Typical data are shown in Fig. 5.

2. *In situ Observations*

Documented LAOS investigations for well-defined triblock copolymers (ABA) systems gives the results for steady-state structure shown in Table 2.

3. *Influence of Molecular Structure*

Many authors found that ABA polymers have considerably higher viscosities (42) or, expressed another way, have much longer maximum relaxation times for the

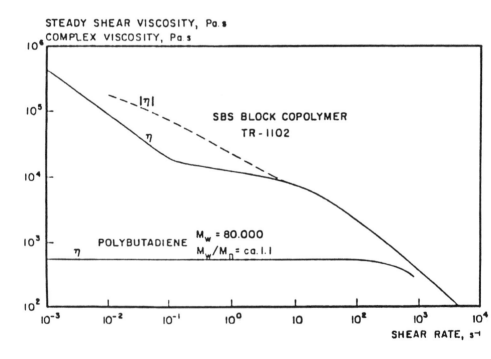

Fig. 5   Viscosity data of Cariflex TR-1102 and polybutadiene at 150°C. (Reproduced from Ref. 42 by courtesy of the publisher, the International Union of Pure and Applied Chemistry.)

Table 2   Structures Observed in Triblock Copolymers After Oscillatory Shear

System	Composition	$\overline{M}_n$ (kg/mol)	Start geometry	Temp (°C)	Angular frequencies (rad/sec)	Steady-state structure	In situ method	Ref.
SIS	51.6 (wt % S)	39.3	Random lamellar	100, 110, 120 130, 140	0.1–10.0	perpendicular	X-ray scattering	30
SIS	17 (vol % S)	120	Cylindrical	$T < 205$	1	=	SAXS TEM	31
SIS	—	120	Spherical	$T > 205$			TEM	31

transition from rubbery to flow behavior than the corresponding homopolymers with the same molecular weights.

This result is confirmed by Futamura and Meinecke (48,49), particularly for many different SXS polymers, where S is styrene and X is varied. They conclude that the maximum relaxation times increase with increasing difference in solubility parameter between the styrene block and the center block, but are not significantly influenced by the glass transition temperature or the molecular weight between entanglements of the center block.

The conclusion reached by Futamura and Meinecke is in excellent agreement with the work of Matzner and coworkers (50,51) on organosiloxane block copolymers. In fact their conclusion was that the differential solubility parameter $| \delta_A - \delta_B |$ [δ measured in $(cal\text{-}cm^{-3})^{1/2}$] should be less than 1 in order to obtain good melt processability. Furthermore, these authors reached the important conclusion that optimum processability and two-phase properties will be obtained in block and graft copolymers if the differential solubility parameter is approximately zero and if at least one of the segments is crystallizable.

The interpretation of structure transitions in block copolymers involves the relatively large number of papers that deal with investigations of dynamic data as preferentially coupled with in situ measurements of morphology as a function of temperature. Block copolymers show thermodynamic transitions from noncompatible at low temperatures to compatible at high temperatures. Both order–order transitions—OOT—and ODT (31,47,52–55,61,64) are observed.

4. Transition Temperatures

The rheological properties of block copolymers deviate from the properties of homogeneous homopolymer melts. Sometimes abrupt changes in rheological properties can be observed. These changes can in many cases be shown to reflect different steady-state microstructures. Several authors have used rheological data as a function of temperature to find the order–disorder transition temperature.

In these analyses master curve representation may be used as a diagnostic tool. Chung and Gale (47) were the first researchers to perform a systematic in-

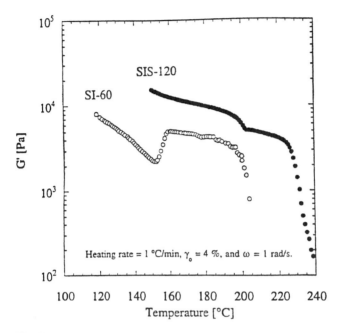

Fig. 6 Plot of log G' against temperature at constant angular velocity. (Reproduced from Ref. 31 by courtesy of the publisher, John Wiley & Sons, Inc.)

vestigation on an SBS sample (S: 7000, B: 43 000), showing a structure transition by increasing the temperature.

Recently many research groups have used plots of log G' against temperature at constant (low) frequency to determine both order–order transition temperatures (OOT) and ODT (31). Figure 6 shows an example.

D. Simple Shear Flow

Han et al. (4) used TEM pictures to evaluate the microstructure of extruded samples of ABA copolymers with cylindrical (C) and spherical morphology. Table 3 reviews some of the findings for these cases. Data measured in cone-and-plate geometry or rotary viscometers (38,42,44,47,56,63–71) and data measured in capillary or slit flow (4,37,39,42,44,46,65–72–89) have been used in investigations of block copolymers.

1. Influence of Shear Rate

One of the first systematic investigations of flow properties of SBS block copolymers at high temperatures, in both cone-and-plate and capillary flow, was reported

Table 3 Morphology-Microstructure of ABA

Sample	Microstructure	$\overline{M}_n$ (kg/mol)	T (°C)	Structure orientation	Measuring equipment	Morphology evaluation	Refs.
SBS (10.5–53.6–10.5)	C	74.4	160–180	=	Capillary rheometer	TEM	4
SIS (10–120–10)	S	140	180–200	S		TEM	4
SBS	C			= Hexagonal packing	Parallel-plate	SAXS	62
SBS (23 wt %)	C			= Coexistence of two ordered cylindrical structures	Cuette	SANS	69, 70
SBS (26 wt %)	C			=	Cone-&-plate Constant shear	SAXS	63

by Holden et al. (65), who observed that one of the most conspicuous differences between the rheological properties of block copolymers and homopolymers is that the block copolymers do not exhibit Newtonian behavior at low shear rates. Furthermore, it was found that materials with a styrene content in the range 39–65% showed two distinct viscosity–shear rate relationships. The authors interpreted this fact as being a result of phase inversion. It is observed that transitions from one state to the other occur at about the same shear stress, which Holden et al. find to be approximately 10^6 dyn-cm^{-2}, independent of temperature.

The IUPAC group reports that low-shear-rate cone-and-plate measurements indicate that shear stress–time curves have very different forms, depending on the shear rate (see Fig. 7). In many cases, no well-defined steady value for the shear stress is found. Consequently, the calculation of viscosity becomes rather arbitrary and depends on how shear stress is evaluated. The scatter between the values from different laboratories was found to be considerable. It was documented that prehistory played a very significant role. At higher shear rates (higher shear stresses), the viscosity was much more reproducible and in agreement with capillary data.

This observation is in accord with the work of Vinogradov et al. (44), who, for the same SBS sample investigated by the IUPAC group, report that the material in the region of low shear stresses exhibits sharply pronounced thioxotrophy. Vinogradov et al. found that in the region 3×10^4 to 3×10^5 dyn-cm^{-2}, nearly Newtonian flow with viscosity η_0 of the material is observed in capillary flow (150°C). This observation is in reasonable accord with the IUPAC data, as shown in Fig. 4. Vinogradov et al. (44) found that capillary-flow data follow a time–temperature position principle of the type log η/η_0 against log $(\eta_0 \dot\gamma)$.

Vinogradov et al. report that "spurt behavior" (near melt fracture) at the capillary wall is observed at $\tau_{cr} = 3.5 \times 10^6$ dyn-cm^{-2} for the sample investigated by them and the IUPAC group. This phenomenon will be discussed in more detail in Sec. V.

Ghijsels and Raadsen (42) interpret the IUPAC result as being caused by structural changes taking place in the melt (breaking of network structures) at low shear stresses. For shear stresses above ~3 × 10^4 dyn-cm^{-2}, star-shaped branched-flow units are taken as responsible for the flow behavior. This star structure may eventually be broken down at high stresses.

Williams and coworkers (13,83,84) have shown that SBS triblock copolymers exhibit yield stress behavior and have developed a special controlled-stress, parallel-plate rheometer for the measurement of yield stresses.

2. Influence of Temperature

The IUPAC group reports that the viscosity at a shear stress 10^4 Pa of the SBS block copolymer (TR-1102) does not follow an Arrhenius type of temperature de-

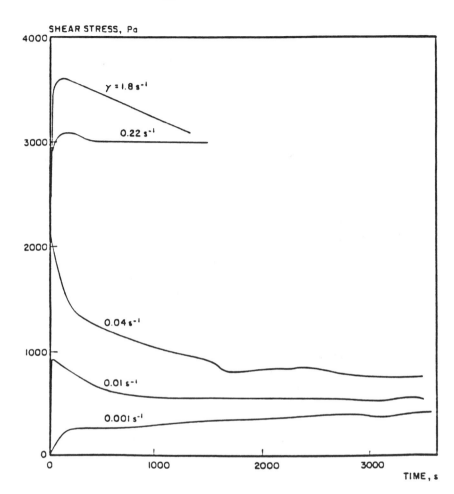

Fig. 7 Typical shear stress–time traces at 150°C. (Reproduced from Ref. 2 by courtesy of the publisher, the International Union of Pure and Applied Chemistry.)

pendence over wide intervals in temperature. The activation energy increases with decreasing temperature; E is about 15 kcal-mol^{-1} at temperatures between 170 and 190°C and about 30 kcal-mol^{-1} at temperatures between 110 and 130°C.

3. Influence of Molecular Structure

The main influence is presented in Table 3. Kraus et al. (46) investigated different samples of BSB and SBS and three- and four-branched samples. The authors show that SBS samples at $\dot{\gamma} = 1 \ s^{-1}$ and $\gamma^{\cdot} = 10 \ s^{-1}$ and $T = 130°C$ and $T = 160°C$ have higher viscosities than BSB samples and that the viscosity of the BSB samples is determined by the butadiene block lengths.

Leblanc (73,74) reports data measured with a slit-die rheometer on a star-shaped butadiene/styrene block copolymer (40% styrene by weight, $\overline{M}_w = 153,000$, $\overline{M}_w/\overline{M}_n = 1.24$). Above a critical shear stress, τ_{cr}, equal to 5.5×10^5 dyn-cm^{-2}, a power law behavior with the same flow index is observed at all temperatures. Below this critical stress, a marked curvature and discrepancy between data from slit dies of different depths are observed. For data above τ_{cr}, a flow activation energy at constant stress is $19\cdot2$ kcal-mol^{-1}.

E. Simple Elongational Flow and Supplementary Measurements (42,85)

The IUPAC group has reported that the elongational viscosity of the IUPAC SBS sample is tension thinning in the stress region 10^5–10^6 Pa at 150°C. Slow stretching of an extrudate at 150°C leads to necking. Die-swell measurements in this sample show a very low value at shear stresses less than approximately 10^5 Pa. Structural changes are investigated by flow birefringence in a slit capillary and indicate a structure transition at approximately 4×10^5 Pa. Furthermore, shear stress relaxation data and birefringence measurements show varying degrees of incomplete relaxation. Steady normal stress data could be obtained only for $\dot{\gamma} > 1 \ s^{-1}$. Transitional investigations of stress and normal stress data are reported in Ref. 86.

V. FRACTURE PHENOMENA

At high deformation rates (or high stresses), a transition from the fluid to a forced high elastic (rubbery) state (FHES) may take place in concentrated polymer melts. This concept is due to Vinogradov and coworkers (87–89). The lowest critical stress corresponding to a transition to FHES, which is a constant for each homologous polymer series, is equal to the maximum value of the loss modulus, determined from dynamic testing. A transition to FHES inevitably leads to fracture of the polymer or loss of adhesion to the surfaces of the measuring equipment and indicates the maximum production rate obtainable with conventional processing methods without melt fracture. Melt fracture phenomena may occur at the inlet to the capillary in a capillary rheometer or in the capillary at the wall and will (normally) result in distorted extrudates.

In capillary flow a transition to FHES for the material in the capillary is revealed by an abrupt change of slope to zero slope in the flow curve (log τ against log γ`). In a double logarithmic delineation of viscosity against shear rate, the spurt is depicted as a straight line with slope of -1. Vinogradov found a critical stress equal to 3.5×10^6 dyn-cm^{-2} for the SBS sample studied by the IUPAC group (44). Thus, from a processing point of view, this spurt phenomenon, which may be observed either in a delineation of shear flow data or in a delineation of dynamic data, is important, one reason being that it is the upper limits for the area where processing of items with smooth surfaces in extrusion may be produced.

The melt fracture phenomenon is often observed at smaller stresses than those corresponding to spurt fracture, where the fracture takes place near the capillary wall. This may be caused by fracture in the entrance zone to the capillary and triggered by the predominantly elongational flow that take place in tapered inlet zones.

The difference between the critical yield stress, τ_{cr}, and the spurt stress seems rather small for SBS samples. For tri- or multiblock copolymers where the yield stress is higher than the melt-fracture stress, a liquid state cannot be reached, at least in shear flows.

The lowest possible viscosities and consequently the easiest processing of a block copolymer are obtained under conditions where the material is close to, or in, a monomolecular melt state. Thus, fast shear rates (production rates) are advantageous in order to obtain structure breakdown. Transition into a forced high elastic state will usually limit the useful range of production rate.

VI. SUMMARY OF ESSENTIAL FEATURES

Earlier sections demonstrate that the rheological behavior of AB, ABA, and A_nB_n block copolymers with noncompatible blocks is considerably different from the behavior of homopolymers as well as random copolymers. The block copolymers are reported to have higher melt viscosities than random copolymers with the same composition and weight-average molecular weight, at least at low shear stresses. The viscosity of multiphase block copolymers does not approach a constant viscosity at low shear rates. Furthermore, abrupt changes of slopes in flow curves have been observed by many research groups.

Summing up: Regular block copolymers form steady-state structures that orient in flow fields. Cox–Merx's rule and master curve principles do not apply to microphase-separated systems. Generally, Newtonian behavior is not observed at low deformation rates. Time effects are very important, and different measuring geometries may give significantly different results, with the last result presumably due to different residence times.

The flow behavior of multiphase block copolymers is indeed complex. As described, microstructure changes do often take place during flow. The subject

Table 4 Structure Parameters and Processing Properties of Block Copolymers

Parameter	Parameter change	Processability
Compatibility (first approximation) $\mid\delta_A - \delta_B\mid$	$\Downarrow$	+
Number average degree of polymerization: $\bar{X}_n$	$\Downarrow$	+
Heterogeneity: H (Mw/Mn)	$\Uparrow$	+
Polymer concentration in solvents for both block components: c	$\Downarrow$	+

has important technological aspects, e.g., the development of a better understanding of the flow behavior of mixtures of block copolymers with blends of homopolymers in order to enhance the interfacial adhesion.

The (limited) knowledge that is available, especially from studies in simple flows, allows some guidelines for the development of new products and a better use of the existing ones.

The conclusions reached by Matzner and coworkers (50,51) that good processability requires $\mid \delta_A - \delta_B \mid < 1$, that optimal processing properties should be obtained for $\mid \delta_A - \delta_B \mid \approx 0$, and that at least one component should be crystalline are roughly in accord with the criterion $\tau_{cr} < \tau_{spurt}$.

Apart from minimizing $\mid \delta_A - \delta_B \mid$, an increasing heterogeneity, H, would tend to give better processing properties for constant $\bar{M}_w$ since τ_{cr} decreases (Eq. 2). Finally, decreasing molecular weight and possibly increased branching ease processing considerably, as has been known for many years.

Thus, in summing up, when developing new block copolymers or choosing block copolymers for given uses, the structure parameters given in Table 4 are the ones most sensitive from a processing point of view.

REFERENCES

1. J Lyngaae-Jørgensen. In: MJ Folkes; ed. Processing, Structure and Properties of Block Copolymers. Elsevier, New York, 1985, pp 75–123.
2. BM Walker, ed. Handbook of Thermoplastic Elastomers. Van Nostrand Reinhold, New York, 1979.
3. NR Legge, G Holden, HE Schroeder. Thermoplastic Elastomers. Hanser, New York, 1987.
4. JH Han, D Feng, C Choi-Feng, CD Han. Polymer, 36(1), 155, 1995.
5. GH Fredrichson, FS Bates. Annu. Rev. Mater. Sci., 26:501, 1996.
6. WP Cox, EH Merz. J. Polym. Sci., 28:619, 1958.
7. RB Bird, RC Armstrong, O Hassager. Dynamics of Polymeric Liquids. Vol. 1: Fluid Mechanics. 2nd ed. Wiley, New York, 1987.

8. CW Macosko. Rheology: Principles, Measurements and Applications. VCH, New York, 1994.
9. K Søndergaard, J Lyngaae-Jørgensen, eds. Rheo-Physics of Multiphase Polymeric Systems: Application of Rheo-Optical Techniques in Characterization. Techomic, Lancaster-Basel, 1995.
10. HH Hull. An Approach to Rheology Through Multi-Variable Thermodynamics. Society of Plastic Engineers, Pittsburgh, 1981.
11. HH Hull. The Thermodynamics of Rheology. 3rd ed. Society of Plastic Engineers, Brookfield, 1995.
12. CP Henderson, MC Williams. J. Polym. Sci. Polym. Letters, Ed., 17:257, 1979.
13. J Diamant, DS Soong, MC Williams. In: WJ Bailey, T Tsuruta, eds. Contemporary Topics in Polymer Science. Plenum, New York, 1984, pp 599–627.
14. M Doi, JL Harden, T Ohta. Macromolecules, 26:4935, 1993.
15. DF Leary, MC Williams. J. Polym. Sci., B., 8:335, 1970; J. Polym. Sci., Phys., 11: 345, 1973; 12:265, 1974.
16. S Krause. Macromolecules, 3:84, 1970.
17. DJ Meier. J. Polym. Sci., C26:81, 1969.
18. E Helfand. Macromolecules, 8:552, 1975; Rubb. Chem. Technol., 49:237, 1976.
19. E Helfand, Z Wasserman. Polym. Eng. Sci., 17:582, 1977.
20. E Helfand, Z Wasserman. Macromolecules, 9:879, 1976; 11:960, 1978: 13:994, 1980.
21. L Leibler. Macromolecules, 13:1602, 1980.
22. J Lyngaae-Jørgensen. Macromol. Chem., Macromol. Symp., 29:109, 1989.
23. J Lyngaae-Jørgensen. ACS Symposium Series, No. 395. American Chemical Society, Washington, DC, Chap 6, pp 128–152, 1989.
24. KA Koppi, M Tirrel, FS Bates, K Almdal, RH Colby. J. Phys. II, 2:1941, 1992.
25. U Wiesner. Macromol. Chem. Phys. 198:3319, 1997.
26. T Tepe, MF Schulz, J Zhao, M Tirrel, FS Bates. Macromolecules, 28:3008, 1995.
27. SS Patel, RG Larsson, KI Winey, H Watanabe. Macromolecules, 28:4313, 1995.
28. V Gupta, R Krishnamoorti, JA Kornfield, SD Smith. Macromolecules, 28:4464, 1995.
29. Y Zang, U Wiesner, HW Spiess. Macromolecules, 28:778, 1995.
30. BL Riise, GH Fredrickson, RG Larson, DS Pearson. Macromolecules, 28:7653, 1995.
31. CY Ryu, MS Lee, DA Hajduk, TP Lodge. J. Polymer Sci., B, Polym. Physics, 35(17):2811, 1997.
32. ME Cates, ST Milner. Phys. Rev. Lett., 62:1856, 1989.
33. KA Koppi, M Tirrel, FS Bates. Phys. Rev. Lett., 70:1449, 1993.
34. FS Bates, KA Koppi, M Tirrel, K Almdal, K Mortensen. Macromolecules, 27:5934, 1994.
35. SD Hudson, KR Amundson, HG Jean, SD Smith. MRS Bull. 20:42, 1995.
36. ME Vigild. Mesomorphic phase behavior of low molar mass PEP-PDMS diblock copolymers. PhD dissertation, University of Copenhagen, Denmark, October 1997 (Risoe-R.998(EU)).
37. FN Cogswell, DE Hanson. Polymer, 16:936, 1975.
38. J Lyngaae-Jørgensen, N Alle, FL Marten. Adv. Chem. Series, 176:541, 1979.
39. VC Nguen, AI Isayev, A Ya Malkin, GV Vinogradov, I Yu Kirchevskaya. Vysoko-mol. Soedin., A17:855, 1975.

40. KI Winey, SS Patel, RG Larsson, H Watanabe. Macromolecules, 26:2542, 1993.
41. EM Friedman, RS Porter. Trans. Soc. Rheol., 19:493, 1975.
42. A Ghijsels, J Raadsen. Pure Appl. Chem., 52 (5):3159, 1980.
43. A Ghijsels, J Raadsen. J. Polym. Sci. Phys., 18:397, 1980.
44. GV Vinogradov, VE Dreval, A Ya Malkin, Yu G Yanovsky, VV Brancheeva, EK Borisenkova, MP Zabugina, EP Plotnikova, O Yu Sabsai. Rheol. Acta, 17:2588, 1978.
45. KR Arnold, DJ Meier. J. Appl. Polym. Sci., 14:427, 1970.
46. G Kraus, FE Naylor, KW Rollman. J. Polym. Sci., A-2, 9:1839, 1971.
47. CI Chung, JC Gale. Polym. Sci. Phys., 14:1149, 1976.
48. S Futamura, E Meinecke. Polym. Eng. Sci., 17:563, 1977.
49. S Futamura. Effect of chemical structure of center block on physical and rheological properties of ABA block-copolymers. PhD dissertation, University of Akron, Ohio, 1975.
50. M Matzner, A Noshay, JC McGrath. ACS, Div. Polym. Chem., Preprint, 14 (1):68, 1973.
51. A Noshay, JE McGrath. Block-Copolymers. Academic Press, New York, 1977, p 408.
52. EV Gouinlock, RS Porter. Polym. Eng. Sci., 17:535, 1977.
53. CJ Chung, MJ Lin. J. Polym. Sci. Phys., 16:545, 1978.
54. CJ Chung, HL Griesbach, L Young. J. Polym. Sci., Phys., 18:1237, 1980.
55. JM Widmaier, GC Meyer. J. Polym. Sci. Phys., 18:2217, 1980.
56. FN Cogswell, DE Hanson. Polymer, 16:936, 1975.
57. PF Erhardt, JJ O'Malley, RG Crystal. In: SL Aggarwal, ed. Block Copolymers. Plenum, New York, 1970, p 195.
58. Bi Le-Khac, L-J Fetters. Macromolecules, 9:732, 1976.
59. R Masuda, P Aroi. Review with 46 references according to Chem. Abstr., 95-133375, 1981.
60. T Masuda, M Kitamura, S Onogi. Kenshu-Kyoto Daigaku Nipon Kagaku Seu i Kenkvusho, 37, 15; Chem. Abstr., 95-188425, 1980.
61. HH Winter, DB Scott, W Gronski, S Okamoto, T Hashimoto. Macromolecules, 26:7236, 1993.
62. FA Morrison, HH Winter, W Gronski, JP Barnes. Macromolecules, 23:4200, 1993.
63. FA Morrison, HH Winter. Macromolecules, 22:383, 1989.
64. CD Han, J Kim, JK Kim. Macromolecules, 22:383, 1989.
65. G Holden, ET Bishop, NR Legge. J. Polym. Sci. C, 26:37, 1969.
66. NP Zoteyev, GM Bartenov. Vysokomol. Soedin, A20:1781, 1978.
67. M Enyiegbulam, DJ Hourston. Polymer, 22:395, 1981.
68. T Masuda, Y Matsumoto, S Onogi. J. Macromol. Sci. Phys., B17:265, 1980.
69. CL Jackson, KA Barnes, FA Morrison, JW Mays, AI Nakatani, CC Han. Macromolecules, 28(3):713, 1995.
70. FA Morrison, JW Mays, M Mathukumar, AI Nakatani, CC Han. Macromolecules, 26:5271, 1993.
71. FA Morrison, HH Winter. Macromolecules, 22:3533, 1989.
72. M Kotamura, M Ishida, T Masuda, S Onogi. Nippon Reoroji Gakkaishi, 9(70); Chem. Abstr., 95-133739, 1981.
73. JL Leblanc. Rheol. Acta, 15:654, 1976.

74. JL Leblanc. Polymer, 17:235, 1976.
75. DR Han, DA Rao. J. Appl. Polym. Sci., 24:225, 1979.
76. G Kraus, JT Gruver. J. Appl. Polym. Sci., 11:2121, 1967.
77. HE Railsback, G Kraus. Kautschuk und Gummi-Kunststoffe, 22(9):497, 1969.
78. VS Al'tzitser, LB Kandyrin, BN Anfimov, VN Kuleznev. Kuch. Rezina, No. 12:16, 1979.
79. ON Sarakuz, BN Timchenko, AG Sinaiskii, AG Makhmurov. Plast. Massy, No. 10:37, 1979.
80. G Ferrando, E Diani. Technol. Plast. Rubber Interface, 2nd Euro. Conf. Plast. Rubber Inst., (Prepr.), 1976, p. 1.
81. Z Horák, Z Krulis, F Vecerka, J Kovar. Macromol. Chem., Macromol. Symp., 56:161, 1992.
82. I Mathew, KE George, DJ Francis. Kautschuk und Gummi-Kunststoffe, 44(5):450, 1991.
83. PJ Hansen, GS Hugenberger, MC Williams. Proceedings 28th IUPAC Macromolecular Symposium, Amherst, MA, 1982, p 781.
84. PJ Hansen, MC Williams. SPE NATEC, Miami, October 25–27, 1982, p 268.
85. TS Ng. Rheol. Acta, 26:453, 1987.
86. CP Han. Macromol. Symp., 118:303, 1997.
87. GV Vinogradov. Polymer, 18:1275, 1977.
88. GV Vinogradov. J. Polym. Sci. Lett., 16:433, 1978.
89. GV Vinogradov, AY Malkin, VV Volosevitch. J. Appl. Polym. Sci. Appl. Polym. Symposia, 27:47, 1975.

15
Orientation Behavior in Axial Tension of Half-Melt Poly(ether-b-ester) Copolymers

Georg Broza and Karl Schulte
Technical University Hamburg-Harburg, Hamburg, Germany

I. INTRODUCTION

Multiblock copoly(ether-b-ester)s as thermoplastic elastomers (PEE) are based on semicrystalline poly(butylene terephthalate) blocks (PBT) and amorphous non-crystalline oxytetramethylene blocks (PTMO). The domains of the crystalline hard segments are interconnected by the soft matrix segments (1). The crystalline phase has the same structure as the α-form of poly(butylene terephthalate), whereas the amorphous phase is a mixture of the uncrystallized ester segments and the ether segments (2).

PEEs are used as an engineering thermoplastic elastomer due to their attractive combination of strength, cold-temperature flexibility, high entropy elasticity, creep resistance, melt stability, and high crystallization rates (3). The morphology and degree of phase separation have been found to be important to the strength and elasticity of the polymer (4).

In recent years, the structure and properties of these multiblock copolymers have been investigated using standard techniques such as differential scanning calorimetry (DSC) (4–5), small angle x-ray scattering (SAXS) (6,7), wide-angle x-ray scattering (WAXS) (8,9), birefringence measurements and various scattering techniques (10). However, only a few studies were performed using optical light microscopy (11) and transmission electron microscopy (TEM) (4,12) in connection with mechanical tests (1,10).

The results of all these investigations are unambiguous in multiblock copolymers, which have a two-phase morphology; nevertheless, as exhibited from electron microscopy, the details of the morphology differ, while fibrils, spherulites, and lamellae have been assumed (1,7,8,12,13). Both TEM and SAXS investigations on PBT/PTMO confirmed the previous conclusions for a lamellae morphology (1,14–16) over the entire composition range investigated. Transmission electron microscopy and SAXS studies generally suggest a morphology similar to a lamella habit, characteristic of semicrystalline homopolymers (1,2). The morphology depends on several factors, such as the nature and concentration of the crystallized hard segments, solvents, and the crystallization conditions. For example, Cella observed shish-kebab structures, during the crystallization of the melt under shearing (1). Further insight has been gained studying model poly(ether-*b*-ester) block copolymers (1,17).

The micro-multiphase structure with crystalline domains may have an application for high functional materials. Similar to semicrystalline polymers, the hard blocks of PEE copolymers crystallize under high strain with a degree of high orientation and a relatively high amount of phase separation without further treatments. Polymers, which have been strained during the manufacturing process, normally show both a high orientation and an increased crystallinity (18,19). With an external load acting on the molecules, it is possible to stretch them, respectively to move or shear them relative to each other (20,21).

It is the purpose of this chapter to study the effect of uniaxial orientation of PBT/PTMO films, the morphological features, the microphase separation of the hard/soft segments, and the degree of crystallinity of poly(ether-*b*-ester) copolymers. This was performed via an observation and recording of the supermolecular structure of highly oriented melt-spun PEEs (under various annealing conditions) using electron microscopy and correlating them with the properties determined by DSC, SAXS and WAXS.

The samples studied varied in their compositions, containing, between 50 and 80 wt.% hard segments and in terms of their morphological structures.

II. DESCRIPTION OF POLY(ETHER-*b*-ESTER) COPOLYMERS

A. Materials Description

Copoly(oxytetramethylene-*b*-butylene) terephthalate was used and synthesized by a two-stage process: transesterification followed by a polycondensation in the melt, as described in Refs. 22 and 23. The materials used in this investigation were three block copoly(ether-*b*-ester)s (PBT/PTMO) with three different hard-segment compositions. Some important characteristics of the three materials are summarized in Table 1.

Table 1 Characteristics of the Copolymers

Designation	χ_H	ω_H	$p_{n,H}$	$p_{n,s}$	T_m
El-1	0.95	0.80	20.6	14	217
El-2	0.86	0.60	7.7	14	210
El-3	0.82	0.50	5.1	14	197

χ_H, ω_H = mole fraction and weight fraction of hard segments; $p_{n,H}$ $p_{n,s}$ = average degree of polymerized hard segments and soft segments, respectively; T_m = highest observed melting point in DSC.

B. Preparation and Testing

Annealing of the thin films (already mounted on copper grids) was carried out on a hot stage (180°C for 15 s), annealing of the bulk specimens was performed at 180°C for 1 hour in an oven. They were fixed with both ends on a copper block.

The TEM observations were carried out with a Philips EM 400 T model with an acceleration voltage of 100 kV. In order to determine the morphology of the samples, bright-field, dark-field, and diffraction techniques were applied.

Nickel-filtered Cu-K radiation was obtained from a Philips PW 1730 generator operated at 40 kV and 35 mA. A Kratky camera, equipped with a position-sensitive detector (Braun), was used to record the SAXS data.

Thermal tests were carried out with loose and fixed ends of the samples with a Perkin-Elmer DSC-7. The heating rate in all cases was 10°C/min. Birefringence measurements were made with a ZEISS E tilting compensator in an optical microscope (Reichert and Jung, Polyvar) with polarized light (λ = 546 nm).

Tensile tests were performed on a Zwick 1445 tensile test machine, at a cross-head speed of 1 mm/min. The width of the specimens was 5 mm and the thickness between 50 and 60 µm. The gauge length was 30 mm.

III. UNORIENTED POLY(ETHER-*b*-ESTER) COPOLYMERS

Polarized optical microscopy has shown that the morphology of PBT/PTMO block copolymers is basically spherulitic (Fig. 1). Figure 2 shows an x-ray diffraction pattern (WAXS) from the unoriented specimen. Only a diffuse pattern was observed for this sample.

Transmission electron micrographs were prepared from the supplied granulates using the following method: Blocks of 0.5 × 2 × 20 mm were prepared by compression molding under a pressure of 120 bar and a temperature of 40°C above the melting point of the respective polymer, followed by a rapid quenching in ice water. Subsequently, stripes with a thickness of about 0.5 mm were cut from these blocks with an ultramicrotom (Reichert and Jung, at −80°C). Figure 3 shows the TEM micrograph. The dark parts represent the crystalline domains.

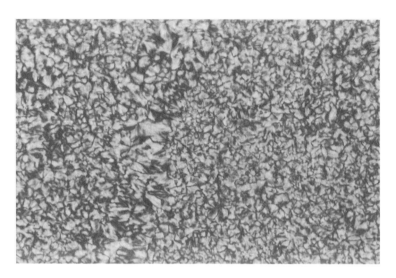

Fig. 1 Light micrograph of PBT/PTMO cross-polarized.

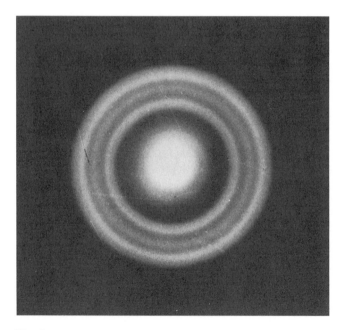

Fig. 2 WAXS of unoriented sample of PBT/PTMO.

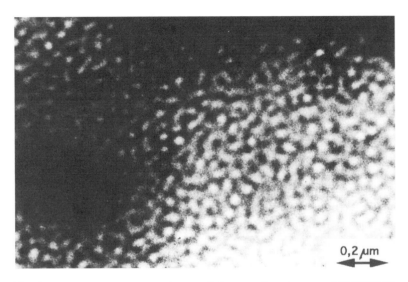

Fig. 3 Transmission electron micrograph from an ultramicrotomed PBT/PTMO sample.

The thin-film technique was developed by Petermann and Gleiter (24) using 1,1,1,3,3,3-hexafluor-3-propanol as a solvent. Spherical textures were observed in all samples. The light areas are the continuous crystalline phases, while the brighter regions represent the continuous amorphous matrix. The amorphous phase is a mixture of the uncrystallized hard segments and the elastomeric segments. Figure 4 shows the TEM micrographs of an El-2 type specimen.

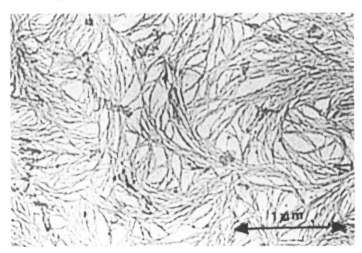

Fig. 4 Transmission electron micrograph of a PBT/PTMO film crystallized from 1,1,1,3,3,3-hexafluor-3-propanol.

IV. ORIENTATION IN AXIAL TENSION OF HALF-MELT

Very soft stretching of the copolymer PBT/PTMO causes differences in the spherulitic form of the hard segments. This can be noticed by comparing it with the neat PBT (25). At the initial stretching, the spherulites change into an ellipsoid form. The typical "Maltese cross" disappears; however, the center and the frames of the spherulites are still visible during the stretching procedure (Fig. 5).

A. Stretching of the Films (For the Structure Investigation)

For high-oriented copolymers, samples were prepared according to the method developed by Petermann and Gohil (26) as follows: A small amount of 0.3 wt. % PEE in a mixture of 1,1,1,3,3,3-hexafluor-3-propanol and 1,2-dichlorethane in a ratio of 1:2 vol. % was poured onto a hot glass plate ($T = 170°C$) so that the solvent could evaporate in a few seconds. The remaining thin polymer film then was picked up with a motor-driven cylinder before crystallization occurred (winding speed 10 cm/s), as shown in Fig. 6. The obtained highly oriented electron-transmissible film exhibited a thickness of about 100 nm.

B. Stretching of the Foils (For the Mechanical Tests)

The PBT/PTMO copolymers were produced in a hot press (Weber Co.; Germany) at 10°C above the melting point. From these foils, small stripes with a

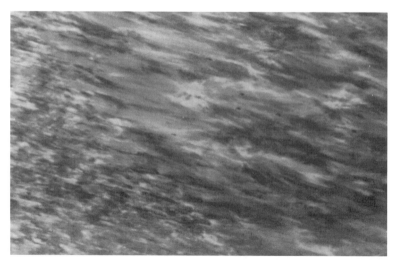

Fig. 5 Light micrograph of oriented PBT/PTMO cross-polarized.

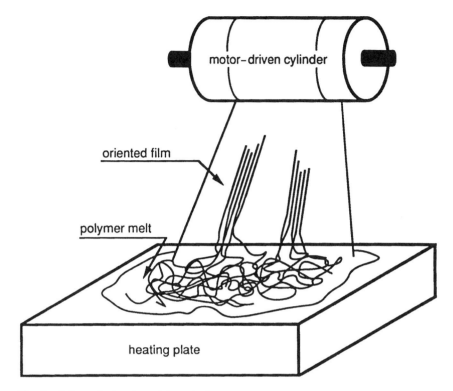

Fig. 6   Schematic drawing of the spinning apparatus.

width of 5 mm and a length of 50 mm were taken. The stripes were fixed in a stretch apparatus (Fig. 7) that consists of two rotating bars with diameters of 10 mm and 15 mm respectively. A small heating wire was located between the two bars (diameter 15 mm). In order to stretch it, the foil was wrapped around the two bars while gliding across the heating wire, where the foil is heated to the appropriate temperature. Because both bars have the same rate of rotation (10 per minute), the different diameters of the rotating bars produce a stretch rate of λ_v = 1.5 in each stretch cycle. The stretch rate was calculated by assuming that λ_v = L_v/L_o, where L_v is the length of the foil after stretching and L_o the initial length.

The total stretch rate of the foils investigated was: λ_v = 4.2 for El-1; λ_v = 5.2 for El-2; λ_v = 5.4 for El-3. The temperature during the stretching was in all cases 125°C.

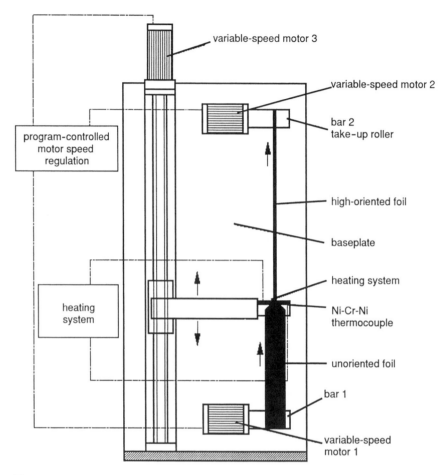

Fig. 7   Schematic of the stress apparatus.

V. MORPHOLOGICAL STRUCTURE

A. Crystalline Phase

The development of the microcrystalline structure at various annealing temperatures is shown in Figs. 8a–c (TEM bright-field graphs). A very fine-stripe structure is observed in the drawn state of the PEE samples. It arises from very small
crystalline lamellae, which are oriented mainly in the drawing direction. The inserted diffraction patterns (Figs. 9a–c (WAXS)) show a relatively high degree of
chain orientation in the crystalline blocks of PBT/PTMO copolymers. This can be
estimated from well-split equatorial reflections and an intense (100) reflection.

Certainly, with increasing fraction of soft segments, the diffraction reflexes are not as sharp as in samples with a high fraction of crystalline hard segments. With increasing annealing temperature, an increase in the lateral crystallite thickness of the lamellae can be detected, accompanied by an enhanced chain orientation, which can be deduced from the sharpening of the equatorial and meridional reflections. However, the formation of the structure is still maintained. A change in structure seems to occur at annealing temperatures above 150°C, in

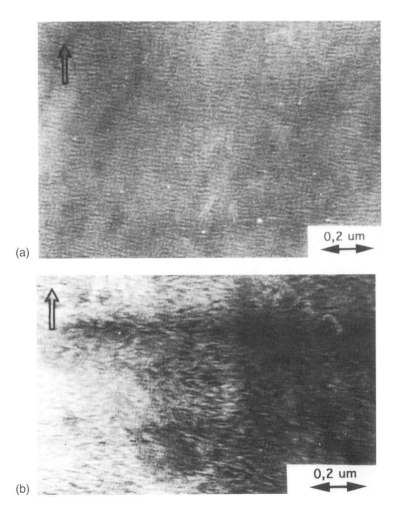

(a)

(b)

Fig. 8 Bright-field electron micrograph of melt-spun PBT/PTMO. (a) As-drawn state; (b) annealed at 150°C; (c) annealed at 180°C. *(figure continues)*

(c)

Fig. 8 *(continued)*

which the growth of the lateral size of the lamellae is visible (Fig. 8b). The diffraction pattern shows a nearly complete split of the (100) reflection, caused by a very high chain orientation in the drawing direction. Both, crystallite size and orientation reach their maximum at an annealing temperature of 180°C, as revealed in WAXS Fig. 9c.

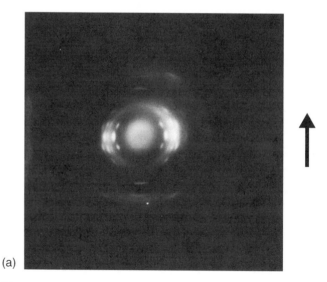

(a)

Fig. 9 WAXS of oriented samples (a) for El-1, (b) for El-2, (c) for El-3. The arrows indicate the direction of uniaxial extension.

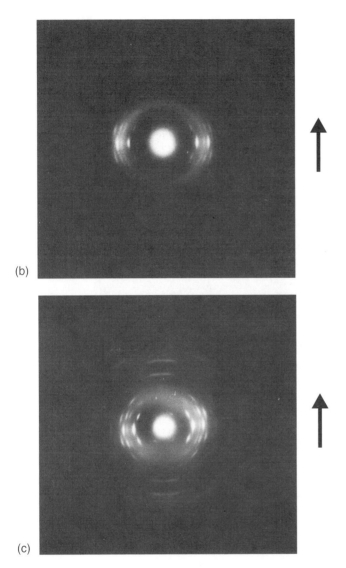

(b)

(c)

Fig. 9 *(continued)*

The diffraction pattern shows a nearly complete split of the (100) reflection, caused by a very high orientation in the drawing direction. Annealing of oriented samples causes an even higher intensity and increases the sharpness of diffraction reflexes. This demonstrates that both crystalline orientation and crystallinity increase. Simultaneously, the crystallinity of the samples drops.

To check the coherence of the lamellae structure (continuous crystals or laterally aligned individual micelle blocks), dark-field imaging was carried out using the (100) reflections. The results are shown in Figs. 10a and b.

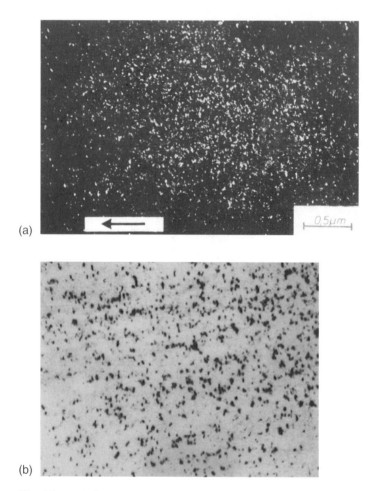

Fig. 10   Dark field electron micrograph using the (100) reflection. (a) As-drawn state; (b) annealed at 180°C.

The TEM and WAXS results lead to the following conclusions:

- In the as-drawn state, PBT/PTMO exhibits individual crystal segments, the lamellae.
- Annealing below 150°C leads to an increase in crystallite size and orientation. The lamellae structure is maintained.
- Annealing above 180°C produces a growth in thickness of the lamellae.
- No evidence for a crystallographic change (neither at $T_a = 150$°C nor at $T_a = 180$°C) is observed.
- The crystals consist of lamellae blocks formed during annealing in the crystallization and recrystallization process.

Figure 11 (SAXS) confirms the previous results and shows the monotonic increase in the long spacing period, L, with increasing temperature. The increase in spacing with temperature is attributed to a thickening of the lamellae, as postulated for crystalline homopolymers (7,17).

The increase in the thickness of the lamellae blocks, when annealing at 180°C, amounts to 68 nm for El-1, to 72 nm for El-2, and to 80 nm for El-3. The increase in the long spacing may result from the thickness of the crystalline phase or the change of the amorphous phase.

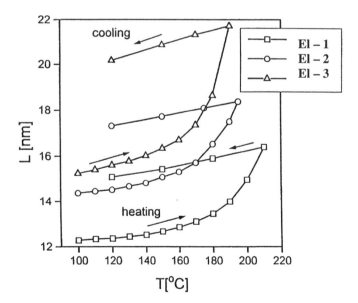

Fig. 11   Long spacing, L, from desmeared SAXS curves vs. the temperature of annealing.

Differential scanning calorimetry reveals the differences occurring in the crystallized phase with changing temperatures. When annealing the samples for 1 hour at 30°C, 20°C, and 10°C below the melting point (Fig. 12), the small, imperfect, and less thermally stable lamellae melt first as long as they do not exceed the critical dimension.

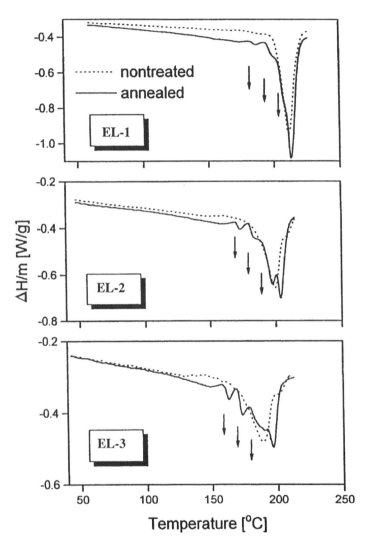

Fig. 12 DSC thermograms for the nontreated and the 1-hour-annealed samples (30°C, 20°C, 10°C below the melting point).

Such phenomena of partly melted samples provide the growth of stabilized lamellae larger than the critical ones. With an increasing amount of soft segments of PBT/PTMO, the number of small lamellae increases, but this increase is restricted by the balling up of the macromolecules of the amorphous phase. The DSC curves show thermal effects shortly before the melting point of PEE. Enthalpy variations are most pronounced for sample El-3 and less for sample El-1, while sample El-2 is in between.

The intensity, taken as the integral below the curve of the electron scattering force, depends mainly on the degree of crystallinity, w_c and the square of the electron density difference: $Q = w_c (\Delta\rho)^2$. With an increasing number of soft segments in the copolymer, the electron scattering force is more intense, which causes a higher electron density in the meridian direction. The force of electron scattering Q also increases with increasing temperature, because imperfectly trained crystals melt. This causes the increase in electron density, as shown in Fig. 13.

Such morphology leads to an increase in optical birefringence, as summarized in Table 2. The measurements were carried out in in situ experiments in which the birefringence was measured from the same area of one specimen but at the various temperatures. Table 2 indicates a slight increase in birefringence at an

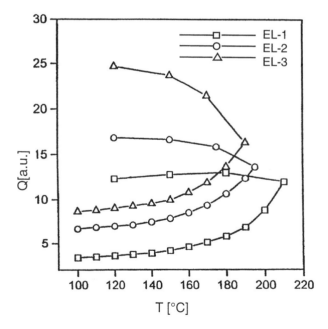

Fig. 13 SAXS curves (Q/T): scattering force vs. the temperature of annealing.

Table 2 Optical Birefringence Measurements

	$\Delta n \times 10^{-3}$		
Sample	$T = 20°C$	$T = 150°C$	$T = 180°C$
El-1	135.7	140.2	150.3
El-2	121.4	130.5	142.9
El-3	115.7	123.3	130.1

annealing temperature of 150°C, which may result from a slight alignment of the lamellea blocks in the orientation direction. In the range from 150°C to 180°C, an even higher orientation is formed. For comparison, see the diffraction patterns of Figs. 9a–c. Optical birefringence measurements are most pronounced for sample El-3 and less for sample El-1, while sample El-2 is in between.

B. Amorphous Phase

Previous investigations (measurements) have been carried out in order to obtain additional information about the amorphous regions. In order to describe the relaxation process in the amorphous phase of the copolymers and to discover the possible differences in the thermal effects between the relaxed amorphous phase and the high-oriented crystalline phase, DSC thermograms were made with copolymer films: in a strained phase (which means with fixed ends) and in a nonstrained (not fixed) phase, where they had free ends. In the samples with fixed ends, there is no possibility of shrinkage during the relaxation of the macromolecules. In the case with free ends, shrinkage is possible during relaxation of the macromolecules. Such differences in the energetic conditions result in, respectively, higher end lower melting points. Figure 14 shows small differences in the specific melting points of just 2°C. This shows that the connecting macromolecules in the amorphous phase are absolutely relaxed.

C. Morphological Model

The results lead to a morphological model, as shown in Fig. 15. The oriented films of copolymers in the as-drawn states have an equiaxial orientation of the lamellae, a statistical size distribution, and a nematic arrangement. The crystals are connected by short tie molecules. Figure 15a shows a schematic picture documenting the features of the proposed morphology. During annealing, melting starts from the small thermodynamically unstable lamellae (partially melted, Fig. 15b), enabling the better trained lamellae (larger in size) to rearrange from groups into blocks (about four lamellae create one block, Fig. 15c). The crystallographic c-direction of the lamellae is nearly perpendicular to the drawing direction. This new

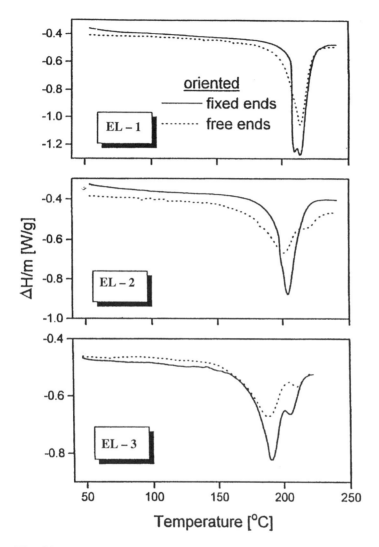

Fig. 14 DSC curves of the copolyether esters. The DSC thermograms lead through copolymer films, in a strained phase, which means fixed endings, whereas in an unstrained phase area, they have free endings.

a)   b)

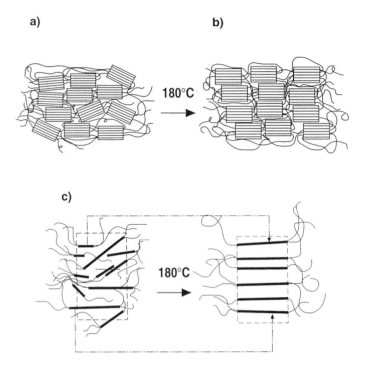

Fig. 15 Morphological model representing amorphous and crystalline regions of PEE annealed below and above 180°C, respectively (deduced from TEM investigations).

reorganization at a temperature of about 180°C can be compared to the smectic orientation, similar to LCP, observed in Ref. 27. The morphological model results in a better separation of the crystalline phase from the amorphous phase.

VI. MECHANICAL BEHAVIOR

A. Tensile Tests

Figure 16a shows the results of tensile tests on the as-drawn highly oriented specimens. With increasing amount of soft segments one can observe an increase in the strain to failure and a decrease in Young's modulus E. In the case of the as-drawn specimens, the dimensions of the crystallites, respectively its aspect ratio is so low that the morphology can be understood to consist of lamellae crystallites rather than a netpoint. (However, netpoints are not existent in this type of copolymer.) The deformation process follows more that of a network, which is elastomer elastic behavior. The relaxation of the soft segments in the network is oriented into

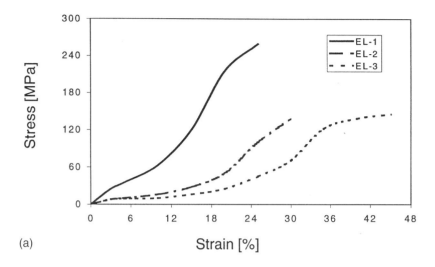

(a) Strain [%]

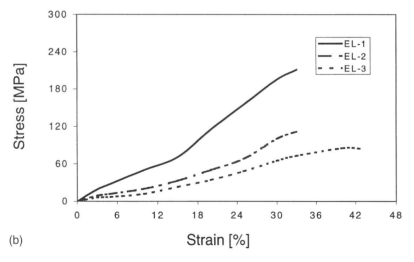

(b) Strain [%]

Fig. 16 (a) Stress–strain diagram of high-oriented as-drawn PBT/PTMO specimens. (b) Stress–strain diagram of the annealed specimens.

the direction of stretching (28). After release of the external load, the amorphous segments return to the energetically favorable entagled form (entropy elasticity) in which they had been before.

The stress–strain curve of the annealed specimens (Fig. 16b) is essentially different from that of the non-annealed ones. They show a nearly linear increase in stress and strain during loading. Two explanations are possible:

1. Due to the high orientation of the hard segments within the specimen, one can assume that the thermoplastic elastomers behave similar to uni-axial short-fiber-reinforced composites. Their behavior can be de-scribed with Eq. (1) (isostrain model), which is simply the rule of mix-tures and describes the influence of the constituents, depending on their volume fraction:

$$E_{GS} = E_A V_A + E_B (1 - V_A) \tag{1}$$

where

E_{GS} = Young's modulus for the total system
E_A = Young's modulus for the A component
E_B = Young's modulus for the B component
V_A = volume fraction of the A component

2. With increasing temperature, a reduction of the shear stresses can be observed in the crystallite/matrix interface, because the stretched bind-ing molecules glide from the crystallite surface (the so-called "Mullins effect") (29,30).

B. Cyclic Deformation

Figure 17(a) shows the behavior of cyclically deformed and extruded specimens, which are characteristic of all the investigated PBT/PTMO materials. The results of a cyclic loading before and after the thermal treatment are shown in Fig. 17(b) and (c). Figure 17(b) shows the deformation behavior of an El-1 specimen in the as-drawn state. The first load cycle was performed within the elastic range, which results 100% reversibility during unloading. One can observe that at each addi-tional load cycle a pronounced reduction in the initial Young's modulus occurs. This behavior is observed for all three types of specimens. As a possible explana-tion, we assume that the stretched binding molecules glide from the crystal sur-face. In the literature a similar effect has been described, the so-called "Mullins effect." This effect was observed mainly during cyclic loading of carbon-black filled elastomers. The typical cyclic loading and the strain response can be ob-served in Fig. 18.

Within the first cycle, the specimens were loaded until point A (curve 1) and than unloaded (curve 2). During the second loading, the rise in stress first follows

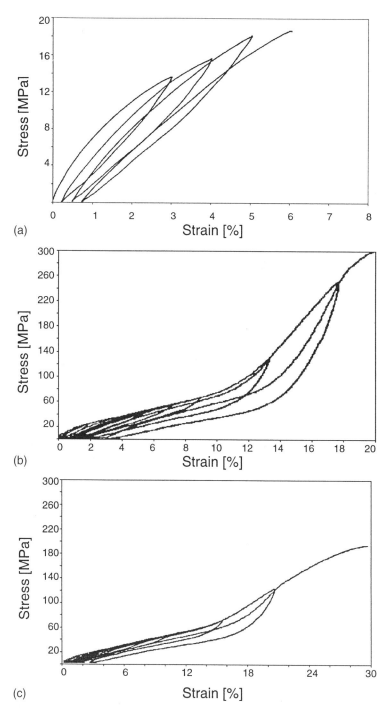

Fig. 17 (a) Cyclic deformation behavior of extruded specimens. (b) Cyclic deformation behavior of as-drawn specimens. (c) Cyclic deformation behavior of annealed specimens.

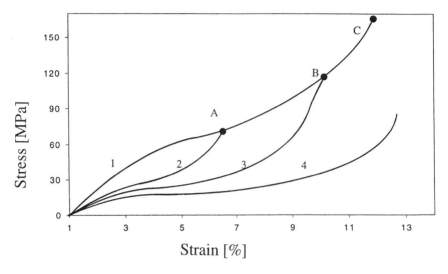

Fig. 18   Mullins effect schematically (From Ref. 29.)

curve 2, until it reaches its higher load level at point *B* and is then followed by a second unloading period (curve 3). The stress–strain curve follows during the consecutive third reloading curve 3 again, until it reaches point *C*. Such a Mullins effect type of deformation can be observed in all as-drawn high oriented copolymers.

In the consecutive load cycle is plotted the value of Young's modulus against the number of load cycles, one can observe the development of the overall deformation behavior [Fig. 19(a–c)]. During the first load cycle the relaxed molecules are stretched, which leads to the observed high Young's modulus. During further loading, Young's modulus first decreases until the maximum deformation capability of the matrix is reached, and then remains on that level because of the continuous stretching of the crystalline hard segments and of the strong intramolecular covalent bonding. Due to the relaxed and entangled molecules, the deformation occurs at the beginning, mainly within the soft segments. The molecules remain to be connected to the hard segments. In each consecutive load cycle the molecules separate from the hard segments and curl up during the unloading cycle. The possibility of load transfer via the connecting molecules is reduced and so therefore is the Young's modulus. This connection or separation can be observed not only during cyclic loading but also during a tensile test. However, it is more pronounced after annealing.

In the annealed specimens, the reduction in Young's modulus after each load cycle is not as pronounced [compare with Fig. 19(a–c)]. This disconnection at high strain rates in the stress–strain curve results in the so-called "strain-softening" effect. This has already been described by Petermann et al. (31) for the high-oriented PE needle morphology. This effect is different from that of the polyurethane block copolymers, which, at higher strain rates, show an increasing Young's modulus during loading. This is called the *strain-hardening effect* and can be explained by the strain crystallization of the soft segments (32).

After annealing at 180°C the imperfect lamellae partially melt and rearrange to thicker lamellae. At the same time, one can observe a rearrangement of the individual lamellae to blocks of lamellae. This results in a decreasing strength and a decreasing Young's modulus when compared to the as-drawn specimens.

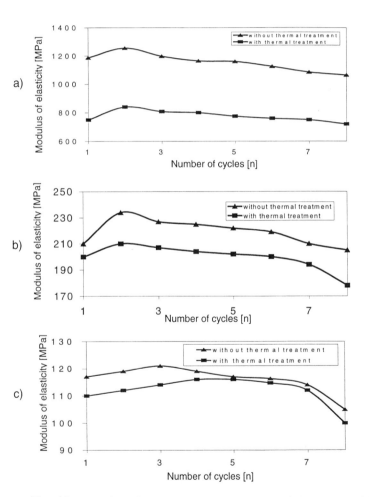

Fig. 19   Variation of Young's modulus as a result of the number of load cycles in as-drawn specimens and annealed specimens: (a) for El-1, (b) for El-2, (c) for El-3.

Table 3 Mechanical Data Calculated from Stress–Strain Curves

Specimen characteristic	Young's modulus (MPa)	Fracture stress (MPa)	Fracture strain (%)
El-1, unannealed	1175	266	23.5
El-1, annealed	775	195	28.4
El-2, unannealed	211	148	32.8
El-2, annealed	209	119	35.6
El-3, unannealed	121	141	46.5
El-3, annealed	112	99	47.9

Table 3 summarizes the results of thermally annealed and thermally anannealed high-oriented poly(ether-*b*-ester) block copolymers. After a thermal treatment, the maximum stresses and Young's modulus are reduced.

C. Elastic Deformation (Reversibility)

As a result of the cyclic deformation experiments, the following conclusions can be drawn on the elastic deformation behavior of highly stretched specimens. The reversibility R follows Eq. (2):

$$R [\%] = \frac{(\varepsilon_G - \varepsilon_P)}{\varepsilon_G} \times 100 \tag{2}$$

$$\varepsilon_G - \varepsilon_P = \varepsilon_E$$

where ε_G is the total strain, ε_P is the plastic strain, and ε_E is the elastic strain. Figure 20 schematically shows how the reversibility was calculated.

Figure 21(a) shows the reversibility of as-drawn specimens dependent on the number of load cycles. After a total strain of about 35%, the El-3 sample still has a reversibility greater than 95%. After a strain value of 20%, specimens with a lower amount of soft segments (El-1) still achieve a reversibility of 90%. Specimens annealed at 180°C have an overall 5% higher reversibility [Fig. 21(b)].

VII. SUMMARY

The morphology of multiblock copolymer poly(ether-*b*-ester) (PBT/PTMO), based on PBT as the hard and PTMO as the soft segments in uniaxially oriented films, were studied by transmission electron microscopy (TEM), differential scanning calorimetry (DSC), and small- and wide-angle x-ray scattering (SAXS and WAXS). Ultra-thin films of copolymers, approximately 100 nm thick, were pre-

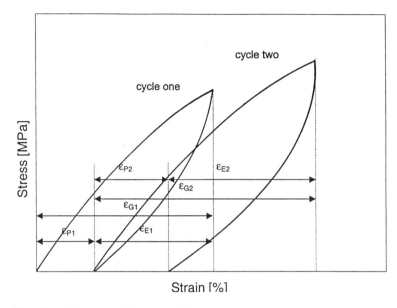

Fig. 20   Schematic of the reversibility.

pared by a special drawing technique, which provides a high extensional flow gradient and a supercooling during crystallization.

Our TEM studies of the PBT/PTMO showed supermolecular structures of different poly(ether-*b*-ester)s with a hard-segment content between 50 and 80 wt. %. The oriented films of copolymers in the as-drawn states have an equiaxial orientation of the lamellae, a statistical size distribution, and a nematic arrangement. Thermal treatment yields an increase in thickness of the crystal segments of the lamellae and their lateral reorganization. The annealing, realized on highly oriented samples with fixed or free ends, created small differences in the specific melting points, for the macromolecules in amorphous phase are absolutely relaxed. An optimal drawing temperature of about 170°C was found to provide the highest degree of crystal orientation. Annealing at a temperature of 180°C for 1 hour led to the transformation of the lamellae from the nematic to the smectic structural order. The morphology is transformed into lamellae crystals, rather then into an agglomeration of individual crystalline blocks, which remain a lamellae substructure (about four lamellas per block).

The mechanical properties determined during tensile tests show the behavior of a polymer filled with hard particles. Fibre reinforced composite materials with a uniaxial orientation of the fibers and a load parallel to the fibers can easily be described by the isotrain model; one can assume that it is also possible to de-

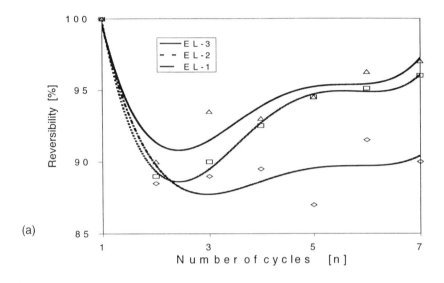

(a)

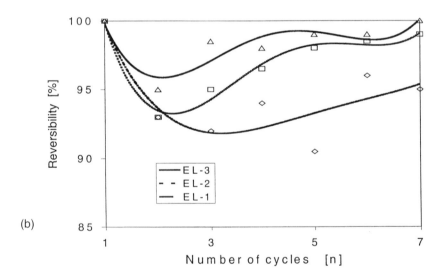

(b)

Fig. 21 Variation of the reversibility due to the number of load cycles: (a) as-drawn specimens; (b) annealed specimens.

scribe the behavior of PBT/PTMO block copolymers with oriented hard segments. It can be especially observed that the dependence of mechanical properties, such as Young's modulus, maximum stress, and fracture strain, follow the morphological structures. The stretched morphology of the block copolymers can be described simply by a model of a composite with blocks of lamellae of PBT as hard segments incorporated into the amorphous matrix. The results of these tests could be correlated with the individual microstructure and the model proposed.

Tensile tests and cyclic tensile tests gave valuable informations about the plastic and elastic behavior of as-drawn and annealed specimens. The results on the cyclic deformation of all investigated poly(ether-b-ester) block copolymers show that the first deformation process was finished within the elastic range, which can be taken from the 100% reversibility during the consecutive unloading. The high reversible-deformation capability of the specimens, higher than 90%, is dependent on the ratio of soft to hard segments.

REFERENCES

1. RJ Cella. J. Polym. Sci. Polym. Symp. 42, 727, 1973.
2. U Bandara, M Dröscher. Colloid Polym. Sci. 261, 26, 1983.
3. P Stahl. Kunststoffe. 82, 749, 1992.
4. WH Buck, RJ Cella, JR, EK Gladding, JR Wolfe. J. Polym. Sci. Symposium No. 48, 47–60, 1974.
5. HW Hässlin, M Dröscher, G Wegner. Makromol. Chem. 179, 1373–1376, 1978.
6. G Wegner, T Fujii, W Meyer, G Lieser. Angew. Macromol. Chem. 74, 295, 1978.
7. S Fakirov, T Gogeva. Makromol. Chem. 191, 603, 1990.
8. MA Vallance, SL Cooper. Macromolecules 17, 1208, 1984.
9. RM Briber, EL Thomas. Polymer 26, 8, 1985
10. A Lilaonitkul, SL Cooper. Paper presented at a meeting of the Rubber Division, American Chemical Society, Minneapolis, Minnesota, 1976.
11. LL Zhu, G Wegner. Makromol. Chem. 182, 3625–3638, 1981.
12. R Bill, M Dröscher, G Wegner. Macromolekules 18, 1727, 1985.
13. G Perego, M Cesari, G della Fortuna. J. Appl. Polym. Sci. 29, 1141–1155, 1984.
14. E Sorta, G della Fortuna. Polymer 21, 728, 1980.
15. A Biggi, G della Fortuna, G Perego, L Zotteri. Kautsch. Gummi Kunststoffe, 34, 349, 1981.
16. S Fakirov, C Fakirov, EW Fischer, M Stamm, AA Apostolov. Colloid Polym. Sci. 271, 811–823, 1993.
17. RW Seymour, JR Overton, LS Corley. Macromolecules 8, 331, 1975.
18. AG Gibson, IM Ward. J. Mater. Sci. 979, 1980.
19. G McCrum, CP Buckley, CB Bucknall. Principles of Polymer Engineering. Oxford University Press, Oxford, 227, 1980.
20. PJ Flory. Principles of Polymer Chemistry. Cornell University Press, Ithaca, NY, 1971.

21. A Keller. Structure–Property Relationship of Polymeric Solids. Ed. Hiltner. Plenum Press, New York, 1983, p. 25.
22. A Szepke-Wrobel, J Slonecki, H Wojcikiewicz. Polimery 27, 400, 1982.
23. R Ukielski, H Wojcikiewicz. Polimery 23, 48, 1978.
24. J Petermann, H Gleiter. Progr. Colloid & Polym. Sci., 64, 122, 1978.
25. L Zhu, G Wegner, U Bandara. Macromol. Chem. 182, 3639, 1981.
26. RM Gohil, J Petermann. J. Pol. Sci. Pol. Phys. Ed. 17, 525, 1979.
27. G Vertogen, WH de Jen. Thermotropic Liquid Crystals, Springer Verlag, New York, 1988.
28. IM Ward. Mechanical Properties of Solid Polymers. Wiley, Chichester, 1983, p. 62.
29. L Mullins. Rubb. Chem. Technol. 42, 339, 1969.
30. JM Schulz. Polymer Material Science. Prentice Hall, Englewood, Cliffs, NJ, 1974, p. 328.
31. J Petermann, RM Gohil, M Massud, D Göritz. J. Mater. Sci. 17, 100, 1982.
32. CHM Jackques. Polymer alloys: blends, blocks, grafts and interpenetrating network. In: D Klempner, KC Frisch, eds. Polymer Science and Technology. Plenum Press, New York, 1977, p. 10.

16
Polyurethane Grafted onto Styrene-Styrene Sulfonic Acid Copolymers

Maria Rutkowska and Mariola Jastrzębska
Gdynia Maritime Academy, Gdynia, Poland

Helena Janik
Technical University of Gdańsk, Gdańsk, Poland

I. INTRODUCTION

Miscibility of polyurethane (PU) with lightly sulfonated polystyrene (PSSSA) in blends has been the subject of several investigation (1,2). In early studies, a proton-transfer mechanism from the PSSA to the hard segment of the PU (which contained a tertiary amine as part of the chain extender) was postulated in analogy with other polymer mixtures employing similar miscibility enhancement techniques (2). The ionic interaction between the sulfonate anions on the styrene and the cations on the PU enhanced the miscibility of the system. More recently, nuclear magnetic resonance (NMR) of the blends of the same PU with PSSSA as well as model compounds in DSMO solutions, along with two-dimensional spectra, showed that the labile protons of the SSA groups were transferred preferentially to the secondary structures (allophanates), followed by the urethane nitrogens and finally by the tertiary nitrogens of the N-methyldiethanoloamine (MDEA) (3).

Since the miscibility enhancement mechanism in these blends is based on interactions between sulfonate anions and several different cations, it was thought convenient to study a system in which the cation was located only on the tertiary amine rather than on the secondary structure. Consequently, a procedure was employed that involves stepwise grafting onto sulfonated polystyrene (PS) chains of MDEA, followed by chain growth on the hydroxyl sites of any polyurethane struc-

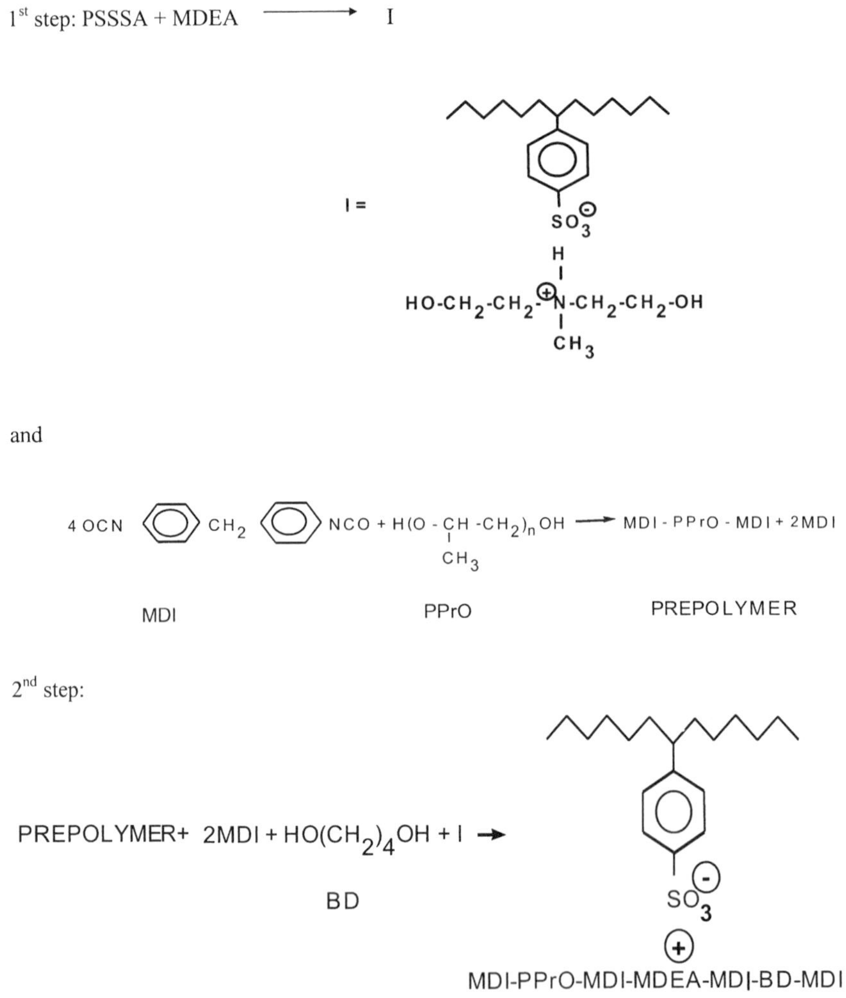

1St step: PSSSA + MDEA ⟶ I

$$I =$$

$$HO\text{-}CH_2\text{-}CH_2\overset{\oplus}{\underset{\underset{CH_3}{|}}{N}}\text{-}CH_2\text{-}CH_2\text{-}OH$$

and

$$4\,OCN \bigcirc CH_2 \bigcirc NCO + H(O\text{-}\underset{\underset{CH_3}{|}}{CH}\text{-}CH_2)_n OH \longrightarrow MDI\text{-}PPro\text{-}MDI + 2MDI$$

MDI   PPro   PREPOLYMER

2nd step:

PREPOLYMER + 2MDI + HO(CH$_2$)$_4$OH + I →

BD

MDI-PPro-MDI-MDEA-MDI-BD-MDI

Scheme 1

tures of interest (4). An example of this approach is shown in Scheme 1. Since this is a condensation polymerization, statistics will dictate the specific nature of the final products, and only the most probable structure is shown here. Further growth is obviously possible. This approach is of interest also because it allows the synthesis of a very wide range of grafts of PU on PS.

The aim of this chapter is to present the structure–morphology relationships of polyurethane grafted onto styrene-styrene sulfonic acid copolymers. Some potential applications of these grafts as compatibilizers will also be described.

II. SYNTHESIS

The PS (Mw = 39,000) was synthesized by radical polymerization and sulfonated by the procedure developed by Makowski et al. (5). The sulfonation was carried out at 50°C in 1,2-dichloroethane solution using a mixed anhydride of acetic acid and sulfuric acid as the sulfonating agent. Polymers containing 4.0, 4.8, 9.9, and 12.0 mol % of styrene sulfonic acid (SSA) were employed in this work.

The PU grafts on PSSSA were prepared by a condensation reaction shown in Scheme 1. 4,4'-Methylene bis (phenylisocyanate) (MDI) was purified by filtration of the liquid at 70°C. Poly(propylene glycol) (PPrO, Mw = 400, 425, 725, 1000, 2000) was dried under vacuum at 60°C for 5 h. N-Methyldiethanoloamine (MDEA), 1,4-butanediol (BD), stannous octoate, and anhydrous N,N-dimethylformamide (DMF) were used as received.

To obtain hard-segment (HS) grafts onto PSSSA, the PSSSA was dissolved in DMF at a weight concentration of about 40%. Then the MDEA was added in an amount needed to maintain an equimolar ratio with that of the SSA groups. The reaction was carried out at room temperature for 2 h. Then the MDI and BD in DMF (40% by weight) along with stannous octoate (0.1% by weight) were added in an appropriate quantity to maintain an NCO:OH molar ratio as shown in Table 1. The reaction was carried out at 70°C for 2 h. The solution was subsequently dried by evaporation of the solvent at progressively higher temperatures up to 80°C and kept under vacuum at that temperature for 1 week.

To obtain a PU with both hard and soft segments grafted onto PSSSA, in the first step of the reaction the prepolymer was prepared as follows: MDI was dissolved in DMF at a weight concentration of 40%. This solution was added to a solution of DMF, PPrO (40% by weight), and stannous octoate (0.1% by weight). The synthesis of the urethane prepolymer was carried out at 70°C for 2 h, maintaining an NCO:OH molar ratio of 3:1 or 4:1. Simultaneously, the PSSSA was dissolved in DMF at a weight concentration of about 40%, and the MDEA was added at an equimolar ratio to that of the SSA groups. The reaction was carried out at room temperature for 2 h to obtain the ammonium salt. In the last step of the reaction, the solution of ammonium salt (PSSSA-MDEA) and the BD were added to the PU prepolymer solution, maintaining the NCO:OH molar ratio shown in Table 1. The reaction was carried out at 70°C for 2 h. The solution was dried by evaporation of the solvent at progressively higher temperatures up to 80°C and kept under vacuum at that temperature for 1 week. The designation of the investigated polyurethane grafts and their chemical composition are shown in Table 1.

Proton NMR spectra were taken in DMSO-d_6 (100% deuterated) solutions. The concentration of the solution was approximately 5% (w/w). The assignments and methods used in the present study follow those of the previous publication (3). The NMR spectra confirmed that the protons of the SSA groups were transferred to the tertiary nitrogen of the MDEA, as was expected (6).

Table 1 Polyurethanes Grafted onto Sulfonated Polystyrene

| | | Glass transition temperature T_g (°C) | | | |
| | | Low | | High | |
Designation	Molar ratio of chemical components	DSC	DMTA	DSC	DMTA
PS-4.8SSA[a]					125
PS-4.8SSA-MDEA	SSA/MDEA = 1/1				86
PS-4.8SSA-HS1[b]	SSA/MDEA/MDI = 1/1/2				92
PS-4.8SSA-HS2	SSA/MDEA/MDI/BD = 1/1/3/1				105
PS-4.8SSA-HS2a	SSA/MDEA/MDI = 1/2/3				103
PS-4.8SSA-HS3	SSA/MDEA/MDI/BD = 1/1/4/2				113
PS-4.0SSA-PU1[c]	SSA/MDEA/MDI/BD/PPrO$_{400}$ = 1/1/4/1/1	−12		87	98
PS-4.0SSA-PU2	SSA/MDEA/MDI/BD/PPrO$_{725}$ = 1/1/4/1/1	−11	~−20	87	100
PS-4.0SSA-PU3	SSA/MDEA/MDI/BD/PPrO$_{1000}$ = 1/1/4/1/1	−13	−38	87	100
PS-4.0SSA-PU4	SSA/MDEA/MDI/BD/PPrO$_{2000}$ = 1/1/4/1/1	−12	−43	92	105
PS-12.0SSA-PU1	SSA/MDEA/MDI/BD/PPrO$_{400}$ = 1/1/4/1/1				102
PS-12.0SSA-PU4	SSA/MDEA/MDI/BD/PPrO$_{1000}$ = 1/1/4/1/1	−12	−32	105	120
PS-4.0SSA-PU5	SSA/MDEA/MDI/BD/PPrO$_{4000}$ = 1/1/3/1/1				90
PS-4.0SSA-PU6	SSA/MDEA/MDI/BD/PPrO$_{725}$ = 1/1/3/1/1	−12	~−30	77	92
PS-4.0SSA-PU7	SSA/MDEA/MDI/BD/PPrO$_{1000}$ = 1/1/3/1/1		−45		95
PS-9.9SSA-PU7	SSA/MDEA/MDI/BD/PPrO$_{1000}$ = 1/1/3/1/1		−38		110
BLEND	SSA/MDEA/MDI/PPrO$_{1000}$ = 1/2/3/1		−8		115
PS-9.9SSA/PU$_n$					

[a] PS-4.8SSA refers to polystyrene containing 4.8 mol % of styrene sulfonic acid.
[b] HS refers to structures containing only hard segments.
[c] PU refers to structures containing both hard and soft segments.

III. MORPHOLOGY

Dynamic mechanical properties and thereby also the phase behavior of polyurethane grafted onto styrene-styrene sulfonic acid copolymers were studied. Dynamic mechanical studies were performed between −100°C and +200°C at 1 Hz using a dynamic mechanical thermal analyzer (DMTA) from Polymer Laboratories at a heating rate of 1°C/min.

The glass transition temperature values of polyurethane grafted onto styrene-styrene sulfonic acid copolymers are shown in Table 1 (4). Figure 1 shows plots of E' vs. temperature for the HS grafts (see Table 1) with different lengths of attached hard segments. The plots show clear evidence of one-phase behavior as detected by mechanical techniques; i.e., a one-step decrease in the modulus is seen. The glass transition temperature (T_g) of the grafts is estimated from the loss tangent peak positions. Looking at the T_g column in Table 1 for the MDEA and HS grafts, it is seen very clearly that T_g values are strongly dependent on the length of attached hard segment. The PSSSA with only MDEA grafted onto it has the lowest T_g value, which increases with increasing length of the hard segment.

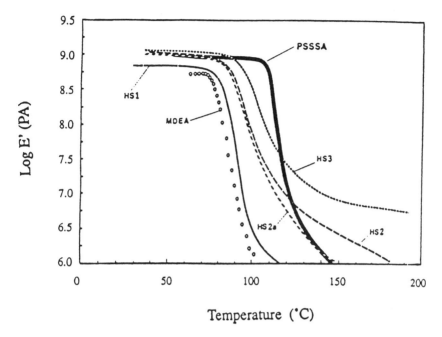

Fig. 1   Variation of the shear storage modulus E′ with temperature for PSSSA containing grafted polyurethane hard segments (HS) of different lengths (see Table 1 with PSSSA omitted). Experimental data are shown only for the MDEA samples; they are omitted in the other plots for the sake of clarity. (From Ref. 4.)

For samples PSSSA-HS2 and PSSSA-HS2a, in which the hard segments have the same length but a slightly different chemical structure, the T_g values are unexpectedly almost the same.

The morphology of HS grafts was investigated by means of transmission electron microscopy (TEM). The solvent-cast film (in DMF) was prepared. After casting (at temperature 70°C), the film was stained with RuO_4. Microscopic investigation of cast films indicates the presence of very small domains in the case when the graft consists of only hard segments of PU (Fig. 2). This suggests the existence of separate domains of PS and of domains of hard segments of PU (Fig. 2) (7). Thus we do not obtain the real miscibility in HS grafts on a nanometer scale. The results confirm that the TEM method is more sensitive for investigation of the morphology of complicated polymer alloys on the 10–50-nm scale than the DMTA method.

Figures 3 and 4 show the plots of $E′$ and tan δ for PU grafts with different lengths of the PPrO soft segment. For the grafts with the shortest soft segment ($M_w = 400$), only one glass transition temperature is observed.

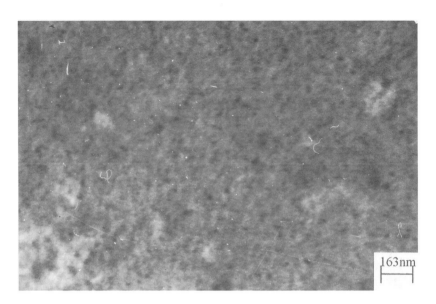

Fig. 2 Transmission electron micrograph of solvent-cast film PS-4.8SSA-HS1. (From Ref. 7.)

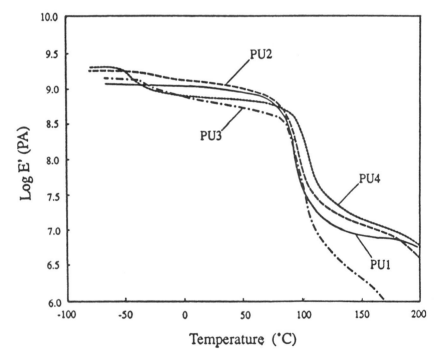

Fig. 3 Variation of shear storage modulus E' with temperature for the PSSSA with grafted PU segments (see Table 1 with PSSSA omitted). (From Ref. 4.)

440

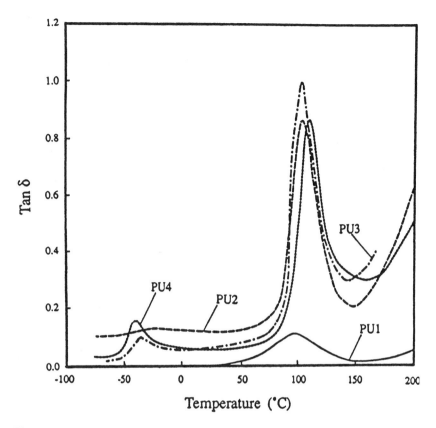

Fig. 4 Loss tangent (tan δ) plots for the same samples as in Fig. 3. (From Ref. 4.)

From Table 1, it is seen that the (high) glass transition temperature for the one-phase system increases with increasing SSA content of the PSSSA (compare PS-4.0SSA-PU1 and PS-12.0SSA-PU1) as well as with increasing length of the hard segment (compare PS-4.0SSA-PU1 and PS-4.0SSA-PU5). However, if a somewhat longer soft segment is used ($M_W = 725$), one begins to see the appearance of a second glass transition (see PS-4.0SSA-PU2 and PS-4.0SSA-PU6). When the soft segment length reaches $M_w = 1000$, the second T_g is seen very clearly in the mechanical tests (Figs. 3 and 4).

The TEM observations of carbon-platinum replica of surfaces of cryogenically fractured samples of the grafts with long soft segments show characteristic grainy aggregates on the fracture surface, indicating a heterogenous system. However, the grafts with short segments also have fine grains distributed homogeneously as separate elements (Fig. 5).

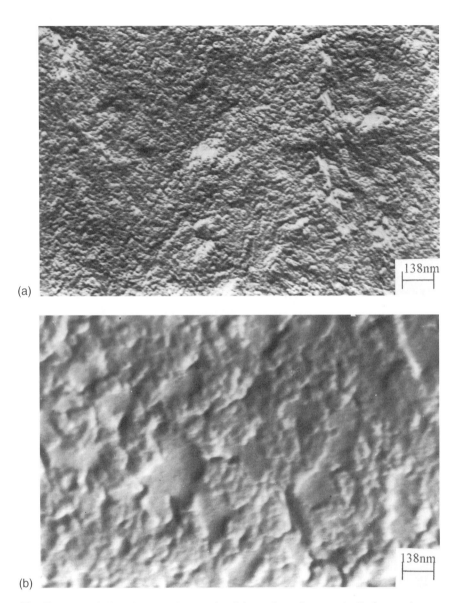

(a)

(b)

Fig. 5   Transmission electron micrographs of the surface of cryogenically fractured samples: (a) PS-4.0SSA-PU1; (b) PS-4.0SSA-PU3. (From Ref. 7.)

In microscopic observations of solvent cast film of grafts with soft segments, small domains rich in PS are distributed in a matrix richer in PU. The domains exibit the shape of short cylinders (18–27-nm width, 36–90-nm length) (Figs. 6a and b).

In regular PU, phase separation occurs when the soft segments exceed a certain length (M_w = 2000) (8). In the case of PU grafts, the soft segments of the PU already exhibit phase separation for soft segment lengths greater than 725. This shows that phase separation in the PU grafts is achieved more easily, presumably because of the single-point attachment. The positions of the low-temperature peaks are dependent on the length of the soft segment. With increasing soft-segment length, the low-temperature peaks move to a lower temperature. The same phenomena were observed for regular PUs and their blends (1).

Differential scanning calorimetry (DSC) at 20°C/min heating rate was also used to determinate the T_g of grafts. The DSC curves show the presence of two T_g in grafts with short and long soft segments and confirm their heterogeneity (9). In DSC investigations there is no relationship between the T_g of the soft phase and the length of soft segment of PU, what is visible in DMTA data. Looking at the data in Table 1 it is seen that the low T_g of the grafts appears faster in the DMTA technique than in DSC. However, the values of high T_g determined by DMTA are lower. This is due to the cohesion of hard domains.

The higher the sulfonate ion concentration in the PSSSA, the higher the T_g (compare PS-4.0SSA-PU4 with PS-12.0SSA-PU4 in Table 1). This finding further confirms the strong existing interactions between the PSSSA and hard segments of the PU. Unexpectedly, however, the low T_g of these systems is higher for higher SSA contents, even if the length of the soft segment is the same (compare PS-4.0SSA-PU4 with PS-12.0SSA-PU4). This suggests that in these syntheses either some of the hard segments that interact with SSA are incorporated into the soft phase or the domain sizes are smaller. Other possible reasons are not excluded.

The PU grafts prepared to yield a high-molecular-weight material with multiple points of attachment per PU chain via the MDEA units were also investigated. Ratios of 1/1/3/1/1 of SSA/MDEA/MDI/BD/PPrO yield, in principle, PUs of very high molecular weight that are attached at multiple sites to the PSSSA chains. As before, soft-segment lengths as low as 725 yield systems that are heterogenous. An increase in the ion concentration of the PSSSA increases both the high and the low T_g (compare PS-4.0SSA-PU7 with PS-9.9SSA-PU7 in Table 1).

Finally, the T_g value of the PS-9.9SSA-PU7 graft was compared with that of the blend of PU_n and PSSSA containing 9.9 mol % sulfonate ions (see the last position in Table 1 and Fig. 7). In the PU_n all the BD is replaced by MDEA. PU_n was obtained by a two-step condensation reaction. The blending at stoichiometric ratio of SSA to tertiary amine was performed by mixing solutions of the high-molecular-weight materials (1).

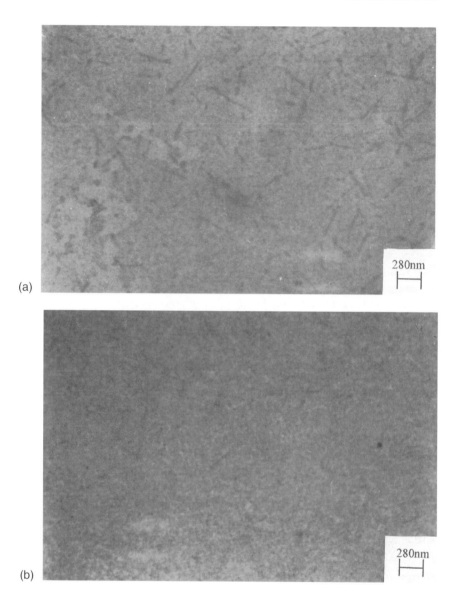

Fig. 6 Transmission electron micrographs of solvent-cast films: (a) PS-4.0SSA-PU1; (b) PS-4.0SSA-PU3; (c) PS-6.6SSA/PU3 blend; (d) PS-6.6SSA/PU7 blend. (From Ref. 7.) *(figure continues)*

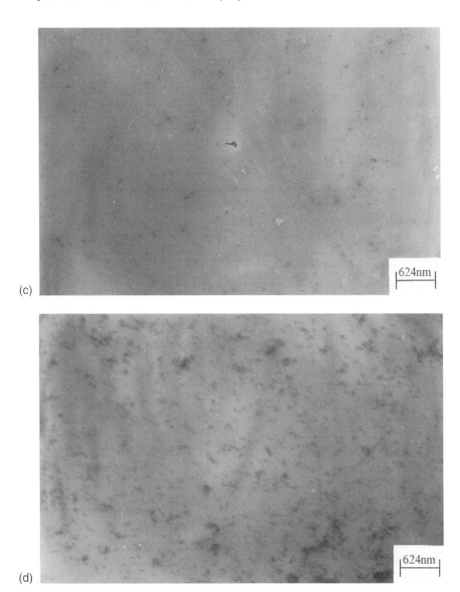

(c)

(d)

Fig. 6 *(continued)*

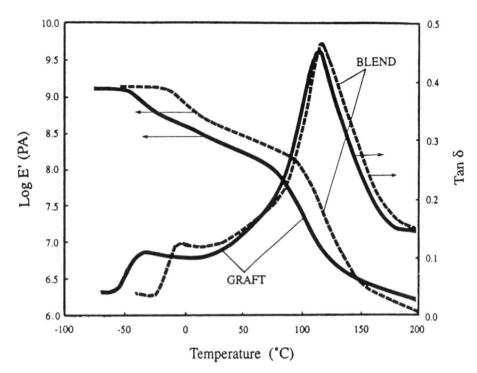

Temperature (°C)

Fig. 7   Variation of shear storage modulus E' and of loss tangent (tan δ) with temperature for the PS-9.9SSA-PU7 graft and a blend of polymers with similar composition. (From Ref. 4.)

It appears that the upper T_g is very similar for both materials, while the low T_g is much lower for the graft than for the blend, suggesting that the phase-separated (low-T_g) regions are purer. This is not unreasonable.

The PU_n used for blending consists of a mixture of soft and hard segments. After blending with the PSSSA, ion interactions occur between the NH groups of secondary structures, as well as the urethanes and tertiary nitrogens of the MDEA (3). These strong coulombic interactions result in a very intimate mixing of the PSSSA copolymer with the hard segments, which in turn leads to the exclusion of the soft segments into a separate phase (Fig. 7). This exclusion, however, is not complete because of the strong coulombic interactions with multiple sites and the statistical nature of the PU. On the other hand, in the grafts the interactions occur only with the tertiary amine of the MDEA, and since the PU is formed after these interactions have occurred, a higher degree of phase purity can be expected. Obviously, this will yield a lower T_g of the low T_g phase.

If the morphology of graft is compared with that of the blend (Figs. 6b and c), the lower degree of phase purity in the blends can also be seen. The grafting method yields materials in which the phase purity in phase-separated systems seems to be better than in the regular blends of similar composition. In the blends, the small black spots richer in PS (size of 19–58 nm) are scattered in a matrix richer in PU. The distance between those spots are in the range of 100 nm. A smaller amount of aggregates of PS (Fig. 6c) is observed for blends with longer hard PU segments (compare Fig. 6c with Fig. 6d). More PS is mixed homogeneously with the PU matrix because of stronger ionic interactions.

IV. APPLICATION

The arms of a graft copolymer are chemically different from the main chain of the copolymer. Thus, some of it could be miscible with one polymer blend component and some with the other (10). As a result, the graft copolymer was used as a compatibilizing agent for blends of regular polystyrene and polyurethane. This system is not miscible from a thermodynamic point of view (11).

Polyurethane grafted onto styrene-styrene sulfonic acid copolymers (PS-4.0SSA-PU1) was used as compatibilizer (C) (0.2% w/w) for the blend PS/PU (90/10% w/w). The blend was obtained by means of injector Anker 8A25 at a temperature 190–210°C. The morphology of regular PS/PU blend and the blend with graft was studied by means of electron microscopy and DSC methods.

The regular PS/PU blend exhibits a two-phase separation structure, with a very clear interface between the polystyrene matrix and the dispersed polyurethane phase, showing spherical globules of diameter $d = 0.65$ μm (Fig. 8). The globules are very weakly bound to the PS matrix, and in some places they can be seen after pulling out the globules from a PS. On the DSC curves, two T_g values for the blend PS/PU can be observed—one for PU about $-30°C$ and the second one for PS about 98°C (Fig. 9). This confirms the idea of a two phase system.

When the compatibilizer is added to the system, the morphology changes from a two-phase system to a multiphase structure. A new phase in the form of the globules appears. These new globules are of higher diameter, $d = 0.74$ μm. They are probably composed of a mixture of PS and PU. The adhesion of PS-PU globules is much better to the PS matrix than in the blend without compatibilizer. This morphology is confirmed by the DSC measurement. After using a compatibilizer, the additional glass temperature of about 50°C on DSC curves is observed. The T_g of polystyrene has not changed, but the signal is much weaker, which may be related to a higher miscibility of polystyrene with polyurethane.

In Table 2 some mechanical properties of PS/PU blend are presented. Looking at the data it could be stated that mechanical properties are altered slightly after using the graft as compatibilizer. The hardness is lower after adding the graft, while the other properties, such as tensile strength and Charpy impact notched strength, are improved.

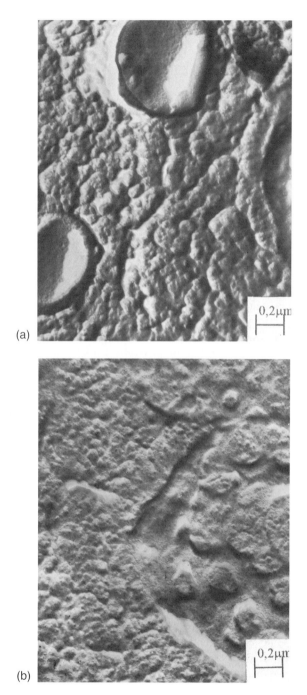

(a)

(b)

Fig. 8 Morphology of PS/PU blends: (a) without and (b) with compatibilizer (PS-4.0SSA-PU1 graft). (From Ref. 6.)

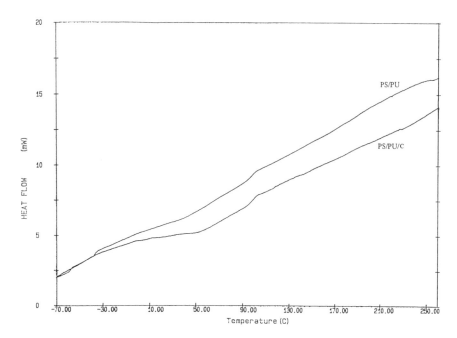

Fig. 9 DSC curves of PS/PU blend and PS/PU/C blend with compatibilizer (C). (From Ref. 6.)

Table 2 Mechanical Properties of PS/PU Blend Without and With Compatibilizer (C)

Mechanical property	PS/PU 90/10	PS/PU/C 89.8/10/0.2
Hardness HR (N/mm^2)	64.35	62.8
Tensile strength (MPa)	24.4	26.1
Elongation (%)	1.71	1.71
Elastic modulus (MPa)	1649	1725
Flexural stress $H = 3.5\%$ (MPa)	29.4	31.9
Flexural modulus (MPa)	1477	1573
Charpy notched impact strength 1 J (kJ/m^2)	5.6	6.0
Izod impact strength 4 J (kJ/m^2)	3.9	4.3

V. CONCLUSION

Polyurethane grafted onto styrene-styrene sulfonic acid copolymers were obtained by stepwise grafting onto PSSSA chains of MDEA followed by chain growth on the hydroxyl sites of a wide range of polyurethane structures. Nuclear magnetic resonance spectra confirm that the protons of the SSA groups are transferred exclusively to the tertiary nitrogens of the MDEA. Dynamic mechanical tests show the occurrence of a single glass transition temperature for the grafts consisting only of hard segments. If long soft segments are added to the grafts, then these phases separate and yield a new glass transition temperature. The values of the high- and low-T_g values depend on length of the soft segments and on the sulfonate ion concentrations within the styrene copolymers. The alloys of PU and PSSSA obtained by the grafting method yield materials in which the phase purity in the phase-separated systems seems to be better than in the regular blend of similar composition.

The TEM investigations indicate the existence of separate domains of PS and of domains of PU—both in blend and in grafts—even those containing only hard segments of PU. The DSC measurements confirm the microscopic observations.

Polyurethane grafted onto copolymer styrene-styrene acid could be applied as a compatibilizer for some blends containing components similar to the arms of the graft.

REFERENCES

1.   M Rutkowska, A Eisenberg. Macromolecules *17*:821–824, 1984.
2.   P Smith, A Eisenberg. J. Polym. Sci. Polym. Lett. Ed. *21*:223–230, 1983.
3.   A Natansohn, M Rutkowska, A Eisenberg. Polymer *28*:885–888, 1987.
4.   M Rutkowska, M Jastrzębska, J-S Kim, A Eisenberg. J. Appl. Polym. Sci. *48*:521–527, 1993.
5.   HS Makowski, RD Lundberg, GH Singhal. U.S. Patent 3,870,841, 1975.
6.   M Jastrzębska. Polyurethane grafted onto styrene-styrene sulfonic acid copolymers. PhD dissertation, Gdynia Maritime Academy, Gdynia, Poland, 1998.
7.   M Rutkowska, M Jastrzębska, H Janik, J-S Kim. Composite Sci. Technol. *57*:1155–1158, 1997.
8.   GM Estes, RW Seymour, SL Cooper. Macromolecules *4*:452–457, 1971.
9.   M Jastrzębska, M Rutkowska. Polimery *43*:540–543, 1998.
10. G Riess, Y Jolivet. Adv. Chem. Ser. A.C.S. Washington *142*:243–256, 1975.
11. PJ Flory. J. Am. Chem. Soc *87*:1833–1838, 1965.

17
Poly(ether-ester) Block Copolymers with LC Segments

Zbigniew Roslaniec
Technical University of Szczecin, Szczecin, Poland

I. INTRODUCTION

The nature of the intermolecular interactions, or the energy level at which the molecules interact with each other, and the manner in which these interactions induce molecular orientation determine polymer properties. The differences in molecular mobility between the macromolecules with various chemical constitutions or segments of the polymer chains may result in phase separation or create the condition for the formation of the various types of supermolecular structures, including the highly oriented forms. As is commonly known, there is a variety of morphologies of semicrystalline polymers that directly influence the physical and functional properties of this group of polymers. The relation between the molecular mobility of macromolecular segments and the structure and intrinsic properties is manifested in block copolymers and liquid crystalline polymers.

Block copolymers can be characterized by a heterophase structure if the differences in the chemical constitution and the physical properties between the blocks are pronounced. The molecular weight of the respective segments (blocks) also has a decisive significance here. Particularly interesting properties of the block copolymers are revealed in the case of large differences in the molecular mobility between the blocks (segments). Polysiloxane, aliphatic polyether, aliphatic polyester, or polybutadiene segments are usually considered flexible segments, since they may form soft (liquid) phases with a relatively low glass transition temperature ($T_g < 0°C$). In contrast, polyamide, polystyrene, aromatic polyester segments, etc., in the presence of aromatic rings in the polymer chains or of groups capable of forming hydrogen bonds, are characterized by a significantly lower mobility; they are called *rigid* segments. There, rigid components form the hard phase with a high glass transition—or melting—temperature (T_g, T_m > 100°C). Such block copolymers exhibit the features of elastomers. The principal factor here is the formation of a matrix of flexible segments (medium, continuous elastomeric phase) in which the rich rigid-segment domains are immersed. As a result of the various types of intermolecular interactions, the rigid-segment domains form so-called "physical crosslinks," ensuring good mechanical properties of the material (1–4).

The strength and types of intermolecular interactions may contribute to the phase separation as well as determining the mechanical properties of the hard phase and of the entire copolymer. In the case of block elastomers, the hard-phase properties usually result from:

The interaction of the van der Waals forces and a high cohesive energy, e.g., in the triblock butadiene-styrene copolymers of the ABA type, the rigid polystyrene segments form the domains of a glassy phase dispersed in the liquid polybutadiene matrix.

Pseudo-crosslinking role of the hydrogen bond, e.g., in urethane copolymers. The ability of hard segments to crystallize, e.g., in ester and olefine elastomers.

The formation of ionic bonds or donor–acceptor complexes (ionic–olefine elastomers, urethane ionomers).

Good mechanical properties of the block copolymers may result not only from the specific intermolecular interactions with the hard phase but also from the possible formation of highly ordered structures. A necessary condition for the formation of such a structure is a close proximity of the rigid and linear macromolecules over a relatively large length. In case of macromolecules composed of aromatic rings, directly connected to each other, the self-ordering could take place due to the interactions arising from the entire chains, if this is permitted by the steric conditions. In reality, the melting temperature of such polymers is significantly higher than the decomposition temperature. The possibility of an ordering of the macromolecule rigid segments is achieved due to the incorporation of bonds between them or groups with enhanced rotation, which enables their displacement and approaching capability. Therefore, an improvement in the packing possibility can be realized via the specific chemical architecture of block elastomers and liquid crystalline polymers.

II. POLY(ETHER-ESTER) BLOCK COPOLYMERS WITH ELASTOTHERMOPLASTIC PROPERTIES

Copoly(ether-block-ester)s (PEE) (Formula 1) comprise a very important group of thermoplastic elastomers with a block constitution.

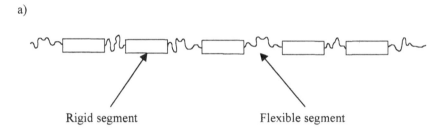

$$\text{PEE} \qquad (1)$$

These copolymers are characterized by elastic properties while they have an heterophase structure. As a result of the aggregation of identical polymer blocks, a matrix-domain structure is formed (Fig. 1). These copolymers combine the properties of elastomers with typical properties of the thermoplastics: flexibility (elasticity), a high impact resistance, outdoor weathering resistance, and chemical durability. In addition, the processability as well as the possibility of recycling,

a)

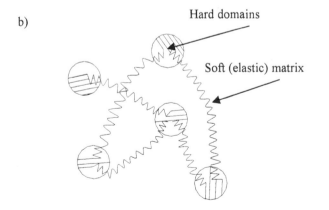

Rigid segment   Flexible segment

b)   Hard domains

Soft (elastic) matrix

Fig. 1 Schematic representation of (a) the chemical constitution and (b) the phase structure of block elastomers.

with regard to the block character of their chemical structure, permits an easy modification of their properties in the synthesis process (1–9).

A number of research works concerning the preparation, chemical constitution, and properties of copoly(ether-ester)s have been published in the scientific literature (1,2,12–27). The most frequently described rigid segments in these copolymers are semicrystalline polyesters (e.g., poly(butylene terephthalate) (1,2,14–16,28,29), whereas for flexible segments-polyethers, the residues of oligo(1,4-oxytetramethylene) are used. Less frequently, the fragments of oligo(1,2-oxypropylene) or oligo(1,2-oxyethylene) (18,30–33) or aliphatic polyesters (34) are also used. The fraction and the type of rigid segment in the copoly(ether-ester)s have a decisive influence on the thermal and mechanical properties of the copolymers. Therefore chemical modifications of the ester segments are performed in order to develop polymeric materials corresponding to particular functional needs.

Stevenson and Cooper (35) have investigated the effect of the modification of poly(butylene terephthalate) rigid segments by using isophthalic acid. They concluded that chemical modification results in depression of the melting temperature, reduction of the degree of copolymer crystallinity, and an enhancement of its flexibility. Another modification method having practical significance is that of incorporating branches in the polyester rigid segment by means of polyalcohols. These branches disturb the crystalline structure of the copolymer and, to a significant degree, influence the increase in the molten polymer viscosity (on extruded products). The modification by pentaerythrite was performed in order to determine the application of such systems for the manufacture of products subjected to high compressive stresses with regard to their applications as the active elements in shock absorbers (36,37). Modification of poly(butylene terphthalane) segments was also carried out by the use of 1,4-butene-2-diol (BD-2) (38–39). Unsaturated PEEs are characterized by a lower melting temperature and a smaller degree of crystallinity as compared to the saturated PEEs.

Poly(ether-ester) elastomers were introduced as a commercial product in 1968 under the trademark of Hytrel (3,10,11); the technology was developed by Du Pont. Other products of this type were launched in subsequent years and include, among others, Pelprene (Toyobo), Lomod (General Electric), Arnitel (DSM), Gaflex (GAF Corporation), and Elitel (Elana, Poland).

III. LIQUID CRYSTALLINE COPOLYMERS

It is known that polymers may exhibit liquid crystalline behavior (40–42) if chain segments with a considerable structural rigidity (mesogens) are incorporated into the chain molecules. Many well-known and functional synthetic polymers hitherto used have relatively flexible macromolecules which, in the molten state, exhibit a random coil configuration with the absence of a mesophase. In order to obtain liquid crystallinity, molecular segments with appropriate rigidity are built in.

A typical example of such structure is poly(*p*-phenylene), but only with a degree of polymerization of $n = 5$–7, since above this number it does not melt but undergoes decomposition. Polyesters are so for the most widely described group of liquid crystalline polymers (40–45). This results from the fact that via polycondensation, chains containing the segments with a high rigidity separated by the flexible segments can be readily obtained.

Groups with relatively high mobility permit the formation of ordered structures in which the mesogenic chain segments can easily be incorporated into the polyesters. As a result, the melting temperature of the polymer is depressed in this manner, yielding a practical significance in the processing. Four principal methods exist for polyester modification:

1. Incorporation of bonds into the macromolecules that change the direction of the main chain (refracting the macromolecule linearity) (46–50) (Formula 2): Either monomers with meta-substituted aromatic rings, e.g., *m*-phenylene links (Formula 3) or monomers with "hinged" atoms or groups (Formula 4) can be used.

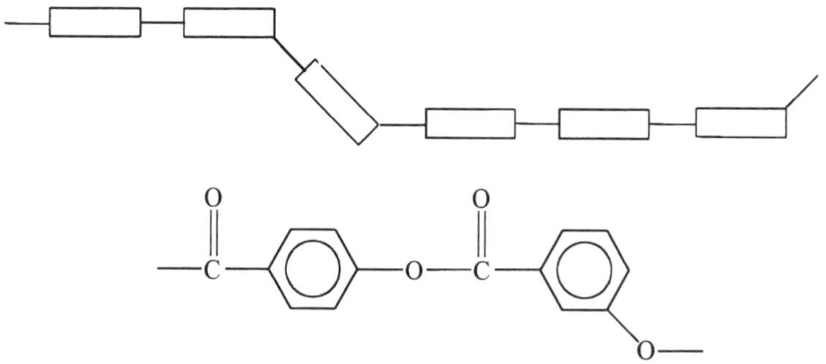

(2)

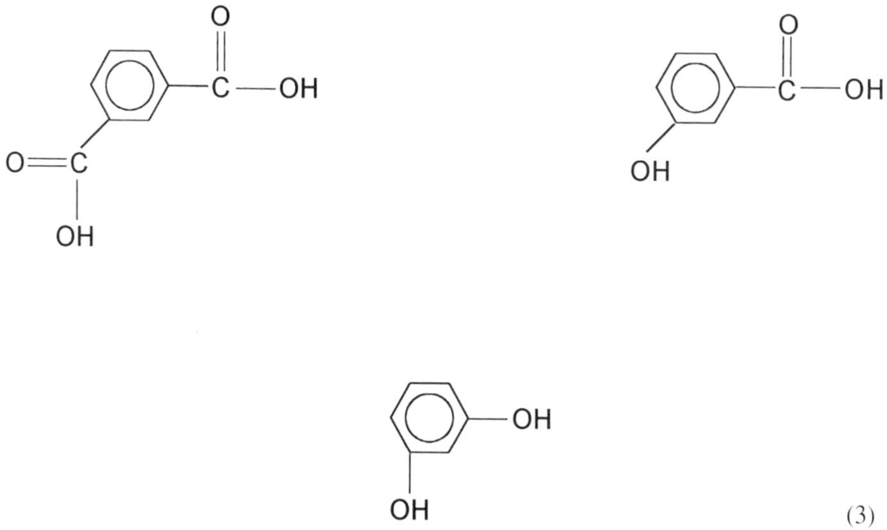

(3)

$$HOOC—\bigcirc—R—\bigcirc—COOH \quad \text{Where:} \quad R = \overset{|}{\underset{|}{C}}H_2 \quad \overset{|}{\underset{|}{CO}} \quad \overset{|}{\underset{|}{O}}$$

$$HO—\bigcirc—R—\bigcirc—OH \quad \text{Where:} \quad R = \overset{|}{\underset{|}{S}} \quad \overset{|}{\underset{|}{O}} \quad \overset{|}{\underset{|}{SO_2}}$$

$$\tag{4}$$

2. Incorporation of aliphatic groups into the main chain (51–56). Such a separating link, and simultaneously linking (coupling) the mesogens (so-called coupling bridge, or spacer) is represented by a linear hydrocarbon or aliphatic ether chain, composed of—CH_2— or —$O(CH_2)_n$—groups, which enables the appropriate locations of the mesogens (Formula 5).

$$\tag{5}$$

$$—\overset{O}{\overset{\|}{C}}—\bigcirc—O—(CH_2—CH_2—O)_n—$$

3. The synthesis of polymers with irregular constitution of the main chain: the irregularity is achieved by the incorporation of side chains (sterical hindrance) into the phenylene rings (57–60), which affect the sterical structure of the polymeric chain, hindering it and changing the molecular mobility. Also, irregularities can be achieved through the building in of different structural elements, e.g., heterocyclic rings (61) near aromatic rings, in addition to the groups—CH_2—and—O—CH_2—CH_2—(62) in the amounts of 5–10 and 2–4, respectively, or nonaromatic cyclic rings (63) (Formula 6).

$$\tag{6}$$

$$—\overset{O}{\overset{\|}{C}}—O—\underset{R}{\bigcirc}—O—\overset{O}{\overset{\|}{C}}—\bigcirc—\overset{O}{\overset{\|}{C}}—O—\overset{R}{\bigcirc}—$$

4. Incorporation of monomers disordering the crystalline structure of the polymer (57,60,64–70). This may be achieved through the incorporation of monomers containing para-aromatic substituents. For example: 4,4-bisphenylene

groups, 2,6-naphthylene substituents or condensed aromatic systems. The incorporation of this type of monomer impedes a close packing of the chains (Formula 7).

$$(7)$$

The structures that ensured the formation of the mesogens in the condensation polymers comprise:

Ring aromatic groups (Formula 8)
Noncyclic elements containing mainly the double bonds (Formula 9)

$$(8)$$

$$(9)$$

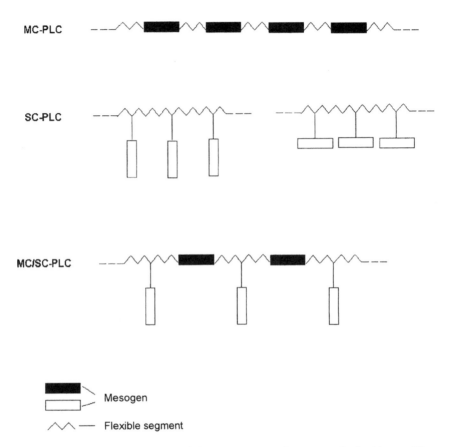

MC-PLC

SC-PLC

MC/SC-PLC

Mesogen

Flexible segment

Fig. 2 Schematic representation of the arrangement of mesogens in liquid crystalline polymers.

The role of mobile elements (segments) are most frequently performed by the alkyne chains—$(CH_2)_n$—or polyoxyethylene—$(CH_2$—CH_2—$0)_n$—, with length usually corresponding to $n = 2$–10. The role of mobile segments may also be accomplished by the polysiloxane chains (71).

Mesogenic structures are incorporated into the polymers either in the main chain of the macromolecule (main-chain polymer-MC-PLCs) or as side chains (side-chain polymers, comb-shaped polymers SC-PLCs) and either in the main chain or in the side chain (combined polymers MC/SC-PLCs) (Fig. 2).

A variety of mesogen structures influence molecular orientation. The best known are the four types of molecular arrangements (the types of mesophases) in

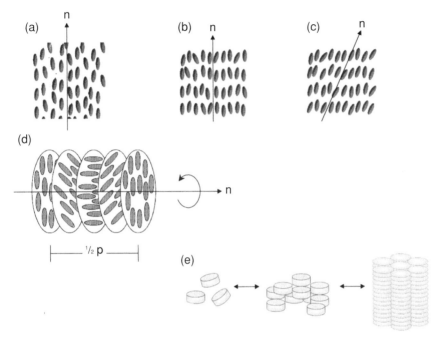

Fig. 3   Schematic representation of the different types of molecular arrangement (order) in liquid crystalline polymers: (a) nematic; (b) smectic A; (c) smectic C; (d) cholesteric; (e) discotic.

the liquid crystalline states (42,94) (Fig. 3):

Nematic order (one-dimensional orientation)
Smectic order (two-dimensional orientation, the best known being the A-type order and the C-type order)
Cholesteric order (these are helical, twisted nematic structures)
Columnar order (mesogens are arranged in layers and hexagonally packed)

Liquid crystalline polyesters have a main-chain constitution and form exclusively nematic and smectic structures. The main-chain constitution of liquid crystalline polyesters makes them high-strength engineering polymer materials (termed *self-reinforced*) processed by the method of extrusion or injection. Processing induces an orientation in the direction of material flow, which results in the formation of a material with high mechanical properties in the oriented direction.

To this group of polymers belong the thermoplastic aromatic polyesters; from them, products can be produced that exhibit high mechanical properties reminiscent of the aromatic polyamides of the Kevlar type. The excellent properties of these products result from the high anisotropy of their mechanical properties.

Suitable treatments and molding of these aromatic polyesters leads to the formation of materials with a very high strength (so-called *modular*). The main-chain
constitution of liquid crystalline polyesters bears a similarity to the constitution of
segmented (block) copoly(ether-ester) elastomers.

A number of companies throughout the world manufacture liquid crystalline polyesters, e.g., Vectra (Hoechst, Germany), Xydar (Amoco, United
States), Utrax (BASF, Germany), Ekkcel, Ekonol (Carborundum, United States)
Rodron (Unitika, Japan), Zenite (Du Pont, United States), Novoaccurate (Mitsubishi, Japan), Granlar (Montedison, Italy), Victrex (ICI, England).

The incorporation of highly oriented liquid crystalline elements as the rigid
segments in the block elastomers opens up possibilities for the development of
new types of polymeric materials.

IV. POLY(ETHER-ESTER) BLOCK COPOLYMERS WITH ESTER LIQUID CRYSTALLINE SEGMENT

A. Multiblock Copoly(ether-ester)s with Poly(p,p'-bibenzoate alkylene) Segment

Much information on the synthesis of copoly(ether-ester)s with liquid crystalline
segments deals with the application of poly(p,p'-bibenzoate alkylene) (BB)
(72–75) as rigid segment. The addition of BB is responsible for the appearance of
the thermotropic liquid crystalline properties of the copolymer.

Hong-Bing Tsai and coworkers (76–78) have synthesized and examine the
physical properties of copoly(ether-ester)s (Formula 10) using BB and
oligo(oxytetramethylene)diol (PTMO, $\overline{M}n$ = 2000 g/mol). Copoly(ether-ester)s
were prepared by the melt polycondensation method using tetrabutyl titanate as a
catalyst.

$$(10)$$

The molar composition of synthesized copolymers in amount (BB-alkylene
glycol-PTMO) is 1:0.9:0.1. The synthesized polymers are characterized by a wide
range of viscosities (between 0.18 and 1.50 dl/g). The influence of alkylene chain

segments—$(CH_2)_n$—on the liquid crystalline properties of the copolymers has been examined. The existence of smectic mesophases has been observed in copolymers containing alkylene chains with length corresponding to $n = 3$–8. An odd–even effect on the melting temperature (T_m), transition temperatures (T_a) of the isotropic-liquid—to-smectic-mesophase and smectic-mesophase-to-solid-crystal transition (T_b) was found. Similar properties are shown by the BB homopolyester (72) and by other polyesters (53,79). Figure 4 illustrates the dependence of the phase transition temperatures $(T_m, T_a,$ and $T_b)$ on the amount of the methylene units (n) in copoly(ether-ester)s, with a molar composition of BB–ethylene glycol–PTMO of 1:0.9:0.1. The temperature range of a mesophase and its type depend on the alkylene-chain length. The mesophase range diminishes amount of—$(CH_2)_n$—groups increases, but for $n = 3 \div 6$ is the largest. For $n = 9$ and $n = 10$ during heating and cooling, the existence of only one peak of the solid-crystal-to-isotropic-liquid transition was observed, and the existence of the liquid crystalline phase was absent.

Copoly(ether-ester)s have been also synthesized and investigated using: 1,6-hexamethylene glycol $(n = 6)$ or 2-methyl 1,4 butanediol (4(2Me)); oligo(oxytetramethylene)diol (PTMO, $\overline{M}n = 2140$ g/mol); dimethyl terphthalate

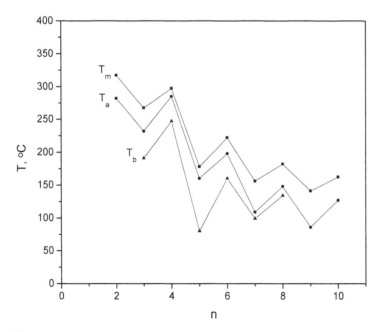

Fig. 4 Dependence of the transition temperatures (T_m, T_a, T_b) on the number of methylene units of alkylene diol (n) for the block copoly(ether-ester)s with a molar composition of BB-alkylenediol-PTMO = 1:0.9:0.1. (From Ref. 76, with permission from Society of Polymer Science, Japan.)

(*T*) or diethyl ester of 4,4-bifenylene dicarboxylic acid and (*B*) and dimethyl ester of 2,6-naphthalenedicarboxylic acid (*N*) (80). In the foregoing block copolymers B6-PTMO(x) (Formula 11), the existence of the smectic mesophase of the A type was found, but only when the fraction of rigid segments in the copolymer was no higher than 50 wt %.

With BN6-PTMO(x) (Formula 12), the obtained material is characterized by a high degree of phase separation. The occurrence of smectic of the A type was also observed in this case, while the content of rigid segments in the copolymer is smaller than 50 wt %.

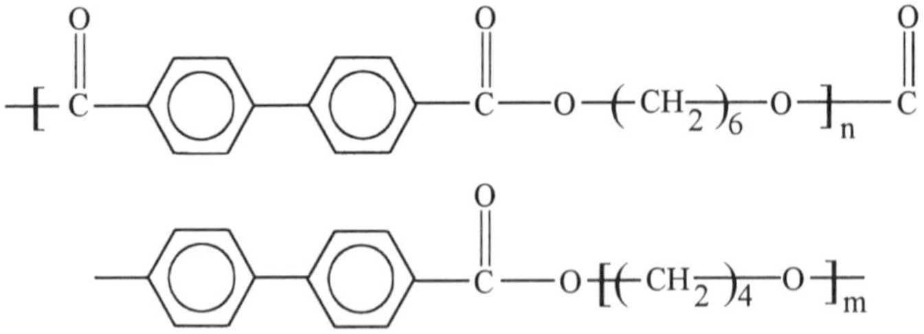

$$+ \overset{O}{\overset{\|}{C}} - \!\!\bigcirc\!\!-\!\!\bigcirc\!\!- \overset{O}{\overset{\|}{C}} - O + CH_2)_6 - O]_n - \overset{O}{\overset{\|}{C}}$$

$$-\!\!\bigcirc\!\!-\!\!\bigcirc\!\!- \overset{O}{\overset{\|}{C}} - O [(CH_2)_4 - O]_m$$

(11)

B6 - PTMO(x)

$$+ \overset{O}{\overset{\|}{C}} - Ar - \overset{O}{\overset{\|}{C}} - O + CH_2)_6 - O]_n - \overset{O}{\overset{\|}{C}} - Ar - \overset{O}{\overset{\|}{C}} - O [(CH_2)_4 - O]_m$$

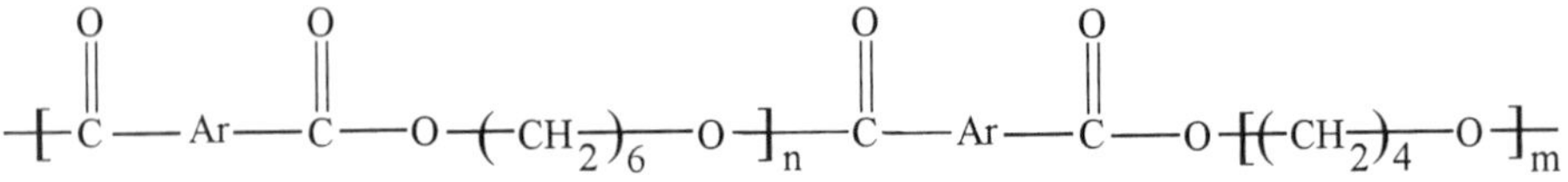

BN6 - PTMO(x)

Where:

$$Ar = + \bigcirc\!\!-\!\!\bigcirc \,]_{0.75} \, , \, + \bigcirc\!\!\!\bigcirc]_{0.25}$$

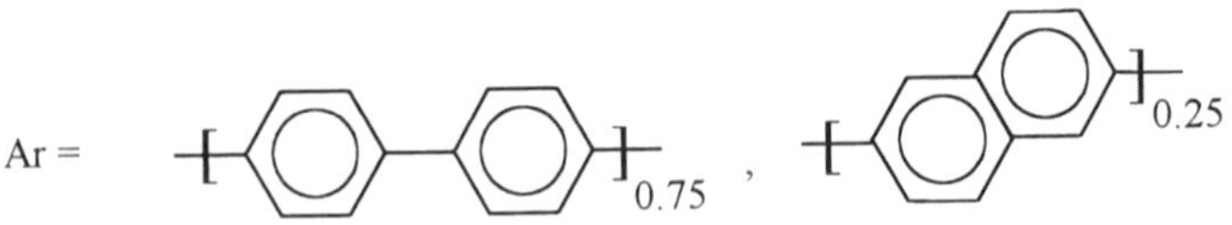

(12)

The distinct effect of the flexible-segment content on the type of mesophase formed was observed in the copolymers of the B4(2Me)-PTMO(x) (Formula 13) type. In the systems containing less than 20 wt % of flexible segments, the existence of an enantropic nematic was observed, whereas for the systems containing an amount over 20 wt % of flexible segments, the monotropic phase transition takes place.

$$+ \overset{O}{\overset{\|}{C}} - \text{\textcircled{}}-\text{\textcircled{}}- \overset{O}{\overset{\|}{C}}-O-CH_2-\overset{CH_3}{\overset{|}{CH}}(CH_2)_2-O\rbrack_n C$$

$$-\text{\textcircled{}}-\text{\textcircled{}}- \overset{O}{\overset{\|}{C}}-O\lbrack(CH_2)_4-O\rbrack_m \quad (13)$$

B4(2Me) - PTMO(x)

B. Multiblock Copoly(ether-ester) with Polyester Segment of Terephthalic Acid and Hydroquinone

Sonpatki et al. (81) have investigated the main-chain block copoly(ether-ester)s (Formula 14) obtained from terephthalic acid, hydroquinone, methylhydroquinone or chlorohydroquinone, and oligo(oxytetramethylene)diol (PTMO, $\overline{M}n = 250$ g/mol).

$$\left[-O-(CH_2CH_2CH_2CH_2-O-)_m \overset{O}{\overset{\|}{C}}-\text{\textcircled{}}-\overset{O}{\overset{\|}{C}}-(-O-\right.$$

$$\left.\text{\textcircled{R}}-O-\overset{O}{\overset{\|}{C}}-\text{\textcircled{}}-\overset{O}{\overset{\|}{C}}-)_n\right] \quad (14)$$

where: $R = H,$ $n = 1, 2, 3, 4, 5$

 $R = CH_3, Cl,$ $n = 2, 3$

The studies performed with copoly(ether-ester)s possessing different substituents in the aromatic rings validate the strong dependence of the thermotropic properties and the crystallinity of copolymers on the substituent features (type, size). The foregoing authors have found that an increase of rigid-segment length results in an increase in the transition temperature, inducing the change of the copolymers from thermoplastic to thermotropic character and even leading to infusible systems.

The microcalorimetric examinations (DSC) of the resulting nonsubstituted copolymers demonstrate the existence of a low-temperature glass transition of flexible segments (T_g) in the temperature range characteristic of the thermoplastic elastomer of the Hytrel type (T_g from $-30°C$ to $-60°C$) (82). These copolymers display a high degree of phase separation and elastothermoplastic properties. The microscopic observations permitted the authors to reach the conclusion that in both series of substituted copoly(ether-ester)s, the nematic mesophase, reminiscent of that observed in the respective substituted thermotropic homopolyesters (83), exists.

C. Copoly(ether-ester)s of Terephthalic Acid and p-hydroxy-benzoic Acid

The synthesis of copoly(ether-ester)s based on terephthalic acid and p-hydroxybenzoic acid may be realized by one of two methods: polycondensation in solution or melt polycondensation.

Mitrach et al. (84) have synthesized copoly(ether-ester)s using high-temperature polycondensation in solution in which oligo(oksytetra-methylene)diol (PTMO, $\overline{M}n$ = from 250 to 4500 g/mol) was used as the flexible segment. The mesogenic units were formed in the reaction of terephthalic chloride with p-hydroxybenzoic acid. The formation process for thermotropic block copoly(ether-ester)s from PTMO as flexible segment is shown in Formula 15. The application of the esters of terphthalic acid and p-hydroxybenzoic acid as mesoelements in blocks of liquid crystalline poly(ether-ester)s with polyoxyethylene or polyoxypropylene as the flexible segment has been previously described in Bilibin's works (85–87).

Polarizing light microscopy observations and DSC measurements have demonstrated that copoly(ether-ester)s are liquid crystalline polymers only when they contain flexible PTMO segments with $\overline{M}n$ = 650 g/mol ($n \approx 9$). In the case of copolymers from PTMO with molecular weight above 1000 g/mol, the existence of the phase separation was also observed. The appearance of two melting endotherms in the DSC scans confirms the heterophase structure of these systems: one peak corresponding to the crystalline polyether (PTMO) and the second peak corresponding to the crystalline aromatic units

Polycondensation in the molten state is an interesting synthesis method for copoly(ether-ester)s that are the derivatives of terephthalic acid and p-hydroxybenzoic acid. The advantage here is the technological simplicity of the removal of acetic acid formed, in comparison with the removal of HCl generated from acidchloride anhydride. The modifications of polyethylene terephthalate (PET) and polybutylene terephthalate (PBT) by p-hydroxybenzoic acid are well known from the literature (51–54). The addition of p-hydroxybenzoic acid results in the liquid crystallinity of polymers and enhances the mechanical properties as compared to the homopolymer. Therefore this addition method improves the qualitative parameters of products and facilitates processing (51,54). Hence, the possibility of combining these two methods for the purpose of forming block copoly(ether-es-

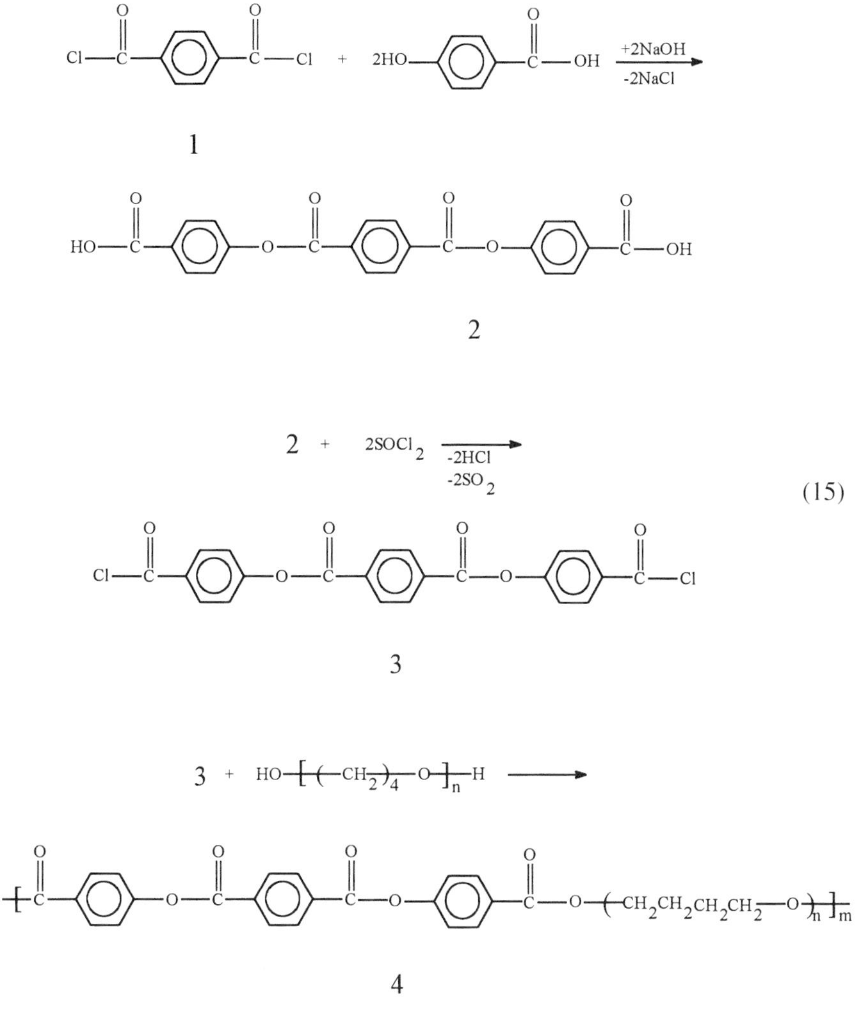

1

2

(15)

3

4

ter) elastomers with liquid crystalline segments seems possible. Most of the synthesis methods for liquid crystalline elastomers described up to now in the literature concern rubbers crosslinked by diisocyanate $R(NCO)_2$ or α, ω-bifunctional oligosiloxanes (88–90). Therefore, the combination of the heterophase structure of the block copolymers with the liquid crystalline properties creates a novel possibility for preparing new group of elastomers.

Copoly(ether-ester)s of the PTMO-*block*-PBT/PHB (Formula 16) type (91–94) were obtained by the melt polycondensation method, with the use of dimethyl terephthalate (DMT), 1,4-butanediol (BD-1,4), acetyl derivative of *p*-hydroxybenzoic acid (*p*-acetoxybenzoic acid), and oligo(oxytetramethylene)diol

(PTMO, $\overline{M}n$ = 1000 g/mol):

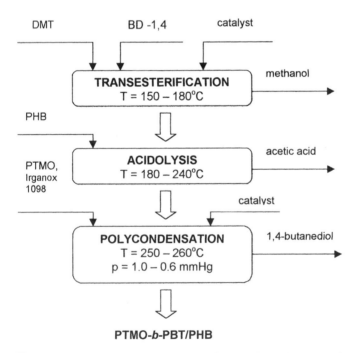

$$\left[\left(C-\bigcirc-O\right)_a\left(C-\bigcirc-C-O-(CH_2)_4-O\right)_b\right]_x C$$

$$-\bigcirc-C-O\left[(CH_2)_4-O\right]_n- \tag{16}$$

PTMO-*b*-PBT/PHB

The course of synthesis of PTMO-*block*-PBT/PHB copolymeris is shown in Fig. 5. The crystalline phase was not obtained with copolymers of this type; however, the glassy solidification of the mesophase was observed. This state characterizes only the glass transition temperature on (Fig. 6). The segmented copoly(ether-ester)s based on (PTMO-*block*-PBT/PHB) copolymers show nematic texture, which

DMT   BD -1,4   catalyst

TRANSESTERIFICATION
T = 150 – 180°C   methanol

PHB

PTMO,
Irganox
1098

ACIDOLYSIS
T = 180 – 240°C   acetic acid

catalyst

POLYCONDENSATION
T = 250 – 260°C
p = 1.0 – 0.6 mmHg   1,4-butanediol

PTMO-*b*-PBT/PHB

Fig. 5   Block diagram for the synthesis of PTMO-*b*-PBT/PHB copolymers.

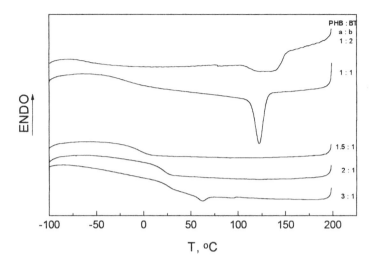

Fig. 6 DSC cooling curves for PTMO-*b*-PBT/PHB copolymers.

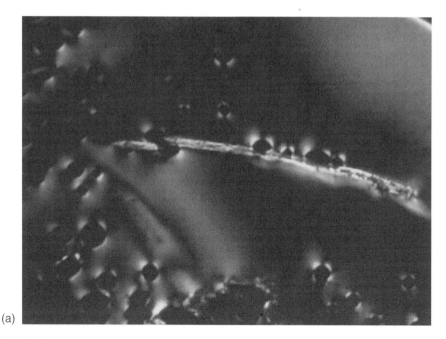

(a)

Fig. 7 Polarizing microscopy photomicrographs of PTMO-*b*-PBT/PHB copolymers with molar composition PHB-BT = 2:1. (a) Droplet texture; (b) Schlieren texture. (From Ref. 101.) *(figure continues)*

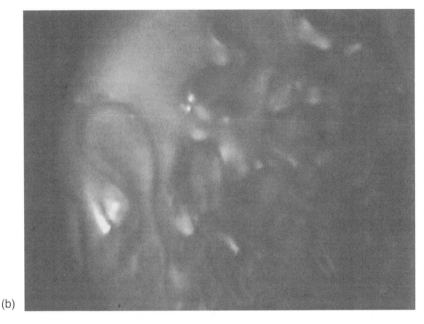

(b)

Fig. 7 *(continued)*

is illustrated on Figs. 7–9, (101). A similar structure has already been observed by
Bhownik P.K. for LC polyester (53).

D. Photoreactive Block Copoly(ether-ester)s

Photo-crosslinked thermotropic liquid crystalline block copoly(ether-ester)s represent new materials, from both the synthesis and properties points of view as well
as for its practical utilization (95). The copolymers were obtained using melt polycondensation, by application of PTMO diol with varied molecular weight
($\overline{M}n = 1000$–3000 g/mol) as a constituent of flexible segments and poly(hexam-
ethylene *p*-phenylene diacrylate) as photoreactive rigid segments. The synthesis
route of photo-crosslinked thermotropic liquid crystalline block copoly(ether-ester)s is shown in Formula 17:

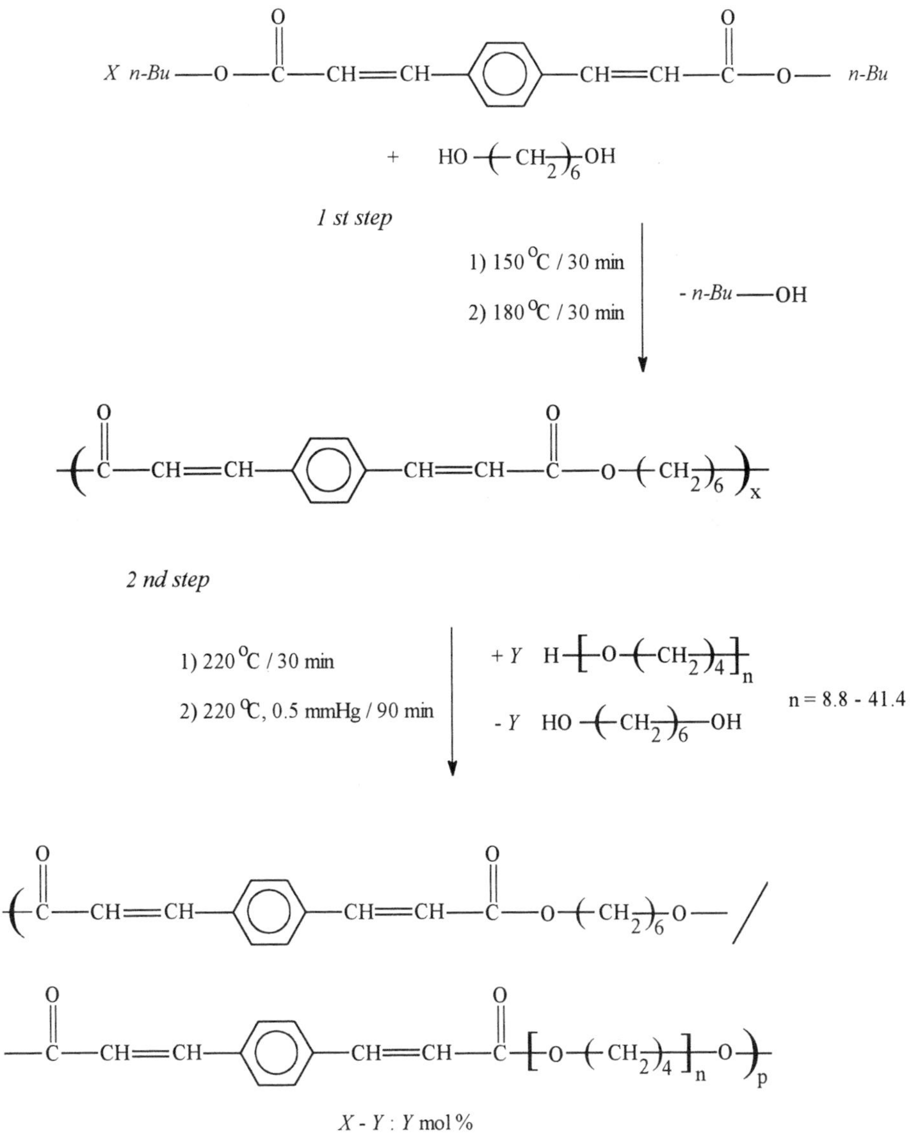

$$X \quad n\text{-}Bu - O - \overset{\overset{\displaystyle O}{\|}}{C} - CH = CH - \langle\bigcirc\rangle - CH = CH - \overset{\overset{\displaystyle O}{\|}}{C} - O - \quad n\text{-}Bu$$

$$+ \quad HO - (CH_2)_6 - OH$$

1 st step

$$\begin{array}{c} 1)\ 150\,^{\circ}C\ /\ 30\ min \\ 2)\ 180\,^{\circ}C\ /\ 30\ min \end{array} \Bigg| \quad - n\text{-}Bu - OH$$

$$\left(\overset{\overset{\displaystyle O}{\|}}{C} - CH = CH - \langle\bigcirc\rangle - CH = CH - \overset{\overset{\displaystyle O}{\|}}{C} - O - (CH_2)_6 \right)_x$$

2 nd step

$$\begin{array}{c} 1)\ 220\,^{\circ}C\ /\ 30\ min \\ 2)\ 220\,^{\circ}C,\ 0.5\ mmHg\ /\ 90\ min \end{array} \Bigg| \quad \begin{array}{l} + Y \quad H\!\left[O - (CH_2)_4 \right]_n \\ - Y \quad HO - (CH_2)_6 - OH \end{array} \quad n = 8.8\ \text{-}\ 41.4$$

$$\left(\overset{\overset{\displaystyle O}{\|}}{C} - CH = CH - \langle\bigcirc\rangle - CH = CH - \overset{\overset{\displaystyle O}{\|}}{C} - O - (CH_2)_6 - O - \right/$$

$$- \overset{\overset{\displaystyle O}{\|}}{C} - CH = CH - \langle\bigcirc\rangle - CH = CH - \overset{\overset{\displaystyle O}{\|}}{C}\!\left[O - (CH_2)_4 \right]_n O - \right)_p$$

X - Y : Y mol %

$$(17)$$

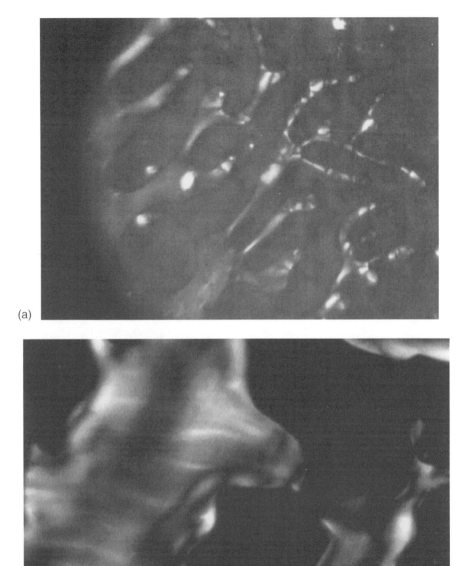

(a)

(b)

Fig. 8 Polarizing microscopy photomicrographs of PTMO-*b*-PBT/PHB copolymers with molar composition PHB-BT = 3:1. (a) and (b) displaying the coexistence of an LC phase and an isotropic phase. (From Ref. 101.)

Fig. 9 Polarizing microscopy photomicrographs of PTMO-*b*-PBT/PHB copolymers with (a) molar composition PHB-BT = 2:1 and (b) molar composition PHB-BT = 3:1, displaying oriented fibrillar morphology. (From Ref. 101.)

The occurrence of the nematic mesophase was confirmed in the present copoly(ether-ester)s. It is noteworthy that through the incorporation of photoreactive mesogenic constitutional units into the copolymer, one may influence the physical properties by the photo-crosslinking reaction.

E. Triblock Copoly(ether-ester)s with Liquid Crystalline Ester Segment

Schulze and Schmidt (96) synthesized triblock copolymers ABA from polyoxyethylene (block A) and liquid crystalline polyester (block B). Formula 18 shows the preparation method for the copolymers:

(18)

Either liquid crystalline block copolymers that do not show the phase separation or the block copolymers with liquid crystalline polyester segment (B) exhibiting the phase separation can be obtained, depending upon the respective block lengths. From the preceding example it is clear that the type and length of the respective blocks significantly affect the properties of the materials obtained. Triblock copolymers of the ABA type represent a new type of liquid crystalline thermotropic material having application as nonionic surfactant active agent.

V. REACTIVE BLENDS OF BLOCK COPOLY(ETHER-ESTER) WITH LIQUID CRYSTALLINE POLYESTER

The blends of thermotropic polyesters with the technical polymers (PET, PBT, PP, PE, PC, PA, PS, PI) have several technical applications. The addition of liquid crystalline polymer acts as the reinforcing fiber, and improvement in the mechanical properties of modified thermoplastic in such a manner can be obtained with its slight participation. Additionally, a substantial lowering of its viscosity creates the possibility of the manufacture of complicated injection moldings. Anisotropy of the mechanical properties was observed with an increase in the liquid crystalline polymer content in the thermoplastic matrix.

With regard to the attractive functional properties, noteworthy are the mixtures of liquid crystalline polyesters with the block elastomers. The occurrence of the liquid crystalline phase in the form of oriented fibers in the thermoplastic matrix, which leads to the formation of self-reinforced composites, is observed in the blends. The blend of thermotropic liquid crystalline polyester Vectra A900 and thermoplastic elastomer Kraton G1650 has been investigated (97,98). The blends were produced in the molten state using a laboratory extruder at a temperature of 300°C. During extrusion, the orientation of the liquid crystalline phase occurs under the influence of shear force in the direction of material flow. This leads to an enhancement of rigidity and strength of the elastomer.

DSC and DMTA examinations demonstrate the thermodynamic immiscibility of liquid crystalline polymer with the elastomer. An elevation of the glass transition temperature of the soft phase of Kraton elastomer was also observed. This phenomenon can be explained by the utilization of a fraction of butadiene from this phase for the formation of the intermediate phase upon the liquid crystalline microregions. The occurrence of a fibrilar structure is observed in the blends from adding a small amount of Vectry A950 (below 20 vol %), whereas in the blends containing a higher fraction (above 20 vol %) of liquid crystalline polyester, the occurrence of laminar forms is observed.

Jang et al. (99) have investigated the blend of copoly(ether-ester) (KOPEL KP 3355) with liquid crystalline polyester Vectra A950. Based on the examination of the thermal properties, they demonstrated that the LCP/PEE blends prepared in the melt state exhibit a partial miscibility between the polymers. The morphological constitution has a fibrilar character.

The properties of "in situ" composites formed from the aromatic polyester Vectra A950 and the thermoplastic poly(ether-ester) elastomer Arnitel 630 have also been investigated (100). The blends were manufactured by feeding the components from separate extruders into a Ross-type mixer. The blends obtained had a phase structure varying from the fibrilar morphology to the core-shell structure. A different phase structure through fibrilar to the *skin-core* structure. The strength increase and the modulus of elasticity in tension for all the obtained blends are induced through an increase in orientation of the liquid crystalline phase during extrusion. Such materials may find applications for the preparation of elastic laminates or films with high performance mechanical properties.

VI. CONCLUSION

The differences in molecular mobility between the macromolecules or polymer chain segments may result in phase separation or various types of the supermolecular structure up to the highly oriented forms. The existence of rigid and flexible segments of the macromolecules is characteristic of block elastomers and liquid crystalline polymers. These systems are readily constructed based on the polyesters. The properties of liquid crystalline block copoly(ether-ester)s depend on the contents and the types of rigid segments (mesogens). Polymer materials can be achieved by means of blending block elastomers with liquid crystalline polyesters. The research on copoly(ether-ester)s containing soft and liquid crystalline microphase may lead to the development of a novel type of liquid crystalline elastomers.

REFERENCES

1. NR Legge, G Holden, HE Schroeder (eds.). Thermoplastic Elastomers. Hanser, New York, 1987.
2. AK Bhowmick, HL Stephens (eds.). Handbook of Elastomers: New Developments and Technology. Marcel Dekker, New York, 1988.
3. BM Walker, ChP Rader (ed.). Handbook of Thermoplastic Elastomers. Van Nostrand Reinhold, New York, 1988.
4. JC Salamone. Polymeric Materials Encyclopedia. CRC Press, Boca Raton, FL, 1996, p. 8343.
5. MJ Folkes. Processing, Structure and Properties of Block Copolymers. Elsevier Appl. Sci., London, 1985.
6. Z Roslaniec. Eko-Plast (Szczecin) 1994, No. 5, 96.
7. JJ Zeilstra. J. Appl. Polym. Sci. 1986, 31, 1977.
8. A Higashiyana, Y Yanamoto, R Chûjô, M Wu. Polymer J. 1992, 24, 1345.
9. GK Hoeschele. Angew. Makromol. Chem. 1977, 58/59, 299.
10. Pat. USA 3 023 192 (1968).
11. Pat. USA 3 766 146 (1973).
12. Pat. PL 150278, 1986 (1990).
13. Pat. PL 156412, 1987 (1992).
14. Pat. USA 3 651 014 (1972).
15. Pat. USA 3 801 547 (1974).
16. H Schroeder, RJ Cella. Encyclopedia of Polymer Science and Engineering. 2nd ed. Wiley, New York, 1988, vol. 12, pp. 75–117.
17. WH Buck, RJ Cella, EK Gladding, JR Wolfa. J. Polym. Sci. Polym. Symp. 1974, 48, 47.
18. GK Hoeschele, WK Witsiepe. Angew. Makromol. Chem. 1973, 29/30, 267.
19. RJ Cella. J. Polym. Sci. Polym. Symp. 1973, 42, 727.
20. R Ukielski, H Wojcikiewicz. Polimery (Warsaw) 1983, 28, 141.
21. A Lilaonitkul, JC West, SL Cooper. J. Makromol. Sci. Phys. 1976, B12(4), 563.
22. K Dietrich, J Hipp, J Schulze, Ch Versäumer, O Metz. Acta Polym. 1980, 31, 34.
23. JL Castles, MA Vallance, JM Mc Kenne, SL Cooper. J. Polym. Sci. Polym. Phys. 1985, 23, 2119.
24. JC Stevenson, SL Cooper. J. Polym. Sci. Polym. Phys. 1988, 26, 953.
25. J Slonecki. Polimery (Warsaw) 1991, 36, 225.
26. J Slonecki. Polimery (Warsaw) 1995, 40, 572.
27. J Slonecki. Polimery (Warsaw) 1996, 41, 31.
28. W Hofmann. Gummi, Fasern, Kunstst. 1987, 40, 650.
29. H Domininghaus. Plastverarbeiter 1989, 40, nr 1, 39.
30. G Wegener, T Fuji, W Meyer, G Lieser. Angew. Makromol. Chem. 1978, 74, 295.
31. J Slonecki. Acta Polym. 1991, 42, 655.
32. S Fakirov, T Gogeva. Makromol. Chem. 1990, 191, 603.
33. F Lembicz, J Slonecki. Makromol. Chem. 1990, 191, 1363.

34. RJ Spontak, MC Williams. Macromolecules 1988, 21, 1377.
35. JC Stevenson, SLJ Cooper. Polym. Sci. Polym Phys. 1988, 26, 953.
36. R Ukielski, H Wojcikiewicz. Polimery (Warsaw) 1988, 33, 9.
37. J Magryta, L Pysklo, Z Roslaniec, E Kapko. Polimery (Warsaw) 1988, 33, 464.
38. SB Nelsen, SJ Gromelski, JJ Charles. J. Elast. Plast. 1983, 19, 256.
39. J Siemiński, Z Roslaniec. Synthesis and characteristics of unsaturated poly(ether-es-
     ter) block copolymers. 12th Sci. Conference "Polymer Modification," Kudowa
     Zdrój (Poland) Sept. 11–15, 1995, p. 207.
40. JC Salomane. Polymeric Materials Encyclopedia. CRC Press, Boca Raton, FL,
     1996, p. 3645.
41. JI Jin, ChS Kang. Prog. Polym. Sci. 1997, 22, 937.
42. AM Donald, AH Windle. Liquid Crystalline Polymers. Cambridge Solid State Sci-
     ence Series, 1992.
43. K Grzebieniak, A Michalski. Wlokna Chem. 1991, 17, 33.
44. A Michalski, K Grzebieniak. Wlokna Chem. 1991, 17, 28.
45. H Han, PK Bhownik. Prog. Polym. Sci. 1997, 22, 1431.
46. Pat. Eur. 220737 (1987).
47. Pat. Eur. 257598 (1988).
48. Pat. Ger. 3535452 (1979).
49. Pat. USA 4156070 (1979).
50. VN Cwietkov, LN Andreeva, PN Lavrenko. Vysokomol. Soed. Ser. A. 1988, 30, 1263.
51. WJ Jackson, HF Kuhfuss. J. Polym. Chem. 1976, 14, 2043.
52. L Strzelecki, D Luyen. Eur. Polym. J. 1980, 16, 299.
53. D Luyen, L Strzelecki. Eur. Polym. J. 1980, 16, 303.
54. MS Chen, C Lee, NH Chang, BC Chang, HB Tsai. Polymer 1989, 29, 1472.
55. BZ Volczek, NS Chalmuratov. Vysokomol. Soed. Ser. A 1986, 28, 924.
56. BZ Volczek, NS Chalmuratov. Vysokomol. Soed. Ser. A 1987, 29, 1097.
57. Patent USA 4159365 (1979).
58. Patent Bryt. 993272 (1965).
59. J Majnusz, JM Catala, RW Lenz. Eur. Polym. J. 1983, 11, 1043.
60. HR Dicke, RW Lenz. J. Polym. Chem. 1983, 21, 2581.
61. DA Gusjeva. Chim. Volokna 1986, No. 4, 13.
62. D Caretti, AS Angeloni, M Laus. Makromol. Chem. Macromol. Chem. Phys. 1989,
     190, 1655.
63. B Reck, H Ringsdorf. Makromol Chem. Macromol Chem Phys. 1989, 190, 2511.
64. Patent USA 4728719 (1987).
65. W Heitz. Makromol Chem. Macromol Symp. 1989, 26, 1.
66. R Sinta, RA Minss, RE Gaudiana, HG Rogers. J. Polym. Sci. Polym. Lett. 1987, 25,
     11.
67. VN Cvietkov, LN Andreeva, IB Biszin. Vysokomol. Soed. Ser. A, 1988, 30, 713.
68. WJ Jackson. Macromolecules 1983, 7, 1027.
69. Pat. USA 3991014 (1976).
70. Pat. USA 4083829 (1978).
71. CK Ober, JI Jin, QF Zhou, RW Lenz. Adv. Polym. Sci. 1984, 59, 103.
72. WR Krigbaum, J Asrar, H Toriumi, A Ciferri, J Preston. J. Polym. Sci. Polym. Lett.
     1982, 20, 109.
73. WJ Krigbaum, J Watanabe. Polymer 1983, 24, 1299.

74. J Watanabe, M Hayashi. Macromolecules 1988, 21, 278.
75. J Watanabe, M Hayashi. Macromolecules 1989, 22, 4083.
76. HB Tsai, C Lee, NS Chang. Polym. J. 1992, 24, 157.
77. HB Tsai, DK Lee, JL Liu, YS Tsao, RS Tsai. J. Appl. Polym. Sci. 1996, 59, 1027.
78. HB Tsai. J. Macromol. Sci. Phys. 1997, B36 (2), 175.
79. J Asrar, H Toriumi, J Watanabe, WR Krigbaum, A Ciferri, J Preston. J. Polym. Sci. Polym. Phys. 1983, 21, 1119.
80. L Rozes, M Tessier, E Marechal. Synthesis and characterization of thermotropic poly(ether-block-ester)s. International Symposium on Polycondensation "Polycondensation 96," Paris Sept. 23–26, 1996, p. 334.
81. MM Sonpatki, K Ravindranath, S Ponrathnam. Polymer J. 1994, 26, 804.
82. KB Wagener. In: Polymer Handbook. 3rd ed. Braudup J., Immergut E. H., eds. Wiley Interscience, New York, 1989.
83. QF Zhou, RW Lenz. J. Polym. Sci. Polym. Chem. 1983, 21, 3313.
84. K Mitrach, D Pospiech, L Häuler, D Voigt D Jehnichen, M Rätzsch. Polymer 1993, 34, 3469.
85. A Yu Bilibin, AV Tenkovtsev, ON Piraner, EE Pashkovsky, SS Skorokhodov. Makromol. Chem. 1985, 186, 1575.
86. A Yu Bilibin, AV Tenkovtse, SS Skorokhodov. Makromol. Chem. Rapid Commun. 1985, 6, 209.
87. A Yu Bilibin, ON Piramer. Makromol. Chem. 1991, 192, 201.
88. R Zentel. Adv. Mater. 1989, 28, 1407.
89. J Franek, ZJ Jedlinski, J Majnusz. Synthesis of polymeric liquid crystals. In: Kricheldorf HR (ed.): Handbook of Polymer Synthesis, Part B; p. 1281; Marcel Dekker, New York, 1972.
90. Y Ikeda, T Yonezawa, K Urayama, S Kohjiya. Polymer 1997, 38, 3229.
91. D Pietkiewicz, Z Roslaniec. Segmented copoly(ether-ester) elastomers with PBT/PHB-LC hard segments—synthesis and characteristics. International Symposium on Polycondensation "Polycondensation '96," Paris, Sept. 23–26, 1996, p. 330.
92. D Pietkiewicz, Roslaniec Z. Synthesis and properties of copoly[ether-block-ester] elastomer with liquid crystalline segments. 4th International Symposium on Polymers for Advanced Technologies, Lipsko, Poland, Sept. 1997, P III 17.
93. D Pietkiewicz, Z Roslaniec. Influence of PHB on physical properties of poly(ether-ester) block copolymers. 13th Sci. Conf. "Polymer Modification," Kudowa-Zdrój, Poland, Sept. 11–14, 1997, p. 175.
94. D Pietkiewicz, Z Roslaniec. Polimery (Warsaw) 2000, 45, 69.
95. N Kawatsuki, A Kikai, R Fukae, T Yamamoto, O Sangen. J. Polym. Sci. Polym. Chem. 1997, 35, 1849.
96. U Schulze, HW Schimdt. Polymer Bull. 1998, 40, 159.
97. H Verhoogt, HC Langelaan, J Van Dam, A Posthuma De Boer. Polym. Eng. Sci. 1993, 33, 453.
98. H Verhoogt, CRJ Willens, J Van Dam, A Posthuma De Boer. Polym. Eng. Sci. 1994, 34, 754.
99. SH Jang, MH Jung, YS Gal, W Ch Lee. J. Macromol. Sci. Pure Appl. Chem. 1997, A34(9), 1645.
100. AGC Machiels, KFJ Denys, J Van Dam, A Posthuma De Boer. Polym. Eng. Sci.

1996, 36, 2451.

101. D Pietkiewicz. Synthesis and properties of liquid crystalline poly(ether-ester) elastomers, PhD dissertation, Technical University of Szczecin, Poland, 1999.

18
Triblock Copolymers for Toughness Modification of Immiscible Engineering Polymer Blends

Volker Altstädt, Thorsten Kirschnick, and Harald Ott
Technical University Hamburg-Harburg, Hamburg, Germany

Miroslawa El Fray
Technical University of Szczecin, Szczecin, Poland

I. OVERVIEW

A. Introduction

Polymer blends can be placed into two widely used categories: miscible and immiscible (1–4). Miscible blends involve thermodynamic solubility and are characterized by the presence of one phase and a single glass transition temperature. Immiscible blends are phase separated, exhibiting the glass transition temperatures and/or melting temperatures of each blend component. The latter blends are of great interest because they can combine some of the important characteristics of each constitutive component. The immiscibility problem of most polymer blends is caused not only by differences in the chemical structure of thermodynamically immiscible components, but also by differences in the phase behavior of these polymers. Two polymers or copolymers can be blended to give a miscible polymer blend only when certain temperature and composition conditions are met. The focus in research on the miscibility of polymers is usually the determination of the Flory interaction parameter χ. From the classical Huggins–Flory relation the free energy of mixing can be expressed as follows:

$$\frac{\Delta G_m}{RTV} \cong \chi_{12}\phi_1\phi_2 \tag{1}$$

where R, T, V, ϕ_i, and χ_{12} are, respectively, the gas constant, temperature, (molar) volume of the system, the volume fraction of component $i = 1, 2$, and the polymer–polymer interaction parameter. Equation (1) states that for the miscibility of the polymer blend system $\chi_{12} < 0$ is required (the condition of $\Delta G_m < 0$ for the miscibility of polymer blends can exist only for negative values of the binary polymer–polymer interaction parameter). But in fact for most mixtures, the enthalpy of mixing ΔH_m is positive. The miscibility of polymers is determined largely by the value of χ, which is the dominating parameter for the equilibrium-phase diagram of the blend. The magnitude of χ depends on the reference volume V, and thus one cannot compare the values from one blend with those from another unless the reference volumes are the same. As a result, the cohesive (interaction) energy density B was applied as a parameter to describe the miscibility of a blend $B = \chi_{12}RT/V$. The advantage of the cohesive energy density approach is that it can easily be compared between blends, and it can be estimated from solubility parameters by using the equation: $B = (\delta_1 - \delta_2)^2$, where δ_i is the solubility parameter of component i.

The resulting multiphase morphology has a considerable influence on the mechanical properties of the blend. In general a blend of immiscible polymers is a material with poor dispersion, low interfacial adhesion, and consequently poor mechanical properties (5). Satisfactory performance in immiscible blends is usually achieved by a compatibilization, which minimizes interfacial tension and improves adhesion between the two phases.

B. Compatibilization Methods

Block copolymers and graft copolymers (Fig. 1) have been widely used as compatibilizers, since block and graft copolymers are very effective in reducing the interfacial tension and improving interfacial adhesion by polymer chain entanglements and therefore bridging the interface. These compatibilizing agents were found both to reduce the dispersed phase dimensions and to enhance the mechanical properties.

The common approach for compatibilization of polymer blends is to decrease the interfacial tension between the blend partners, causing an increase in interfacial width or mixing, thereby enhancing the phase-size uniformity and stability (7). From this point of view, the following routes of compatibilization of immiscible polymer blends can be considered:

1. Compatibilization by reactions between the blend partners or by using specific interactions between them, e.g., hydrogen bonding and ion–ion and ion–dipole interactions (8,9).
2. Compatibilization by block copolymers containing segments chemically similar to the blends partners. Diblock, multiblock and graft copolymers are composed of segments that either are miscible or have

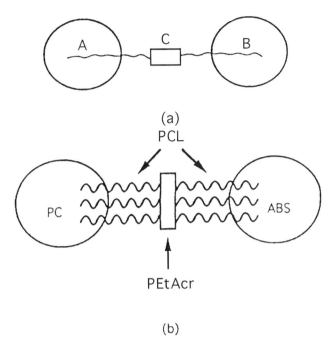

(a)

PCL

PEtAcr

(b)

Fig. 1 Localization of (a) block copolymers and (b) graft copolymers at the interface between immiscible polymers. PC: polycarbonate; ABS: poly(acrylonitrile-*co*-butadiene-*co*-styrene); PCL: polycaprolactone; PEtAcr: poly(ether acrylate). (Reprinted from Ref. 6 with permission from Elsevier Science.)

strong affinities to one of the two homopolymer phases. As a consequence, each block localizes itself in the corresponding phase, thus reducing interfacial tension and promoting phase adhesion (10,11).

The effectiveness of block copolymers as compatibilizers depends, on one hand, on the molecular architecture, molecular weight, and block composition, and on the other it is important to anchor these interfacial agents at the interface (12). The practical importance of diblock copolymers as compatibilizing agents has been extensively studied (3,4,13,14). The most popular are A-B diblock copolymers, for blends of A and B components, and C-D diblocks, in which each one of the blocks is thermodynamically miscible with one of the blend components but chemically different.

Taking into consideration that diblock copolymers are well established, we have focused on triblock copolymers as a new class of compatibilizers for multicomponent polymer systems. The following section will give an overview in the field of general performance and application of those systems in polymer blends.

C. Styrenic SBS Triblock Copolymers

A-B-A triblock copolymers consisting of polystyrene end blocks and a long poly-butylene, polyisoprene, or poly(ethylene-*co*-butylene) midblock are thermoplas-tic elastomers that exhibit rubberlike physical properties due to the midblock elas-ticity but are melt processable like conventional thermoplastics (15). The phase-separated glassy polystyrene microdomains act as physical crosslinks for the elastomeric sequences, providing high strength.

Addition of these A-B-A block copolymers as a compatibilizer to a blend system results in finer particle dispersion in the final blend. This was shown, for example, for blends based on high-impact polystyrene/polypropylene (HIPS/PP). In this case the triblock copolymer polystyrene-*block*-polybutadiene-*block*-polystyrene (SBS) was much more efficient with regard to the mechanical prop-erties of the (HIPS/PP) blend than a polystyrene-*block*-polybutadiene (SB) di-block copolymer (16).

D. Styrenic SEBS Triblock Copolymers

Polystyrene-*block*-poly(ethylene-*co*-butylene)-*block*-polystyrene (SEBS) poly-mers have been shown to act as compatibilizers for different polymer blends, es-pecially for those of polystyrene or polyesters with polyolefins, where they bridge the separated phases in the blend through physical or chemical interactions (17–22). It has been found by Yang et al. (23) that low-molecular-weight SEBS triblock copolymers (Kraton G 1652) are able to effectively compatibilize polystyrene (PS) in a blend with linear low-density polyethylene (LLDPE). The superiority of the SEBS triblock copolymer over the polystyrene-*block*-poly(ethy-lene-*co*-propylene) (SEP) diblock copolymer was observed. The proposed com-patibilization mechanism is shown in Fig. 2.

The addition of SEBS triblock copolymer not only induces finer phase dis-persion, but also results in an interlocked structure. The strong immiscibility of PS and LLDPE blends without compatibilizer results in poor mechanical properties (Fig. 3) as compared to a compatibilized system. The blends modified with 2 wt. % and 5 wt. % of SEBS showed improved mechanical properties over the entire composition range (approximately twice higher tensile strength and improved elongation at break).

SEBS was also found to be a good compatibilizer in binary polystyrene/high-density polyethylene (PS/HDPE) blends (17). Results from dynamic me-chanical thermal analysis suggest that the copolymer tends to be located prefer-entially at the interface, with a partial penetration into the polymer phases consid-erably improving the toughness and impact properties. The addition of a small amount of SEBS yielded a finer dispersion of the polymer phases and a higher toughness due to the phase-adhesion-promoting effect of SEBS. Small quantities

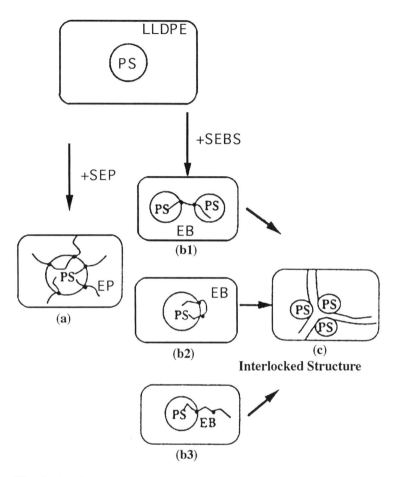

Fig. 2 Possible compatibilization mechanism of the SEBS triblock and SEP diblock copolymers within PS/LLDPE blends. (Reprinted from Ref. 23 by permission of John Wiley & Sons, Inc.)

(up to 2 wt. %) of SEBS also gave good mechanical properties in PS/LLDPE blends (24).

Plawky et al. (21) found that the addition of SEBS to an isotactic polypropylene/linear low-density polyethylene (IPP/LLDPE) blend increased the adhesion between the iPP matrix and the PE domains (a) by modifying the interface and (b) by reinforcing the matrix through its role as a physical crosslinker. In these systems SEBS may act as an interfacial agent, considerably enhancing the impact resistance of such a blend.

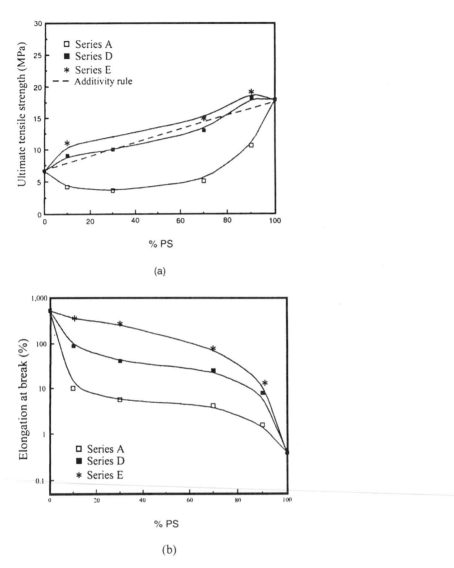

Fig. 3 Effect of SEBS triblock copolymer addition on mechanical properties of PS/LLDPE blends. (a) Ultimate tensile strength; (b) elongation at break. Series A: unmodified PS/LLDPE blend; Series D: PS/LLDPE blends modified by 2% SEBS copolymer; Series E: PS/LLDPE blends modified by 5% SEBS copolymer. (Reprinted Ref. 24 with permission from VSP.)

As previously mentioned, the molecular weight of the interfacial agent might have a strong influence on the mechanical properties of polymer blends. Cigana et al. (25) used two SEBS block copolymers of different molecular weight as interfacial agents for PS/poly(ethylene-*co*-propylene) rubber (EPR) blends. They found that the high-molecular-weight SEBS did not effectively migrate to the interface; furthermore, the stress could not be effectively transferred from the PS to the rubber phase, resulting in a low toughness. The shorter chains of the lower-molecular-weight SEBS saturated the interface and provided an effective stress transfer across the interface, resulting in a ductile fracture mechanism at the saturation level (20 wt. % of SEBS) (see Fig. 4). The arrow in Fig. 4 indicates the transition between brittle and ductile fracture mechanisms (25).

The versatility of SEBS block copolymers was significantly improved by introducing functionalities like maleic anhydride or epoxy groups to the midblock. The modified block copolymers improved mechanical properties of blends consisting of polar engineering polymers like polyamides, polyesters, or polyester-type liquid crystalline polymers (LCPs) with nonpolar polymers like polyolefins (15). Maleic anhydride or epoxy functionalized SEBS polymers were evaluated as potential compatibilizers for poly(ethylene terephthalate)/polypropylene

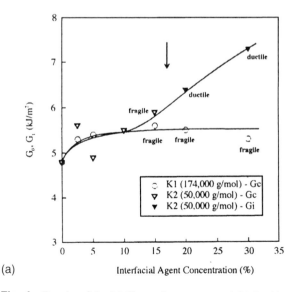

(a)

Fig. 4 Results of the (a) Charpy impact test and (b) Izod impact tests for blends of 80% PS and 20% EPR. K1: SEBS compatibilizer of high molecular weight; K2: SEBS of low molecular weight; Gc: brittle fracture; Gi: ductile fracture. (Reprinted from Ref. 25 with permission from the American Chemical Society.) *(figure continues)*

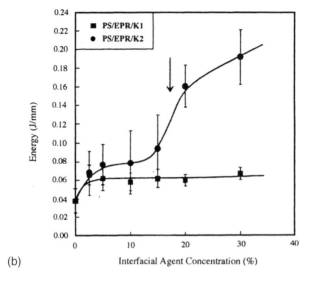

(b)   Interfacial Agent Concentration (%)

Fig. 4 *(continued)*

(PET/PP) blends. Especially epoxy functionalities (glycidyl methacrylate, GMA) have shown to improve toughness properties in combination with relatively high values of strength and modulus (15).

SEBS functionalized with maleic anhydride was found to improve toughness of polyamide 6/polycarbonate (PA6/PC) blends (26–28). PA6 and PC are thermodynamically immiscible and incompatible over the whole temperature and composition range (29,30). As a reactive compatibilizer, anhydride-functionalized SEBS can form chemical linkages through the reaction of anhydride groups with the polyamide chain end groups. Figure 5 schematically shows the possible location of SEBS grafted maleic anhydride (SEBS-gMA) as proposed by Horiuchi et al. (26). As shown in Fig. 6, the addition of SEBS-gMA improves the tensile properties from brittle to ductile, resulting in the yielding and elongation with necking of the specimens. The improvement of tensile properties suggests that the encapsulation of SEBS-gMA around PC domains enhances the interfacial adhesion, and therefore SEBS-gMA acts as an impact modifier for the PA6 matrix (see Fig. 7) (26).

Another class of important industrial polymer blends is based on polyamides and polyolefines. If properly compatibilized, they offer a wide range of desired properties, such as chemical resistance and low water absorption combined with low production costs (31,32). The toughening of polyamide 6/polypropylene (PA6/PP) blends can be achieved by using SEBS-gMA block copolymer as a compatibilizer and impact modifier (Fig. 8).

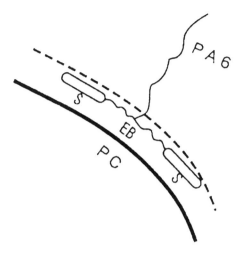

Fig. 5 Suggested illustration of the location of SEBS-gMA on the interface between PA6 and PC. (Reprinted from Ref. 26 with permission from Elsevier Science.)

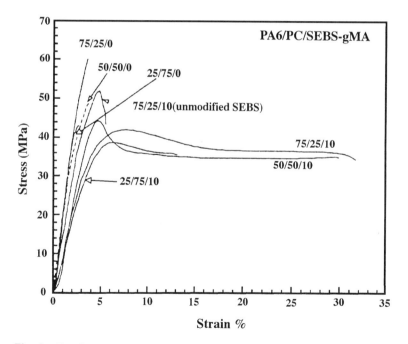

Fig. 6 Tensile stress–strain curves of the PA6/PC uncompatibilized blends and compatibilized with SEBS-gMA and with unfunctionalized SEBS. (Reprinted from Ref. 26 with permission from Elsevier Science.)

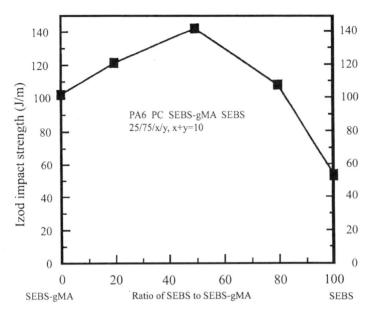

Fig. 7 Notched Izod impact strength of PA6/PC blends as a function of the SEBS-gMA-to-SEBS ratio. (Reprinted from Ref. 26 with permission from Elsevier Science.)

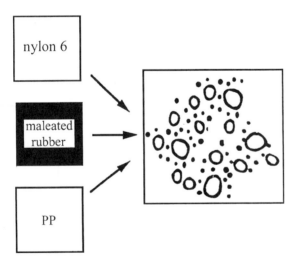

Fig. 8 Schematic representation of the morphology of PA6/PP blends compatibilized by maleated rubber, SEBS-gMA, or EPR-gMA, respectively. (Reprinted from Ref. 31 with permission from Elsevier Science.)

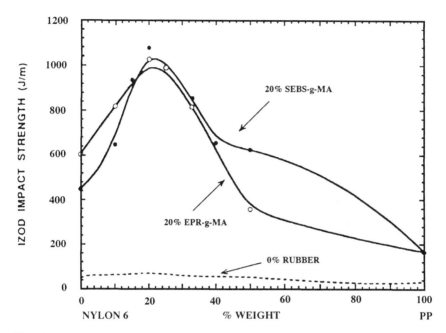

Fig. 9 Izod impact strength of PA6/PP blends modified with 20% of SEBS-gMA and 20% of EPR-gMA. (Reprinted from Ref. 31 with permission from Elsevier Science.)

Mechanical test results (Fig. 9–11) indicated that the addition of maleated SEBS block copolymer improved the impact strength in the range of 0–50 wt. % polypropylene concentration in the blend. The addition of this copolymer to a PA6/PP blend reduced the yield stress together with the modulus.

SEBS was also found to be an effective compatibilizer for PS/PE (33–35), poly(phenylene oxide) (PPO)/PP (36,37), PPO/HDPE (38), and HDPE/PPO/PS (39–41) blends.

E. Different Types of Triblock Copolymer Compatibilizers

Hydrogenated polystyrene-*block*-polyisoprene-*block*-polystyrene (SEPS) triblock copolymer is another specific compatibilizer for the HDPE/PPO/PS blend system (42). The toughness of this polymer blend considerably increases with an increasing amount of the block copolymer at lower temperatures (Table 1). This is due to a successful incorporation of the elastomeric block copolymer in the HDPE/PPO/PS system. A significant decrease in both the tensile strength and the flexural tests was observed, which is often accompanied by an improved toughness in the polymer blend (42).

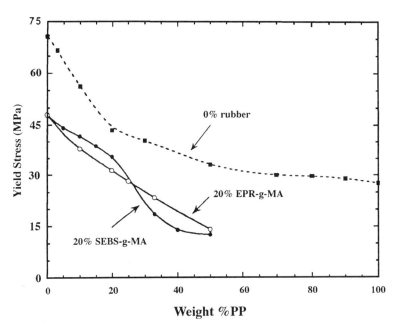

Fig. 10 Yield strength of PA6/PP blends modified with 20% of SEBS-gMA or 20% of EPR-gMA. (Reprinted from Ref. 31 with permission from Elsevier Science.)

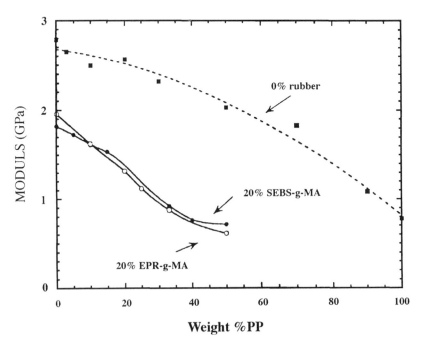

Fig. 11 Modulus of PA6/PP blends modified with 20% of SEBS-gMA or 20% of EPR-gMA. (Reprinted from Ref. 31 with permission from Elsevier Science.)

Table 1 Properties of PE/PPO/PS Blend (Weight Ratio 50:27.5:22.5) as a Function of the Triblock Copolymer Concentration

Property	Unit	Block copolymer concentration (% by weight)	
		7	14
Density	g/cm³	0.985	0.978
Notched Izod impact strength			
23°C	kJ/m²	3.8	12.5
−20°C	kJ/m²	3.7	9.2
Flexural modulus	MPa	800	450
Flexural stress	MPa	19.7	12.3
Tensile modulus of elasticity	MPa	874	453
Yield stress	MPa	12.9	12.1
Ball indentation hardness	N/mm²	31	23
Vicat softening temperature	°C	129	127
Heat deflection temperature	°C	54	51

Source: Reprinted from Ref. 42 by permission of John Wiley & Sons, Inc.

The compatibilizing effect of polystyrene-*block*-poly-*t*-butyl-methacrylate-*block*-polymethylmethacrylate (PS-PtBMA-PMMA) block copolymer was evaluated for immiscible polymer blends of PS/PMMA and PPO/PMMA (43). The use of a triblock copolymer for compatibilization resulted in micelle formation and a significant reduction of the blend particle size. The examined blends were stable during thermal treatment (43). A summary of different triblock copolymers and compatibilized polymer blend systems is presented in Table 2.

Auschra and Stadler have synthesized (59) polystyrene-block-poly(ethylene-co-butylene)-block-polymethylmethacrylate (SEBM) triblock copolymers representing a new design of block compatibilizers successfully combining high modulus with toughness for effective compatibilization of glassy polymer blends (60,62–64).

II. TOUGHNESS MODIFICATION OF PPO/SAN-BLENDS

A. Introduction

Since the 1960s, a homogenous PPO/PS blend (Noryl®) has been commercially available. The heat-distortion temperature (HDT) as well as the processing behavior of this blend can be adjusted by varying the PS content. In comparison to PS, SAN has the advantage of a higher HDT, environmental stress corrosion resistance, chemical resistance, stiffness, and fracture toughness.

Table 2 Different Polymer Blend Systems and Useful Compatibilizers

Compatibilizer	Blending system (matrix ratio + wt% comp.)	Characteristic feature	Ref.
SBS	PS/PP (80/20+6)	Impact, elongation	16
	HIPS/PP (45-70/0/40-5+5-25)	Thermal resistance	44
	iPP/ABS (80/20+5)	Thermal, tensile, morphology	45
	PS/LLDPE (+10)	Impact	35
SEBS	HDPE/PS (45-80/20-6+2-50)	Impact, toughness, tensile	17, 33
	LDPE/PS (30/70 +1-10)	Morphology	18, 34
	PP/PS (71/24-5)	Phase structure	19
	PS/EPR (80/20 +2-4)	Morphology, impact	20, 25
	PBT/PPO	Impact, elongation	46
	iPP/LLDPE (80/20+5-20)	Impact, adhesion	21, 47
	PS/LLDPE (19-20/76-80+5)	Morphology	23
	PS/PE (50-20/80-45+2-50)	Interfacial tension, tensile, electrical properties	22, 24 48, 49
	PPO/PP (5-90/95-10+10-20)	Chemical, thermal, impact	37
	PPO/HDPE (48/32+2-50)	Impact	38
	PPO/PA (5-94/94-5+10-90)	Impact	50
	PPO/PP (30-70/70-30+10)	Impact, tensile, elongation	65
SEBS-gMA	PA6/PC (25-75/75-25+5-15)	Tensile	26
	PA6/PP (9.5/76+5)	Impact	32
	PP/SAN (5-95/95-5+5)	Impact, tensile	51
	PA6/SEBS (85/15-11+1-4)	Toughness	28
SIS	PE/PS	Rheology, morphology	52
	LDPE/PS	Morphology	53

Table 2 *(continued)*

SEPS	HDPE/PPO/PS (50/27.5/22.5+7.14)	Young's modulus, impact	42
	iPP/LLDPE	Morphology	54
	SB-g/LLDPE (5-94/4-95+1-20)	Toughness, stiffness	55
PPO-PSF-PPO	PS/PSF (65/5+30)	Morphology	56
Ionomer SBS	PC/HDPE (80/20+5)	Young's modulus, tensile	57
PS-PtBMA-PMMA	PPO/PMMA (39/70+5)	Morphology	43
PA6-PES-PA6	PES/PA6	Young's modulus, thermal resistance	58
SEBM	PPO/SAN (30-40/70-60+15-30)	Morphology	60–62
	PPO/SAN (40/60+6-38)	Morphology, tensile	64

Like most polymer blends, PPO and SAN are immiscible, and the blend has unsatisfactory mechanical properties due to a coarse phase morphology and weak interfacial adhesion.

In order to enhance the mechanical performance, compatibilizers and toughness modifiers have to be added to control the phase morphology during the blend processing. In the case of PPO and SAN this can be done either by a polystyrene-*block*-polymethylmethacrylate diblock copolymer (SM) compatibilizer and polybutadiene block copolymers containing polymers that are miscible with the corresponding homopolymer phases, or by adding triblock copolymers combining the compatibilizing effect with a toughness modification utilizing an elastomeric midblock. The end blocks of an A-B-C block copolymer have to penetrate into the corresponding miscible matrix components and form entanglements with the constitutive chains to achieve a good interfacial adhesion essential for stress transfer from one phase to the other.

B. Unmodified PPO/SAN Blends

An unmodified PPO/SAN blend (40:60 wt. %) exhibits a coarse two-phase morphology. The diameter of the dispersed phase (in this case PPO) is in the range of 1 μm. Depending on the preparation method, this morphology can be either of a cocontinuous or disperse-like appearance. The mechanical properties of the blend therefore depend on the morphology. In the case of a cocontinuous morphology, for example, the mechanical performance of the blend above the T_g of SAN is much better compared to a disperse morphology. The development of thermal stresses along the interface, due to different thermal expansion coefficients, is also morphology dependent.

C. General Aspects of Compatibilization

In order to compatibilize a PPO/SAN blend with block copolymers, these have to contain components, which are miscible with either the PPO or the SAN. Polystyrene is miscible with PPO, whereas PMMA is miscible with SAN with an acrylonitrile content between 11% and 35% by weight.

For instance, an SM block copolymer would localize itself as an amphiphilic species at the PPO/SAN interface. It would lower the interfacial tension and therefore promote a finer dispersion of the minority component of the blend (here PPO).

In this context, the entanglement molecular weight M_e (66) of the components is of critical importance, because only above this molecular weight are the chains anchored by entanglements (Fig. 12) that bridge and strengthen the interface. Otherwise only a dispersing effect can be expected. The entanglement molecular weights of the components of the block copolymer relative to the corresponding blend partners (i.e., PPO, SAN) are different. In a blend, M_e of the corresponding mixture is a combination of the component M_e values.

The molecular weight also influences the rheological behavior of the components, particularly the viscosity. Concerning melt-mixing techniques, it is easier for lower-molecular-weight compatibilizers, e.g., SM, to reach the interface because of their higher mobility in the melt compared to higher-molecular-weight

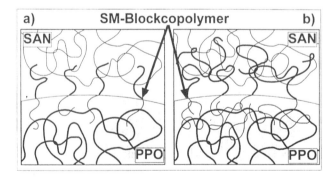

Fig. 12 Schematic representation of the molecular weight influence of SM copolymer at SAN/PPO interface: (a) no entanglements at low molecular weight M_W (SM) $< M_e$; (b) SM entanglements with itself and the matrix components at higher M_W ($\gg M_e$).

SM compatibilizers. Considering these two effects, one has to adjust the molecular weights to match mechanical properties as well as processing properties.

Impact modifiers (see Sec. II.D) can be added to the PPO/SAN blend to further improve the toughness of the otherwise brittle blend. In general, impact modifiers form rubbery domains that initiate multiple crazing in response to an external overloading. Such a crazing helps to dissipate the energy within the sample. It is worth recalling that impact modifiers lower the elastic modulus of the matrix.

D. Conventional Modification Concept

A well-established way to modify immiscible PPO/SAN blends would implicate the use of a diblock compatibilizer (e.g., SM copolymer) and perhaps an additional impact modification by incorporating rubber-toughening agents.

The impact modification can be achieved in the PPO phase (e.g., with SB block copolymers or HIPS) and/or in the SAN phase (e.g., with BM block copolymers). These modifiers are normally block copolymers containing an elastomeric component and one that is miscible with its corresponding matrix.

The elastomeric component of the block copolymer is phase separated from the matrix and forms rubbery domains (micelles in the case of an SB or BM block copolymer) with a low elastic modulus. If a force is applied to the matrix (Fig. 13), the stress at the rubber domains concentrates at their equatorial regions. Ideally each rubber particle affected by this stress concentration initiates crazes so that multiple crazing dissipates the energy in the stressed volume element. The generation and growth of crazes absorbs energy by increasing the surface area. The rubber domains themselves contribute to a smaller extent to the energy dissipation by undergoing an elastic deformation. This requires a certain phase adhesion between

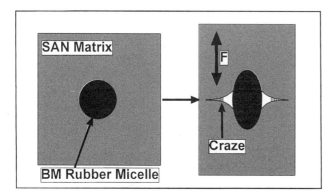

Fig. 13 Schematic representation of conventionally toughness-modified SAN and multiple crazing upon deformation.

the rubber phase and the matrix. The advantages of impact modification by rubber particles are an increased fracture strain, higher fracture energy, and therefore a better damage tolerance. Negative consequences are the lowered overall modulus of the material as well as the often-decreased maximum strength.

As already mentioned, the SM compatibilizer lowers the interfacial energy, and consequently a finer dispersion is achieved. If the molecular weights of the block components exceed a critical value (see Sec. II.C), then an additional strengthening effect at the interface can be expected.

An example of a PPO/SAN blend compatibilized with SM block copolymers and impact modified with BM and SB block copolymers is shown in Fig. 14. The BM block copolymer forms rubber micelles in the SAN, whereas the SB forms the rubber micelles in the PPO. The PPO/SAN interface is enriched with SM block copolymer, resulting in a better dispersion of the PPO. The sizes of these PPO domains are in the range of one micron.

E. SEBM and SBM Block Copolymers as Compatibilizers

Auschra and Stadler (64) used SEBM block copolymers to combine the compatibilizing effect of the end blocks with the toughening effect of the elastomeric midblock. They found a significant improvement in the mechanical properties in comparison to a conventionally modified PPO/SAN blend. The loss especially in the elastic modulus was found to be very small. The high elongation at break combined with the high energy at break proved the efficiency of this concept compared to an unmodified blend.

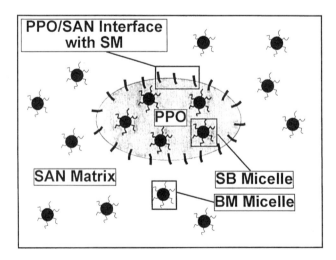

Fig. 14 Schematic representation of a conventionally modified SAN/PPO blend.

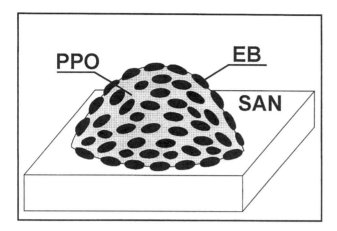

Fig. 15 Schematic representation of a raspberry particle in a SAN/PPO blend with SEBM. EB-microdomains at PPO particle about 10–20 nm thick and 30–70 nm in diameter.

A morphological characterization revealed a special morphology termed raspberry morphology (Fig. 15), with the elastomeric phase discontinuously located at the PPO/SAN interface. This localization provides an additional toughening mechanism at the interface that seems to effectively dissipate the stresses acting on the interface.

The fact that the elastomeric component was located almost exclusively at the PPO/SAN interface and not inside the PPO and SAN phases contributed to the small loss in modulus compared to conventional rubber-toughening modifications.

Successive work (67) on this concept was performed with SBM block copolymers, which are easier to obtain because no hydrogenation step is necessary. Those blends also exhibited enhanced mechanical properties compared to conventionally modified blends, and the investigated samples revealed the raspberry morphology (Figs. 16 and 17).

So far the preparation of the blends modified by SEBM/SBM copolymers was done by coprecipitation of the dissolved blend components followed by a time-consuming workup and subsequent compression moulding of plates. In order to prepare the blends in a manner suitable for technical-scale production, the means of melt processing and extrusion techniques were investigated.

F. Melt Processing of PPO/SAN Blends Containing SBM

Prior to a blend preparation in an extruder, the stability and rheology of the components at the processing temperature have to be considered. The T_g of PPO is

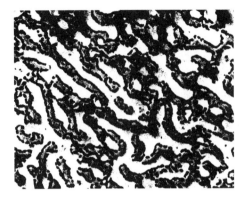

Fig. 16 TEM micrograph of a raspberry morphology from a SAN/PPO blend with SBM (stained with OsO_4).

above 210°C (68) and therefore it is the highest of all components even if blended with a certain quantity of PS. Two components of the SBM copolymer require special attention during melt processing. The PB midblock is sensitive to crosslinking at elevated temperatures as well as in the presence of oxygen. The MMA block has a ceiling temperature of 164°C at atmospheric pressure with its monomer. A processing temperature beyond 240°C should therefore be avoided in order to prevent substantial depolymerization.

Due to their phase separation even in the melt, triblock copolymers show higher melt viscosities compared to homopolymers (69). The high difference in the melt viscosity of SAN and PPO has to be taken into account when considering how the components should be added in a melt-mixing process, e.g., extrusion starting from the pure components.

Experiments with a twin-screw compounder at 240°C revealed that the SBM block copolymer first became dispersed in the SAN component. Cylindrical structures of phase-separated PS surrounded by a perforated PB layer were distributed in the SAN phase. Thermodynamically the interaction parameter between the PS block and the dispersed PPO phase is more favorable than between the PMMA block and the SAN. However, in the early stage of mixing, both the SAN and the SBM melt prior to PPO, causing the SBM to become dispersed in the

SAN. This SBM dispersion increases the viscosity of the SAN, making the subsequent dispersion of the delayed softened PPO easier. This is due to the fact that two phase-separated components of a comparable melt viscosity will give better dispersion results (70).

During this process the SBM chains can reach their most favorable thermodynamic location at the SAN/PPO interface, resulting in the desired raspberry morphology.

It is also important to apply high shear deformations to quickly disperse the SBM domains before the PB block undergoes crosslinking reactions that would complicate or even prevent the dispersion of the SBM chains.

Formation of the raspberry morphology was also accomplished by blend preparation using twins-crew extruders that provide higher shear deformations than single-screw extruders. Subsequent injection molding of these samples yielded specimen that displayed superior mechanical properties in stress–strain experiments with regard to elongation at break and fracture energy (Fig. 18). The elastic modulus decreased only marginally as compared to reference samples.

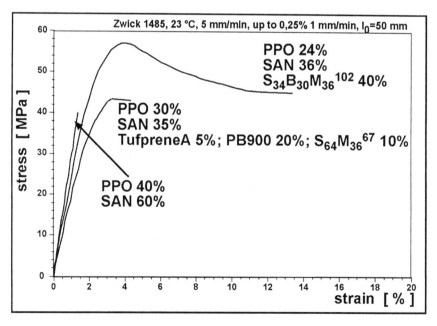

Fig. 17 Stress–strain diagram of a raspberry-morphology sample prepared by coprecipitation and compression moulding.

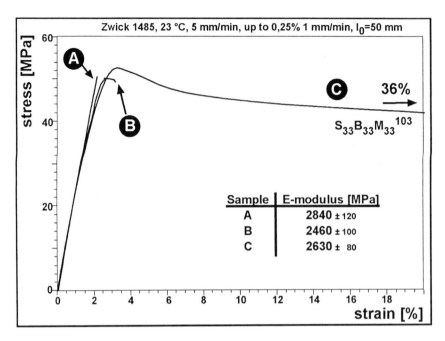

Fig. 18   Stress–strain diagram of a raspberry-morphology sample prepared by extrusion and subsequent injection moulding. Sample A: SAN/PPO (HIPS) 60:40; sample B: SAN/PPO (HIPS)/SB-Rubber 48:32:20; sample C: SAN/PPO (HIPS)/$S_{33}B_{33}M_{33}{}^{103}$ 40:27:33.

G.   Conclusions and Outlook

Compatibilization of immiscible polymer blends by block and graft copolymers has proven to be an efficient strategy for achieving high-performance polymer alloys. The suitability of this approach for industrial developments depends on the availability of copolymers prepared at low costs and the use of existing processing equipment. Polymer blends processed in the melt often result in nonequilibrium situations, where the processing conditions and the addition order of the components accounts for differences in the properties. Further insight in the thermodynamics of mixing in relation to the effects of composition, temperature, and specific interactions at the interface with regard to interfacial tension and thickness are highly desirable.

Triblock copolymers in general have proven to be useful compatibilizers and modifiers for several blend systems. The mechanical properties of the corresponding blend systems can be adjusted by varying the amount of triblock copolymer. Presently a breakthrough in commercial applications cannot be expected because of the relatively high production costs of the triblock copolymers compared to those of most major blend components.

The SEBM and SBM block copolymers in particular have been shown to be effective compatibilizers and impact modifiers at the same time for the PPO/SAN system. In contrast to conventional impact modifiers, they decreased the modulus only marginally by a discontinuous distribution of the elastomeric midblock along the interface. This results in a pronounced strengthening of the interface that could not be achieved by the use of SM copolymer. The latter did not provide such an efficient mechanism to dissipate the stresses acting on the interface upon cooling and mechanical loading.

The SBM content in the blend as well as the SBM composition have to be optimized for the desired property profile. For a commercial application also, some processing parameters may have to be adjusted.

The substitution of one SBM end-block component by another polymer may allow this concept to be adopted for other immiscible blend systems, e.g., polyolefins.

ACKNOWLEDGMENTS

The authors would like to thank Bundesministerium für Bildung und Forschung, BASF AG (Ludwigshafen, Germany) and the working group of Prof. Stadler at the University of Bayreuth (Germany) for their collaboration.

ABBREVIATIONS

ABS	poly(acrylonitrile-*co*-butadiene-*co*-styrene)
BM	polybutadiene-*block*-polymethylmethacrylate
EB	poly(ethylene-*co*-butylene)
EPR	poly(ethylene-*co*-propylene) rubber
EPR-gMA	maleic anhydride grafted EPR
GMA	glycidyl methacrylate
HDPE	high-density polyethylene
HDT	heat distortion temperature
HIPS	high-impact polystyrene
IPP	isotactic polypropylene
LCPs	liquid crystalline polymers
LDPE	low-density polyethylene
LLDPE	linear low-density polyethylene
PA	polyamide
PA6	polyamide 6 (Nylon 6)

PBT	poly(butylene terephthalate)
PC	polycarbonate
PCL	polycaprolactone
PE	polyethylene
PES	poly(ether sulfone)
PET	poly(ethylene terephthalate)
PEtAcr	poly(ether acrylate)
PMMA	polymethylmethacrylate
PP	polypropylene
PPO	poly(2,6-dimethyl-1,4-phenylene oxide)
PS	polystyrene
PSF	polysulfone
PS-PtBMA-PMMA	polystyrene-*block*-poly-*t*-butyl-methacrylate-*block*-polymethylmethacrylate
SAN	poly(styrene-*co*-acrylonitrile)
SB	polystyrene-*block*-polybutadiene
SBM	polystyrene-*block*-polybutadiene-*block*-polymethylmethacrylate
SBS	polystyrene-*block*-polybutadiene-*block*-polystyrene
SEBM	polystyrene-*block*-poly(ethylene-*co*-butylene)-*block*-polymethylmethacrylate
SEBS	polystyrene-*block*-poly(ethylene-*co*-butylene)-*block*-polystyrene
SEBS-gMA	maleic anhydride grafted SEBS
SEP	polystyrene-*block*-poly(ethylene-*co*-propylene)
SEPS	polystyrene-*block*-poly(ethylene-*co*-propylene)-*block*-polystyrene
SIS	polystyrene-*block*-polyisoprene-*block*-polystyrene
SM	polystyrene-*block*-polymethylmethacrylate

REFERENCES

1. O Olabisi, LM Roberson, MT Shaw. Polymer-Polymer Miscibility. New York: Academic Press, 1979, p. 65.
2. LA Utracki. Polymer Alloys and Blends, Thermodynamics and Rheology. New York: Carl Hanser Verlag, 1990, pp. 30–42.
3. S Datta, DJ Lohse. Polymeric Compatibilizers. Uses and Benefits in Polymer Blends. New York: Carl Hanser Verlag, 1996, pp. 10–15.
4. T Ouhadi, R Fayt, Ph Teyssie. J. Appl. Polym. Sci. *32*:5647, 1986.
5. M Taha, V Frerejean. J. Appl. Polym. Sci. *61*:969–979:1996.
6. CG Cho, TH Park, YS Kim. Polymer *38*(18):4687–4696, 1997.

7. AA Adedeij, AM Jamieson. Compos. Interf. *3:*51, 1995.
8. L Häubler, D Pospiech, K Eckstein, A Janke, R Vogel. J. Appl. Polym. Sci. *66:*2293–2309, 1997.
9. NC Liu, WE Baker. Adv. Polym. Techn. *11*(4):249–262, 1992.
10. J Devaux, P Godard, JP Mercier. Polym. Eng. Sci. *22:*229, 1982.
11. S Endo, K Min, JL White, T Kyu. Polym. Eng. Sci. *26:*45, 1986.
12. S Haaga, C Friedrich. Polym. Networks Blends *4*(2):61–67, 1994.
13. CE Koning, R Fayt, W Bruls, L vd Vonderwoort, T Rauch, Ph. Teyssie. Makromol. Chem. Makromol. Symp. *75:*159, 1993.
14. TA Vilgis, J Noolandi. Macromolecules *23:*2941, 1990.
15. M Heino, J Kirjava, P Hietaoja, J Seppälä. J. Appl. Polym. Sci. *65:*241–249, 1997.
16. D Hlavata, Z Horak, V Fort. Polym. Networks Blends *6*(1):15–19, 1996.
17. S Karrad, J-M Cuesta, J Lopez, A Crespy. Plast. Rubber Compos. Process. Appl. *26* (5):193–198, 1997.
18. H Eklind, FHJ Maurer. Polymer *37*(20):4465–4471, 1996.
19. I Fortelny, D Michalkova. Polym. Networks Blends *7*(3):125–131, 1997.
20. P Cigana, V Benoit, A Tremblay, BD Favis. Annu. Tech. Conf. Soc. Plast. Eng. 55th *(Vol. 2):*1527–1531, 1997.
21. U Plawky, M Schlabs, W Wenig. J. Appl. Polym. Sci. *59*(12):1891–1896, 1996.
22. M Bousmina, P Bataille, S Sapieha, HP Schreiber. J. Rheol. (N.Y.) *39*(3):499–517, 1995.
23. L-Y Yang, TG Smith, D Bigio. J. Appl. Polym. Sci. *58*(1):117–127, 1995.
24. M Bousmina, P Bataille, S Sapieha, HP Schreiber. Compos. Interfaces *2*(3):171–86, 1994.
25. P Cigana, BD Favis, C Albert, T Vu-Khanh. Macromolecules *30*(14):4163–4169, 1997.
26. S Horiuchi, N Matchariyakul, K Yase, T Kitano, HK Choi, YM Lee. Polymer *37*(14):3065–3078, 1996; Polymer *38*(1):59–78, 1997; Polymer *38*(26):6317–6326, 1997.
27. CJ Wu, JF Kuo, CY Chen. Polym. Prepr. *34*(1):858–859, 1993.
28. CJ Wu, JF Kuo, CY Chen. Adv. Polym. Technol. *14*(2):129–136, 1995.
29. E Gattiglia, A Turturro, FP La Mantia, A Valenza. J. Appl. Polym. Sci. *46:*1887, 1992.
30. FP La Mantia, A Valenza, E Gattiglia, A Turturro. Intr. Polym. Processing *3:*240, 1994.
31. A Gonzales-Montiel, H Keskkula, DR Paul. Polymer *36*(24):4587–4603, 1995.
32. US Pat. US 5278231, Inv. D. Chundury (Ferro Corp., USA), 1994.
33. S Haaga, HA Schneider, C Friedrich. Polym. Prepr. *34*(2):799–800, 1993.
34. KU Kim. Pollimo *13*(3):229–237, 1989.
35. RM Ruiz, DR Vivas, JL Feijoo, AJ Muller. Mem. Simp. Latinoam. Polim. 3rd: 394–400, 1992.
36. H-T Chiu, D-S Hwung. Eur. Polym. J. *30*(10):1191–1195, 1994.
37. Jap. Pat. JP 07018173 A2, Inv. Y. Fukaya, H Chitoku, S Mya (for Mitsubishi Gas Chemical Co. Japan), 1995.
38. Jap. Pat. JP 07003083 A2, Inv. Y Kurasawa, K Nishida, H Sano, T Miwa (for Mitsubishi Kagaku Kk, Japan), 1995.
39. MC Schwarz, JW Barlow, DR Paul. J. Appl. Polym. Sci. *37:*403, 1989.
40. MC Schwarz, H Keskulla, JW Barlow, DR Paul. J. Appl. Polym. Sci. *35:*653, 1988.

41. MC Schwarz, JW Barlow, DR Paul. J. Appl. Polym. Sci. *35:*2053, 1988.
42. J Schellenberg. J. Appl. Polym. Sci. *64*(9):1835–1842, 1997.
43. P Guegan, AK Khandpur, ChW Macosko. Polym. Prepr. *36*(1):188–189, 1995.
44. Pat. WO 9313168 A1, Inv. SB Swartzmiller, RJ Donald, JE Bonekamp (for Dow Chemical Co., USA), 1993.
45. M Frounchi, RP Burford. Iran. J. Polym. Sci. Technol. *2*(2):59–68, 1993.
46. D Kim, J-H Park, J Kim, J Ahn, H Kim, S-S Lee. Pollimo *20*(4):611–621, 1996.
47. V Flaris, W Wenig, ZH Stachurski. Mater. Forum *16*(2):181–184, 1992.
48. S Haaga, C Friedrich. Polym. Networks Blends *4*(2):61–67, 1994.
49. A Gustafsson, R Salot, UW Gedde. Polym. Compos. *14*(5):421–429, 1993.
50. Pat. EP 297633 A1, Inv. R Vander Meer (for General Electric Co., USA), 1989.
51. Pat. EP 518447 A2, Inv. A Kobayashi, T Kawamura, T Teraya, E Kuchiki, Y Fujita (for Tonen Corp., Japan), 1992.
52. G Kim, M Libera, P Potluri, CG Gogos. Mater. Res. Soc. Symp. Proc. *461:*141–146, 1997.
53. R Potluri, CG Gogos, MR Libera, SS Dagli. Annu. Tech. Conf. Soc. Plast. Eng. *53* (2):3172–3176, 1995.
54. V Flaris, A Wasiak, W Wenig. Annu. Tech. Conf. Soc. Plast. Eng. (1992) *50*(1): 1265–1266, 1992.
55. Pat. EP 580051 A1, Inv. S Seelert, H Hoenl, HD Schwaben (for BASF A.G., Germany), 1994.
56. CJG Plummer, JL Hedrick, JG Hilborn, H-H Kausch. J. Polym. Sci., Part B: Polym. Phys. *34*(13):2177–2192, 1996.
57. N Mekhilef, A Ait Kadi, A. Ajji. Polym. Eng. Sci. *32*(13):894–902, 1992.
58. TO Ahn, S Ch. Hong, HM Jeong. Pollimo *18*(4):521–527, 1994.
59. C Auschra, R Stadler. Polym. Bull. *30:*257, 1993.
60. A Gottschalk, K Muehlbach, F Seitz, R Stadler, C Auschra. Macromol. Symp. *83:* 127–146, 1994.
61. R Stadler, C Auschra. Vysokomol. Soedin., Ser. A Ser. B *35*(11):1802–1807, 1993.
62. C Auschra, R Stadler. Vortr. Poster Symp. Materialforsch. 1991, 2nd (1991), Vol. 1, pp. 213–228.
63. Germ. Pat. DE 4240445 A1, Inv. A. Gottschalk, F Seitz, R Stadler, C Auschra (for BASF A-G, Germany), 1994.
64. C Auschra, R Stadler. Macromolecules *26*(24):6364–6377, 1993.
65. MK Akkapedi, B vanBuskirk. Adv. Polym. Technol. *11*(4):263–275, 1992.
66. GH Michler. Kunststoffmikromechanik: Morphologie, Deformations- und Bruch-mechanismen. Munich: Hanser Verlag, 1992, pp. 12–13.
67. T Kirschnick, G Broza, V Altstädt, C Mehler, A Gottschalk, R Stadler. The effects of adding polystyrene-*block*-polybutadiene-*block*-polymethyl-methacrylate triblock copolymers on the mechanical properties of PPO/SAN blends. 6th European Polymer Federation Symposium on Polymeric Materials, Greece, Oct. 1996.
68. J Brandrup, EH Immergut. Polymer Handbook. 3rd ed. New York: Wiley, 1989.
69. DJ Walsh, JS Higgins, A Maconnachie. Polymer Blends and Mixtures. NATO ASI Series E: Applied Sciences, No. 89. Boston: Matinus Nijhoff, 1985, pp. 190–191.
70. CD Han. Multiphase Flow Behavior in Polymer Processing. New York: Academic Press, 1981.

19
Poly(ether-*b*-ester) Copolymers, Blends, Composites, and Their Applications

Regina Jeziórska, Witold Zielinski, and Edward Grzywa
Industrial Chemistry Research Institute, Warsaw, Poland

I. INTRODUCTION

Poly(ether-*b*-ester) copolymers (PEE-*b*-EC) belong to the group of multiblock thermoplastic elastomers characterized by high elasticity and good mechanical properties combined with thermoplastic properties. This unique combination of properties is due to the specific chemical structure and distribution of the segments along the macromolecule. In these systems, the hard segments of high T_g are easily crystallizable blocks that form domain structures, whereas the soft segments of low T_g contribiute to the formation of the amorphous continuous matrix. This immiscibility of segments and their two-phase microstructure results in a unique behavior of the multiblock elastomers. In PEE-*b*-EC, the hard segments are oligoester chains of various molecular weights (MW), whereas the soft segments are usually aliphatic oligoethers of MW ranging from 300 to 6000. The chemical nature of the hard segments imparts a good mechanical strength, hardness, and high temperature and solvent resistance to the material. The nature of the soft segments is responsible for hydrolysis resistance, elastic properties, and behavior at low temperatures of PEE-*b*-EC. Compounds with electrical and a variety of other properties are also available. Because of the unique combination of properties, PEE-*b*-EC have found several commercial applications in the automotive, machine, wire, and cable industries and other areas, despite their high price, ca. three times that of styrenic elastomers (1,2).

In general, commercial poly(ether-*b*-ester) copolymers are based on poly(butylene terephthalate) (PBT), which forms the hard phase, and poly(tetramethylene oxide) (PTMO), which forms the soft one. The companies that specialize worldwide in this type of elastomer production are: Du Pont, United States (Hytrel), Akzo-Plastics, Holland (Arnitel), Sumitomo, Japan (Sumitex), and Elana, Poland (Elitel).

The PEE-*b*-EC copolymers have hardness values in the 34–72 Shore D range and are available in pellet, granular, or powder form. The grades are offered for extrusion, injection, and rotational molding on thermoplastic processing equipment. Typical applications include hoses, tubing, wire coating, film, injection molded mechanical goods, and flexible couplings (3).

Properties are said to include good weatherability (grades stabilized with carbon black) and high impact strength and minimum loss of flexibility at low temperatures, as well as the ability to perform well at elevated temperatures. The PEE-*b*-EC copolymers also offer high resistance to abrasion, flex fatigue, deformation under moderate stress, and a broad range of chemicals, solvents, oils, and greases. Toughness, low specific gravity, and low-temperature properties make them suitable for use in agricultural and garden equipment, recreational vehicles, sporting goods, toys, etc. (3–6).

The softest materials of the PEE-*b*-EC copolymers have a high degree of flexibility and impact resistance at low temperatures. Their rubberlike qualities and high elongations make them particularly suitable for shoe soling. The grades with hardness values of 50–60D have a medium melt viscosity and are particularly useful for molding complex and/or thin-walled parts. Further up along the hardness scale there are also grades with Shore hardness values of 63–72D. These materials are said to offer the best combination of flexibility at high and low temperatures, as well as resistance to impact and creep (6,7).

Poly(ether-*b*-ester) copolymers possess mechanical properties that qualify them, uniquely, for a number of demanding applications. These properties are achieved only when the polymers are correctly processed. Owing to its inherent melt stability, PEE-*b*-EC can be processed over a broad temperature range, depending on polymer type and process. Melt temperatures for most grades should be 177–232°C. Higher melt temperatures, 260–271°C, have been used for thin-walled parts to aid filling the mold, while thicker parts can be molded with melt temperatures very close to the polymer melting points. Injection pressures of 40–90 MPa are typical (7).

To achieve intermediate properties, standard grades can be blended together or with other polymers, and also with various fillers in order to improve their mechanical properties. The ability to upgrade a commodity made of standard materials by including special fillers or by specific process techniques means that such plastics can be used in many "engineering applications." Furthermore, blends of PEE-*b*-EC and thermoplastic polymers and their filled or reinforced grades cost less than the pure polymer. This is a very important aspect from an economic point of view.

This chapter describes in detail the poly(ether-*b*-ester) (Elitel™, Elana SA, Poland) copolymer blends with thermoplastic polymers in order to improve their mechanical properties, and also modified products with various fillers.

II. EXPERIMENTAL PART

A. Preparation of Polymer Blends

The blends of poly(ether-*b*-ester) copolymers and thermoplastic polymers and/or fillers were obtained in the melt by using a ZE-25 twin-screw extruder (screw diameter $D = 25$ mm, $L/D = 33$) made by Berstorff, Germany. The blend components were mixed together by using gravimetric and volumetric dosing units. In the last step of extrusion molding, the molten polymers were vacuum degassed. The temperature in each barrel section of the extruder was 190–230°C, and it was highly dependent on the grade of poly(ether-*b*-ester) copolymer used and the thermoplastic polymer content.

B. Investigational Method

The mechanical properties of the injection molded specimens were tested with an Instron 4505 tensile tester and a Zwick apparatus operated in the tensile and flexural mode.

The RDS-2 rheospectrometer Rheometrics was used for dynamic rheological measurements. The dynamic viscosities η^*, η' and η'' and dynamic shear moduli G^*, G', and G'' were measured as a function of frequency in the range from 10^0 to 5×10^2 rad/s at constant strain, at three selected temperatures: 165, 205 and 215°C. Test temperatures close to the melting range of PEE-*b*-EC components were chosen for preliminary investigations. The measurements were performed in the range of linear viscoelasticity that was determined beforehand from the plots of G' vs. strain (8).

Differential scanning calorimetry (DSC) was performed on a Perkin-Elmer (DSC-2) apparatus. The process was carried out in a triple cycle "heating-cooling-heating" in the temperatures range -50–250°C. The rate of heating and cooling was 10°C min^{-1}.

III. BLENDS OF POLY(ETHER-*b*-ESTER) COPOLYMERS AND THERMOPLASTIC POLYMERS

A number of thermoplastic polymers have been studied in blends with poly(ether-*b*-ester) copolymers. Some of them may be of greatest commercial value, e.g., atactic polystyrene (PS) and poly(butylene terephtalate) (PBT).

A. Polystyrene Blends

Polystyrene is compatible with soft segments and immiscible with the hard seg-ments of PEE-*b*-EC (9). It can increase hardness and improve tensile and flex properties (10,11). The PEE-*b*-EC/PS blends consist of hard polyester blocks and hard polystyrene phase interspersed with amorphous regions that impart elas-tomeric qualities (11). This phase-separated microstructure is a typical morpho-logical feature of block copolymers with incompatible sequences and is due to the crystallization ability of the hard segments in the molten state. This may be also explained by the immiscibility of PS and the hard segments of PEE-*b*-EC (11–13).

In blends with PEE-*b*-EC, PS offers a wide range of mechanical properties and stiffness. The results of the investigations show that the increase of mechani-cal properties is connected with a growing amount of PS. It is also highly depen-

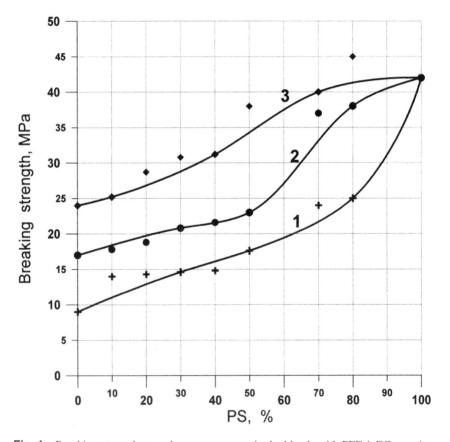

Fig. 1 Breaking strength vs. polystyrene amount in the blends with PEE-*b*-EC contain-ing (1) 50%, (2) 60%, and (3) 80% of the hard segments.

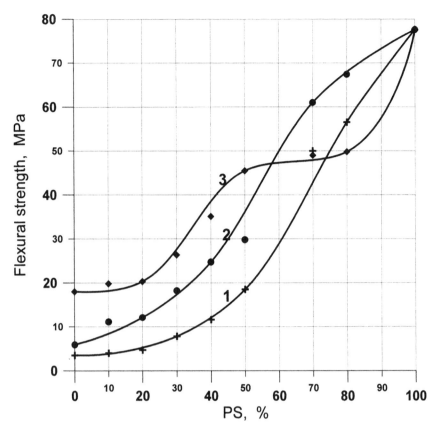

Fig. 2 Flexural strength vs. polystyrene amount in the blends with PEE-*b*-EC containing (1) 50%, (2) 60%, and (3) 80% of the hard segments.

dent on the grade of poly(ether-*b*-ester) copolymer used. The dependence of breaking and flexural strength vs. PS content in the blends is shown in Figs. 1 and 2, respectively. For PS contentration below 30%, PEE-*b*-EC/PS blends retain elastic properties and exhibit high impact strength (no cracks at −40°C). The mechanical properties of the blends are comparable with these of elastomers even up to 40% of PS when PEE-*b*-EC containing 50% of the hard segments is used. Thermal stability and shape stability at elevated temperatures are also higher (Figs. 3 and 4), and stiffness is improved as compared with PEE-*b*-EC (Tables 1–3). When the PS content is increased further, the mechanical properties of the blends change into the thermoplastic behavior. The blends show good impact strength, small elongation at break, and high tensile and flexural modulus. The stress/strain characteristics of the blends as compared with thermoplastics, vulcanized rubbers, and thermoplastic rubbers are shown in Fig. 5.

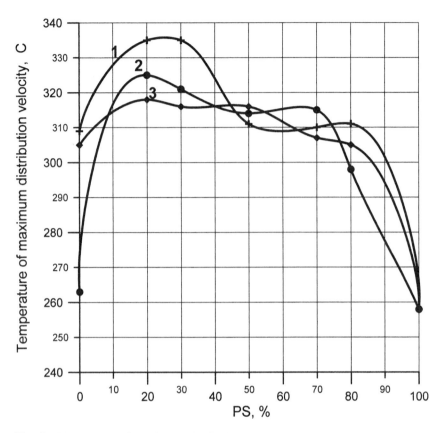

Fig. 3 Temperature of maximum distribution velocity vs. polystyrene amount in the blends with PEE-*b*-EC containing (1) 50%, (2) 60%, and (3) 80% of the hard segments.

As evident from Tables 1–3, an increase of the hard-segment amount in PEE-*b*-EC results in higher breaking and flexural strength, tensile and flexural modulus, hardness and deflection temperature (Fig. 4), on the one hand, and lower elongation at break and impact strength on the other (11,12,14).

The material is designed for typical applications of thermoplastic elastomers. Therefore its flow properties are essential for characterizing the processing behavior. The dependence of the dynamic complex viscosity η^* vs. frequency for the PEE-*b*-EC/PS blends is shown in Figs. 6–8. The shear thinning effect, as well as an influence of the PS component on the values of η^*, is observed for all blends. It was found that the grades show viscoelastic properties in the frequency range from 10^0 to 10^2 rad/s, similar to the unmodified PEE-*b*-EC (11).

All grades exhibit a sharp crystalline melting point that increases (and becomes sharper) with increasing hardness and crystallinity. Melt viscosity is strongly dependent upon melt temperature, and this dependency increases with increasing hardness. The crystallization rate increases with increasing hardness. The moisture content of all grades of PEE-*b*-EC/PS blends must be 0.1% in order to achieve an acceptable processing characteristic with minimum degradation (11).

The PEE-*b*-EC/PS blends show hardness values in the range from 35 to 78D and are available in granular form. They can be injection molded and extruded. All grades can be molded on all types of screw-injection molding machines, but screws with gradual transition sections of the type used for polyethylene are recommended. The *L/D* (length-to-diameter) ratio can vary from 18:1 to 24:1. Longer

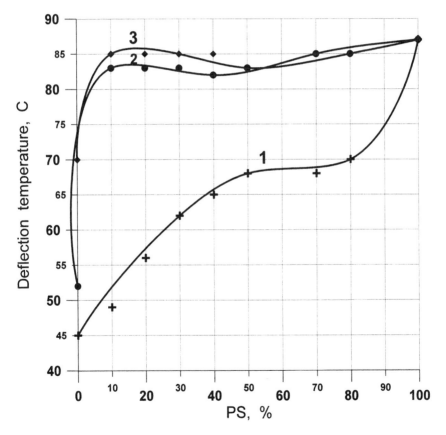

Fig. 4   Deflection temperature vs. polystyrene amount in the blends with PEE-*b*-EC containing (1) 50%, (2) 60%, and (3) 80% of the hard segments.

Table 1   Properties of PEE-*b*-EC/PS Blends with PEE-*b*-EC Containing 50% of the Hard Segments

Property	PS, %								
	0	10	20	30	40	50	70	80	100
Breaking strength, MPa	9.0	14.0	14.3	14.6	14.8	17.6	24.0	25.0	42.0
Elongation at break, %	>590	>490	>490	222	84	26.0	23.0	21.0	2.6
Tensile modulus, MPa	135	180	224	236	396	736	2000	2330	3010
Flexural modulus, MPa	120	187	255	281	435	792	1980	2450	3000
Flexural strength, MPa	3.5	4.0	4.7	7.8	11.6	18.5	50.0	56.5	77.6
Impact strength Charpy at 23°C, kJ/m²	nb	nb	nb	nb	nb	nb	50% nb	39.4	24.8
Impact strength Charpy at −40°C, kJ/m²	nb	nb	nb	nb	nb	nb	42.5	36.2	17.6
Impact (notched) Charpy at 23°C, kJ/m²	nb	nb	nb	nb	26.0	12.7	5.2	4.2	1.9
Impact (notched) Charpy at −40°C, kJ/m²	nb	nb	nb	nb	16.0	9.1	5.0	3.2	1.5
Hardness Shore, °ShD	34	35	40	45	50	54	65	73	84
Deflection temperature under load of 0.45 MPa, °C	45	49	56	62	65	68	68	70	87

Source: Refs. 11, 12, and 14.
nb = no break.

Table 2   Properties of PEE-*b*-EC/PS Blends with PEE-*b*-EC Containing 60% of the Hard Segments

Property	PS, %								
	0	10	20	30	40	50	70	80	100
Breaking strength, MPa	17.0	17.8	18.8	20.8	21.6	23.0	37.0	38.0	42.0
Elongation at break, %	>450	>270	63	21	8.7	7.1	6.0	4.0	2.6
Tensile modulus, MPa	230	318	418	589	765	1110	2230	2410	3010
Flexural modulus, MPa	210	345	433	633	885	1220	2450	2575	3000
Flexural strength, MPa	5.9	11.1	12.1	18.2	24.7	29.8	61.0	67.4	77.6
Impact strength Charpy at 23°C, kJ/m²	nb	nb	nb	nb	nb	nb	29.9	28.5	24.8
Impact strength Charpy at −40°C, kJ/m²	nb	nb	nb	nb	40% nb	45.1	22.5	20.7	17.6
Impact (notched) Charpy at 23°C, kJ/m²	nb	nb	19.3	7.5	5.0	4.8	4.0	2.5	1.9
Impact (notched) Charpy at −40°C, kJ/m²	nb	18.4	10.0	6.4	3.2	2.5	2.0	1.9	1.5
Hardness Shore, °ShD	54	58	58	64	66	69	76	77	84
Deflection temperature under load of 0.45 MPa, °C	52	83	83	83	82	83	85	85	87

Source: Refs. 11, 12, and 14.
nb = no break.

Table 3 Properties of PEE-*b*-EC/PS Blends with PEE-*b*-EC Containing 80% of the Hard Segments

Property	PS, %								
	0	10	20	30	40	50	70	80	100
Breaking strength, MPa	24.0	25.2	28.7	30.8	31.2	38.0	40.0	45.0	42.0
Elongation at break, %	>330	>450	>390	19.0	12.0	5.7	4.0	2.6	2.6
Tensile modulus, MPa	570	625	830	1120	1490	1830	2700	2740	3010
Flexural modulus, MPa	460	498	693	939	1270	1610	2500	2620	3000
Flexural strength, MPa	18.0	19.8	20.3	26.3	35.1	45.5	49.0	49.8	77.6
Impact strength Charpy at 23°C, kJ/m^2	nb	nb	nb	nb	nb	nb	53.9	48.3	24.8
Impact strength Charpy at −40°C, kJ/m^2	nb	nb	nb	nb	nb	60% nb	32.5	26.2	17.6
Impact (notched) Charpy at 23°C, kJ/m^2	nb	80% nb	21.3	11.8	9.8	6.6	5.0	4.0	1.9
Impact (notched) Charpy at −40°C, kJ/m^2	7.5	12.2	11.0	8.9	5.3	4.6	3.8	3.0	1.5
Hardness Shore, °ShD	64	65	66	68	71	73	75	78	84
Deflection temperature under load of 0.45 MPa, °C	70	85	85	85	85	83	85	85	87

Source: Refs. 11, 12, and 14.
nb = no break.

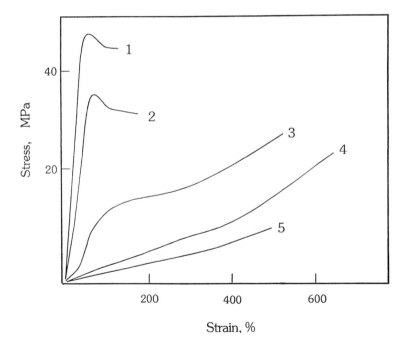

Fig. 5 Stress/strain curves of various polymers (tensile test). 1 = thermoplastic (PS); 2 = PEE-*b*-EC/PS (30/70) blends; 3 = PEE-*b*-EC/PS (70/30) blends; 4 = vulcanized rubbers; 5 = thermoplastic rubbers.

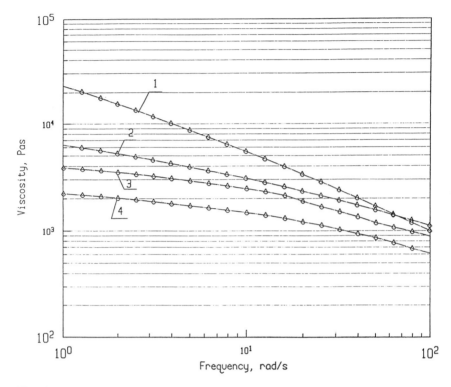

Fig. 6 Complex viscosity η^* vs. frequency. PEE-*b*-EC/PS blends with PEE-*b*-EC containing 50% of the hard segments at 165°C. 1 = PS; 2 = PEE-*b*-EC/PS 70/30; 3 = PEE-*b*-EC/PS 90/10; 4 = PEE-*b*-EC containing 50% of the hard segments.

L/D ratios are preferred to provide a more uniform melt temperature and better mixing. A compression ratio for the screw between 3.0:1 and 3.5:1 is suggested. Higher ratios will cause excessive machine amperage and polymer shear, while lower compression ratios may not produce a uniform melt (11).

Owing to inherent melt stability (11,13), PEE-*b*-EC/PS blends can be processed over a broad temperature range, depending on polymer type and process. Typical injection molding conditions for these blends are shown in Table 4. They are based on studies made with a screw machine Model ES 25/50 manufactured by Engel. Melt temperatures for most grades should be 146–222°C. Melting points determined by differential scanning calorimeter are shown in Table 5. Mold-cooling temperatures from 25°C to 50°C have been used. A lower temperature is better for the lowest PS contents. Higher temperatures are compared with lower elongation at break. Injection pressures of 60–95 MPa are typical (11).

Extruders suitable for use with common thermoplastics such as nylon or polyolefins are also suitable for extrusion of PEE-*b*-EC/PS blends, but screws with gradual transition sections of the type used for polyethylene are recommended. Length-to-diameter ratios of at least 20:1, and preferably 24:1, provide the best melt quality for precision extrusions. A compression ratio for the screw between 3.0:1 and 3.5:1 is recommended. The three sections of the screw are the feed, transition, and metering zones. Feed sections comprising up to 50% of the total screw length have been used successfully. The length of the transition section should be at least 25% of the total screw length. The screw should also have a rounded head to eliminate dead spots in front of the screw, where the polymer can degrade (11).

Selected PEE-*b*-EC/PS blends are recommended for many applications, such as high-impact machine parts working at temperatures from −40 to 100°C, hydraulic hoses working at low temperatures down to −40°C, parts of winter

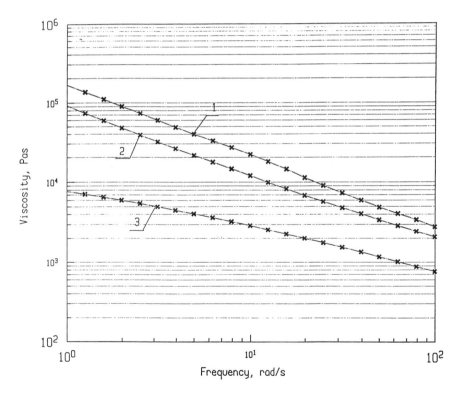

Fig. 7 Complex viscosity η^* vs. frequency. PEE-*b*-EC/PS blends with PEE-*b*-EC containing 60% of the hard segments at 205°C. 1 = PEE-*b*-EC containing 60% of the hard segments; 2 = PEE-*b*-EC/PS 70/30; 3 = PEE-*b*-EC/PS 20/80.

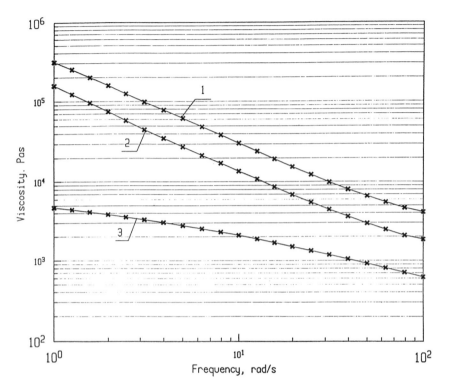

Fig. 8   Complex viscosity η* vs. frequency. PEE-*b*-EC/PS blends with PEE-*b*-EC containing 80% of the hard segments at 215°C. 1 = PEE-*b*-EC containing 80% of the hard segments; 2 = PEE-*b*-EC/PS 70/30; 3 = PEE-*b*-EC/PS 20/80.

sports equipment, housings and parts of domestic appliances, electrical and electronic flexible parts and cases, flexible couplings, and silent-running gears.

B. PBT Blends

PEE-*b*-EC/PBT blends were investigated mainly in order to develop new thermoplastic elastomers of high impact strength for use at low temperatures (below −40°C). The results proved that this aim could be achieved—no break was observed in unnotched specimens even at −50°C, and the impact strength for notched specimens at this temperature was very high (20 kJ/m^2).

Selected properties of these blends are presented in Table 6. The mechanical strength and hardness are seen to increase, and impact strength and elongation at break decrease as the PBT content in the blend is increased. A blend with opti-

Table 4 Typical Injection Molding Conditions for PEE-*b*-EC/PS Blends

	Hard-segment amount in PEE-*b*-EC					
	50		60		80	
	PS content in the blend, %		PS content in the blend, %		PS content in the blend, %	
Condition	30	70	30	70	30	70
Feed throat jacket	Cooled	Cooled	Cooled	Cooled	Cooled	Cooled
Cylinder temperature, °C						
Rear	172–177	152–172	185–200	185–198	194–209	193–208
Center	177–184	172–179	200–211	198–211	209–219	208–216
Front	177–189	172–184	200–219	198–219	209–225	208–222
Nozzle	177–195	172–190	200–219	198–219	209–225	208–222
Melt temperature, °C	176–195	171–190	205–219	203–219	214–225	213–222
Screw speed, rev/min	70–100	70–100	70–100	70–100	70–100	70–100
Screw retraction time	Equal to hold time, if possible	Equal to hold time, if possible	Equal to hold time, if possible	Equal to hold time, if possible	Equal to hold time, if possible	Equal to hold time, if possible
Mold temperature, °C	25	40	25	50	25	50
Cycle time*, s						
Injection	12	12	12	12	12	12
Booster	3	3	3	3	3	3
Hold	20	20	15	15	15	15

Source: Ref. 11
* Cycle time depends on part dimension and shape.

Table 5 Melting Points and Beginning of Rapid Distribution of PEE-*b*-EC/PS Blends Determined by DSC

	Temperatures, °C		
PS, %	Peak of endotherm	Melt complete	Beginning of rapid distribution
PE-*b*-EC containing 50% of the hard segments			
30	146	161	223
50	152	152	216
70	156	156	199
PE-*b*-EC containing 60% of the hard segments			
30	205	214	220
50	206	212	220
70	203	214	220
PE-*b*-EC containing 80% of the hard segments			
30	214	220	229
50	214	221	229
70	213	219	227

Source: Ref. 11.

Table 6 Properties of PEE-*b*-EC/PBT Blends with PEE-*b*-EC Containing 50% of the Hard Segments

	PBT, %				
Property	0	50	75	90	100
Breaking strength, MPa	9	22	29	33	36
Elongation at break, %	>590	24.0	3.9	3.2	3.0
Flexural strength, MPa	3.5	30	46	62	70
Flexural modulus, MPa	120	900	1400	2100	2500
Impact strength Charpy at 23°C, kJ/m^2	nb	nb	nb	25	25
Impact strength Charpy at −20°C, kJ/m^2	nb	nb	41	25	20
Impact Charpy at −50°C, kJ/m^2	nb	nb	28	11	16
Impact (notched) Charpy at 23°C, kJ/m^2	nb	22.4	12.1	5.6	1.4
Impact (notched) Charpy at −50°C, kJ/m^2	nb	16.7	10.6	4.7	1.4
Hardness, °ShD	34	59	69	72	75

nb = no break.

mum properties for a application can easily be selected with reference to these results supplemented with impact strength measurements at very low temperatures.

Available in pellet or granular form, PEE-*b*-EC/PBT blends are engineering thermoplastics designed primarily for injection molding on all types of screw-injection molding machines, but screws with gradual transition sections of the type used for polyethylene are recommended.

Varying in hardness from 59 to 72 Shore D, these products are intended for a broad range of applications, including gaskets for piping and refrigeration machinery, as well as for any machinery working at extremely low temperatures, vacuum cleaners casings, utensils, plugs, telephone elements, computer keyboards, headlight casings, lamp elements, car body elements, safety belt elements, ski boot elements and buckles, slide bearings, gears, optical and medical instrument elements, tape recorder elements.

IV. COMPOSITES BASED ON POLY(ETHER-*b*-ESTER) COPOLYMERS

Filler geometry expressed by an *L/D* (length-to-diameter) ratio with interfacial adhesion is the principal agent for qualifying properties of filled polymers. Table 7 includes some of the filler characteristics and their effect on the properties of modified polymers. Filled grades have a particularly higher modulus of elasticity, hardness, and utilization temperatures as compared with properties of the matrix polymer.

As a rule, high interfacial adhesion is difficult to obtain because of the lack of affinity between an organic matrix and generally an inorganic filler. To increase adhesion, fillers are usually coated with chemically active coatings (silanes, stearates, and titanates derivatives).

The consumption structure of fillers indicates a predominance of chalk and glass fibers. Both account for around 70% of the total use of fillers. Chalk is used because of accessibility, low cost, and improvement in mechanical properties.

Adding fillers to poly(ether-*b*-ester) copolymers can modify properties, particularly rigidity, impact strength, resistance to deformation under heat, and dimensional stability, and reduce material costs. These properties make filled copolymers far more suitable for specific applications than unfilled. And they can be used in many different industries: automotive, aerospace, building construction, furniture, sporting goods, marine, oil, electronics, among many more (15–18).

The following fillers have been investigated in blends with PEE-*b*-EC: cut glass fibers (3-mm length) coated with aminosilane; very short cotton fibers (maximum length 0.5 mm); calcium metasilicate (length <35 microns, diameter 3.5 microns); mica powder (diameter <50 microns); graphite (diameter <60 microns); glass microballs coated with aminosilane (diameter <100 microns); carbon black powder (diameter <20 microns); quartz powder (diameter <30 microns); chalk (diameter <30 microns); and MoS_2 powder (diameter 4.3 microns). The effect of an increase in filler content on the properties of PEE-*b*-EC varied with the filler and coating.

Table 7 Inorganic Fillers

Fillers	Aspect ratio (*L/D*)	Improved properties
Carbon black	1	Weatherability, mechanical and heat resistance, conductivity
Quartz powder	1	Chemical resistance, surface properties
Chalk	1	Mechanical properties
Glass microballs	1	Mechanical properties, rigidity
Mica	1:20–1:30	Mechanical properties, heat distortion, temperature, chemical, and water resistance, dielectric and surface properties
Graphite	<10	Friction, heat conductivity, rigidity
MoS_2	<10	Abrasion resistance, coefficient of friction
Wollastonite	3–15	Mechanical and heat resistance
Cotton fiber	>10	Stiffness, rigidity, deflection temperature
Glass fiber	10–100	Mechanical properties, mechanical and heat resistance

Adding glass fibers to PEE-*b*-EC results in a considerable expansion of the range and fields of application of these materials, thanks to an improvement in their strength characteristics, above all the elastic modulus and bending strength (Table 8). Simultaneously, a high impact strength is retained. It is also obvious that glass fiber causes a drastic reduction of elasticity of poly(ether-*b*-ester) copolymers connected with a decreasing elongation at break. With Shore hardnesses ranging from 58 to 70D, these materials are designed primarily for injection molding. The results (Table 8) indicate that the most useful properties of these reinforced materials are obtained for a glass fiber content in the range 20–30%. These properties are achieved when the glass fiber is coated with suitable chemically active coatings; aminosilane coating is especially recommended.

Reinforced grades can be used mainly in the field of electronics, the electric and power industries, and the automotive and aircraft industries. These materials can be particularly useful for the manufacture of parts with required high strength properties and impact resistance at low temperatures (15,16,18).

Cotton-fiber-filled PEE-*b*-EC have the lowest specific gravity of any filled or reinforced PEE-*b*-EC. It is understood that cotton-fiber-filled poly(ether-*b*-ester) copolymers have higher stiffness than unfilled ones (Table 9). The improvement in rigidity and deflection temperature was dependent on the amount of filler and the grade of PEE-*b*-EC (17).

Table 10 indicates the influence of calcium metasilicate (Wollastonite) concentration on the mechanical properties of PEE-*b*-EC blends. The grades filled with calcium metasilicate showed improved properties, such as impact strength, rigidity, and elasticity. This unique combination of properties is due to the acicular shape of the filler. It is important that filled PEE-*b*-EC have very high elongation at break, even up to 30% of calcium metasilicate contents (16,18).

The effect of the amount of calcium metasilicate on the properties of PEE-*b*-EC/PBT blends is illustrated in Table 11. The grade comprising 40% of PBT and modified with 10% of the filler offers the highest increase in rigidity and the minimal loss of elasticity. Because of the unusual combination of high impact strength at low temperatures and improved tensile and flexural properties, this blend can be used as an engineering elastomer.

Table 12 presents the influence of lamellar materials such as mica and graphite on the mechanical properties (including hardness) of PEE-*b*-EC containing 50% of the hard segments. According to the foregoing results, the improvement in breaking and flexural strength, flexural modulus, and hardness was observed. It was highly dependent on the contents of the filler and their shape (16,18).

To improve the weatherability of PEE-*b*-EC, carbon black can be incorporated (4). The properties of PEE-*b*-EC containing 50% of the hard segments modified with different amounts of carbon black are presented in Table 13. As compared with the unmodified copolymers, filled grades have better tensile and

Table 8 Properties of PEE-*b*-EC Reinforced with Short Glass Fiber

	Hard-segment amount in PEE-*b*-EC											
	50				60				80			
	Glass fiber, %				Glass fiber, %				Glass fiber, %			
Property	0	10	20	30	0	10	20	30	0	10	20	30
Breaking strength, MPa	9	28	60	78	18	38	64	78	24	47	69	79
Elongation at break, %	>590	20	11	10	>450	15	10	8	>330	11	9	7
Flexural strength, MPa	3.5	12.1	34.0	55.0	5.9	34.0	46.0	71.0	18.0	45.0	76.0	80.0
Flexural modulus, MPa	120	450	1350	2530	210	1300	1800	3200	460	1650	3100	4100
Impact strength Charpy, kJ/m^2	nb	nb	nb	68	nb	nb	71	60	nb	86	68	58
Impact (notched) Charpy, kJ/m^2	nb	46	25	25	nb	36	25	25	nb	25	20	17

Source: Refs. 15, 16, and 18.
nb = no break.

Table 9 Properties of PEE-b-EC Filled with Cotton Fiber

Property	Hard-segment amount in PEE-b-EC											
	50				60				80			
	Cotton fiber, %				Cotton fiber, %				Cotton fiber, %			
	0	5	10	20	0	5	10	20	0	5	10	20
Specific gravity, g/cm³	1.19	1.21	1.23	1.25	1.21	1.23	1.25	1.27	1.22	1.25	1.27	1.29
Breaking strength, MPa	9.0	13.0	14.9	19.2	18.0	17.9	19.3	23.2	24.0	28.3	28.5	28.7
Elongation at break, %	>590	290	48	18	>450	181	39	19	>330	135	32	16
Flexural strength, MPa	3.5	5.7	8.7	17.0	5.9	10.0	13.6	19.4	18.0	20.3	20.3	20.5
Flexural modulus, MPa	120	200	300	690	210	340	480	730	460	480	560	760
Impact strength Charpy at 23°C, kJ/m²	nb	nb	nb	nb	nb	nb	nb	nb	nb	nb	nb	nb
Impact strength Charpy at −40°C, kJ/m²	nb	nb	nb	42	nb	46	37	40	nb	40	38	35
Impact strength Charpy at −60°C, kJ/m²	nb	nb	nb	42	nb	46	37	30	35	34	33	27
Impact (notched) Charpy at 23°C, kJ/m²	nb	nb	nb	19.1	nb	nb	14.6	9.8	nb	nb	13.0	10.4
Impact (notched) Charpy at −40°C, kJ/m²	nb	36.0	16.0	13.2	nb	25.0	12.4	9.8	7.5	7.3	7.0	6.8
Impact (notched) Charpy at −60°C, kJ/m²	4.8	4.8	4.8	4.6	4.4	4.2	4.2	3.8	4.1	4.1	4.0	3.6
Hardness, °ShD	34	45	47	53	54	57	59	62	64	64	64	65
Deflection temperature under load of 1.86 MPa, °C	48	50	53	68	54	55	59	78	59	62	68	80

Source: Ref. 17.
nb = no break.

Table 10 Properties of PEE-*b*-EC Filled with Calcium Metasilicate (Wollastonite)

	Hard-segment amount in PEE-*b*-EC														
	50					60					80				
	Calcium metasilicate, %					Calcium metasilicate, %					Calcium metasilicate, %				
Property	0	10	20	30	40	0	10	20	30	40	0	10	20	30	40
Breaking strength, MPa	9.0	15.0	15.0	15.0	14.0	17.0	34.0	31.0	28.0	23.0	24.0	36.0	33.0	29.0	25.0
Elongation at break, %	>590	450	430	215	80	>450	445	300	153	31	>330	300	270	120	17
Flexural strength, MPa	3.5	4.5	5.8	6.2	8.6	5.9	8.6	9.0	12.2	17.5	18.0	11.3	14.5	18.0	23.7
Flexural modulus, MPa	120	150	180	210	315	210	280	340	460	635	460	400	550	660	950
Impact strength Charpy, kJ/m^2	nb	nb	nb	nb	nb	nb	nb	nb	nb	nb	nb	nb	nb	nb	nb
Impact (notched) Charpy, kJ/m^2	nb	nb	nb	nb	nb	nb	nb	nb	nb	25	nb	nb	nb	27	21
Hardness, °ShD	34	48	50	51	53	54	57	57	59	62	64	63	64	66	67

Source: Refs. 16 and 18.
nb = no break.

524   Jeziórska et al.

Table 11 Properties of PEE-*b*-EC/PBT Blends Filled with Calcium Metasilicate (Wollastonite), with PEE-*b*-EC Containing 80% of the Hard Segments

			PBT, %					
			10		25		40	
			Calcium metasilicate, %		Calcium metasilicate, %		Calcium metasilicate, %	
Property	PEE-*b*-EC	PBT	10	20	10	20	10	20
Breaking strength, MPa	24.0	36	30	29	32	28	33	30
Elongation at break, %	<330	3.0	300	280	320	76	200	42
Flexural strength, MPa	18	70	16	18	22	25	29	34
Flexural modulus, MPa	460	2500	550	600	800	900	1100	1300
Impact strength Charpy at 23°C, kJ/m^2	nb	25	nb	nb	nb	nb	nb	nb
Impact strength Charpy at −20°C, kJ/m^2	nb	20	43	—	38	—	28	—
Impact strength Charpy at −40°C, kJ/m^2	nb	18	39	—	34	—	21	—
Impact strength Charpy at −60°C, kJ/m^2	nb	15	33	—	26	—	16	—
Impact (notched) Charpy, kJ/m^2	nb	1.4	nb	nb	25	21	19	16
Hardness, °ShD	64	75	64	65	67	69	71	72

nb = no break.

Table 12 Properties of PEE-*b*-EC Containing 50% of the Hard Segments Filled with Lamellar Materials

| | | Mica, % | | | | Graphite, % | | |
Property	PEE-*b*-EC	10	15	20	30	5	10	20
Breaking strength, MPa	9.0	12.7	12.7	12.7	14.8	11.1	13.2	15.5
Elongation at break, %	>590	280	72	31	25	590	23	16
Flexural strength, MPa	3.5	6.6	8.4	10.3	11.8	5.5	8.3	14.0
Flexural modulus, MPa	120	200	220	360	520	185	295	580
Impact strength Charpy, kJ/m^2	nb	nb	nb	nb	nb	nb	nb	nb
Impact (notched) Charpy, kJ/m^2	nb	nb	26.8	17.1	16.1	nb	nb	nb
Hardness, °ShD	34	50	52	53	58	45	48	53

Source: Refs. 16 and 18.
nb = no break.

Table 13 Properties of PEE-*b*-EC Containing 50% of the Hard Segments Filled with Carbon Black and MoS$_2$ Powder (Molykote)

Property	PEE-*b*-EC	Carbon black, %				MoS$_2$, %			
		5	10	20	30	2.5	10	20	40
Breaking strength, MPa	9.0	17.7	17.8	16.7	14.1	17.4	15.3	11.9	13.2
Elongation at break, %	>590	>590	590	520	345	>590	>590	117	23
Flexural strength, MPa	3.5	4.2	4.3	5.0	5.5	3.9	4.5	5.7	8.2
Flexural modulus, MPa	120	140	140	170	190	125	145	195	290
Impact strength Charpy, kJ/m^2	nb	nb	nb	nb	nb	nb	nb	nb	nb
Impact (notched) Charpy, kJ/m^2	nb	nb	nb	nb	nb	nb	nb	nb	15.8
Hardness, °ShD	34	45	45	46	47	44	44	44	47

nb = no break.

flexural strength. In addition, their resistance to degradation by oxygen, ozone, and UV light is excellent. Good weatherability combined with mechanical properties are said to be a key property.

Adding MoS$_2$ powder to PEE-*b*-EC can improve abrasion resistance. The properties of PEE-*b*-EC containing 50% of the hard segments modified with different amounts of MoS$_2$ powder (Molykote) are presented in Table 13. The filled grades offer improved breaking strength and flexural properties as well as high impact strength and very high elongation at break, even up to 20% of MoS$_2$ powder contents.

Typical mechanical properties of PEE-*b*-EC containing 50% of the hard segments filled with spherical materials like glass microballs, quartz powder, and chalk are shown in Table 14. Filled PEE-*b*-EC have good tensile and flexural properties and unusual combinations of hardness and resilience. These properties are highly dependent on the kind and contents of the filler (16,18).

Table 15 illustrates the properties of PEE-*b*-EC containing various amounts of glass fiber in combination with calcium metasilicate (Wollastonite). As evidenced from these investigations, an increase in glass fiber amount in blends results in higher breaking and flexural strength, flexural modulus, and hardness and lower elongation at break and impact strength. To retain flexural properties of calcium-metasilicate-filled PEE-*b*-EC (see Table 10), the glass fiber content had to be no higher than 4%, although sometimes it reached 7%).

From the foregoing investigations, the properties of filled PEE-*b*-EC are seen to be generally better than those for unmodified polymers. This is a very interesting fact, because such composites are much cheaper as compared to the PEE-*b*-EC materials. Filling with various fillers brings maximum weight reduction, adds great strength, and results in a significantly lowered applied cost.

Table 14 Properties of PEE-*b*-EC Containing 50% of the Hard Segments Filled with Spherical Materials

Property	PEE-*b*-EC	Glass microballs, %				Quartz powder, %				Chalk, %			
		10	15	20	30	10	20	30	40	10	15	20	30
Breaking strength, MPa	9.0	15.3	14.2	13.3	11.8	15.0	15.0	15.0	14.0	15.2	15.8	15.9	12.3
Elongation at break, %	>590	390	300	300	255	450	450	275	82	590	690	420	230
Flexural strength, MPa	3.5	4.1	4.5	5.0	6.0	4.1	4.8	5.5	7.6	3.7	4.1	4.6	5.3
Flexural modulus, MPa	120	140	155	170	200	130	160	190	285	120	130	142	175
Impact (notched) Charpy, kJ/m²	nb	nb	nb	nb	nb	nb	nb	nb	nb	nb	nb	nb	nb
Hardness, °ShD	34	49	51	53	54	45	47	50	54	45	45	46	48

Source: Refs. 16 and 18.
nb = no break.

Table 15 Properties of PEE-*b*-EC Filled with 20% Calcium Metasilicate (Wollastonite) and 4 or 7% Glass Fiber

	Hard-segment amount in PEE-*b*-EC			
	50		60	
	Glass fiber, %		Glass fiber, %	
Property	4	7	4	7
Breaking strength, MPa	22	26	28	41
Elongation at break, %	61	36	50	18
Flexural strength, MPa	10	13	15	24
Flexural modulus, MPa	380	480	560	960
Impact strength Charpy, kJ/m^2	nb	nb	nb	nb
Impact (notched) Charpy, kJ/m^2	nb	29	29	18
Hardness, °ShD	49	53	53	62

nb = no break.

The wide processing range of filled poly(ether-*b*-ester) copolymers permits the polymers to be used for injection molded products varying in shape, size, and thickness. They can be injection molded in all types of injection molding machines. A compression ratio for the screw from 3.0:1 to 3.5:1 is recommended. Higher ratios will cause excessive machine amperage and polymer shear; lower compression ratios may not produce a uniform melt.

The *L/D* (length-to-diameter) ratio can vary between 18:1 and 24:1. Screws with gradual transition zones comprising at least 25% of screw length are recommended. Nylon reverse-taper nozzles can be used. Shut-off nozzles are not required, because filled poly(ether-*b*-ester) copolymers have high viscosity and do not drool at normal operating temperatures. The grades need to be predried in a hopper dryer for 2–3 h at 100–105°C.

Molding conditions can be varied over a broad range of temperatures, since filled PEE-*b*-EC have excellent melt stability. For PEE-*b*-EC containing 50% of the hard phase modified with fillers, barrel temperatures of 180–210°C are recommended. Injection pressures are in the range of 80–90 MPa. Mold-filling rates vary with part thickness and geometry. Thin-wall moldings (<3 mm) should be filled rapidly, before the polymer viscosity becomes too high. However, a less-than-maximum fill rate may be required to prevent flow lines. Slow filling rates are needed for part thicknesses >6 mm and to prevent jetting into the mold cavity. Such jetting can cause turbulence and produce rough surfaces.

Molding cycle time is dependent on part size and on polymer melt and mold cavity temperatures. Cycle time ranges from 15 seconds for thin parts to 3 minutes for thick parts. For a simple part of 5-mm thickness, a cycle time of 1–1.2 minutes is a good starting point.

For filled PEE-*b*-EC, a medium-to-fast screw speed of 100 rev/min, low back pressure of 3.5 MPa, and clamp capacity of 40–70 MPa are usually adequate. Mold-cooling temperatures from 15°C to 60°C should be used. Lower temperatures reduce the cycle time and improve the ejection of parts. Higher temperatures improve flow and surface finish. Part shrinkage can be minimized by maintaining the lowest mold temperature possible while still allowing complete filling of the cavity without premature freeze-off of the gate.

Many applications can be considered for these composites, including gears and clutches, self-lubricating slide bearings, gaskets and housings, automobile bumpers, and parts of sport shoes, i.e., areas where their unique features, e.g., higher mechanical strength (glass microballs), self-lubricating properties (graphite, cotton fibers), or very low density (cotton fibers), can be favorably combined with lower price (14–17).

V. PLASTICIZERS FOR POLY(ETHER-*b*-ESTER) COPOLYMERS

A number of plasticizer types have been studied in blends with PEE-*b*-EC of 34 Shore D hardness, such as methylpentachlorostearate, methyldihydroabietate, butyl phthalylbutylglycolate, dipropyleneglycoldibenzoate (19), *p*-toluenesulfonamide, and *N*-butylbenzoatesulfonamide. Two of them may be of big commercial value, i.e., *p*-toluenesulfonamide (Ketjenflex 9) and *N*-butylbenzoatesulfonamide

Table 16 Physical Properties of Plasticized PEE-*b*-EC Containing 50% of the Hard Segments

Properties	PEE-*b*-EC	Ketjenflex 9, %		Plastmol BMB, %	
		10	25	10	25
Breaking strength, MPa	9	<4	<4	<4	<4
Elongation at break, %	>590	>1000	>1000	>1000	>1000
Flexural strength, MPa	3.5	1.6	1.5	1.8	1.5
Flexural modulus, MPa	120	47	40	58	40
Impact strength Charpy, kJ/m^2	nb	nb	nb	nb	nb
Impact (notched) Charpy, kJ/m^2	nb	nb	nb	nb	nb
Hardness, °ShA	88	86	85	87	86
Hardness, °ShD	34	28	27	28	28

nb = no break.

(Plastmol BMB). Nevertheless, a judicious choice of plasticizers can be made based on considerations of cost, volatility, color, flame retardancy, and oil and water extractability.

Because the harder grades of PEE-*b*-EC, e.g. the 54 and 64D grades, have much greater fluid resistance than the 34D grade, plasticizers are generally less compatible with the harder polymers. Some physical properties of 34D PEE-*b*-EC containing 10% and 25% of plasticizers are indicated in Table 16. Plasticizer content effects a decrease in breaking and flexural strength and flexural modulus with increasing elongation at break.

Presently, softer grades of PEE-*b*-EC are designed for a variety of molded and extruded parts, such as seals and sound and vibration absorbers.

VI. CONCLUSIONS

Modification of poly(ether-*b*-ester) copolymers to gain special properties is frequently described in the literature. Nevertheless, these investigations have failed to gain technical significance when PS and PBT are used as thermoplastics as described for applications in this chapter. The modified products have the characteristic engineering thermoplastic properties of strength, stiffness, and resilience coupled with excellent processability. The properties of filled PE-*b*-E copolymers depend largely on filler shape, filler size, and filler type. The most important effect of fillers is the increase in stiffness and dimensional stability over a wide range of temperatures. In many cases, other mechanical and thermal properties are changed. The properties are seen to be generally better than those for unmodified polymers. This is a very interesting fact, because such blends and composites are much cheaper as compared to the PE-*b*-E copolymers. Furthermore, fillers bring maximum weight reduction and add great strength.

Many applications can be considered for poly(ether-*b*-ester) copolymers, blends, and composites, including high-impact machine parts working at temperatures from −40 to 100°C, electrical and electronic flexible parts and cases (PEE-*b*-EC/PS blends), gaskets for piping and refrigeration machinery, as well as for any machinery working at extremely low temperatures (PEE-*b*-EC/PBT blends), gears and clutches, self-lubricating slide bearings, gaskets and housings, automobile bumpers, and parts of sport shoes, i.e., areas where their unique features, e.g., higher mechanical strength (glass microballs), self-lubricating properties (graphite, cotton fibers), or very low density (cotton fibers), can be favorably combined with lower price.

REFERENCES

1. J Slonecki. Szczecin: Prace Nauk. Polit. Szczecinskiej No. 479, 1992.
2. A Szepke-Wróbel, J Slonecki, H Wojcikiewicz. Polimery *10:*400–403, 1982.

3. Du Pont. Prospectus No. HYT-001A: Types, properties and uses of Hytrel.
4. L Silin-Boranowska. Polimery 9:349–351, 1986.
5. A Szepke-Wróbel, J Slonecki, H Wojcikiewicz. Inter. Polym. Sci. Technol. 10: 14–18, 1983.
6. A Szepke-Wróbel, W Zielinski. Polimery 11–12:455–456, 1981.
7. W Zielinski, A Szepke-Wróbel. Przeglad Mechaniczny 16:4–6, 1985.
8. Z Dobkowski, M Zielonka. Polimery 42:321–326, 1997.
9. AFM Barton. Handbook of Solubility Parameters and Other Cohesion Parameters. 2nd ed. Boca Raton, FL: CRC Press, 1985.
10. Pol. Pat. Appl. 311 022 (1995).
11. R Jeziorska. PhD dissertation, Industrial Chemistry Research Institute, Warsaw, 1998.
12. R Jeziorska, E Grzywa, W Zielinski. Annual Report '96. Warsaw: Industrial Chemistry Research Institute, 1997, pp 31–36.
13. R Jeziorska, E Grzywa, W Zielinski, T Jaczewska. Proceedings of XIII Science Conference on Modification of Polymers, Wroclaw, Poland, 1997, pp 149–151.
14. R Jeziorska, E Grzywa, W Zielinski. Papers of 10th International Conference on Deformation, Yield and Facture of Polymers, Cambridge, England 1997, pp 561–564.
15. W Zielinski, R Jeziorska. Annual Report '93. Warsaw: Industrial Chemistry Research Institute, 1994, pp 69–70.
16. R Jeziorska, W Zielinski. Papers of XIX International Conference on Reinforced Plastics, Karlovy Vary of Czech Republic, 1997, pp 170–176.
17. W Zielinski, R Jeziorska. Annual Report '94. Warsaw: Industrial Chemistry Research Institute, 1995, pp 46–47.
18. R Jeziorska, W Zielinski. Polimery 5:331–336.
19. Du Pont. Prospectus No. HYT-302.: Plasticizers for Hytrel.

20
Future Trends in Block Copolymers

Ernest Maréchal
University P. M. Curie (Paris VI), Paris, France

I. GENERAL CONSIDERATIONS

The literature relative to block copolymers splits into two very different parts. A majority of the publications, mainly patents, are relative to mixtures of block copolymers and homopolymers; they are poorly characterized, with the exception of elementary analysis, and at the best a limited spectroscopic investigation that does not permit one to distinguish between the block copolymers and the corresponding homopolymers. On the other hand, an increasing number of articles and reviews describe well-defined products, and often their preparation associates different techniques, for instance, living cationic and mediated radical polymerization.

The syntheses carried out in solution or in the melt remain the most important way to prepare block copolymers; however, some other techniques are used, although their development is rather slow: reactive processes, reactions in the solid state, and block copolymer epitaxy (1).

The nature of block copolymers is no longer limited to synthetic products such as polyamides and polystyrene, they also include segments derived from natural compounds, with a rapid development in medical applications. The same trend is observed in analytical techniques, where new and efficient methods lead to a thorough knowledge of block copolymers.

This chapter will survey the lastest developments and the expected trends in the following domains: syntheses, characterizations, applications.

II. SYNTHESES

Two main trends characterize the preparation of block copolymers: the development of new techniques, for instance, mediated radical polymerization, and their

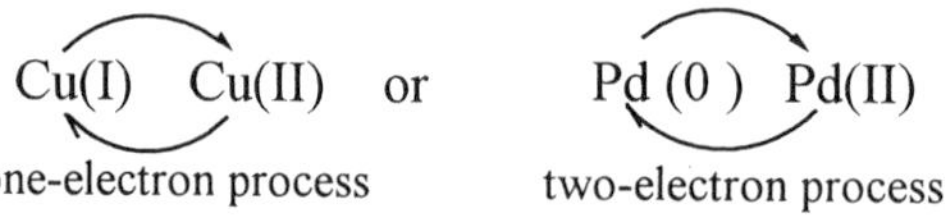

Cu(I)   Cu(II)   or   Pd (0)   Pd(II)

one-electron process   two-electron process

Scheme 1

association in sequential reactions, which is a new strategy described in several reviews and books (2–4)

A. Polycondensation

The synthesis of block copolymers by polycondensation of α,ω-difunctional-oligomers (Pebax®) or the polycondensation of an α,ω-difunctional-oligomer with the precursors of another block (Hytrel®) remain important industrial processes. However, polycondensation probably will not be the dominant technique in block copolymer synthesis in upcoming years even if new catalysts can greatly improve the control of these reactions. No revolutionary technique emerged in polycondensation, in contrast to what is observed in chain polymerization, such as mediated radical polymerization or living cationic polymerization.

Classical block polycondensation is applied to new structures. For instance, Jedlinski and Brandt (5) prepared copolymers associating soft blocks and crosslinkable blocks and Ziegast and Pfannemüller (6) coupled oligosaccharides with several linear synthetic polymers by condensing, for instance, maltooligosaccharides with an amino-group-ended oligomer.

The most significant breakthroughs have been in catalysis. The use of metal derivatives and of enzymes is very promising, even though they are little applied to the synthesis of block copolymers, at least for the time being.

1. Metal-Catalyzed Polycondensations and Polyadditions

Metal-catalyzed polycondensations and polyadditions are characterized by a change of the valence of the metal, and one or two electrons are exchanged (Scheme 1) (7).

Palladium derivatives are used in the preparation of difunctional monomers (Scheme 2), oligomers, or high polymers. There are permanent ex-

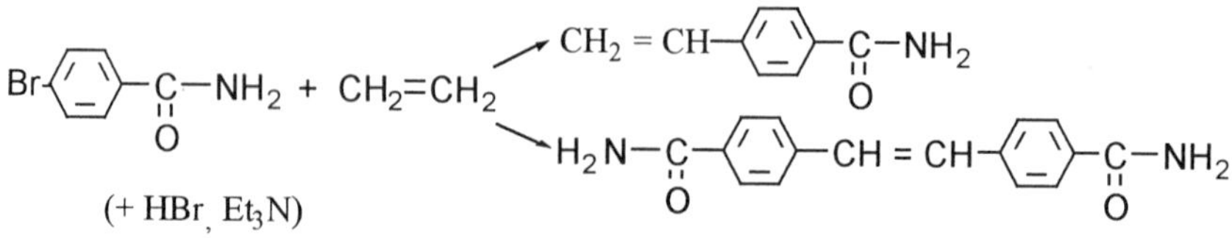

Scheme 2

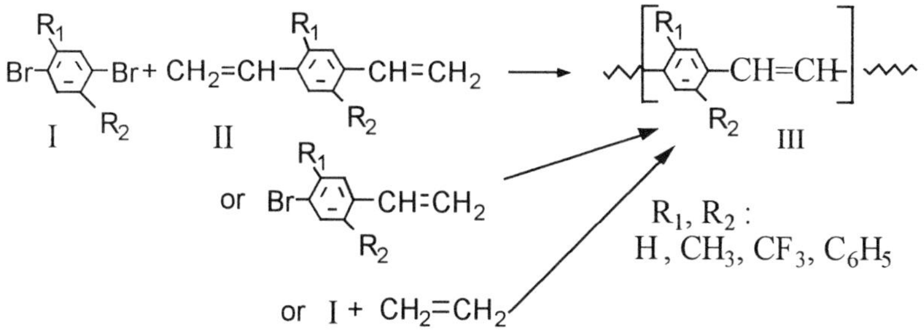

Scheme 3

changes between Pd(0) and Pd(II); Pd(0) is used as complexes and Pd(II) as salts.

These reactions are efficient in the preparation of α,ω-difunctional oligomers. Pd(II) derivatives catalyze reactions leading to polymers that could not be prepared in another way, at least in one step. Telechelic oligomers III or IV (8) were obtained by the Heck reaction, which can be the first step of a block copoly-condensation (Schemes 3 and 4).

Nickel derivatives have also seen an important development. Takagi et al. (9) block copolycondensed alkoxyallenes with phenylallene, catalyzed by [(p-al-lyl)NiOOCCF$_3$]$_2$ and proceeding through sequential additions (Scheme 5). Even though these monomers have rather different reactivity, the corresponding block copolymers are characterized by narrow weight distributions.

2. Enzyme-Catalyzed Polycondensations and Polyadditions

Since the 1980s, more and more chemists have been using the enormous possibil-ities of enzymes in organic chemistry. They exhibit a much higher selectivity than classical catalysts: an enzyme catalyzes only one type of reaction, and it recog-nizes only one compound (the substrate). For a long time, enzymatic catalysis was

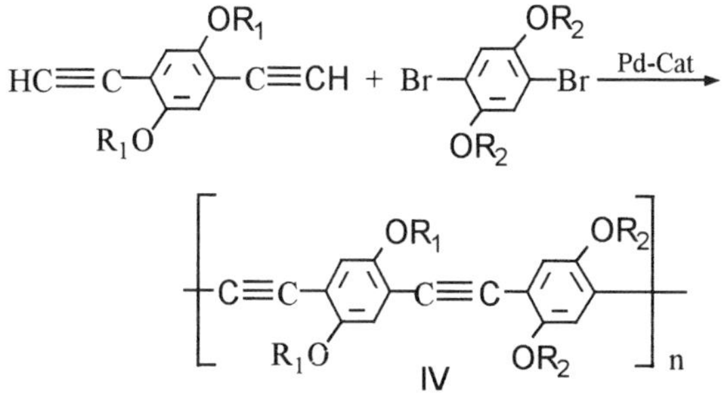

Scheme 4

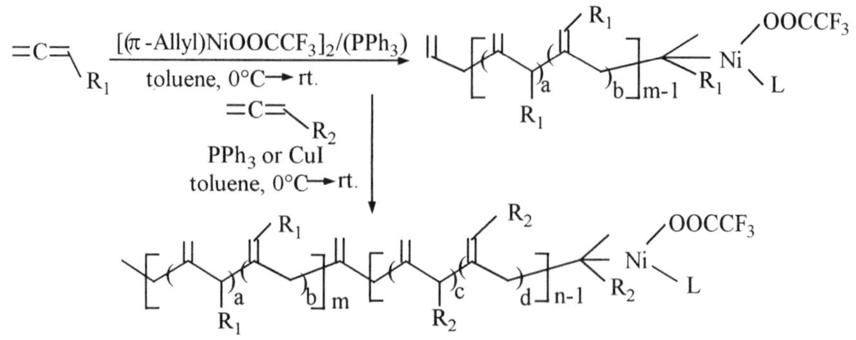

L is PPh_3 R_1 = OR or Ph R_2 = OR or Ph R is n-octyl or n-butyl

Scheme 5

used mainly in aqueous media, and only a few reactions were carried out in organic solvents. Now this situation has completely changed, and enzymatic catalysis in organic media has seen a very rapid development (10). Enzyme-catalyzed synthesis can be carried out in mild conditions, which permits the efficient control of reactions and structures. Wallace and Morrow (11) describe the polycondensation of bis(2,2,2-trichloroethyl)-3,4-epoxyadipate with 1,4-butanediol in anhydrous ether at room temperature and in the presence of porcine pancreas lipase; they observed that the epoxy group content is not modified during the polycondensation (Scheme 6), and only one of the eniantiomers was polycondensed.

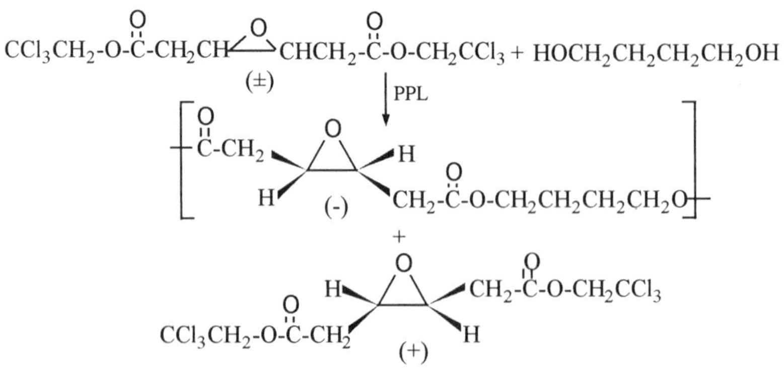

Scheme 6

Even though there are few examples relative to the enzyme-catalyzed synthesis of block copolymers, some of them are particularly interesting, and these examples are the beginning of an important development. Copolymers associating two lignin blocks with a poly(*p*-cresol) block were prepared by the polycon-

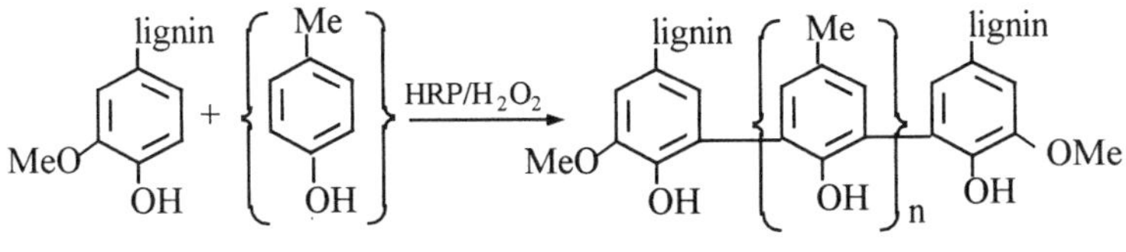

Scheme 7

densation of *p*-cresol with *o*-cresol bearing lignin catalyzed by horse radish peroxidase (HRP) (Scheme 7) (12).

Kobayashi et al. (13,14) prepared several oligocellulose samples by polycondensing β-D-cellobiosyl fluoride catalyzed by cellulase and carried out in a mixture of acetonitrile and water with acetate buffer (Scheme 8). Oligomers II and I have similar end groups. Oligomer II can be block polycondensed with another β-D-cellobiosyl fluoride.

Loos and Stadler (15) prepared amylose-*block*-polystyrene (III); α-amino-polystyrene is modified by maltoheptaose moities (I) leading to oligomer (II), which initiates the polymerization of α-D-glucose-1-phosphate catalyzed by potato phosphorylase in water, at room temperature (Scheme 9). The amylose block has a very narrow distribution.

B. Chain Polymerization

Chain polymerization has seen the arrival of powerful new tools: mediated radical polymerization, quasi-living cationic polymerization, the use of metallocene as initiators, etc. Their use in block copolymerization promises fantastic developments, particularly when several techniques are associated in sequential reactions.

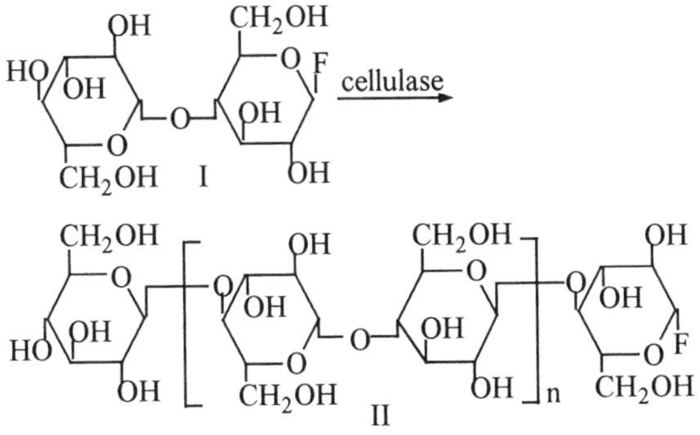

Scheme 8

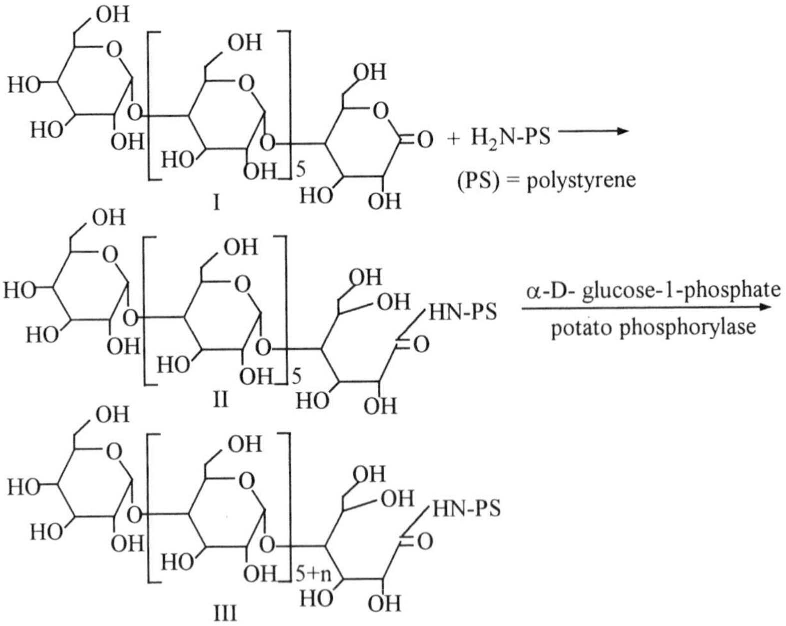

Scheme 9

1. Radical Polymerization

Radical block polymerization can be carried out in different ways, for instance, telomerization, the use of oligomeric radicals prepared from macromolecular initiator such as azo-containing polymers (16), and photoinduced block copolymerization (17–19). Mediated radical polymerization is probably the most promizing technique for preparing block copolymers, and it will probably be the dominant process in the immediate future (20–22). These different techniques were critically analyzed by Améduri et al. in a comprehensive review (23).

Li et al. (24) prepared azo-function-containing initiators with two 2,2,6,6-tetramethylpiperidinyl-1-oxy (TEMPO) as end groups (Scheme 10); this was applied to prepare copolymers associating various blocks, for instance, polystyrene, polyacrylates, polymethacrylates, polyacrylonitrile, and polyisoprene. Gravert and Janda (25) prepared initiators containing an azo group and ester or ether groups. Wang and Matyjaszewski (26) proposed a new free-radical polymerization, the Atom Transfer Radical Polymerization (ATRP), where the initiator is a Cu(I)/bipyridinyl system. This is an essential contribution; however, it requires significant amounts of Cu(I) and bipyridinyl. In order to eliminate this technical problem, Benoit et al. (27) proposed N-tertiobutyl-1-diethylphosphono-2,2-dimethylpropyl nitroxyl (DPEN), which is a new nitroxyl radical. This asymmet-

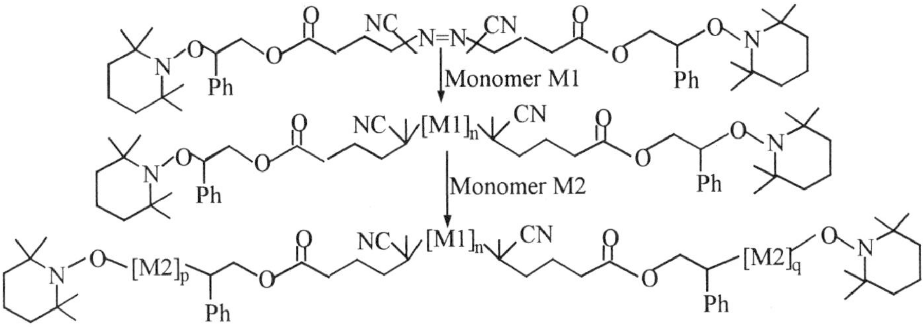

Scheme 10

ric radical leads to a fast and well-controlled polymerization of styrene, and it can be used to prepare diblock or triblock copolymer (Scheme 11).

2. Ionic Polymerizations

Two important reviews were recently published: anionic polymerization, by Hsieh and Quirk (2), and cationic polymerization, by Sawamoto (3). Most of the articles concern classic living anionic and cationic polymerization (28–31); Franta and Rempp critically analyzed the use of these techniques (32). Zundel et al. (33) succeeded in circumventing the classic rule as to why the nucleophilic reactivity of the initiating species must be equal to or larger than that of the (co)monomer. For instance, "living" potassium poly(ethylene oxide) is unable to initiate styrene or methyl methacrylate polymerization; however, its oxyanionic end was reacted

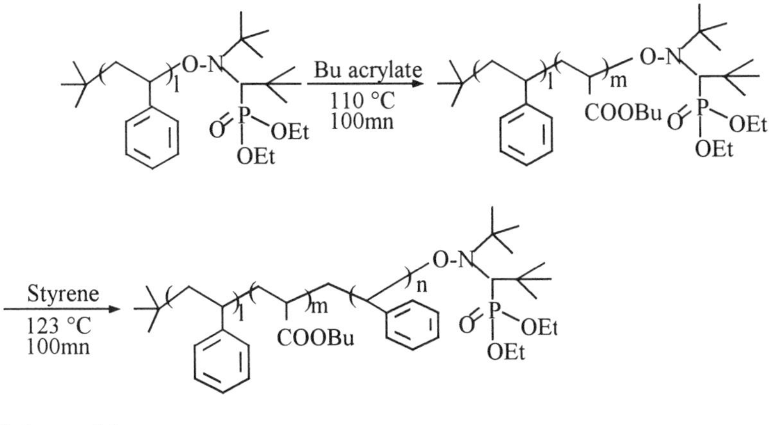

Scheme 11

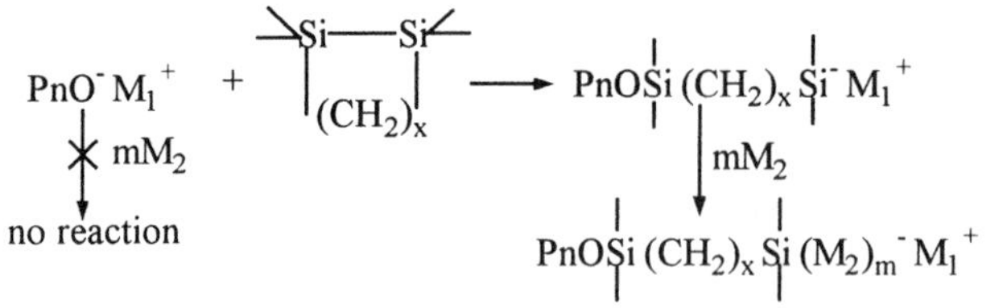

$$\text{PnO}^- \text{M}_1^+ \quad + \quad \underset{(CH_2)_x}{\overset{\displaystyle \text{Si}\!-\!\text{Si}}{|\quad|}} \quad \longrightarrow \quad \text{PnOSi}\,(CH_2)_x\,\overset{|}{\underset{|}{\text{Si}}}^-\text{M}_1^+$$

$$\underset{\displaystyle \text{no reaction}}{\overset{\displaystyle \times\ m\text{M}_2}{\big\downarrow}} \qquad\qquad\qquad \overset{\displaystyle \big\downarrow m\text{M}_2}{}$$

$$\text{PnOSi}\,(CH_2)_x\,\overset{|}{\underset{|}{\text{Si}}}\,(M_2)_m^-\,\text{M}_1^+$$

Pn can be poly(ethylene oxide) and M_2 can be styrene or methyl methacrylate

Scheme 12

with a cyclic disila derivative, leading to a silylanionic end, which is able to initiate the sequential copolymerization of styrene and methyl methacrylate (Scheme 12). The disila derivative can be, for instance, 1,1,2,2-tetramethyl-1,2-disilacyclopentane.

Cramail and Deffieux (34) used a weak Lewis, such as $ZnCl_2$, to initiate the "living" cationic polymerization of vinyl ether, and this was applied to an α,β-dihydroxy-polybutadiene (scheme 13) to obtain poly(butadiene-*block*-ethyl vinyl ether) and poly(butadiene-*block*-chloroethyl vinyl ether). In fact, the complete synthesis implies sequential reactions: anionic polymerization, chemical modification, and then cationic polymerization.

Classic cationic polymerization was applied to the formation of copolymers prepared directly on a specific solid support (1); the copolymers obtained contain either two crystalline blocks or a crystalline block and an amorphous one. Figure 1 shows the principle of epitaxy: a substrate (for instance, graphite) is introduced in a THF-containing solution where $BF_3(OC_2H_5)_2$ is introduced and an epitaxial layer of crystalline poly(THF) is formed. In second step, a mixture of THF and 3,3-bis(chloromethyl)oxycyclobutane (BCMO) is added, which results in an amorphous block. The amorphous block forms a viscous layer anchored to the support through epitaxial adsorption of the crystalline layer; it can be modified by

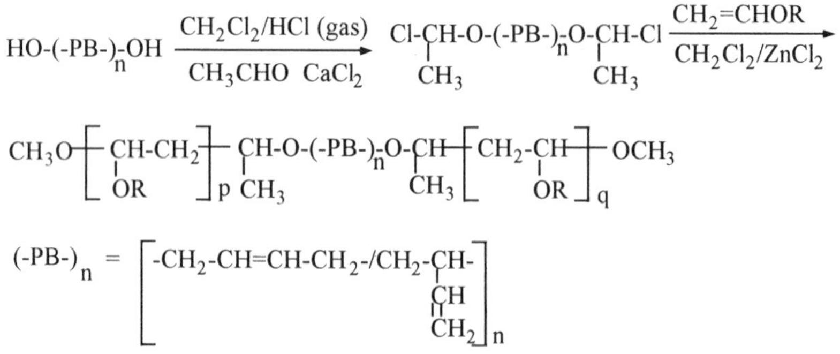

$$\text{HO-(-PB-)}_n\text{-OH} \xrightarrow[\text{CH}_3\text{CHO} \;\; \text{CaCl}_2]{\text{CH}_2\text{Cl}_2/\text{HCl (gas)}} \underset{\text{CH}_3}{\text{Cl-CH-O-(-PB-)}_n\text{-O-CH-Cl}} \xrightarrow[\text{CH}_2\text{Cl}_2/\text{ZnCl}_2]{\text{CH}_2=\text{CHOR}}$$

$$\text{CH}_3\text{O}\!\!\left[\!\!\underset{\text{OR}}{\overset{}{\text{CH-CH}_2}}\!\!\right]_p\!\!\underset{\text{CH}_3}{\overset{}{\text{CH-O-(-PB-)}_n\text{-O-CH}}}\!\!\left[\!\!\underset{\text{OR}}{\overset{}{\text{CH}_2\text{-CH}}}\!\!\right]_q\!\!\text{OCH}_3$$

$$(\text{-PB-})_n = \left[\!\underset{\displaystyle \overset{|}{\underset{\displaystyle \overset{\text{CH}}{\underset{\text{CH}_2}{\|}}}{\text{CH}}}}{\text{-CH}_2\text{-CH=CH-CH}_2\text{-/CH}_2\text{-CH-}}\!\right]_n$$

Scheme 13

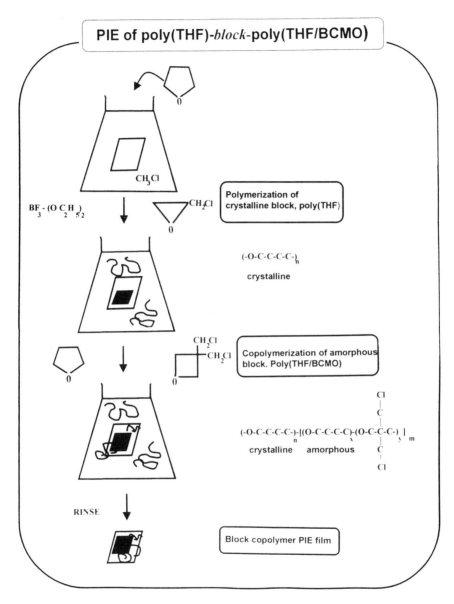

Fig. 1 Schematic diagram of polymerization-induced epitaxy (PIE) process applied to the diblock copolymer. (From Ref. 1.)

different reactants, such as sodium azide (Fig. 2). This work paves the way to new syntheses and products.

3. Organometallic Complexes as Initiators

Mashima et al. (35) studied the polymerization of α-olefins catalyzed by organometallic complexes in early transition metals, and Yasuda and Ihara (36) reviewed the rare-earth-metal-initiated polymerizations, which give high polymers with low polydispersity index. Shiomi et al. (37) report that the block copolymerization of MMA with lactones gives copolymers with index ranging between 1.11 and 1.23 when the monomers are introduced in the following order: first MMA, then the lactone (Scheme 14).

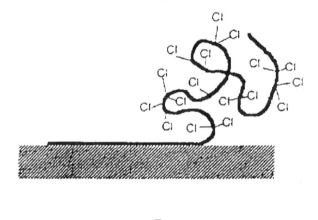

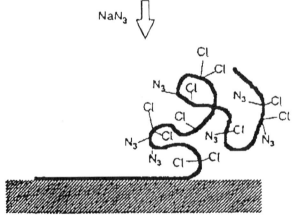

Fig. 2 Schematic diagram showing a substitution reaction of chlorine on the noncrystalline block by azide. Due to the irreversibility-adsorbed crystalline block, the polymerization-induced epitaxy film remains soluble during the reaction. (From Ref. 1.)

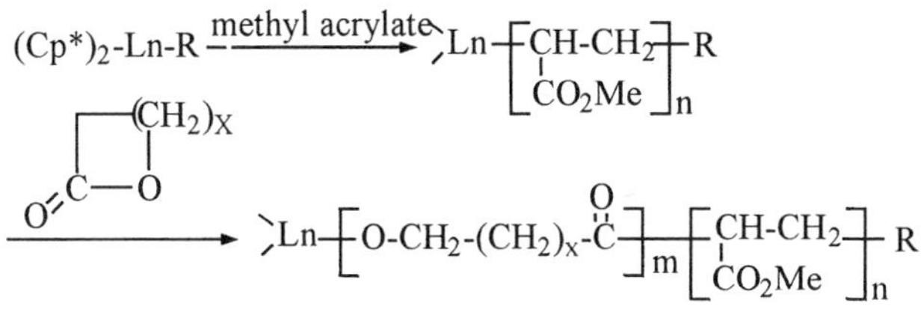

$$(Cp^*)_2\text{-Ln-R} \xrightarrow{\text{methyl acrylate}} {}_{\searrow}\text{Ln} \underset{}{\overset{}{\left[\underset{CO_2Me}{\overset{}{CH\text{-}CH_2}}\right]_n}}\!\!\!R$$

$$\xrightarrow{\underset{O^{\diagdown}}{\overset{\overset{\displaystyle(CH_2)_x}{|}}{\underset{}{C\text{-}O}}}} {}_{\searrow}\text{Ln}\underset{}{\left[O\text{-}CH_2\text{-}(CH_2)_x\text{-}\overset{\overset{\displaystyle O}{\|}}{C}\right]_m}\underset{}{\left[\underset{CO_2Me}{\overset{}{CH\text{-}CH_2}}\right]_n}\!\!\!R$$

CP* is pentamethyl-cyclopentadiene; Ln is a lanthanide; R is H or alkyl

Scheme 14

Several other block copolymers were prepared: poly[poly(THF)-*block*-poly(caprolactone)], with SmI_2 as initiator (38), and poly[poly(L.lactide)-*block*-poly(caprolactone)] and poly[poly(L.lactide)-*block*-poly(valerolactone)] (39). The most important part of Yasuda and Ihara's review (36) concerns the block copolymerization of ethylene with polar monomers such as methyl methacrylate, methyl acrylate, ethyl acrylate, valerolactone, and caprolactone using $LnR(Cp^*)_2$ complexes, with Ln = Sm, Yb, Lu, and R is H or Me; two examples are reported in Schemes 15 and 16. This initiator permits control of the relative ratio of polyethylene and poly(MMA) or poly(caprolactone) blocks in the copolymer.

Yasuda et al. (40–42) prepared high copolymers of methyl methacrylate (MMA) and *n*-butyl acrylate (BuA) with narrow molecular distribution, such as Ln-poly(BuA)-*block*-poly(MMA) and poly(MMA)-*block*-poly(BuA)-*block*-poly(MMA) (Schemes 17 and 18). The SEC traces of poly(MMA)-*block*-poly(BuA)-*block*-poly(MMA) show the distribution of the products obtained after each step (Fig. 3). The molar mass of the triblock is 144,000, and its polydispersity index is 1.09.

C. Sequential Reactions

Sequential reactions represent a new strategy, a first block is prepared and then its active center is modified to initiate another polymerization. Several authors have provided valuable information on this topics, such as Schué (4) and Rempp and Franta (43).

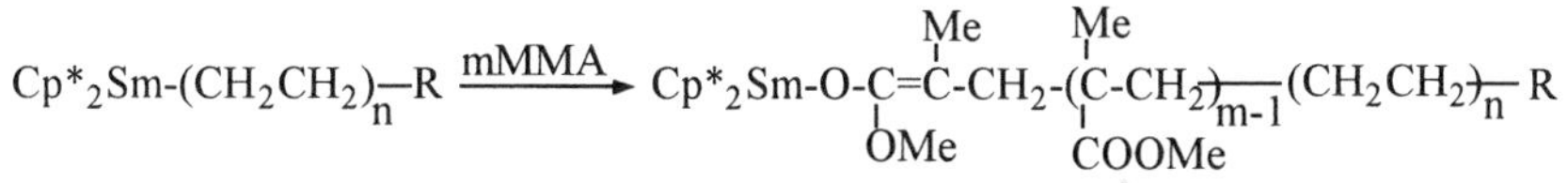

$$Cp^*_2Sm\text{-}(CH_2CH_2)_{\overline{n}}R \xrightarrow{mMMA} Cp^*_2Sm\text{-}O\text{-}\underset{OMe}{\overset{\overset{\displaystyle Me}{|}}{C}}=\underset{}{C}\text{-}CH_2\text{-}\underset{COOMe}{\overset{\overset{\displaystyle Me}{|}}{(C\text{-}CH_2)_{m\text{-}1}}}(CH_2CH_2)_{\overline{n}}R$$

R = H or Me

Scheme 15

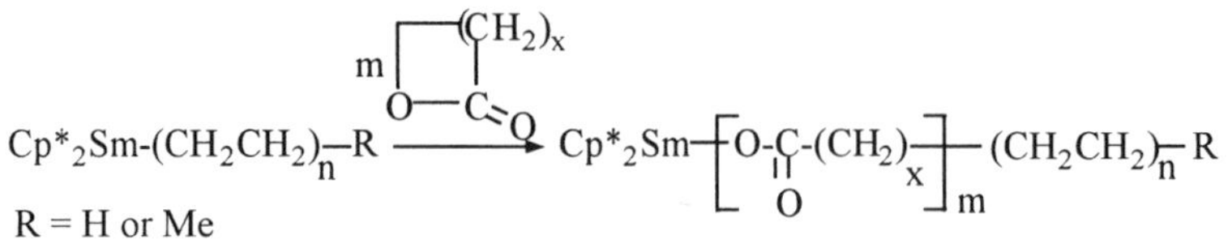

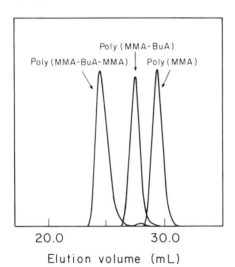

$$Cp*_2Sm\text{-}(CH_2CH_2)_n\text{-}R \xrightarrow{\quad m \overset{(CH_2)_x}{\underset{O-C=O}{\boxed{\quad}}} \quad} Cp*_2Sm\text{-}\left[O\text{-}\underset{O}{\overset{\|}{C}}\text{-}(CH_2)_x\right]_m (CH_2CH_2)_n\text{-}R$$

R = H or Me

Scheme 16

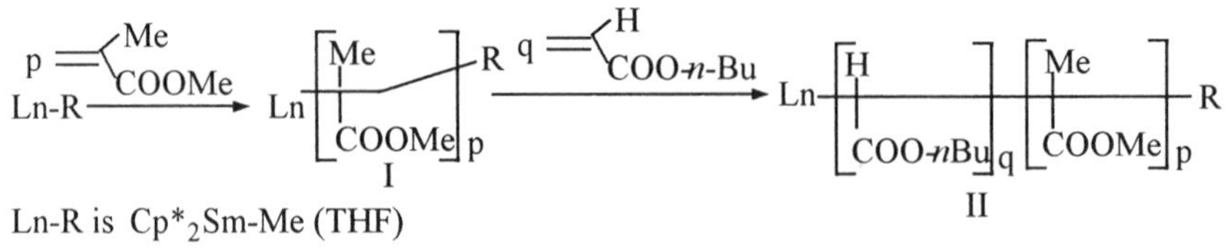

$$Ln\text{-}R \xrightarrow{\quad p \overset{Me}{\underset{COOMe}{=}} \quad} Ln\left[\underset{COOMe}{\overset{Me}{|}}\right]_p R \xrightarrow{\quad q \overset{H}{\underset{COO\text{-}n\text{-}Bu}{=}} \quad} Ln\left[\underset{COO\text{-}nBu}{\overset{H}{|}}\right]_q \left[\underset{COOMe}{\overset{Me}{|}}\right]_p R$$
                                                  I   II

Ln-R is Cp*$_2$Sm-Me (THF)

then

Scheme 17

$$II \xrightarrow{\quad r \overset{H}{\underset{COO\text{-}Me}{=}} \quad} Ln\left[\underset{COO\text{-}Me}{\overset{Me}{|}}\right]_p\left[\underset{COOnBu}{\overset{H}{|}}\right]_q\left[\underset{COOMe}{\overset{Me}{|}}\right]_r R$$

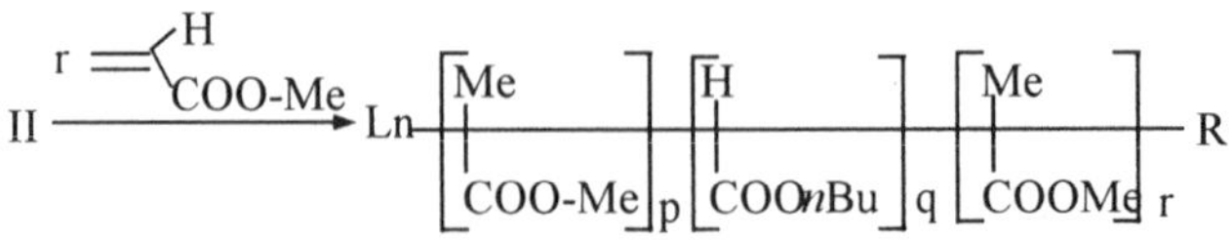

Scheme 18

Poly (MMA-BuA)

Poly (MMA-BuA-MMA)   Poly (MMA)

20.0   30.0

Elution volume (mL)

Fig. 3 SEC traces of poly(MMA), poly(MMA-*block*-BuA), poly(MMA-*block*-BuA-*block*-MMA) prepared by sequential polymerization. (From Ref. 40.)

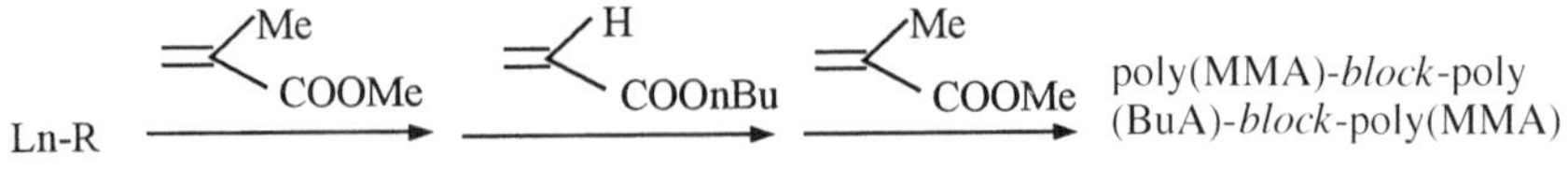

$$Ln\text{-}R \xrightarrow{\quad \overset{Me}{\underset{COOMe}{=}} \quad} \xrightarrow{\quad \overset{H}{\underset{COOnBu}{=}} \quad} \xrightarrow{\quad \overset{Me}{\underset{COOMe}{=}} \quad} \begin{array}{l} poly(MMA)\text{-}block\text{-}poly \\ (BuA)\text{-}block\text{-}poly(MMA) \end{array}$$

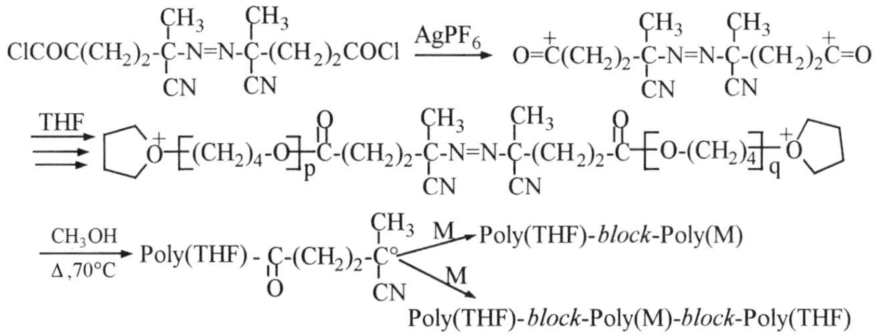

$$\text{ClCOC(CH}_2)_2\text{-}\underset{\underset{CN}{|}}{\overset{\overset{CH_3}{|}}{C}}\text{-N=N-}\underset{\underset{CN}{|}}{\overset{\overset{CH_3}{|}}{C}}\text{-(CH}_2)_2\text{COCl} \xrightarrow{\text{AgPF}_6} \text{O=}\overset{+}{C}\text{(CH}_2)_2\text{-}\underset{\underset{CN}{|}}{\overset{\overset{CH_3}{|}}{C}}\text{-N=N-}\underset{\underset{CN}{|}}{\overset{\overset{CH_3}{|}}{C}}\text{-(CH}_2)_2\overset{+}{C}\text{=O}$$

$$\overset{\text{THF}}{\rightleftharpoons} \overset{+}{\underset{}{\bigcirc}}\text{O-}[(CH_2)_4\text{-O}]_p\overset{O}{\overset{||}{C}}\text{-(CH}_2)_2\text{-}\underset{\underset{CN}{|}}{\overset{\overset{CH_3}{|}}{C}}\text{-N=N-}\underset{\underset{CN}{|}}{\overset{\overset{CH_3}{|}}{C}}\text{-(CH}_2)_2\text{-}\overset{O}{\overset{||}{C}}[\text{O-(CH}_2)_4]_q\overset{+}{\bigcirc}$$

$$\xrightarrow[\Delta,70°C]{CH_3OH} \text{Poly(THF) - }\underset{\underset{CN}{|}}{\overset{\overset{CH_3}{|}}{C}}\text{-(CH}_2)_2\text{-}\overset{}{\underset{O}{\overset{|}{C}}}{}^\circ \overset{\nearrow M}{\underset{\searrow M}{<}} \begin{array}{l}\text{Poly(THF)-}block\text{-Poly(M)}\\ \\ \text{Poly(THF)-}block\text{-Poly(M)-}block\text{-Poly(THF)}\end{array}$$

Scheme 19

1. Radical-to-Cationic Polymerization

Yagci published several articles in this field, for instance, the preparation of poly[polystyrene-*block*-polyoxytetramethylene] (44) (Schemes 17 and 18 of Chap. 3). Galli et al. (16) prepared several AB and ABA copolymers (Scheme 19). The cationic step was applied to THF or cyclohexene oxide, and the azo group initiates the radical polymerization of various acrylates or methacrylates (M).

2. Cationic-to-Radical Polymerization

This sequential process is particularly well developed (45–47); Ref. 46 was already analyzed (Scheme 32, Chap. 3). Nguyen et al. (48) prepared an initiator containing an azo group and two vinyl ether moities and associated with HI (Scheme 20); its polydispersity index ranges from 1.1 to 1.4.

3. Anionic-to-Radical Polymerization

Kobatake et al. (49) prepared a polybutadiene-*block*-polystyrene copolymer (III) by associating anionic polymerization, chemical transformation, and mediated

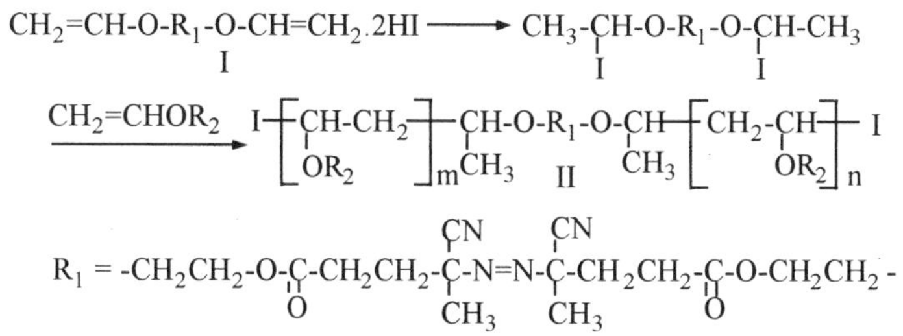

$$CH_2\text{=CH-O-R}_1\text{-O-CH=CH}_2.2HI \longrightarrow CH_3\text{-}\underset{\underset{I}{|}}{CH}\text{-O-R}_1\text{-O-}\underset{\underset{I}{|}}{CH}\text{-CH}_3$$

$$\xrightarrow{CH_2\text{=CHOR}_2} I\left[\underset{\underset{OR_2}{|}}{CH}\text{-CH}_2\right]_m\underset{CH_3}{\overset{}{}}\underset{\underset{CH_3}{}}{CH}\text{-O-R}_1\text{-O-}\underset{II}{CH}\left[\underset{CH_3}{\overset{}{}}CH_2\text{-}\underset{\underset{OR_2}{|}}{CH}\right]_n I$$

$$R_1 = \text{-CH}_2CH_2\text{-O-}\underset{O}{\overset{||}{C}}\text{-CH}_2CH_2\text{-}\underset{\underset{CH_3}{|}}{\overset{\overset{CN}{|}}{C}}\text{-N=N-}\underset{\underset{CH_3}{|}}{\overset{\overset{CN}{|}}{C}}\text{-CH}_2CH_2\text{-}\underset{O}{\overset{||}{C}}\text{-O-CH}_2CH_2\text{ -}$$

Scheme 20

radical polymerization. The polymerization of butadiene is initiated by butyllithium (oligomer I), whose end group is reacted with an epoxy-functionalized nitroxide (oligomer II) and then II initiates a nitroxide-mediated styrene polymerization (Scheme 21). The polydispersity index of I is 1.02 and that of III ranges from 1.4 to 2.2.

4. Cationic-to-Anionic Polymerization

Nomura et al. (50) prepared several poly(THF)-*block*-polylactones. The polymerization of THF is initiated by methyl trifluorosulfonate (the cationic step), and then HMPA and samarium diodide are added to the macrocation, leading to

$$CH_3—[O—(CH_2)_4]_{n+1}—SmI_2$$

(the chemical transformation step); after a time, the lactone is added (the anionic step) (Scheme 22). The polydispersity index ranges from 1.35 to 1.45; the polymerization of the lactone is quantitative.

5. Anionic-to-Cationic Polymerization

Labeau et al. (51) prepared amphiphilic block copolymers of controlled dimensions with hydrophilic glycosidic vinyl ether moities using the following series of reactions: (i) living anionic polymerization of styrene; (ii) quantitative functional termination with chloracetal derivatives; (iii) addition of trimetylsilyl iodide leading to a-iodo ether macroinitiators; (iv) "living" cationic polymerization of protected glycosidic vinyl ethers in the presence of weak Lewis acid (typically $ZnCl_2$) (Scheme 23).

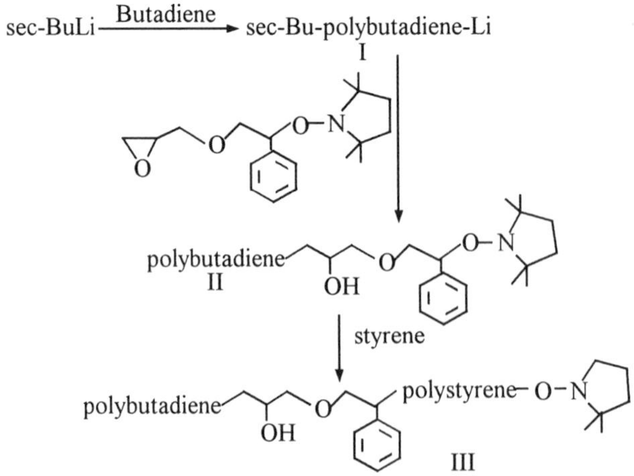

Scheme 21

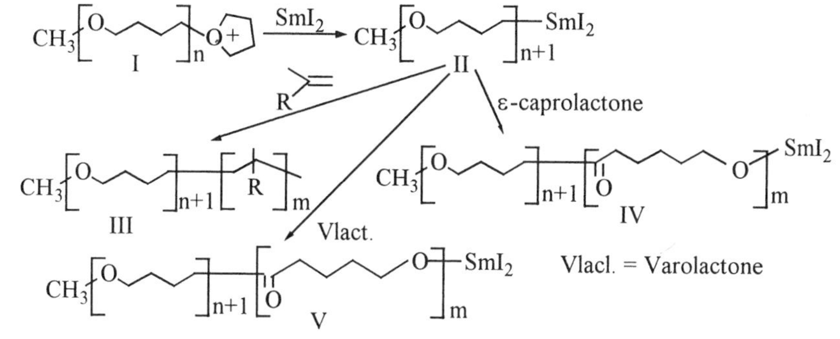

Scheme 22

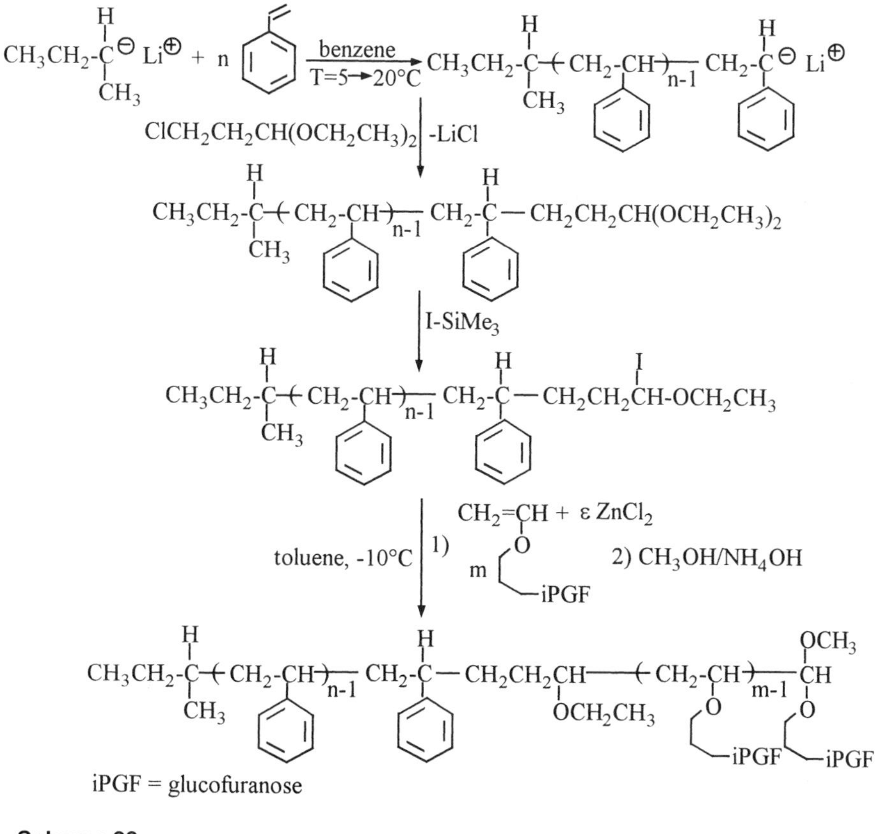

iPGF = glucofuranose

Scheme 23

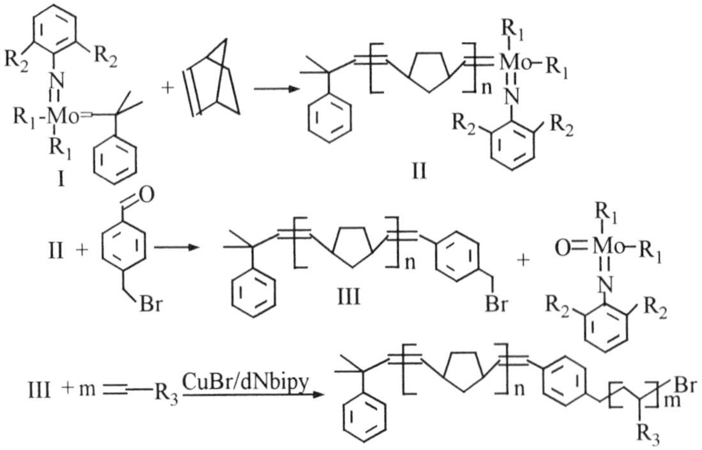

R$_1$ =t-butoxy   R$_2$ = i-propoxy   R$_3$ = C$_6$H$_5$ or CO$_2$CH$_3$

dNbipy = 4,4'-di(5-nonyl)-2,2' bipyridine

Scheme 24

6. Metathesis-to-Radical Polymerization

Coca et al. (52) successively carried out ring-opening metathesis polymerization, chemical transformation, and "living" radical polymerization to prepare diblock copolymers such as polynorbornene-*block*-polystyrene (Scheme 24), polynor-bornene-*block*-poly(methyl actylate), or poly(dicyclopentadiene)-*block*-poly(methyl methacrylate). Their polydispersity index ranges from 1.35 to 1.45. The oligonorbornene is prepared by ring-opening methathesis polymerization initiated by a molybdene-alkylidene complex:

$$Mo—(=CHCPhMe_2)(=NAr)(O-t=Bu)_2$$

where Ar is 2,6-diisopropylphenyl.

7. Radical Polymerization to Chemical Modification

X Huang et al. (53) prepared a poly(vinyl acetate)-*block*-poly(methyl methacry-late) (I) and a poly(vinyl alcohol)-*block*-poly(methyl methacrylate) (II). I was pre-pared by a photo-induced polymerization using a charge-transfer complex composed of aniline and benzophenone (Scheme 28, Chap. 3). The block poly(vinyl

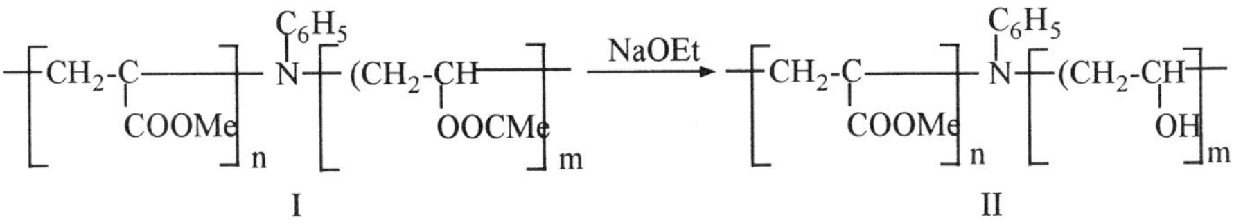

Scheme 25

acetate) was hydrolyzed in the presence of sodium ethoxide (Scheme 25). Only the polyacetate block was hydrolyzed; nuclear magnetic resonance (NMR) showed that the methoxy content of poly(methyl methacrylate) was nearly the same before and after hydrolysis.

D. Coupling Reactions

Coupling reactions (Scheme 26) are applied to many difunctional oligomers, and they were reviewed by Fradet (54), who particularly studied the behavior of bisoxazines, bisoxazolinones, and bis-oxazolines.

The coupling agent can be reacted either with oligomers of the same kind or with different oligomers (55–57). Their end groups are identical or different (Schemes 27 and 28).

A-(OLIGOMER I)-A+B-(OLIGOMER II)-B+C—C ⟶

⟿(OLIGOMER I)-A'C'-C'B'-(OLIGOMER II)⟿

A and B can be different or identical

Scheme 26

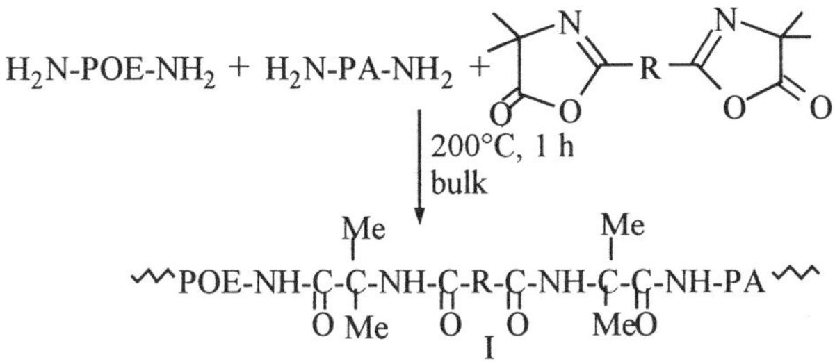

Scheme 27

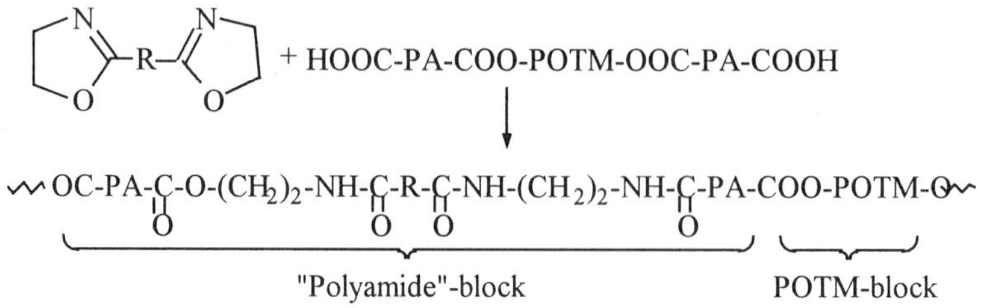

$$\underset{O}{\overset{N}{\bigcup}}\text{-R-}\underset{O}{\overset{N}{\bigcup}} \quad + \text{HOOC-PA-COO-POTM-OOC-PA-COOH}$$

$$\downarrow$$

$$\sim\!\!\sim \text{OC-PA-}\underset{O}{\overset{}{C}}\text{-O-(CH}_2)_2\text{-NH-}\underset{O}{\overset{}{C}}\text{-R-}\underset{O}{\overset{}{C}}\text{-NH-(CH}_2)_2\text{-NH-}\underset{O}{\overset{}{C}}\text{-PA-COO-POTM-O}\!\!\sim\!\!\sim$$

|_____| |_____|
            "Polyamide"-block   POTM-block

PTOM is poly(oxytetramethylene)

Scheme 28

III. CHARACTERIZATION

The progress in block copolymer characterization results from two different factors: the increase in the efficiency of each technique, and their association.

A. Molar Mass and Molar Mass Distribution of the Copolymers and the Blocks

Chromatography, alone or associated with other techniques, permits one to determine not only molar masses and molar mass distributions but, thanks to new developments in this technique, the molar mass distribution of each block, in one run, and the chemical distribution of the blocks.

Determination of molar mass, molar mass distribution, and functionality remains a difficult problem. It requires the contribution of at least two different techniques; one is based upon colligative properties of the polymer, and the other is based on the titration of specific groups of the chain, for instance, the end groups. Unfortunately, the determination of end group concentration is rapidly limited when the molar mass becomes high; the use of NMR is probably one of the most efficient tools in that field. It is astonishing to see that only a very few authors use the chemical modification of the end group, for instance, the reaction of an α,β-dihydroxy-polymer with trifluoroacetic anhydride followed by 19F-NMR analysis or the introduction of a fluorescent group or a group adsorbing in visible or UV light.

Ohishi et al. (58) prepared α,β-dicarboxylic-polystyrene, using radical polymerization; then the acid funtions were modified into phenols. Number average molar mass ($\overline{M}_n$ was determined by size exclusion chromatography and by basic titration of the acid end group concentration; unfortunately, the experimental part does not provide the necessary information to carry out the corresponding calculation. The transformation of the acid functions into phenolic groups and the direct titration of the latter could have given another value of $\overline{M}_n$; unfortunately, phenol concentrations were obtained from the difference between the acid concentrations before and after the chemical modification, which does not provide a confirmation of $\overline{M}_n$ and shows only that the conversion is quantitative.

Improvements in chromatography permit one to obtain more accurate values and additional information on the heterogeneity of the sample. Pash (59–62) proposed a new technique for characterizing polymers already mentioned by Tennikov et al. (63) and Belenkii and Gankina (64). The results of a chromatography depend on several parameters in particular their overlapping; for instance, the molar mass distribution of a functional homopolymer (α,β-difunctional-polymer, macromers) is superimposed by a functionality-type distribution. Copolymers are distributed in molar mass and in chemical composition. The size exclusion chromatogram of a heteropolymer may depend on its chemical composition or on its functionnality; reciprocally, the molar mass distribution must be taken into account in the determination of the chemical heterogeneity. From this point of view, block copolymers are complex macromolecular systems characterized by two distributions in molar mass and chemical composition. Liquid chromatography at the critical point of adsorption permits the determination of different types of molecular heterogeneity. Operating at the transition point of size exclusion and adsorption modes of liquid chromatography, polymers are separated according to their functionality. At the critical point of adsorption the polymer chain behaves like an invisible part of the macromolecule, and only the heterogeneities (functional groups, blocks, grafts) are observed. This technique was applied to the determination of the chemical heterogeneity of block copolymers and of the composition of the blends; in particular, the copolymers can be analyzed with respect to the block lengths of the individual blocks. Pash (62) analyzed a diblock copolymer of styrene and MMA using mixtures of methylethylketone and cyclohexane as eluents. When the eluent contains more than 70% (vol/vol) of methylethylketone, PMMA is separated according to a size exclusion mode; when its volume is rigorously 70%, PMMA is eluted in the critical mode. Figure 4 shows the size exclusion chromatograms of the PS precursors and of the block copolymers; as expected, the copolymers are eluted earlier than PS, since their molar masses are above those of PS.

On the other hand (Fig. 5), critical chromatography gives completely different patterns. The traces of PS and the block copolymers are identical in position and in shape: at the critical point of PMMA, the copolymers are eluted exclusively with respect to the PS blocks. This method was applied to the same copolymers, but they were carried out at the critical mode of polystyrene and the molar mass distribution of the PMMA blocks in the copolymers.

Braud et al. (65) used capillary zone electrophoresis. Neutral phosphate buffer as the background electrolyte was used to analyze water-soluble telechelic oligomers obtained by polycondensation of racemic lactic acid, which can be precursors of block copolymers.

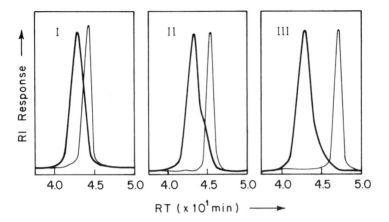

Fig. 4 Size exclusion chromatograms of different PS-PMMA block copolymers (numbered I, II, III). Traces of the block copolymers (▬) and traces of the polystyrene precursors (—). (From Ref. 62.)

Kilz et al. (66) used two-dimensional chromatography for the deformulation of complex copolymers. They combined the advantages of HPLC and SEC using a fully automated, software-controlled, two-dimensional chromatography system for the online analysis of composition, end group functionality, and molar mass distribution. This system consists of two chromatographs, one that separates by

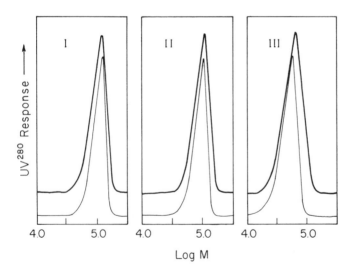

Fig. 5 Critical chromatograms of the the same samples. Eluent: methyl ethyl ketone-cyclohexane (70:30% by volume). (From Ref. 62.)

chemical composition (e.g., a gradient HPLC), and an SEC instrument for subsequent separation by size. The application of a single method, such as multidetector SEC or SEC, with light-scattering detection to a mixture containing block copolymers, cycles, and functional oligomers has some limitations. The two-dimensional chromatography is able to give a good description of such mixtures.

The association of several techniques is very promising. Gores and Kilz (67) studied block copolymers using multiple-detection SEC, online multiangle laser light scattering, and online viscosimetry; multiple detection permitted them to measure the chemical composition distribution, using a detector calibration and fitting the results of NMR.

When a copolymer results from the polycondensation of an α,ω-difunctional-oligomer with the precursors of another block, the molar mass of the latter is not known; however, there are a few exceptions. In some cases the comparison of the NMR signal intensity relative to each block permits one to remove this uncertainty (68–70). In the same way, Khim and Adamson (71) characterized the diblock copolymer polystyrene-block-polylactide by association of size exclusion chromatography and ^{1}H-NMR spectroscopy.

The triblock copolymers polyA-polybutadiene-polyA was prepared by metathesis polymerization catalyzed by tungsten chloride associated with α,β-dihydroxy-polybutadiene; A can be phenylacetylene (Scheme 29), norbornene, cyclooctadiene, and cyclopentene (72). Its characterization shows the efficiency of the association of different techniques. Its molar mass was determined by both SEC and ^{1}H-NMR spectroscopy comparing the intensities of the aromatic and aliphatic peaks; the values obtained are in reasonable agreement. In a second method, W content was determined by elementary analysis and fitting the results of calculation when it is assumed that each chain contains two W atoms per chain. The third method is based on the delinking of the three blocks of the copolymer (Scheme 30) based on a Wittig-like reaction where the product is reacted with benzaldehyde. Delinking method is very efficient, since it gives the molar mass of each block, confirming the values obtained by SEC, ^{1}H-NMR, and tungsten titration.

The selective degradation of one of the blocks permits one to obtain the molar mass and the distribution of the other block; the use of enzymes is particularly efficient due to their selectivity. Valiente et al. (73) studied the enzyme-catalyzed

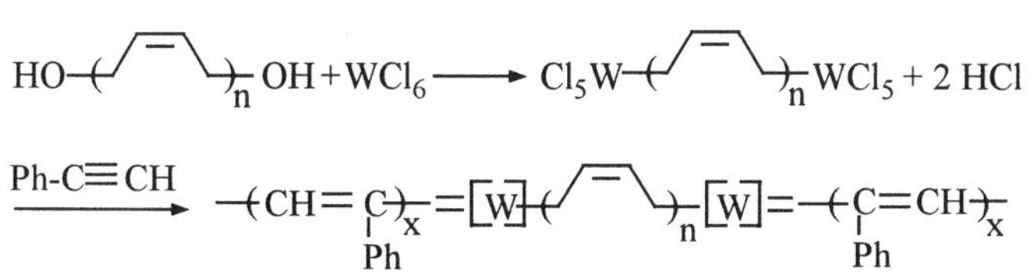

$$HO\!\!-\!\!\left(\!\!\overbrace{}\!\!\right)_{\!\!n}\!\!-OH + WCl_6 \longrightarrow Cl_5W\!\!-\!\!\left(\!\!\overbrace{}\!\!\right)_{\!\!n}\!\!-WCl_5 + 2\ HCl$$

$$\xrightarrow{Ph-C\equiv CH} -\!\!\left(CH\!=\!\underset{\underset{Ph}{|}}{C}\right)_{\!\!x}\!\!=\!\!\left[W\right]\!\!-\!\!\left(\!\!\overbrace{}\!\!\right)_{\!\!n}\!\!-\!\!\left[W\right]\!\!=\!\!-\!\!\left(\underset{\underset{Ph}{|}}{C}\!=\!CH\right)_{\!\!x}$$

Scheme 29

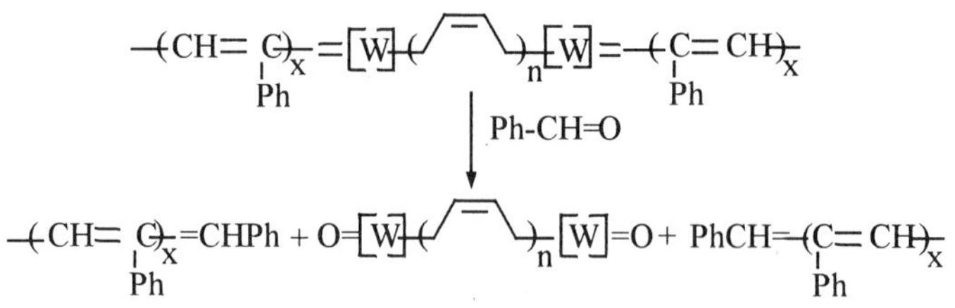

$$-(CH=\underset{\underset{Ph}{|}}{C})_x=[W]-\!\!\!\left(\!\!\!-\!\!\!\right)_n\!\!\![W]=-(\underset{\underset{Ph}{|}}{C}=CH)_x$$

$$\downarrow \;\; Ph\text{-}CH{=}O$$

$$-(CH=\underset{\underset{Ph}{|}}{C})_x=CHPh + O{=}[W]-\!\!\!\left(\!\!\!-\!\!\!\right)_n\!\!\![W]{=}O + PhCH{=}(\underset{\underset{Ph}{|}}{C}=CH)_x$$

Scheme 30

hydrolysis of phthalic-unit-containing copolyesters, which results in an efficient analytical tool permitting one to analyze the structure of the chain, particularly the determination of both length and molar mass distribution of the blocks. The hydrolysis of four polymers—poly(propane-1,2-diyl fumarate), two copolymers containing fumarate and phthalate units (15% and 75%), and poly(propane-1,2-diyl phthalate)—was catalyzed by *Chromobacterium viscosum* in a phosphate buffer (pH = 7.8) at 40°C (Table 1). Each line of Table 1 is relative to one of the copolyesters; it shows that the presence of the enzyme is essential, particularly in the case of poly(propane-1,2-diyl fumarate), which is aliphatic. On the other hand, the efficiency of the enzyme decreases with the increasing number of aromatic rings in the chain, when the polyester is fully aromatic, there is no catalytic effect: the enzymatic catalysis is efficient only in aliphatic parts. When the process is repeated several times, only the aromatic blocks remain, and their molar mass and distribution can be determined.

B. Behavior of Chains in Solution and in the Solid State

Lodge (74) reviewed the use of scattering techniques in the characterization of polymers, particularly block copolymers. Static and dynamic light scattering,

Table 1 Behavior of the Copolymers in Noncatalyzed and Enzymatic-Catalyzed Hydrolysis

	$(m_0 - m)\,/\,m_0\,\%$	
Sample phthalic unit %	Enzyme	No Enzyme
0	100	19
5	89	15
75	50	11
100	8	7

m_0 and m are the mass of the sample before and after hydrolysis, respectively.

small-angle neutron and x-ray reflectivity were analyzed, and emphasis was placed on their similarities and differences. Static light scattering was used to study the expansion of the block copolymers, such as the diblock polystyrene-block-poly(methyl methacrylate) (75). Dynamic light scattering was applied to block copolymers in good solvents or in "selective" solvents, where one of the blocks is preferentially solvated; this is the case of the symetric triblock poly(2-vinylpyridine)-polystyrene-poly(2-vinylpyridine) in toluene, which is a very poor solvent for poly(2-vinylpyridine) (76).

Small-angle neutron scattering (SANS) is particularly adapted to the study of order–order and disorder–order transitions; it is an excellent and nondestructive tool establishing the morphology of block copolymer in the ordered state, in this instance, it is complementary to electron microscopy. Bates et al. (77–79) applied this technique to the diblock copolymer resulting from the hydrogenation of poly(1,4-isoprene)-block-poly(1,2-butadiene), which permitted them to establish their morphology. Munkenbucsh et al. (80) showed the efficiency of SANS and neutron spin-echo in the formation of aggregates formed from diblock copolymers in selective solvent. They particularly studied the system polystyrene-block-polyisoprene/decane, which gives spherical aggregates, whereas the aggregates formed from the system polyethylene-block-poly(ethylene-co-propylene)/decane are platelets resulting from the crystallization of the polyethylene block.

Small-angle x-ray scattering (SAXS) is particularly efficient in the study of amorphous polymers, including microstructured materials, such as block copolymers. The advent of synchotron sources for x-ray scattering provided new information; for instance, it permits the examination of the time evolution of block copolymer microstructures with temporal resolution below 1 second. Singh et al. (81) applied this technique to the response of diblock styrene-butadiene copolymer solutions to a temperature quench from the disordered to the ordered regime.

Neutron reflection is a relative new technique that provides a unique means to characterize the atom composition as a function of depth in a thin film; Foster et al. (82) hydrogenated the diblock copolymer poly(1,4-isoprene)-poly(1,2-butadiene) deposited as films on silicon wafers; they established the corresponding concentration profiles. X-ray reflection is used in the same domains as neutron reflection, particularly the study of block copolymer films.

X-ray and neutron reflection are often complementary techniques associated with other ones, for instance, in the analysis of the structure of films. A number of studies are relative to polystyrenes-block-polyoxyethylenes at the air/water interface; most of them focus on the structure perpendicular to the interface (83,84); on the other hand, Cox et al. (85) characterized the lateral structure of nonionic block copolymers, such as polystyrene-block-polyoxyethylene, spread on water using transmission electronic microscopy and atomic force microscopies, which permitted them to distinguish and to characterize two kinds of ag-glomerate: regular and irregular.

Tsukruk (86) reviewed the applications of scanning probe microscopy to polymer surfaces; block polymers with long-range and well-defined microphase structures of various symmetry are particularly adapted to this technique.

More and more, several techniques are associated. Liu and Jiang (87) stud-ied the crystallization, the morphology, and the crystalline structure of diluted solid solutions of poly(tetrahydrofuran)-poly(methyl methacrylate) diblock copolymer in poly(ethylene oxide) and poly(tetrahydrofuran) using differential scanning calorimetry, x-ray, and optical microscopy. Liu and Jiang's work pro-vides a new insight into the behavior of block copolymers when crystallizing. In the same way, Hashimoto et al. (88) carried out a theoretical investigation of elas-tic small-angle scattering of x-rays, neutrons, and light applied to an assembly of hexagonally packed cylindrical particles with a paracrystalline distortion; this scattering theory is applied to a sphere-forming copolymer and a lamellar-form-ing copolymer.

IV. APPLICATIONS

Thermoplastic elastomers such as Pebax®, Hytrel® and polyurethane remain very important materials; likewise, fibers and implants continue to be the subject of ex-tensive research. The following concerns only applications that more recently ap-peared on the market or in full development: the use of block copolymers in blends, in micellar systems, and above all in medical applications.

A. Applications in Blends

Blends have seen a rapid development in the materials industry, they generate ma-terials with new properties, and their cost is often lower than that of many copoly-mers. Most commercial blends are prepared by melt-mixing, however, many of them are made from incompatible polymers, and compatibility is often achieved only when block or graft copolymers are added to the blend.

The compatibilizing copolymer can be independently prepared and added to the blends (premade copolymer) or produced in the blends during the melt-mix-ing (in situ formed copolymer). Yamane (89) reviewed the compatibizing effect of premade and in situ formed block copolymers in polymer blends. Nakayama et al. (90) compared these two techniques by studying the polystyrene/poly(methyl methacrylate) blends, which are homogeneous only when a diblock copolymer polystyrene-poly(methyl methacrylate is added; it was premade or prepared in situ by reacting a-carboxylic-polystyrene with a-epoxy-polystyrene. The characteriza-

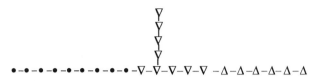

• styrene unit   ∇ butadiene unit

Δ methyl or cylohexyl methacrylate unit

Scheme 31

tion of the blends showed that the reactive system yielded a finer and more stable phase morphology than the premade copolymer; the latter does not completely cover the interface, whereas the coverage increases with time when the in situ technique is used. Fisher and Hellmann (91) studied a diblock polystyrene-polybutadiene whose polybutadiene block is grafted by methyl methacrylate or cyclohexyl methacrylate (Scheme 31). These block copolymers compatibilize blends polystyrene or poly(phenylene ether) with poly(vinyl choride) or poly(styrene-*co*-acrylonitrile); electron micrography clearly shows the layer of block graft copolymer at the interface.

B. Applications in Micellar Systems

Riess and Hurtrez (92) reviewed the use of block copolymers in micellar systems, particularly their amphiphilic properties in solution and in the microdomains of the solid state and their applications, such as dispersants, emulsifiers, wetting agents, foam stabilizers, flocculants, and demulsifiers. The same authors point out the essential applications of block copolymer micellar systems:

> The encapsulation for controlled-release systems
> Viscosity improvers and sludge dispersants in base lubricating oil
> Micellar catalysis
> Oxygen transport systems

Their medical applications are analyzed in Sec. IV.C.

Block ionomer complexes hold great promise in domains such as drug or gene delivery, nanotechnology membrane, and electroconductive polymers. Kabanov et al. prepared block ionomer complexes (93) prepared from ionomers and polyions of opposite charge, such as poly[poly(oxyethylene)-*block*-poly(methyl methacrylate)] anion and poly(*N*-ethyl-4-vinylpyridinium) cation, or prepared from block ionomers and surfactants.

C. Medical Applications

Kataoka et al. (94–102) studied the use of block copolymers in targetable polymeric drugs. Many molecules have a low therapeutic efficiency, and this failure
requires an important increase in the dose, which results in side effects such as
toxicity, unwanted disposition, and degradation. These problems can be overcome
by site-specific drug carriers designed to eliminate toxic side effects and degradation and to avoid a large uptake of the biologically active molecules. Ringsdorf
(103) and Kataoka (102) summarized the different functions of a targetable polymeric drugs in Fig. 6.

Micelle-forming, block copolymer–drug conjugates have seen important
developments in pharmacology; the most used block copolymers are listed in
Table 2.

Copolymers containing a hydrophilic block and a hydrophobic block (PAA)
have a great advantage for drug delivery, for drugs may be covalently coupled to
the PAA blocks and to form micelle-forming block copolymer–drug conjugates
(104–108). In most cases the hydrophilic block is poly(ethylene oxide) (POE), and
PAA can be poly(β-benzyl L-aspartate) or poly(γ-benzyl L-glutamate), for example. The CMC of the PEO block containing copolymers ranges from between
10^{-6} to 10^{-7} instead of 10^{-3} or 10^{-4} for the low-molar-mass surfactants (100).
The use of block copolymer micelles in drug delivery is illustrated in Fig. 7. This
technique was applied to diblock copolymers such as poly(ethylene oxide)-
poly(aspartic acid) (POE-PAsp), poly(ethylene oxide)-poly(γ-benzyl L-glutamate) (POE-PBLA), or poly(ethylene oxide)-poly(lactic acid) (109); the drug is
covalently coupled to the hydrophobic block (for instance, PAsp or PBLA), and

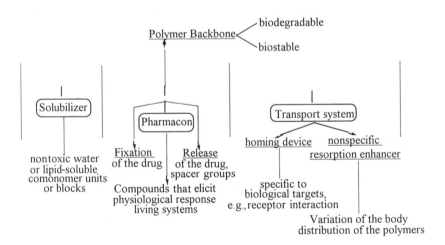

Fig. 6   Model for targetable polymeric drugs. (From Refs. 102 and 103.)

Table 2 Block Copolymers that Form Micelles in Aqueous Medium and Can Be Used in the Preparation of Block Copolymer–Drug Conjugates

AB block copolymers
Poly(ethylene oxide-*block*-styrene)
Poly(methacrylic acid-*block*-styrene)
Poly(ethylene oxide-*block*-β-benzyl L-aspartate)
Poly(ethylene oxide-*block*-γ-benzyl L-glutamate)
ABA block copolymers
Poly(ethylene oxide-*block*-propylene oxide-*block*-ethylene oxide)
Poly(ethylene oxide-*block*-isoprene-*block*-ethylene oxide)
Poly(methacrylic acid-*block*-isoprene-*block*-methacrylic acid
Block copolymer-drug conjugates
Poly(ethylene oxide-*block*-lysine)-cyclophosphamide conjugates
Poly(ethylene oxide-*block*-aspartic acid)-doxorubicin conjugates
Poly(ethylene oxide-*block*-aspartic acid)-cisplatin conjugates

Source: Ref. 101.

they form micelle-forming copolymer–drug conjugates. Scheme 32 shows the preparation of the conjugate POE-PAsp-DOX, where DOX is doxorubicin, a well-known hydrophobic anticancer drug (100,104).

The formation of a copolymer–drug conjugate in water is described in Fig. 8 (110). The copolymer contains a hydrophilic block that surrounds the core and forms a hydrated shell. The core contains the hydrophibic blocks where the drug molecules are covalently or noncovalently linked to the blocks.

Depending on whether the incorporation is chemical or physical, the copolymer–drug conjugate is prepared two different ways: first the drug molecules are linked to the hydrophobic blocks and then the micelles are formed,

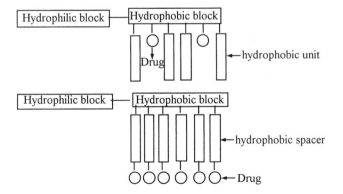

Fig. 7 Models of micelle-forming, block–drug conjugates. (From Refs. 101 and 104.)

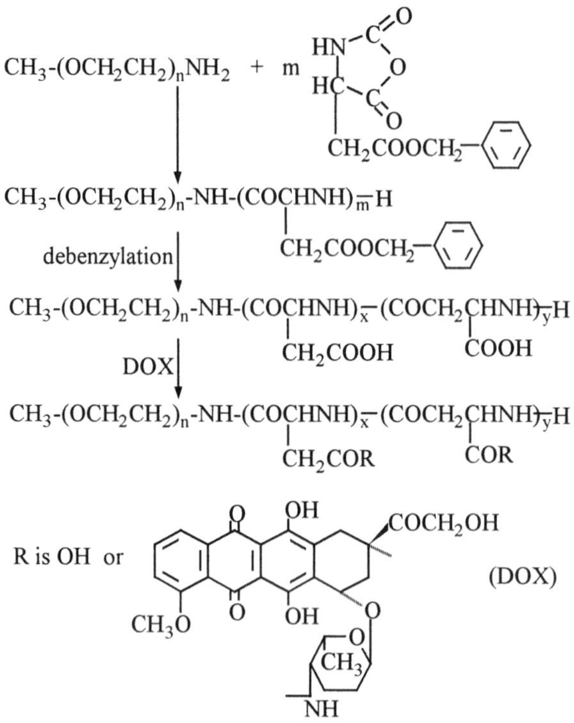

CH₃-(OCH₂CH₂)ₙNH₂ expressed as $CH_3\text{-}(OCH_2CH_2)_nNH_2$

$$CH_3\text{-}(OCH_2CH_2)_nNH_2 \ + \ m \ \begin{array}{c} HN\text{-}C{\diagup}^{O} \\ HC\text{-}C{\diagup}^{O} \\ | \quad \searrow^O \\ CH_2COOCH_2\text{-}\langle\overline{}\rangle \end{array}$$

$$\downarrow$$

$$CH_3\text{-}(OCH_2CH_2)_n\text{-}NH\text{-}(CO\underset{|}{C}HNH)_{\overline{m}}H$$
$$CH_2COOCH_2\text{-}\langle\overline{}\rangle$$

debenzylation $\downarrow$

$$CH_3\text{-}(OCH_2CH_2)_n\text{-}NH\text{-}(CO\underset{|}{C}HNH)_{\overline{x}}(COCH_2\underset{|}{C}HNH)_{\overline{y}}H$$
$$\qquad\qquad CH_2COOH \qquad COOH$$

DOX $\downarrow$

$$CH_3\text{-}(OCH_2CH_2)_n\text{-}NH\text{-}(CO\underset{|}{C}HNH)_{\overline{x}}(COCH_2\underset{|}{C}HNH)_{\overline{y}}H$$
$$\qquad\qquad CH_2COR \qquad COR$$

R is OH or [doxorubicin structure]   (DOX)

Scheme 32

or the micelles are prepared in a first step and then the drug molecules are introduced in the micelles (Fig. 9).

Kataoka et al. (112) extended these systems to copolymers containing a poly(ethylene oxide) and a polyion block. These polyion complex micelles result from interaction between oppositely charged block copolymers, for instance, POE-poly(L-lysine) and POE-poly(α,β-aspartic acid). More recently these au-

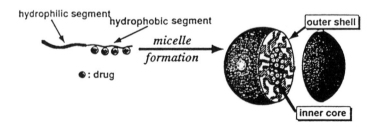

Fig. 8 Architecture of the block copolymer micelles. (From Ref. 110.)

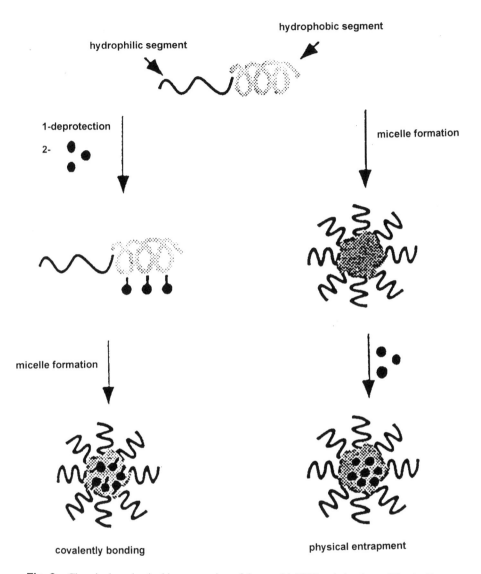

Fig. 9 Chemical or physical incorporation of drugs with POE-poly(amino acid) micelles. (From Ref. 111.)

thors (113) prepared polyion complex micelles entrapping enzyme molecules in the core, such as lisozyme.

Polyether-containing block copolymers have large applications in medical science. For instance, Moghimi et al. (114) used poly[poly(ethylene oxide)-*block*-poly(propylene oxide)] to stabilize polystyrene nanoparticules, which permitted them to control the rate of drainage from the subcutaneous injection site and to manipulate the lymphatic distribution.

The delivery of drugs can result from the degragation of microspheres containing a mixture of drug and block copolymer. Feng and Jia (115) prepared biodegradable block copolymers AB, ABA, AC, and AD, where A, B, C, and D are poly(L-lactide), poly(D,L-lactide), poly(p-dioxanone), and D poly(ε-caprolactone), respectively, which were used to prepare drug-containing microspheres prepared by evaporating the solvent of a solution of drug and copolymer. The release rate is almost constant, even from the initial stage.

The complexes between DNA and cationic polymers are attracting increasing attention as novel synthetic vectors for delivery of genes, either grafted (116) or block copolymers (117). Wolfert et al. (117) used AB-type cationic-hydrophilic block copolymers to introduce a protective surface shielding; the following oriented self-assembly with DNA are poly(ethylene oxide)-*block*-poly(L-lysine) or poly(N-hydroxypropyl-methacrylamide)-*block*-poly(trimethyl-ammonioethyl methacrylate chloride). Poly(ethylene oxide)-*block*-poly(L-lysine)/DNA complexes form discrete spheres, whereas poly(N-hydroxypropyl-methacrylamide)-*block*-poly(trimethyl-ammonioethyl methacrylate chloride)/DNA complexes adopt an extended structure. Their cytotoxicity is limited.

Block copolymers are playing an increasing role in biomaterials: bioadhesive on mucosal surfaces (118), semifast artificial muscle (119), implantable medical devices (120), wound-healing materials (121), biocompatible osteoconductive materials (122), and amphiphilic biosurfaces for prosthetics (123).

D. Miscellaneous Applications

The role of block copolymers in adhesives is the subject of many studies (124). Their use in the electronic and electrochromic industries (125), in lithography (126), and in inkjet technology (127) is rapidly increasing. The antistatic properties of block copolymers have seen an important development. Their efficiency depends on the hydrophily/hydrophoby balance of the blocks; one is responsible of the antistatic properties, whereas the other makes the additive compatible with the material. PEBAX confers a permanent and very stable increase in material surface resistivity; this is the case when it is added to ABS.

V. GENERAL CONCLUSIONS

The rocketing development of block copolymers is both quantitative and qualitative. The number of the relative articles and patents is rapidly increasing; on the other hand, their fundamental content is more and more important.

For a long time, the synthesis of a block copolymer was prepared by a given method (anionic polymerization or polycondensation of telechelic oligomers, for instance). Now many of their preparations call into action several sequential reactions. This new strategy is at its very beginning, and it will expand with the development of the new methods recently discovered, such as "living" cationic polymerization or mediated radical polymerization. Some techniques are almost absent in this strategy; this is the case for chemical modification; we found very few examples (1,53); this will rapidly change with the incoming of new catalytic processes, such as enzymatic-catalyzed reactions, particularly when carried in organic medium.

The characterization of the block copolymers is the subject of significant changes: drastic changes in some methods, such as chromatography (59–62), and the association of several techniques. The introduction of chemical modification in structural analysis should be a powerful tool in the near future, for instance, the separation of the blocks, which permits their independent characterization (72); from this point of view, enzymatic catalysis would be most useful (73).

Many of the analytical techniques we listed are used mainly to characterize the molecules rather than their organization in the solid state or in the melt. However, their interaction and self-organization are the subject of numerous publications, particularly the segregation of the blocks in different phases (128–130). The knowledge of phase organization is at its very beginning; thus Stadler et al. (131) showed that the triblock copolymer polystyrene-*block*-polybutadiene-*block*-poly(methyl methacrylate) gives rise to the formation of a chiral supramolecular assembly in which the chirality is not defined on a molecular level but results from the assembly of the molecules: the polybutadiene phase forms helical strands surrounding the polystyrene cylinders.

The number of articles describing the morphology of the block copolymers is rapidly increasing. Many of them are theoretical; however, they are often coupled with one or more experimental techniques; T. Hashimoto et al. (132) used small-angle scattering of x-rays to investigate the hexagonally packed cylindrical particles with paracrystalline distortion, comparing experimental values to those of scattering theory. In the same way Stocker et al. (133) associated force microscopy with transmission electron microscopy to determine the surface reconstruction of the lamellar morphology in a symmetric poly(styrene-*block*-butadiene-*block*-methyl methacrylate) triblock copolymer or transmission electron microscopy associated to small-angle x-ray scattering angle and theoretical calculations (134). The contribution of theoretical studies will rapidly increase, particularly the use of self-consistent schemes associated with experimental studies (135–137). Many of the calculations are relative to the dynamics of the copolymer chains, to order–disorder, and to diffusion problems. The simulation in block copolymer studies will see a rapid development in the near future, in both static and dynamic analyses (138).

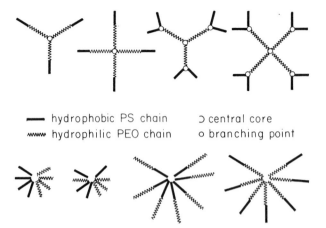

— hydrophobic PS chain   ⊃ central core
〰〰 hydrophilic PEO chain   o branching point

Fig. 10 Dendrimer-like and star architectures. (From Ref. 139.)

For several years, a clear break has been observed in the nature of the applications. Their use in thermoplastic elastomers, electronics, and printing are essentially described in patents, with little attention to characterization. On the other hand, their applications in blends, where they behave as compatibilizers, in industries using micelles, and above all in activities linked to medicine, pharmaceuticals, and cosmetology generate numerous fundamental studies; this is clearly shown in Sec. IV.C.

This chapter is devoted to the linear block copolymers. However, the arrival of copolymers associating different structures will make an important con-

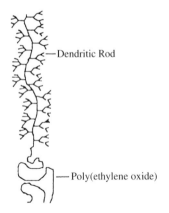

—Dendritic Rod

—Poly(ethylene oxide)

Fig. 11 Linear-dendritic rod diblock copolymers. (From Ref. 141.)

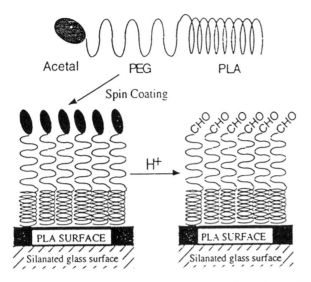

Fig. 12 Schematic representation of an acetal-terminated PEG surface and conversion into an aldehyde-terminated surface using an on-surface reduction, PLA is a polylactide[142].

tribution to polymer science and technology in the near future. Gnanou et al. (139) gathered some dendrimer-like and star architectures made from different blocks (Fig. 10).

Shim et al. (140) described the preparation of star copolymers; each arm of the star is made of a diblock copolymer polystyrene-*block*-polyisobutylene.

More sophisticated architectures are proposed, such as the preparation of a dendritic diblock copolymers (Fig. 11) (141) and a reactive polyoxyethylene-polylactide densely packed layer anchored on the biodegradable polymer surface and forming a brush polymer (Fig. 12) (142).

REFERENCES

1. M Sano. Block copolymer epitaxy. Adv. Mater. 9:509–511, 1997.
2. HL Hsieh, RP Quirk. Anionic Polymerization. Marcel Dekker, New York, 688–710, 1997.
3. M Sawamoto. Controlled polymer synthesis. In: K Matyjaszewski, ed. Cationic Polymerizations. Marcel Dekker, New York, 390–436, 1996.
4. F Schué. Synthesis of block copolymers by transformation reactions. In: SL Aggarwal, S Russo, eds. Comprehensive Polymer Science. 2nd suppl. Pergamon Press: Oxford, UK, 1989, pp. 359–367.
5. Z Jedlinski, K Brandt. Novel segmental polymers containing alternating rigid and soft blocks. Acta Polymerica 39:13–17, 1988.

6. G Ziegast, B Pfannemüller. Linear and star-shaped hybrid polymers. An improved purification procedure for coupling products of oligosaccharides by amide linkage. Makromol. Chem. *185:*1855–1866, 1984.

7. W Heitz. Metal catalyzed polycondensation reactions. Pure Appl. Chem. *67:*1951–64, 1995.

8. R Giesa, RC Schulz. Soluble poly(1,4-phenyleneethynylene). Makromol. Chem. *191:*857–867, 1990.

9. T Takagi, I Tomita, T Endo. Block copolymerization of alkoxyallenes with phenylallene by living coordination system with p-allylnickel catalyst. Polym. Bull. *39:*685–692, 1997.

10. Th Lalot, M Maryvonne, E Maréchal. Enzymatic catalysis in organic media for polymer synthesis. In: SL Aggarwal, S Russo, eds. Comprehensive Polymer Science 2nd suppl. Pergamon Press: Oxford, UK, 1996, pp. 29–70.

11. S Wallace, CJ Morrow. Biocatalytic synthesis of polymers. Synthesis of an optically active, epoxy-substituted polyester by lipase-catalyzed polymerization. J. Polym. Sci. Part A: Polym. Chem. Ed. *27:*2553–2567, 1989.

12. AM Blinkovsky, JS Dordick. Peroxydase-catalyzed synthesis of lignin phenol copolymers. J. Polym. Sci. Part A: Polym. Chem. Ed. *31:*1839–1846, 1993.

13. S Kobayashi, K Kashiwa, T Kawasaki, S-E Shoda. Novel method for polysaccharide synthesis using an enzyme: the first in vitro synthesis of cellulose via a non-biosynthetic path utilizing cellulase as catalyst. J. Am. Chem. Soc. *113:*3079–3084, 1991.

14. S Kobayashi, K Kashiwa, J Shimada. Enzymatic polymerization: the first in vitro synthesis of cellulose via a non-biosynthetic path catalyzed by cellulase. Makromol. Chem. Makromol. Symp. *54/55:*509–518, 1992.

15. K Loos, R Stadler. Synthesis of amylose-*block*-polystyrene rod coil block copolymers. Macromolecules *30:*7641–7643, 1997.

16. G Galli, E Chiellini, Y Yagci, EI Serhatli, M Laus, AS Angeloni. Liquid crystalline block copolymers by sequential cationic or promoted cationic and free-radical polymerizations. Macromol. Symp. *107:*85–97, 1996.

17. Y Yagci, A Onen, W Schnabel, S Denizligil, G Hizal, U Tunca. Novel polymeric photoinitiators and their use in block copolymerization. Photoinitiation: Radcure coat. inks. *149:*151–172, 1993.

18. Y Yagci. Macrophotoinitiators. Synthesis and their use in block copolymerizations. In: MK Mishra, ed. Macromol. Eng. [Proc. Int. Conf. Adv. Polym. Macromol. Eng.] Plenum, New York, 151–161, 1995.

19. XM Yang, KY Qiu. Block copolymerization of vinyl monomers with macrophotoiniferter. J. M. S. Pure Appl. Chem. *A34:*543–549, 1997.

20. D Greszta, D Mardare, K Matyjaszewski. "Living" radical polymerization. 1—Possibilities and limitations. Macromolecules *27:*638–644, 1994.

21. K Matyjaszewski, S Gaynor, D Greszta, D Mardare, T Shigemoto. Synthesis of well-defined polymers by controlled radical polymerization. Macromol. Symp. *98:* 73–89, 1995.

22. K Matyjaszewski, S Gaynor, JS Wang. Controlled radical polymerizations: the use of alkyl iodides in degenerative transfer. Macromolecules, Communications *28:* 2093–2093, 1995.

23. B Améduri, B Boutevin, Ph. Gramain. Synthesis of block copolymers by radical polymerization and telomerization. Adv. Polym. Sci. *127:*88–142, 1997.

24. IQ Li, BA Howell, MT Dineen, PE Kastl, JW Lyons, DM Meunier, PB Smith, DB Priddy. Block copolymer preparation using sequential normal/living radical polymerization techniques. Macromolecules *30:*5195–5199, 1997.

25. DJ Gravert, KD Janda. Bifunctional initiators for free radical polymerization of non-crosslinked block copolymers. Tetrahedron Let. *39:*1513–1516, 1998.

26. JS Wang, K Matyjaszewski. Controlled/"living" radical polymerization. Halogen atom transfer radical polymerization promoted by Cu(I)/Cu(II) redox process. Macromolecules *28:*7901–7910, 1995.

27. D Benoit, S Grimaldi, JP Finet, P Tordo, M Fontanille, Y Gnanou. Controlled free-radical polymerization in the presence of a novel asymmetric nitroxyl radical. Polym. Prep. *38*(1):729–730, 1997.

28. L Baloch, L Samuelson, K Shridara Alva, A Blumstein. Synthesis, homopolymerization, and block copolymerization of *N*-ethyl-2-ethynyl-pyridinium trifluoromethanesulfonate with styrene and butadiene. J. Polym. Sci. Part A, Polym. Chem. Ed. *36:*703–712, 1998.

29. O Nuyken, S Ingrisch. Block copolymers from isobutyl vinyl ether and 2-chloroethyl vinyl ether. Macromol. Chem. Phys. *199:*607–612, 1998.

30. S Penczek, P Kubisa. Living polymers systems. In: Encyclopedia of Polymer Science Engineering. Suppl. Wiley, New York, 1989, pp. 380–399.

31. D Li, R Faust. Polyisobutylene-based thermoplastic elastomers. 3. Synthesis, characterization and properties of poly(a-methylstyrene-*b*-isobutylene-*b*-a-methylstyrene) triblock copolymers. Macromolecules *28:*4893–4898, 1995.

32. E Franta, P Rempp. The block copolymer bag of tricks. Chemtech *26:*24–28, 1996.

33. T Zundel, J Baran, M Mazurek, J-S Wang, R Jérôme, Ph. Teyssié. Climbing back up the nucleophylic reactivity scale. Use of cyclosila derivatives boosters in anionic polymerization. Macromolecules *31:*2724–2730, 1998.

34. H Cramail, A Deffieux. Derivatization of hydroxyl into a-chloro ether: application to the synthesis of block copolymers from hydroxy-telechelic polymers. Makromol. Chem. *193:*2793–2806, 1992.

35. K Mashima, Y Nakayama, A Nakamura. Recent trends in the polymerization of a-olefins catalyzed by organometallic complexes of early transition metals. Adv. Polym. Sci. *133:*2–51, 1997.

36. H Yasuda, E Ihara. Rare earth metal–initiated living polymerizations of polar and nonpolar monomers. Adv. Polym. Sci. *133:*53–101, 1997.

37. M Shiomi, H Shiamara, H Yasuda. Unpublished results reported in Ref. 33, p. 76.

38. R Numora, T Endo. Synthesis of poly(tetrahydrofuran-*b-e* caprolactone) macromer via the SnI$_2$-induced transformation. Polym. Bull. *35:*683–689, 1995.

39. WM Stevels, MJK Ankone, PD Dijstra, J Frijin. A versatile and highly efficient catalyst system for the preparation of polyesters based on lanthanide tris (2,6-di-*ter*-butyl(phenolate)s and various alcohols. Macromolecules *29:*3332–3333, 1996.

40. H Yasuda, E Ihara, M Morimoto, M Nodomo, S Yoshioka, M Furo. Rare earth metal initiated polymerizations to give high polymers with narrow molecular weight distribution. Macromol. Symp. *95:*203–216, 1995.

41. E Ihara, M Morimoto, H Yasuda. Living polymerizations and copolymerizations of alkyl acrylates by the unique catalysis of rare earth metal complexes. Macromolecules *28:*7886–7892, 1995.

42. H Yasuda, E Ihara. Rare earth metal initiated polymerizations of polar and nonpolar monomers to give high molecular weight polymers with extremely narrow molecular weight distribution. Macromol. Chem. Phys. *196:*2417–2441, 1995.

43. P Rempp, E Franta. New developments in the synthesis of block and graft copolymers via anionic and cationic methods. Macromol. Symp. *91:*51–63, 1995.

44. Y Yagci, A Önen, W Schnabel. Block copolymers by combination of radical and promoted cationic polymerization routes. Macromolecules *24:*4620–4623, 1991.

45. S Coca, K Matyjaszewski. Block copolymers by transformation of "living" carbocationic into "living" radical polymerization. Macromolecules *30:*2808–2810, 1997.

46. E Yoshida, A Sugita. Synthesis of poly(styrene-*b*-tetrahydrofuran-*b*-styrene) triblock copolymers by transformation from living cationic into living radical polymerization using 4-hydroxy-2,2,6,6-tetramethylpiperidine-1-oxyl as a transforming agent. J. Pol. Sci. Part A Polym. Chem. *36:*2059–2068, 1998.

47. E Yoshida, A Sugita. Synthesis of poly(tetrahydrofuran) with a nitroxyl radical at the chain end and its application to living radical polymerization. Macromolecules *29:*6422–6426, 1996.

48. O Nguyen, H Kröner, S Aechner. Macro azo initiators via cationic polymerization. Synthesis and application. Makromol. Chem. Rapid Commun. *9:*671–679, 1988.

49. S Kobatake, HJ Harwood, RP Quirk, DB Priddy. Block copolymer synthesis by styrene polymerization initiated with nitroxy-functionalized polybutadiene. Macromolecules *31:*3735–3739, 1998.

50. R Nomura, Y Shibasaki, T Endo. Block copolymerization of tetrahydrofuran with d-valerolactone by samarium iodide-induced transformation. Polym. Bull. *37:*597–601, 1996.

51. MP Labeau, H Cramail, A Deffieux. Amphiphilic block copolymers of controlled dimensions with hydrophilic glycosidic vinyl ether moities. Macromol. Chem. Phys. *199:*335–342, 1998.

52. S Coca, H Paik, K Matyjaszewski. Block copolymers by transformation of living ring-opening metathesis polymerization into controlled "living" atom transfer radical polymerization. Macromolecules *30:*6513–6516, 1997.

53. X Huang, Z Lu, J Huang. Preparation of diblock copolymer of methyl methacrylate and vinyl acetate by successive radical mechanism and selective hydrolysis of the poly(vinyl acetate) block. Polymer *39:*1369–1374, 1998.

54. A Fradet. Coupling reactions in polymer synthesis. In: SL Aggarwal, S Russo, eds. Comprehensive Polymer Science. 2nd suppl. Chap. 4. Pergamon Press, UK, 1996, pp. 133–162.

55. A Douhi, A Fradet. Study of bulk chain-coupling reactions; part IV. Polym. Bull. *36:*455–462, 1996.

56. T Fischer, H Lefèbvre, A Fradet. Synthesis of block copolymers by oligomer-coupling reactions in the bulk: bisimidazoline/COOH and bisoxazolinone/NH$_2$ reactions. Macromol. Symp. *118:*79–87, 1997.

57. H Lefèbvre, A Fradet. Block copolymers by high temperature oligomer-coupling re-

actions: The bis(5-oxazolinone)/OH reaction. Macromol. Symp. *122:*25–31, 1997.

58. H Ohishi, T Ohwaki, T Nishi. A novel process for synthezing polystyrene and polyarylate block copolymers utilizing telechelic polystyrene. J. Polym. Sci: Part A: Polym. Chem. *36:*2839–2847, 1998.

59. H Pash, Y Gallot. Chromatographic investigations of macromolecules in the critical range of liquiq chromatography: 4-analysis of poly(styrene-*b*-methacrylate). Polymer *34:*4100–4104, 1993.

60. H Pash, C Brinkmann, Y Gallot, B Trathnijg. 7—Analysis of the poly(methycralate) block in poly(styrene-*block*-methacrylate). Polymer *34:*4986–4989, 1993.

61. H Pash, M Augenstein. Chromatographic investigations of macromolecules in the critical range of liquiq chromatography: 5—Characterization of block copolymers of decyl and methyl methacrylate. Makromol. Chem. *194:*2533–2541, 1993.

62. H Pash. Liquid chromatography at the critical point of adsorption—a new technique for polymer characterization. Macromol. Symp. *110:*107–120, 1996.

63. MB Tennikov, PP Nefedov, MA Lazavera, SJ Frenkel. Single mechanism of liquid chromatography of macromolecules on porous sorbents. Vysokomol. Soedin. *A19:*657–660, 1977.

64. BG Belenkii, ES Gankina. Thin-layer chromatography of polymers. J. Chromatogr. Chromatogr. Rev. *141:*13–90, 1977.

65. C Braud, R Devarieux, A Atlan, C Ducos, M Vert. Capillary zone electrophoresis in normal or reverse polarity separation modes for the analysis of hydroxy acid oligomers in neutral phosphate buffer. J. Chromatography B *706:*73–82, 1998.

66. P Kilz, RP Krüger, H Much, G Schulz. 2-Dimensional chromatography for the deformulation of complex copolymers. In: T Provder, ed. Polymer Characterization. ACS Advances in Chemistry. *247:*233–241, 1995.

67. TF Gores, P Kilz. Copolymer characterization using conventional size-exclusion chromatography and molar-mass-sensitive detectors. In: T Provder, ed. Chromatography of Polymers. ACS Symp. Ser. *521:*123–148, 1993.

68. A Boularès, L Rozès, M Tessier, E Maréchal. Structure and properties of block copolymers with polyethers as flexible blocks. Europhysics Conference Abstracts *21A:*24, 1997.

69. A Boularès, L Rozès, M Tessier, E Maréchal. Synthesis and characterization of poly(copolyethers-*block*-polyamides). I. Structural study of polyether precursors. J. Macrom. Sci. Pure Appl. Chem. *A35:*933–953, 1998.

70. P Thuillier, M Tessier, Maréchal. Synthesis and characterization of thermotropic polyester-*block*-polyethers. Makromol. Chem. Macromol. Symp. *70/71:*37–45, 1993.

71. YJ Khim, DH Adamson. Synthesis of diblock copolymer of polystyrene and polylactide with well-defined structure. Polym. Prepr. *39(2):*423–424, 1998.

72. G Sundararajan, V Vasudevan, KA Reddy. Synthesis of triblock copolymers—(polyA-polybutadiene-polyA)—via metathesis polymerization. J. Polym. Sci. *36:* 2601–2610, 1998.

73. N Valiente, T Lalot, M Brigodiot, E Maréchal. Enzymatic hydrolysis of phthalic unit containing copolyesters as a potential tool for block length determination. Polymer degradation and stability *61:*409–415, 1998.

74. T Lodge. Characterization of polymer materials by scattering techniques, with applications to block copolymers. Mikrochim. Acta *31*:1–31, 1994.

75. MS Kent, M Tirrell, TP Lodge. Properties of polystyrene-poly(methyl methacrylate) random and diblock copolymers in dilute and semidilute solutions. J. Polym. Sci. Polym. Phys. Ed. *32*:1927–1941, 1994.

76. NP Balsara, M Tirrell. Micelle formation of BAB triblock copolymers in solvents that preferentially dissolve the A block. Macromolecules *24*:1975–1986, 1991.

77. FS Bates, MF Schulz, JH Rosedale, K Almdal. Correlation of binary polyolefin phase behavior with statistical segment length asymmetry. Macromolecules *25*: 5547–5550, 1992.

78. FS Bates, JH Rosedale, GH Frederickson. Fluctuation effects in a symmetric diblock copolymer near the order–disorder transition. J. Chem. Phys. *92*:6255–6270, 1990.

79. K Almdal, JH Rosedale, FS Bates, GD Wignall, GH Frederickson. Gaussian to stretched-coil transition in block-copolymer melts. Phys. Rev. Lett. *65*:1112–1115, 1990.

80. N Munkenbucsh, D Schneiders, D Richter, B Farago, JS Huang, LJ Fetters, AP Gast. Neutron scattering investigation on diblock copolymer aggregates with a polymer brush as outer layer. Polym. Prep. *35*(1):568–569, 1994.

81. MA Singh, CR Harkless, SE Nagler, RF Shannon Jr., SS Ghosh. Time-resolved small-angle x-ray scattering study of ordering kinetics in diblock styrene-butadiene. Phys. Rev. B *47*:8425–8435, 1993.

82. MD Foster, M Sikka, N Singh, FS Bates, SK Satija, CF Majkrzak. Structure of symmetric polyolefin block copolymer thin films. J. Chem. Phys. *96*:8605–8615, 1992.

83. M Niwa, T Hayashi, N Higashi. Surface monolayers of well-defined amphiphilic block copolymer composed of poly(acrylic acid) or poly(oxyethylene) and poly(styrene). Interpolymer complexion at the air–water interface. Langmuir *6*: 263–268, 1990.

84. HD Bijsterbosch, VO de Haan, AW de Graaf, M Mellema, FAM Leermakers, MA Cohen Stuart, AA van Well. Tethered adsorbing chains: neutron reflectivity and surface pressure of spread diblock copolymer monolayers. Langmuir *11*:4467–4473, 1995.

85. JK Cox, K Yu, A Eisenberg, RB Lennox. Characterization of the lateral structure of non-ionic block copolymers spread on water. Polym. Prepr. *39*(2):723–724, 1998.

86. V Tsukruk. Scanning probe microscopy on polymer surfaces. Rubber Chem. Techn. *70*:430–467, 1997.

87. LZ Liu, B Jiang. Crystallization of micelles and matrix in dilute diblock copolymer/homopolymer solutions. J. Polym. Sci. Part B. Polym. Phys. *36*:2961–2970, 1998.

88. T Hashimoto, K Kawamura, M Harada, H Tanaka. Small-angle scattering from hexagonally packed cylindrical particles with paracrystalline distorsion. Macromolecules *27*:3063–3072, 1994.

89. H Yamane. Polymer alloys. SPSJ 42nd Symposium of Macromolecules, Tokyo, 1993, pp. 275–281.

90. A Nakayama. Quoted in Ref. 89.

91. M Fischer, GP Hellmann. Block-graft copolymers on the basis of a styrene-butadiene diblock copolymer. Polymer *37*:4547–4554, 1996.

92. G Riess, G Hurtrez. Block copolymers: syntheses, colloidal properties and application possibilities of micellar systems. In: SE Weber et al., eds. Solvents and Self-Organization of Polymers. Kluwer Academic, Dordrecht, 1996, pp. 33–51.

93. AV Kabanov, TK Bronich, K Yu, A Eisenberg, AA Lysenko, VA Kabanov. Block ionomer complexes. Polym. Prep. 39(2):388–389, 1998.

94. K Kataoka, GS Kwon, M Yokoyama, T Okano, Y. Sakurai. Polymeric micelles as novel drug carriers and virus mimicking vehicles. In: Y Kahovec, ed. Macromolecules. VSP: Utrecht, Netherlands, 1992.

95. K Kataoka, GS Kwon, M Yokoyama, T Okano, Y Sakurai. Block copolymer micelles as vehicles for drug delivery. J. Controlled Release 24:119–132, 1993.

96. K Kataoka. Design of nanoscopic vehicles for drug targeting based on micellization of amphiphilic block copolymers. JMS Pure Appl. Chem. 1759–1769, 1994.

97. M Yokoyama, GS Kwon, T Okano, Y Sakurai, K Kataoka. Development of micelle-forming polymeric drug with superior anticancer activity. In: M Ottenbrite, ed. Polymeric Drugs and Drug Administration. ACS Symposium Series, 545:126–134, 1994.

98. K Kataoka. Design of nanoscopic vehicles for drug targeting based on micellization of amphiphilic block copolymers. J. M. S. Pure Appl. Chem. 1759–1769, 1994.

99. K Kataoka. Preparation of novel drug carrier based on the self-association of block copolymers. Drug Delivery System 10:363–370, 1995.

100. GS Kwon, K Kataoka. Block copolymer micelles as long-circulating drug vehicles. Advanced Drug Delivery Rev. 16:295–309, 1995.

101. S Cammas, K Kataoka. Site-specific drug-carries: polymeric micelles as high-potential vehicles for biologically active molecules. In: SE Weber et al., eds. Solvents and Self-Organization of Polymers. Kluwer Academic, Dordrecht, Netherlands, 1996, pp. 83–113.

102. K Kataoka. Targetable polymeric drugs in controlled drug delivery: challenges and strategies. In: K Park, ed. ACS Prof. Ref. Book. Washington, DC: Am. Chem. Soc., 1997, pp. 49–71.

103. H Ringsdorf. Structure and properties of pharmacologically active polymers. J. Polym. Sci. Symp. 51:135–153, 1975.

104. K Dorn, G Hoerpel, H Rindsdorf. Polymeric anti-tumor agents on a molecular and cellular level. In: CG Gebelein, CE Carraher, Jr., eds. Bioactive Polymeric Systems. Plenum, New York, 1985, pp. 531–585.

105. M Yokoyama, S Inoue, K Kataoka, N Yui, Y Sakurai. Preparation of adriamycin-conjugated poly(ethylene glycol)-poly(aspartic) block copolymer: a new type of polymeric anticancer drug. Makrom. Chem. Rapid Commun. 8:431–435, 1987.

106. M Yokoyama, M Miyauchi, N Yamada, T Okano, Y Sakurai, K Katoaka, S Inoue. Characterization and anticancer activity of the micelle-forming polymeric anticancer drug adriamycin-conjugated poly(ethylene glycol)-poly(aspartic) block copolymer. 50:1693–1700, 1990.

107. M Yokoyama, T Okano, Y Sakurai, H Ekimoto, C Shibazaki, K Kataoka. Toxicity and tumor-activity against solid tumor and micelle-forming polymer anticancer drug and its extremely long circulation in blood. Cancer Res. 51:3229–3236, 1991.

108. M Yokoyama, GS Kwon, GS Naito, T Okano, Y Sakurai, T Seto, K Kataoka. Prepa-

ration of micelle-forming polymeric drug conjugates. Bioconj. Chem. *3:*295–301, 1992.

109. JH Kim, M Iijima, Y Nagasaki, K Kataoka. Synthesis and properties of PEG/ PDLLA block copolymer micelle with polymerizable methacryloyl end. Polymer Prep. *39*(2):130–131, 1991.

110. M Yokoyama, GS Kwon, T Okano, Y Sakurai, K Kataoka. Development of micelle-forming polymeric drug with superior anticancer activity. ACS Symp. *545:*126–134, 1994.

111. GS Kwon, M Naito, K Kataoka, M Yokoyama, Y Sakurai, T Okano. Block copolymer micelles as vehicles for hydrophobic drugs. Colloids Surf. B: Biointerfaces *2:* 429–434, 1994.

112. A Harada, K Kataoka. Formation of polyion complex micelles in an acqueous milieu from a pair of oppositely charged block copolymers with poly(ethylene glycol) segments. Macromolecules *28:*5294–5299, 1995.

113. A Harada, K Kataoka. Novel polyion complex micelles entrapping enzyme molecules in the core: preparation of narrowly distributed micelles from lysozyme and poly(ethylene glycol)-poly(aspartic acid) block copolymer in acqueous medium. Macromolecules *31:*288–294, 1998.

114. SM Moghimi, AE Hawley, NM Christy, T Gray, L Illum, SS Davis. Surface engineered nanospheres with enhanced drainage into lymphatics and uptake by macrophages of the regional lymph modes. FEBS Lett. *344:*25–30, 1994.

115. Xin-De Feng, Y Jia. Synthesis and drug controlled release of block copolymers of poly(L-lactide) with poly(D, L-lactide) and related monomers. Macromol. Symp. *118:*625–630, 1997.

116. V Toncheva, MA Wolfert, PR Dash, D Oupicky, K Ulbrich, LW Seymour, EH Schacht. Novel vectors for gene delivery formed by self-assembly of DNA with poly(L-lysine) grafted with hydrophilic polymers. Biochim. Biophys. Acta *1380:* 354–368, 1998.

117. MA Wolfert, EH Schacht, V Toncheva, K Ulbrich, O Nazarova, LW Seymour. Characterization of vectors for gene therapy formed by self-assembly of DNA with synthetic block copolymers. Human Gene Therapy *7:*2123–2133, 1996.

118. AS Hoffman, G Chen, X Wu, Z Ding, B Kabra, K Randeri, M Schiller, E Ron, AN Pappas, C Brazel. Graft copolymers of POE-POP-POE triblock polyethers on bioadhesive polymer backbones: synthesis and properties. Polym. Prepr. *38:*524–525, 1997.

119. PG de Gennes. A semi-fast artificial muscle. CR Acad. Sci. Ser. IIb. *324:*343–348, 1997.

120. JB Kohn, B Yu, C Yu. Block copolymers of tyrosine-based polycarbonate and poly(alkylene oxide) for biocompatible articles. PCT Int. Appl. WO 97 19,996, 5 June 1997.

121. A Morikawa, K Shiho, N Kawahashi, Y Yamakawa, K Kuruda (Japan Synthetic Rubber Co., Ltd.). Block copolymers and wound-healing materials. Jn Kokai Tokkyo Koho JP 09,268,227 [97,268,227], 14 Oct. 1997.

122. BM Silberstein, S Rabinovich, ME Sizikov. Experimental studies of combined use of biocompatible osteoconductive polymers block (BOP-B) and superelastic porous

TiNi alloy implants in spine. Shape Mem. Mater. 94 Proc. Int. Symp. 657–658, 1994.

123. B Wesslen, C Freij-Larsson, M Kober, A Ljungh, M. Pauisson, P Tengvall. Amphiphilic biosurfaces. Mater. Sci. Eng. C. C1:127–131, 1994.

124. EB Brandes, TS Coolbaugh, FC Loveless, F Shirazi. Liquid elastomers provide versality for adhesive and sealant design. Adhesives Age, 1998, pp. 24–28.

125. C Michot, D Baril, M Armand. Polyimide-polyether mixed conductors as switchable materials for electronic devices. Sol. Energy Mater. Sol. Cells *39:*289–299, 1995.

126. HA Gabor, DR Allen, P Gallagher-Wetmore, CK Ober. Block and random copolymer resists for 193-nm lithography and environmentally friendly supercritical CO_2 development. Proc. SPIE Int. Soc. Opt. Eng. In: Adv. Resist Techn. and Process. XIII ed. The International Society for Optical Engineering *2724:*410–417, 1996.

127. D Sentilli (Eastman Kodak Company, USA). Inkjet inks containing block copolymer of polyethylene oxide and polypropylene oxide for printing high-quality images. Eur. Pat. Appl. EP 780,451, 25 June 1997.

128. M Olvera de la Cruz, AM Mayes, BW Swift. Transition to lamellar-catenoid structure in block-copolymer melts. Macromolecules *25:*944–948, 1992.

129. I Erukhimovich, V Abetz, R Stadler. Microphase separation in ternary ABC block copolymers: ordering control in molten diblock AB copolymers by attaching a short, strongly interacting C block. Macromolecules *30:*7435–7443, 1997.

130. L de Luca Freitas, MM Jacobi, G Goncalves, R Stadler. Microphase separation induced by hydrogen bonding in poly(1,4-butadiene)-*block*-poly(1,4-isoprene) diblock copolymer—an example of supramolecular organization via tandem interactions. Macromolecules *31:*3379–3382, 1998.

131. R Stadler, U Krappe, I Voigt-Martin. Chiral nanostrutures based on amorphous ABC-triblock copolymers. Polym. Prepr. *35(2):*542–543, 1994.

132. T Hashimoto, T Kawamura, M Harada, H Tanaka. Small-angle scattering from hexagonally packed cylindrical particles with paracrystalline distorsion. Macromolecules *27:*3063–3072, 1994.

133. W Stocker, J Beckmann, R Stadler, JP Rabe. Surface reconstruction of the lamellar morphology in a symmetric poly(styrene-*block*-butadiene-*block*-methyl methacrylate) triblock copolymers: A tapping mode scanning force microscopy study. Macromolecules *29:*7502–7507, 1996.

134. U Breiner, U Krappe, T Jakob, V Abetz, R Stadler. Spheres on spheres. A novel spherical multiphase morphology in symmetric polystyrene-*block*-polybutadiene-*block*-poly(methyl methacrylate) triblock copolymers. Polym. Bull. *40:*219–226, 1998.

135. M Benmouna, U Maschke, B Ewen. Properties of polymer blends and triblock copolymers obtained by neutron scattering. J. Polym. Sci. Part B: Polym. Phys. *34:* 2161–2168, 1996.

136. MW Matsen, FS Bates. Conformationally asymmetric block copolymers. J. Polym. Sci. Part B: Polym. Phys. *35:*945–952, 1997.

137. RR Netz, M Schick. Liquid-crystalline phases of semiflexible diblock copolymer melts. Phys. Rev. Lett. *77:*302–305, 1996.

138. T Pakula. Simulation of copolymers by means of the cooperative-motion algorithm. J. Computer-aided Mater. Design *3:*329–340, 1996.
139. Y Gnanou, S Angot, D Taton, V Héroguez, M Fontanille. Novel branched amphiphiles with dendrimer-like and star architectures. Polym. Mater. Sci. Eng. 80:59–60, 1999.
140. JS Shim, S Asthana, N Omura, JP Kennedy. Novel thermoplastic elastomers. I. Synthesis and characterization of star-block copolymers of Pst-*b*-PIB arms emanating from cyclosiloxane cores. J. Polym. Sci. Part A: Polym. Chem. *36:*2997–3012, 1998.
141. CM Bambenck, TA Hatton, PT Hammond. Synthesis and characterization of linear-dendritic rod diblock copolymers. Polym. Mater. Sci. Eng. *79:*449–501, 1998.
142. H Otsuka, Y Nakasaki, K Kataoka, T Okano, Y Sakurai. Reactive-PEG-polylactide block copolymer for tissue engineering. Polym. Prepr. *39(2):*128–129, 1998.

Index